| | Multiply | By | To Obtain |
|---|---|---|---|
| PRESSURE | millimeters mercury (mm Hg) at 0° C | 133.32 | pascals (Pa; N per m$^2$) |
| | pounds per square inch (psi) | 6.895 | kilopascals (kPa; 1000 pascals) |
| | pascals | 0.0075 | millimeters mercury at 0° C |
| | kilopascals | 0.1450 | pounds per square inch |
| | bars | 1000 | millibars (mb) |
| | bars | 100000 | pascals |
| | bars | 0.9869 | atmospheres (atm) |
| ENERGY* | joules (J) | 0.2389 | calories (cal) |
| | kilocalories (kcal) | 1000 | calories |
| | joules | 1.0 | watt-seconds (W-sec) |
| | kilojoules (kJ) | 1000 | joules |
| | calories | 0.00397 | Btu (British thermal units) |
| | Btu | 252.0 | calories |
| POWER* | joules per second | 1.0 | watts (W) |
| | kilowatts (kW) | 1000 | watts |
| | megawatts (MW) | 1000 | kilowatts |
| | kilocalories per minute | 69.93 | watts |
| | watts | 0.0013 | horsepower (hp) |
| | kilowatts | 5.7 | Btu per minute |
| | Btu per minute | 0.235 | horsepower |
| TEMPERATURE | degrees Fahrenheit (°F) plus 459.67 | 0.5555 | kelvins (K) |
| | degrees Celsius (°C) plus 273.15 | 1.0 | kelvins |
| | kelvins | 1.80 | degrees Fahrenheit minus 459.67 |
| | kelvins | 1.0 | degrees Celsius minus 273.15 |
| | degrees Fahrenheit minus 32 | 0.5555 | degrees Celsius |
| | degrees Celsius | 1.80 | degrees Fahrenheit plus 32 |

It is important to understand the distinction between *energy* and *power* as well as the corresponding metric and English units.

**Energy** in its various forms (for example, heat, chemical energy, radiant energy) represents the ability to perform work. Energy is therefore expressed in units of *work,* which is defined as a force applied over some distance. In the metric system, units of energy are the joule and the erg. By definition, 1 *joule* is the energy required when 1 newton acts through 1 meter. (One *newton* is the force that accelerates a 1-kilogram mass by 1 meter per second per second.) Alternately, 1 *erg* is the energy required when 1 dyne acts through 1 centimeter. (One *dyne* is the force that accelerates a 1-gram mass by 1 centimeter per second per second.) One joule is the equivalent of 10 ergs, and 1 calories of heat (defined in Chapter 3) equals 4.186 joules.

**Power** is the rate at which energy is used, that is, released or converted. One *watt* is defined as 1 joule per second, and 1 *kilowatt* is 1000 joules per second. One *megawatt* equals 1000 kilowatts.

When solar radiation is absorbed by something (air, water, or land, for example), radiant energy is converted to heat. We can describe the amount of heat in terms of calories or joules. We may also be concerned about the rate at which solar radiation travels through some cross-sectional area or the rate at which it is absorbed by some surface. The flux of energy is then described in terms of watts per square meter, joules per square meter per minute, or calories per square centimeter per minute.

# Meteorology

## The Atmosphere and the Science of Weather

# Meteorology

## The Atmosphere and the Science of Weather

**Fourth Edition**

**Joseph M. Moran** *University of Wisconsin–Green Bay*

**Michael D. Morgan** *University of Wisconsin-Green Bay*

Chapter 11 written by
**Patricia M. Pauley** *Naval Postgraduate School, Monterey, California*

MACMILLAN COLLEGE PUBLISHING COMPANY
*New York*

MAXWELL MACMILLAN CANADA
*Toronto*

Editor: Robert McConnin
Production Supervisor: Elisabeth Belfer
Production Manager: Su Levine
Text designer: Andrew Zutis
Cover designer: Tom Mack
Cover photograph: Warren Faidley/International Stock
Frontispiece: Woody Woodworth/Superstock


This book was set in Times Roman by The Clarinda Company and was printed and
bound by Von Hoffmann Press. The cover was printed by Philips Winson, Inc.

Macmillan College Publishing Company
866 Third Avenue, New York, New York 10022

Macmillan College Publishing Company is part of the
Maxwell Communication Group of Companies.

Maxwell Macmillan Canada, Inc.
1200 Eglinton Avenue East
Suite 200
Don Mills, Ontario M3C 3N1

Library of Congress Cataloging-in-Publication Data

Moran, Joseph M.
     Meteorology: the atmosphere and the science of weater / Joseph
M. Moran, Michael D. Morgan : chapter 11 written by Patricia M.
Pauley. -- 4th ed.
        p.   cm.
     Includes index.
     ISBN 0-02-383341-6
     1. Meteorology.  2. Weather.  3. Atmospheric physics.  I. Morgan.
Michael D.  II. Title.
QC861.2.M625   1994
551.5--dc20                                              93-1035
                                                         CIP

Printing:  1  2  3  4  5  6  7  8        Year:  4  5  6  7  8  9  0  1  2  3

# Preface

This book is intended to introduce the college nonscience major to the fundamentals of atmospheric science. It is appropriate for an introductory course on the atmosphere, weather, or weather and climate. Our primary goal is to demonstrate how scientific principles govern the circulation of the atmosphere and the day-to-day sequence of weather events. In so doing, we also introduce the nonscience student to the nature of scientific inquiry and the methodology of science. Atmospheric science is especially well suited to these goals because it is an applied science that readily lends itself to familiar illustrations.

Our approach is based on our combined 50 years of teaching science to nonscience majors who have little or no background in the sciences or mathematics. Our aim in writing *Meteorology* is to present the principles of meterology as simply as possible in the context of everyday examples but without sacrificing scientific integrity. Clear and logical explanations of the principles underlying meteorological phenomena and observations make it unnecessary to write around or avoid even the more sophisticated ideas. We break down basic concepts into elementary components and arrange them so that one concept builds logically on another. The geostrophic wind, for example, is introduced only after a detailed examination of the separate forces contributing to that wind. The atmosphere gradually emerges as a complex and interactive system that is governed by physical laws.

Our emphasis on scientific methodology provides the student with a perspective on the accomplishments of atmospheric scientists and the challenges still facing them. The reader soon understands that weather is not an arbitrary act of nature, and yet weather forecasting has limits and the climatic future is uncertain. We have integrated topics of special contemporary interest including acid deposition, the potential for global climatic change, and threats to the stratospheric ozone shield.

Chapters are arranged in a traditional sequence. Chapter 1 introduces the basic properties and structure of the atmosphere. Chapters 2, 3, and 4 cover radiation and energy conversions within the atmosphere and focus on the global radiation balance. The main theme in these chapters is that weather is a response to imbalances in radiational heating and cooling within the Earth—atmosphere system. Air pressure is discussed in Chapter 5 and is related to other atmospheric properties through the gas law. Water in the atmosphere is the subject of Chapters 6, 7, and 8, with special emphasis on saturation and precipitation processes. Having established the basis for atmospheric circulation, we then examine the forces governing weather systems (Chapter 9). There follow descriptions of planetary-scale circulation (Chapter 10), synoptic-scale weather systems (Chapter 11), and local and regional weather systems (Chapter 12). Chapter 13 deals with the genesis and characteristics of thunderstorms and associated weather hazards including, for example, lightning and flash floods. We cover the life cycles and impacts of tornadoes in Chapter 14 and hurricanes and tropical storms in Chapter 15. Our treatment of weather closes with Chapter 16 on weather forecasting, which is structured to integrate and apply the key concepts developed earlier in the book. A special feature is the inclusion of four final chapters on topics that are of particular contemporary interest: Chapter 17 on air pollution meteorology, Chapter 18 on world climates, Chapter 19 on the climate record, and Chapter 20 on the causes of climate variability.

Although it is desirable to cover Chapters 1 through 16 in sequence, certain sections of most chapters may be dropped without loss of continuity. For example, sections on weather instruments or atmospheric optics may be deleted if necessary. Chapters 17, 18, 19, and 20 may be covered in any order, and any one or all may be omitted to fit the available time.

Special sections, set apart from the main flow of the chapters, provide considerable flexibility with regard to depth and breadth of topic coverage. Each chapter contains one or two supplementary Special Topics, such as "Why Is the Sky Blue?" and "Solar Power," which are related to a major theme of the chapter. Mathematical Notes at the ends of eight chapters provide quantitative discussions of concepts in the chapters, with the basic meteorological equations for those who are interested.

In addition, each chapter features the following elements, designed to guide student understanding:

- chapter outline
- key term list
- summary statements
- review questions
- questions for critical thinking
- selected readings with annotations

New to this edition, many chapters also have quantitative questions; this is part of our effort to more properly represent meteorology as a quantitative science. Furthermore, key words are boldfaced and defined at their first use in the chapter and in the Glossary at the back of the book.

Metric units are used throughout the book, with the English equivalents in parentheses. Unit conversions are given inside the front cover. Psychrometric tables appear inside the back cover. The appendixes feature the history of meteorology, the standard atmosphere, weather map symbols, weather extremes, and climate data for the United States and Canada.

## The Revision

In this fourth edition, we have significantly revised and/or updated our treatment of the global radiation balance, ozone shield, the greenhouse effect, atmospheric stability, clouds, precipitation processes, centripetal force, El Niño and La Niña, vorticity, desert winds, mountain and valley breezes, Doppler radar, hurricane dynamics, weather forecasting, air pollution potential, acid deposition, climate controls, global climate models, and climate change. We are very pleased that Patricia M. Pauley of the Naval Postgraduate School agreed to revise and update Chapter 11, "Air Masses, Fronts, Cyclones, and Anticyclones." In addi-

tion, we have added Special Topics on wind profilers, the link between West African rainfall and the frequency of Atlantic hurricanes, and the problem of chaos in predicting future climate.

We have retained the same basic organization as in the third edition. However, in response to suggestions by many reviewers, we now treat tornadoes and hurricanes in separate chapters. We also cover climatology in three chapters rather than two, although the length of topic coverage is substantially unchanged. We have added an *Introduction* that is designed to encourage students to become involved in observing and following weather events from the outset of the course. This edition features full-color to improve the physical appearance and utility of photographs and line drawings. Brief notes on weather phenomena, known as *Weather Facts,* appear in many chapters.

## Supplements

Supplements available to adopters of this book include an *Instructor's Manual* (with a set of transparency masters of selected line drawings from the text), a Computerized Test Bank, and a set of 50 35-mm color slides of atmospheric phenomena. Also available are a companion *Study Guide* by Joseph M. Moran and *Meteorology Exercise Manual and Study Guide* by Robert A. Paul. Both are published by Macmillan College Publishing Company.

## Acknowledgments

In preparing the four editions of this book, we have profited greatly from the creative ideas and constructive criticisms of many reviewers. We are especially grateful to Gaylen L. Ashcroft, Utah State Climatologist; David D. Houghton, University of Wisconsin–Madison; John E. Oliver, Indiana State University–Terre Haute; Robert A. Ragotzkie, University of Wisconsin–Madison; and Clayton H. Reitan, Northern Illinois University. We also thank L. Dean Bark, Kansas State University; Bradley Bramer, University of Wisconsin–Madison; Arnold Court, California State

University, Northridge; William A. Dando, University of North Dakota; Russell L. DeSouza, Millersville State University; Lee Guernsey, Indiana State University–Terre Haute; B. Ross Guest, Northern Illinois University; Edward J. Hopkins, University of Wisconsin–Madison; William C. Kaufman, University of Wisconsin–Green Bay; Bruce E. Kopplin, University of Nebraska–Lincoln; Garrick B. Lee, Butte College; William H. Long, Florida State University, Tallahassee; Thomas H. McIntosh, University of Wisconsin–Green Bay; David W Marczely, Southern Connecticut State University, New Haven; Shamin Naim, Illinois State University; T. R. Oke, University of British Columbia; Robert A. Paul, Northern Essex Community College; Peter J. Robinson, University of North Carolina at Chapel Hill; Charles C. Ryerson, University of Vermont; Russell Schneider, University of Wisconsin–Madison; William M. Smith, University of Wisconsin–Green Bay; Gregory E. Taylor, Creighton University; Charles L. Wax, Mississippi State University; Wayne M. Wendland, Illinois State Water Survey; James H. Wiersma, University of Wisconsin–Green Bay; Thomas B. Williams, Indiana University at Indianapolis; George Wooten, Hillsborough Community College.

In addition, the following reviewers were helpful in the preparation of the Fourth Edition: Richard A. Anthes, The University Corporation for Atmospheric Research; Terrill R. Berkland, Central Missouri State University; Joanne Bowers, Wayland Baptist University; Donald B. Cruikshank, Jr., Anderson University; Rick Dellinger, Auburn University; Robert Eisenson, SUNY Maritime College; Kenneth Hatfield, Lake Superior State University; Rex J. Hess, University of Utah; Julie Laity, California State University–Northridge; Karl K. Leiker, Westfield State College; Gong-Yuh Lin, California State University–Northbridge; Frank A. Lombardo, Daytona Beach Community College; Max A. Malquist, Anoka Ramsey Community College; Steven B. Newman, Central Connecticut State University; Kurt Nielsen, Naval Postgraduate School; James Norwine, Texas A & I University; Harold Reese, Wayland Baptist University; Carl R. Snipes, Southeastern Community College (Iowa); Alfred Stamm, SUNY Oswego; Allen E. Staver, Northern Illinois University; Douglas D. Streu, National Weather Service; Lynn R. Thomson, Ricks College; Anastasios Tsonis, University of Wisconsin–Milwaukee; Norman K. Wagner, University of Texas–Austin; Jill Wiesman, University of Wisconsin–Green Bay.

We give special thanks to Professor Reid A. Bryson of the University of Wisconsin–Madison for his insight, guidance, and example over the years.

We also have been fortunate in working with some of the most talented and enthusiastic professionals in college publishing. We thank Robert A. McConnin of Macmillan for guiding this project to completion. And we extend a special note of appreciation to Elisabeth Belfer, whose outstanding editorial skills and commitment to quality contributed much to the second, third, and fourth editions.

We are grateful to our typist, Pattie Dimmer, for her fine attention to detail. We thank our wives, Jennifer and Gloria, for their understanding and encouragement. Finally, we acknowledge with gratitude the contributions of our students and colleagues at the University of Wisconsin–Green Bay.

**Joseph M. Moran**
**Michael D. Morgan**
*Green Bay, Wisconsin*

# Brief Contents

# Contents

# 3 Heat and Temperature

58

# 4 Heat Imbalances and Weather

78

# 5 Air Pressure

98

# 6 Humidity and Stability 114

# 7 Dew, Frost, Fog, and Clouds 142

# 8 Precipitation, Weather Modification, and Atmospheric Optics 166

# 9 The Wind

# 10 Planetary-Scale Circulation

# 11 Air Masses, Fronts, Cyclones, and Anticyclones *by Patricia M. Pauley*

# 12 Local and Regional Circulation Systems    276

# 13 Thunderstorms    296

# 14 Tornadoes    322

# 15   Hurricanes   338

# 16   Weather Analysis and Forecasting   356

# 17   Air Pollution Meteorology   382

# 18 World Climates                                            404

# 19 The Climatic Record                                       430

# 20 Causes of Climatic Variability                            450

## Appendixes

# Meteorology

## The Atmosphere and
## the Science of Weather

Meteorology, the subject of this book, is the study of the atmosphere and the processes that cause weather. [Photograph by Mike Brisson]

# Introduction

We are about to embark on a study of the atmosphere, weather, and climate. As with most scientific disciplines, this study requires us to first become familiar with some basic physical principles and ideas. We then apply what we learn in exploring a wide variety of atmospheric phenomena ranging from rainbows to thunderstorms to hurricanes. This means that we are well into the book before actually getting into the more dynamic, practical, and exciting aspects of weather.

Those of us who were counting on studying storms, tornadoes, fronts, hailstorms, and the like right away need not be disappointed, however. There are some activities that we can begin now—today, in fact—and continue thoughout the course that will immediately involve us in weather events, and enrich and enliven our study of the atmosphere. The purpose of this introduction is to suggest and describe some of these activities.

Just about everyone who enrolls in an introductory weather and climate course already has considerable experience with (and understanding of) the subject matter. After all, each of us has been living with weather all our lives. No matter where we live or our walk of life, we are well aware of weather's far-reaching influence. To a large extent, weather dictates the clothes we wear, how we drive, and even our choice of recreational activities (Figure I.1). Before setting off in the morning, most of us check the latest weather forecast on radio or TV and then glance at the sky or read the thermometer. Hence, every day we acquire information on weather through the media, our senses, and perhaps even our own weather instruments. And from that information, we derive a basic understanding of how weather and the atmosphere function.

Until now, for many of us, keeping track of the weather has been a casual but relatively important part of daily life. Our first suggestion is to make weather

**FIGURE I.1**
The variability of weather makes possible a wide variety of recreational activities. [Photographs by J. M. Moran (left) and Mike Brisson (right)]

1

observation a more formal and regular part of the day. From now on, or at least for the duration of this course, tune to a televised weathercast at least once a day. Usually, weathercasts are routine segments of the local morning, noon, and evening news reports. For those of us with access to cable television, we may choose to watch the *Weather Channel,* a 24-hour-a-day telecast devoted exclusively to weather reports. (Comparable cable weather service is available in parts of Canada.)

If television is not available, weather information may be obtained from local or national newspapers or by listening to radio broadcasts. Newspapers usually include a weather page that features maps and statistical summaries. Although local radio stations provide the latest weather conditions and forecasts, most of the time they do not present the big picture, that is, a summary of weather across the nation.

Another valuable source of local weather information is the NOAA weather radio. The National Oceanic and Atmospheric Administration (NOAA) operates low-power, VHF-FM radio transmitters that broadcast continuous weather information 24 hours a day. A taped message is repeated every 4 to 6 minutes and is revised every 2 to 3 hours. Special weather radios are designed to receive this signal and some are even equipped with alarms in the event that severe weather threatens (Figure I.2).

Televised and newspaper weather reports may include (1) national and regional weather maps, (2) satellite photographs (or movie loop) depicting large-scale cloud patterns, (3) data on current and past (24-hour) weather conditions, and (4) weather forecasts for the short term (24 to 48 hours) and long term (up to 5 days or even longer). So that your weather watching is more meaningful and useful, we devote the remainder of this introduction to a description of what to watch for. Let's begin with the national weather map.

National weather maps typically use symbols that represent the principal weathermakers, that is, pressure systems and fronts (Figure I.3). Pressure systems are of two types, highs (or anticyclones) and lows (or cyclones). We can think of atmospheric pressure as the weight per unit area of a column of air that stretches from the Earth's surface to the top of the atmosphere. On any given day, air pressure at the Earth's surface varies from one place to another across the continent. On a weather map, H or HIGH symbolizes regions where the air pressure is relatively high compared to

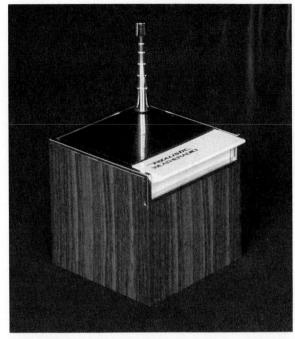

**FIGURE I.2**
At the push of a button, this portable weather radio issues NOAA weather reports and forecasts. [Photograph by J. M. Moran]

that of the surrounding areas and L or LOW symbolizes regions where the air pressure is relatively low compared to that of the surrounding areas.

As you examine weather maps, note the following about pressure systems.

1. Usually highs are accompanied by fair weather and hence are often described as *fair-weather* systems. Highs that originate in northwestern Canada bring cold weather in winter, whereas highs that develop further south bring hot, dry weather in summer.

2. Winds surrounding a high-pressure system blow in a clockwise and outward spiral as shown in Figure I.4A. It is usual, however, for winds to be calm or nearly so over a large area surrounding the center of a high.

3. Most lows are accompanied by cloudy, rainy, or snowy weather and hence are often described as *stormy-weather* systems. An exception may occur over arid or semiarid terrain, especially in summer,

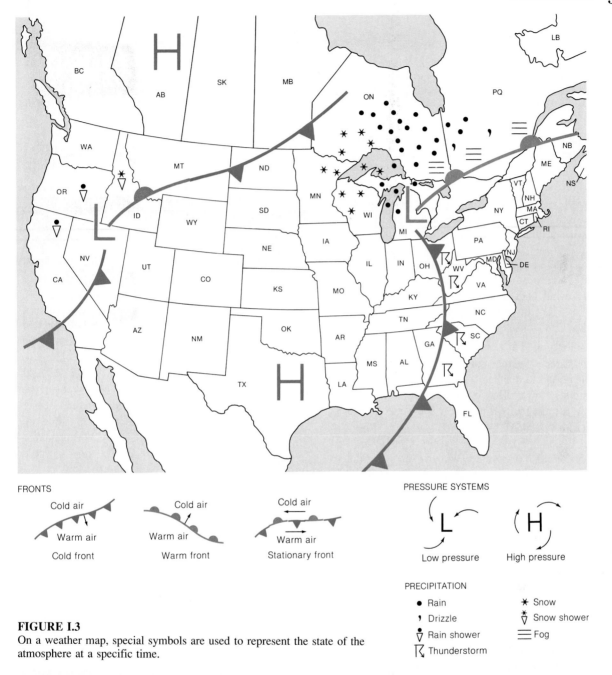

**FIGURE I.3**
On a weather map, special symbols are used to represent the state of the atmosphere at a specific time.

where intense heating by the sun may lower the air pressure and produce a stationary low that is not accompanied by stormy weather.

4. Winds surrounding a low-pressure system blow in a counterclockwise and inward spiral as shown in Figure I.4B.

5. Both lows and highs move with the prevailing winds from west to east across North America and as they do, the weather changes. Lows are followed by highs, and highs are followed by lows. As a rule, lows track toward the east and northeast, whereas highs track toward the east and southeast.

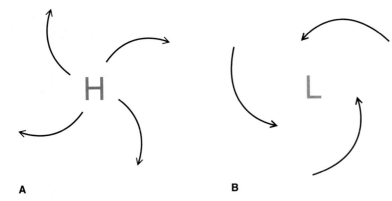

**FIGURE I.4**
Near the Earth's surface, horizontal winds blow (A) in a clockwise and outward spiral about a high-pressure system and (B) in a counterclockwise and inward spiral about a low-pressure system.

6. In general, lows that track across the northern United States or southern Canada produce less rain or snowfall than lows that track further south, for example, out of eastern Colorado or up the eastern seaboard.

7. Weather to the left (west and north) of a storm track tends to be relatively cold, whereas weather to the right (east and south) of a storm track tends to be relatively warm. Hence, for example, in winter, snow is most likely to the west and north of a low-pressure system.

Fronts are also important weathermakers. A front is a narrow zone of transition between air masses that differ in temperature, humidity, or both. An air mass, in turn, is a huge volume of air, covering hundreds of thousands of square kilometers, that is relatively uniform in temperature and humidity. The specific characteristics of an air mass depend on the type of surface over which the air mass forms and travels. Hence, cold air masses form at high latitudes over ground that is often snow covered, whereas warm air masses form in the tropics where the Earth's surface is relatively warm year-round. Humid air masses form over moist surfaces (e.g., Pacific Ocean, Gulf of Mexico), and dry air masses develop over dry surfaces (e.g., desert southwest, northwestern Canada). Thus, the four basic types of air masses are cold and dry, cold and humid, warm and dry, and warm and humid.

Where contrasting air masses meet, a front forms; along a front, cloudiness and precipitation often develop. The most common fronts are stationary, cold, or warm. Weather map symbols for all three are shown in Figure I.3. As the name implies, a stationary front is just that, stationary. On either side of a stationary front, winds blow parallel to the front but in opposite directions. A change in wind direction may cause a portion of a stationary front to become active and advance northward as a warm front or southward as a cold front. Warm air is less dense than cold air so that warm air advances by gliding up and over cold air. The cold air thus forms a wedge under the warm air (Figure I.5A). On the other hand, denser cold air advances by sliding under and pushing up the less dense warm air (Figure I.5B). Consequently, a warm front slopes more gently than does a cold front.

As you examine weather maps, note the following about fronts.

1. The map symbol for a front is plotted where the front intersects the Earth's surface.

2. Most cloudiness and precipitation associated with a warm front occur over a broad band, often hundreds of kilometers wide, to the north and east of the surface front. Precipitation ahead of a warm front generally is light to moderate in intensity and may persist for upward of 12 to 24 hours.

3. Most cloudiness and precipitation associated with a cold front occur in a narrow band along or just ahead of the front. Often the precipitation is showery.

4. Wind direction shifts from one side of a front to the other.

5. Some fronts are not accompanied by cloudiness and/or precipitation.

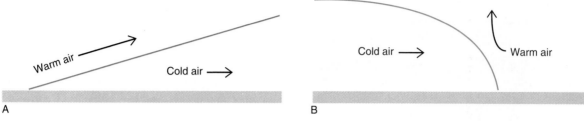

**FIGURE I.5**
The two most common types of fronts are (A) a warm front, which marks the boundary between advancing warm air and retreating cold air, and (B) a cold front, which marks the boundary between advancing cold air and retreating warm air. Both diagrams are vertical cross sections.

6. In summer, air temperatures are sometimes very nearly the same both ahead of and behind a cold front. In that case, the air masses differ primarily in humidity, that is, the air mass ahead of the front is more humid and the air mass behind the front is drier (and more comfortable).

7. On a weather map, cold and warm fronts usually form broad arcs that are anchored at the center of a low-pressure system. In fact, the counterclockwise and inward circulation about a low-pressure system brings contrasting air masses together to form fronts.

8. Low-pressure systems may form along a stationary front and travel rapidly along that front.

9. Thunderstorms and severe weather (e.g., tornadoes) often occur to the east of a sharply defined cold front and south of a warm front.

In addition, as you monitor national and regional weather maps, watch for

1. Cool sea breezes or lake breezes that push inland and bring relief on a hot summer afternoon.

2. In late fall and throughout much of the winter, heavy lake-effect snows that fall in narrow belts on the downwind (east and south) shores of the Great Lakes and Great Salt Lake.

3. The relatively high frequency of tornadoes and severe thunderstorms in spring across the central United States (from Texas north to Nebraska and Iowa east to Indiana).

4. The low frequency of thunderstorms along the Pacific coast and the high frequency of thunderstorms in Florida and the western High Plains.

5. Tropical storms and hurricanes off the Atlantic and

Gulf coasts, primarily from August through October.

In telecasts and in newspapers, statistical summaries of present and past weather conditions use a number of parameters. It is useful to briefly comment on their meaning.

1. *Maximum temperature.* The highest air temperature recorded over a 24-hour period, typically from midnight to midnight. Usually, but not always, the maximum temperature occurs in the early to mid afternoon.

2. *Minimum temperature.* The lowest air temperature recorded over a 24-hour period, typically from midnight to midnight. Usually, but not always, the minimum temperature occurs around sunrise.

3. *Dew-point (or frost-point) temperature.* The temperature to which air must be cooled for dew (or frost) to form on a relatively cold surface.

4. *Relative humidity.* A measure of the water vapor concentration in air, expressed as a percentage (from 0% to 100%). The relative humidity varies with temperature so that on most days, the relative humidity is highest when it is coldest (around sunrise) and lowest when it is warmest (in the early afternoon).

5. *Precipitation amounts.* Rainfall or snowfall over a 24-hour period, from midnight to midnight, usually expressed in inches. On average, 10 inches of snow melt down to 1 inch of water.

6. *Air pressure.* The weight of a column of air over a unit area of the Earth's surface. In the traditional instrument used to monitor air pressure (mercury barometer), air pressure supports a column of mer-

**FIGURE I.6**
A wind vane points in the direction *from which* the wind blows.

cury in a glass tube. The mercury column fluctuates up and down as the air pressure rises and falls. Hence, air pressure is commonly expressed as the height of the mercury column in inches (e.g., 29.97 in.). Falling air pressure often indicates the approach of a low-pressure system and a turn to stormy weather. Rising air pressure, on the other hand, signals the approach of a high-pressure system and clearing skies or continued fair weather.

7. *Wind direction* and *wind speed*. Wind direction is the compass direction *from which* the wind blows (Figure I.6). As a general rule, a wind shift from east to northeast to north is accompanied by falling temperatures. On the other hand, a wind shift from east to southeast to south usually brings warmer weather. Within a high, winds are light or calm. Wind speeds tend to increase when a front approaches and are particularly gusty in thunderstorms.

8. *Sky conditions*. Based on the fraction of the sky that is covered by clouds, the sky is described as clear (less than 1/10 cloud-covered), scattered clouds or partly cloudy (1/10 to 5/10), broken clouds or cloudy (6/10 to 9/10), or overcast (more than 9/10 cloud-covered). All other factors being equal, nights are coldest when the sky is clear.

9. *Weather watch*. Issued when atmospheric conditions are favorable for the development of severe weather such as tornadoes or heavy snowfall.

10. *Weather warning*. Issued when severe weather conditions have developed and are approaching your location.

At this beginning stage of our study of the atmosphere and weather, it is also a good idea to develop the habit of observing the sky and changes in clouds and cloud cover. Here are some things to watch for.

1. Clouds are made up of tiny water droplets, ice crystals, or some combination of these. Ice-crystal clouds occur at relatively high altitudes where temperatures are low; they exhibit a fibrous appearance such as shown in Figure I.7A. Water-

**FIGURE I.7**
Clouds that are composed of mostly ice crystals appear fibrous (A), whereas clouds that are mostly water droplets have more sharply defined edges (B). [Photographs by J. M. Moran]

A

B

**FIGURE I.8**
Low, dark clouds accompany a fast-moving, well-defined cold front. [Photograph by J. M. Moran]

**FIGURE I.9**
Fair-weather cumulus clouds. [Photograph by J. M. Moran]

droplet clouds occur at lower altitudes where temperatures are higher; they are characterized by more sharply defined edges such as shown in Figure I.7B.

2. As a warm front approaches your locality, clouds gradually lower and thicken so that eventually they block the sun during the day or the moon at night.

3. A well-defined cold front may be accompanied by towering clouds that produce bursts of heavy rain, lightning, and thunder (Figure I.8).

4. The day may begin clear but after several hours of bright sunshine, small puffy clouds often develop (Figure I.9). These are fair-weather cumulus clouds that dissipate rapidly as the sun sets.

5. Sometimes cumulus clouds build both vertically and laterally and eventually form cumulonimbus (thunderstorm) clouds (Figure I.10).

In summary, we can learn much about weather systems by keeping track of local, regional, and national weather patterns via television, radio, and newspapers. In addition, we are well advised to develop the habit of watching the skies for changing conditions and to monitor weather instruments if they are available. In this way, we are able to get involved with the subject matter of the course from the beginning and what we learn becomes more meaningful and practical.

**FIGURE I.10**
Cumulus clouds may build into cumulonimbus (thunderstorm) clouds. [Photograph by J. M. Moran]

# 1 Atmosphere: Origin, Composition, and Structure

*A generation goes, and a generation comes. But the earth remains forever.*

*The sun rises and the sun goes down, and hastens to the place where it rises.*

*The wind blows to the south, and goes round to the north; round and round goes the wind, and on its circuits the wind returns.*

*Ecclesiastes 1:4–6*

The atmosphere is a thin envelope of gases and aerosols surrounding the planet. [Photograph by Tranquality/Philip Chaudoir]

**TABLE 1.1**
**Approximate Annual Losses to Selected Weather Hazards in the United States**

| Hazard | Fatalities | Cost (millions of dollars) |
|---|---|---|
| Flood | 163 | 3175[a] |
| Hurricane | 33 | 796[b] |
| Tornado | 98 | 300 |
| Lightning | 97 | 200 |
| Hail | | 750 |
| Drought | | 800 |

*Source:* W. E. Riebsame et al. "The Social Burden of Weather and Climate Hazards," *Bulletin of the American Meteorological Society* 67 (1986):1379.

[a]In 1985 dollars.
[b]In 1982 dollars.

E
VERYONE SEEMS TO BE INTERESTED in the weather, probably because it affects virtually every aspect of daily life—our clothing, our outdoor activities, the price of oranges and coffee in the grocery store, and even the outcome of a football game. Tranquil, pleasant weather allows us to enjoy a variety of outdoor activities. A turn to stormy weather may bring mixed blessings: heavy rains wash out picnics but also benefit crops wilting under the searing summer sun. Occasionally, the weather is extreme, and the impact may range from mere inconvenience to a disaster that is costly in human lives and property (Table 1.1). Heavy snowfall snarls commuter traffic (Figure 1.1), thick fog causes flight delays and cancellations, and a night of subfreezing temperatures takes its toll on Florida citrus. But these impacts are minor when compared to the death, injury, and property damage that may attend a hurricane or tornado (Figure 1.2).

Regardless of where we live in the world, each of us is well aware from personal experience that weather is variable. This variability prompted Mark Twain—not one to shy from exaggeration—to quip of spring weather in New England: "I have counted one hundred and thirty-six different kinds of weather inside of four-and-twenty hours."* Of course, weather's variability is not the same everywhere. For example, the temperature contrast between winter and summer is much more pronounced on Canada's prairies, where summers are very warm and winters are bitter cold, than in south Florida, where the weather is usually subtropical year-round.

Any reasonable definition of weather must reflect its geographical and temporal variability. Hence, we define **weather** as the state of the atmosphere at some

*Excerpted from "Address of Mr. Samuel L. Clemens: The Weather in New England," *New England Society in the City of New York, Annual Report,* 1876, p. 59.

**FIGURE 1.1**
Heavy snowfall can disrupt motor vehicle traffic and cause considerable inconvenience. [Photograph by Arjen and Jerrine Verkaik/SKYART]

**FIGURE 1.2**
Occasionally, the weather turns violent and may take lives and cause considerable property damage. Hurricane Camille caused this damage in Biloxi, Mississippi, in August 1969. [NOAA photograph]

place and time, described in terms of such variables as temperature, cloudiness, precipitation, and wind speed and direction. **Meteorology** is the study of the atmosphere and the processes that cause weather.

**Climate** is often defined as weather conditions at some locality averaged over a specified time period, but climate encompasses more than this. Departures from long-term averages and extremes in weather are also important aspects of climate. For example, farmers are interested in knowing not only the average rainfall for July, but also the frequency of extremely wet or dry Julys. Climate is the ultimate environmental control in that it governs, for example, what crops can be cultivated, the fresh water supply, and the average heating and cooling requirements for homes. **Climatology** is the study of climate, its controls and variability.

The **atmosphere,** where weather takes place, encircles the globe as a relatively thin envelope of gases and tiny, suspended particles. In fact, 99% of the atmosphere's mass is confined to a layer that in thickness is only about 0.25% of the Earth's diameter.

Hence, the planet's atmosphere is comparable in thickness to the thin skin of an apple. Yet, the thin atmospheric skin is essential for life and the orderly functioning of physical and biological processes on Earth. The atmosphere shields organisms from exposure to hazardous levels of ultraviolet radiation;* it contains the gases necessary for the life-sustaining processes of cellular respiration and photosynthesis; and it supplies the water needed by all forms of life.

## Understanding the Atmosphere

Our current understanding of the atmosphere, weather, and climate is the culmination of centuries of painstaking inquiry by scientists from many disciplines. Physicists, chemists, astronomers, and others have ap-

*Radiation and its various forms are discussed in detail in Chapter 2.

plied basic principles in unlocking the mysteries of the atmosphere. The roots of modern meteorology, in fact, reach back to the fourth century B.C. and Aristotle's *Meteorologica,* the first treatise on atmospheric science. Other milestones in the history of meteorology are listed in Appendix I.

Although much progress has been made, many questions remain regarding the workings of the atmosphere, weather, and climate. Modern atmospheric scientists (meteorologists and climatologists) continue the efforts of their predecessors, and, although armed with more sophisticated tools such as satellites, radar, and electronic computers, they still rely on the scientific method of investigation.

## THE SCIENTIFIC METHOD

The **scientific method** is a systematic form of inquiry involving observation, speculation, and reasoning. An acid rain problem illustrates how scientists apply the scientific method.

Lakes in the Adirondack Mountain region of upstate New York have long been noted for their abundance of game fish and other aquatic life, and good fishing had made the lakes a popular attraction for sportsmen and other vacationers. More than two decades ago, however, populations of fish in some Adirondack lakes began to decline. Local businesses became concerned that poor fishing would hurt the region's recreational industry, and environmentalists speculated that the lakes were becoming yet another casualty of pollution.

Biologists initially proposed that toxic substances (poisons) were entering the lakes and killing aquatic life, but they did not know the nature of the toxins or their origins. Subsequent chemical testing of water samples failed to find hazardous concentrations of toxins, but did reveal that lake waters were abnormally acidic. Scientists hypothesized that acidic rainwater (and snowmelt) was responsible for acidifying the lakes. And laboratory studies have shown that excessively acidic waters are lethal to young fish.

Why was the rainwater (and snowmelt) so acidic? Rain (and snow) is normally slightly acidic, because raindrops (and snow flakes) dissolve some of the carbon dioxide ($CO_2$) in the air and form the same weak and harmless acid found in carbonated beverages. Rain-

water samples collected in the vicinity of the Adirondack lakes, however, were at least one hundred times more acidic than expected. Additional laboratory tests identified sulfuric acid ($H_2SO_4$) and nitric acid ($HNO_3$) in the rainwater, which scientists proposed were formed from oxides of sulfur and nitrogen—common industrial air pollutants—dissolving in rainwater.

The next question concerned the source of the sulfur and nitrogen oxides. Because the Adirondack Mountains are downwind of some major industrial sources of those air pollutants, such as coal-fired electric power plants, scientists reasoned that the loss of fish in Adirondack lakes was linked, at least circumstantially, to industrial air pollution. Today, the hypothesis that excessively acidic rainwater (and snowmelt) led to the fish kills is generally accepted because it is consistent with our knowledge of (1) the limited tolerance of fish to acidic waters and (2) the chemical reactions that involve rainwater and air pollutants.

From our acid rain example, it is evident that the scientific method involves a sequence of steps in which scientists (1) identify specific questions related to the problem at hand; (2) propose an answer to one of these questions in the form of an educated guess; (3) state the educated guess in such a way that it can be tested, that is, formulate a **hypothesis;** (4) predict what the outcome of the test would be if the hypothesis were correct; (5) test the hypothesis by checking to see if the prediction is correct; and (6) reject or revise the hypothesis if the prediction is wrong.

In actual practice, scientists do not follow this scheme cookbook style, and discrete steps are often thoroughly integrated as a single avenue of inquiry. Furthermore, the scientific method is not a formula for creativity because it does not provide the key idea, the hunch, or the educated guess that spontaneously springs to mind and forms the original basis of the hypothesis. Rather, it is a technique for assessing the validity or worth of a creative key idea however and wherever it originates.

As in the acid rain example, a hypothesis is a tool that suggests new experiments or observations or opens new avenues of inquiry. Hence, even an erroneous hypothesis may be fruitful. Above all, scientists must bear in mind that a hypothesis is merely a working assumption that may be accepted, modified, or rejected. They must be objective in evaluating a hypothesis and not allow personal biases or expectations to cloud that evaluation. In fact, scientists actively search for observations or information that could disprove their beliefs. If they find such evidence, they revise their hypothesis to incorporate the disparate observations or information. Inquiry, creative thinking, and imagination are stifled when a hypothesis is considered to be immutable.

A new hypothesis (or an old, resurrected one) may be hotly debated within the scientific community. History warns us of a natural human resistance to new ideas that threaten to displace long-held notions. Also, disagreements among scientists on a particularly controversial issue sometimes receive considerable media attention, which may confuse the general public. As of this writing, for example, climatologists are engaged in public debate as to whether burning coal and oil for power is leading to global warming.

In some cases, the prevailing public reaction may be, "Well, if the so-called experts can't agree among themselves, whom am I to believe? Is there really a problem after all?" However, debate and disagreement are essential steps in the process of reaching scientific understanding; they generate useful suggestions, stimulate new thinking, and uncover errors. In fact, such debate and skepticism buffer the scientific community from too hastily accepting new ideas. If a hypothesis survives the scrutiny and skepticism of scientists and the public, it is probably accurate.

## ATMOSPHERIC MODELS

In applying the scientific method, scientists often find that models aid their investigation. This is certainly the case in the atmospheric sciences. Because we will be using models throughout this book, we consider at the outset some of the general objectives and limitations of scientific models, especially as they apply to meteorology and climatology.

A **scientific model** is defined as an approximate representation or simulation of a real system.* A model eliminates all but the essential variables, or characteristics of that system. For example, to learn how to improve the fuel efficiency of automobiles, we might examine a model automobile in a wind tunnel. An automobile can be designed to reduce its air resistance, which, in turn, can increase its miles per gallon. The shape of the model automobile is the critical variable and is the focus of study. Other variables, such as the color of the model automobile or whether it is equipped with whitewall tires, are irrelevant to the experiment and are ignored. Often what is or is not relevant is determined by trial and error.

Sometimes models are used to organize information. Because they are not cluttered with extraneous and distracting details, they may provide important insights as to how things interact, or they may trigger creative thinking about complex phenomena. Models can also be used to make predictions. For example, in a model composed of many variables, one variable may be perturbed in order to assess its effect on the other variables.

Depending on their particular function, scientific models are classified as conceptual, graphical, physical, or numerical. A **conceptual model** describes the general relationships among components of a system. For example, the geostrophic wind (described in Chapter 9) is a conceptual model that relates the interaction of certain forces operating in the atmosphere to straight, horizontal winds that blow at altitudes of 1000 meters (m) (3280 ft)* or higher. A **graphical model** assembles and displays data in an organized format that can readily be interpreted. For example, a weather map (Figure I.3) integrates weather observations taken simultaneously at hundreds of locations into a coherent representation of the state of the atmosphere.

A **physical model** is a miniaturized version of some system. For example, atmospheric scientists at Purdue University simulate a tornadic circulation in a specially-designed Tornado Vortex Chamber (Figure 1.3). Precision instruments monitor the detailed motions within vortices (swirls) such as the one shown in Figure 1.4. In this way, scientists learn more about the internal characteristics of tornadoes.

Scientific modeling is greatly assisted by electronic computers that can accommodate enormous quantities of data and perform calculations very rapidly. Typically, a computer is programmed with a **numerical model** consisting of one or more mathematical equations that portray the behavior of a particular physical system such as the atmosphere. Variables in the numerical model (such as temperature or humidity) are manipulated, individually or in groups, in order to assess their impact on the system.

Computerized numerical models of the atmosphere have been used to forecast the weather since the 1950s.

---

*A *system* is composed of parts that interact in an orderly manner according to some governing principles.

*Metric units, followed by the English-unit equivalents, are used throughout this book. For unit conversions, see the inside front cover.

**FIGURE 1.3**
Exterior view of Purdue University's Tornado Vortex Chamber used to simulate the circulation within tornadoes. The apparatus is 8 m high, 5 m wide, 5 m deep, and features a work area/viewing platform. [Photograph by the Center for Instructional Services, Purdue University, courtesy of Dr. John T. Snow and C. R. Church]

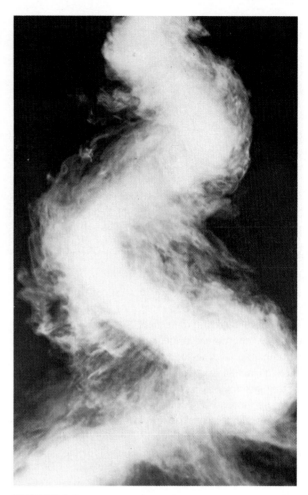

**FIGURE 1.4**
An example of a vortex generated by Purdue University's Tornado Vortex Chamber. [Photograph courtesy of Dr. John T. Snow and C. R. Church, Purdue University]

More recently, they have also been used to predict the potential impact on the global climate of rising levels of atmospheric carbon dioxide. As discussed in more detail in Chapters 2 and 20, the atmospheric carbon dioxide concentration has been increasing for many decades, primarily as a byproduct of fossil fuel (coal, oil, and natural gas) burning. Higher air temperatures may be the consequence because carbon dioxide slows the escape of the Earth's heat to space.

Atmospheric scientists have employed numerical models of the atmosphere in experiments to predict the temperature rise that might accompany a continued in-crease in atmospheric carbon dioxide concentration (Figure 1.5). Three steps are followed. First, they design a numerical model of the atmosphere that accurately depicts the present worldwide air temperature pattern, given the current level of atmospheric carbon dioxide. Second, holding all other variables (except temperature) constant, they elevate the carbon dioxide concentration (typically, it is doubled), and the numerical model computes a new worldwide air temperature pattern. Third, they subtract the initial temperature pattern from the final, predicted temperature pattern. Presumably, the net temperature change can be attributed to elevated carbon dioxide levels.

It is important to emphasize that models are only

**FIGURE 1.5**
Numerical model prediction of the temperature change that would accompany a doubling of atmospheric carbon dioxide concentration. Temperature changes greater than 4 Celsius degrees are shaded deep red and cross-hatched; those greater than 8 Celsius degrees are shown in bright red; black depicts areas that showed a warming of 0 to 4 Celsius degrees. [National Center for Atmospheric Research/University Corporation for Atmospheric Research/National Science Foundation]

simulations of reality and, hence, subject to error. For example, one potential difficulty with numerical models concerns the accuracy of their component equations. Typically, the equations are only approximations of the way a system really works in nature, and they may not adequately account for all the relevant variables. This is one reason why long-range weather forecasting, based on numerical models of the atmosphere, declines in accuracy as the forecast period lengthens.

## Evolution of the Atmosphere

Now that we have a general idea of how meteorologists, climatologists, and other scientists go about the job of studying the atmosphere, it is time to begin our investigation of the basic properties of the atmosphere. In this section we review what is presently understood about the origins of the atmosphere.

The Earth's atmosphere is the product of a lengthy evolutionary process that began at the Earth's birth about 4.6 billion years ago. Astronomers scanning the solar system and geologists analyzing evidence obtained from meteorites, rocks, and fossils have given us a reasonable, albeit as yet incomplete, scenario of the origins of the atmosphere.

### PRIMEVAL PHASE

Earth as well as the sun and the entire solar system is believed to have developed out of an immense cloud of dust and gases within the Milky Way galaxy. In the beginning, Earth was an aggregate of dust and meteorites surrounded by a gaseous envelope of mostly hydrogen and helium. For millions of years, the Earth's mass grew by accretion as the planet swept up cosmic dust in its path. Bombardment by meteorites heated up the Earth's surface and its atmosphere, eventually driving off most of the original gases.

Earth became geologically active as volcanoes spewed forth huge quantities of lava, ash, and a variety of gases. Then as now the principal gaseous emissions of volcanoes consisted of water vapor ($H_2O$), carbon dioxide ($CO_2$), and some nitrogen ($N_2$). Initially, free oxygen (O or $O_2$) was absent, although oxygen was combined with other elements in various chemical compounds such as carbon dioxide. Millions of years of meteorite bombardment and volcanic activity produced a dense atmosphere rich in carbon dioxide and some nitrogen. Intense solar radiation in the upper atmosphere caused some water vapor to break down into its constituent atoms, thereby contributing minor amounts of oxygen to the early atmosphere. Radioactive decay of an isotope of potassium (potassium-40) in the planet's bedrock added argon (Ar), an inert gas, to the evolving atmosphere.

This early atmosphere was perhaps 10 to 20 times denser than the present atmosphere. The abundance of carbon dioxide meant higher surface temperatures than now because, as mentioned previously, carbon dioxide slows the escape of the Earth's heat to space. Numerical models that simulate this early atmosphere predict an average surface temperature in the range of 85 to 110 °C (185 to 230 °F).* This phase of the planet's primeval atmosphere probably lasted for several hundred million years.

With the subsequent formation of seas and the coming of life, the atmosphere continued to evolve. Perhaps 3.8 billion years ago, volcanic activity and meteorite bombardment subsided somewhat, and the Earth's surface and atmosphere gradually cooled. Cooling caused some of the water vapor to condense into clouds, and rains gave rise to the first rivers, lakes, and seas. The global water cycle had begun. In those seas, the first primitive forms of life appeared about 3.5 billion years ago, and with the coming of the first marine plants, photosynthesis began. **Photosynthesis** is the process whereby plants use sunlight, water, and carbon dioxide to manufacture their food. A byproduct of this process is oxygen, which is released to the atmosphere. Through subsequent millions of years, photosynthesis added oxygen to the atmosphere so that eventually oxygen became the second most abundant atmospheric gas after nitrogen.

For reasons not totally understood, the concentration of atmospheric oxygen has fluctuated significantly since the arrival of photosynthetic organisms. Geological evidence suggests a dramatic rise in oxygen levels about 2.1 to 1.7 billion years ago and again about 1.1 billion years ago to 700 million years ago. The latter pulse in oxygen concentration apparently triggered the evolution of large animals. Also, scientists found that ancient air bubbles trapped in 85-million-year-old amber (hardened tree resin) contained 50% more oxygen than modern air.

While atmospheric oxygen concentrations trended upward, the concentration of carbon dioxide declined, albeit with some important reversals. Photosynthesis removed some $CO_2$, but much of it dissolved in ocean waters. Marine organisms used some of the dissolved carbon dioxide in building their shells, and when they

*The various temperature scales are described in detail in Chapter 3.

died their remains accumulated as sediment on the sea floor. In time these sediments converted to marine carbonate rocks such as limestone ($CaCO_3$). With the global carbon cycle established, atmospheric carbon dioxide declined to only a small fraction of its original concentration.

From time to time in the geologic past, the concentration of atmospheric carbon dioxide has fluctuated with important implications for climate. All other factors being equal, a more $CO_2$-rich atmosphere means a warmer global climate. Geological evidence points to a burst of volcanic activity on the Pacific ocean floor about 100 to 120 million years ago. Some of the carbon dioxide released by that activity eventually reached the atmosphere and elevated the temperature of the lower atmosphere. The result was tropical conditions at high latitudes. During the Ice Age, also, atmospheric carbon dioxide levels fluctuated, decreasing during episodes of glacial expansion and increasing during episodes of glacial recession (Chapter 19).

## MODERN PHASE

Ultimately, these gradual evolutionary processes produced the modern atmosphere, which is a mixture of many different gases. Because the lower atmosphere continually circulates and mixes, the principal atmospheric gases occur almost everywhere in about the same relative proportions up to an altitude of about 80 kilometers (km) (50 mi). That portion of the atmosphere is called the **homosphere.** Above 80 km, gases are stratified such that concentrations of the heavier gases decrease more rapidly with altitude than do concentrations of the lighter gases. The region of the atmosphere above 80 km is known as the **heterosphere.**

Nitrogen and oxygen are the chief gases of the homosphere. Not counting water vapor (which has a highly variable concentration), nitrogen ($N_2$) occupies 78.08% by volume of the homosphere, and oxygen ($O_2$) is 20.95% by volume. The next most abundant gases are argon (0.93%) and carbon dioxide (0.035%). As shown in Table 1.2, the atmosphere also contains small quantities of helium (He), methane ($CH_4$), hydrogen ($H_2$), ozone ($O_3$), and several other gases. Unlike the atmosphere's principal gases, the percent volume of

**TABLE 1.2**
**Relative Proportions of Gases Composing Dry Air in the Lower Atmosphere (below 80 km)**

| Gas | % by Volume | Parts per Million |
|---|---|---|
| Nitrogen | 78.08 | 780,840.0 |
| Oxygen | 20.95 | 209,460.0 |
| Argon | 0.93 | 9,340.0 |
| Carbon dioxide | 0.035 | 350.0 |
| Neon | 0.0018 | 18.0 |
| Helium | 0.00052 | 5.2 |
| Methane | 0.00014 | 1.4 |
| Krypton | 0.00010 | 1.0 |
| Nitrous oxide | 0.00005 | 0.5 |
| Hydrogen | 0.00005 | 0.5 |
| Ozone | 0.000007 | 0.07 |
| Xenon | 0.000009 | 0.09 |

some of these trace gases varies with time and location within the homosphere.

Within the heterosphere, above about 150 km (93 mi), oxygen is the chief atmospheric gas but occurs primarily in the atomic (O) rather than diatomic ($O_2$) form. Ultraviolet radiation from the sun photodissociates $O_2$ into its constituent atoms. **Photodissociation** is the breakdown of molecules by radiation. Two oxygen atoms can recombine into a molecule only by colliding with another atom or molecule, but air density (mass per unit volume) at these high altitudes is so low that such collisions are infrequent. At lower altitudes the intensity of incoming solar ultraviolet radiation is less, thereby reducing the rate of photodissociation of $O_2$. Also, with greater air density at lower altitudes, molecular collisions are more frequent. Hence, below about 100 km (62 mi), the rate of recombination of oxygen atoms exceeds the rate of photodissociation of oxygen molecules and oxygen occurs mostly as $O_2$.

The Earth's nitrogen/oxygen-dominated atmosphere is in striking contrast to the carbon dioxide rich atmospheres of neighboring planets, Venus and Mars. The atmosphere of Venus is almost 100 times denser than the Earth's atmosphere and features an average surface temperature of about 475 °C (890 °F). The Martian atmosphere, on the other hand, is much thinner than the Earth's atmosphere and has an average surface temperature of about −53 °C (−63 °F). This contrast occurs even though all three planets likely started out with very similar atmospheres. The atmospheres of Earth,

Mars, and Venus evidently followed different evolutionary paths. For more on this, refer to the Special Topic "The Martian Atmosphere."

In addition to gases, the Earth's atmosphere contains minute liquid and solid particles, collectively called **aerosols.** Some aerosols—water droplets and ice crystals—are visible as clouds. Others are too small to be visible. Most aerosols occur in the lower atmosphere near their source, the Earth's surface. They originate through forest fires, from wind erosion of soil, as tiny sea-salt crystals from ocean spray, in volcanic emissions, and from industrial and agricultural activities. Also some aerosols, such as meteoric dust, enter the atmosphere from above.

It may be tempting to dismiss as unimportant those substances that make up only a small fraction of the atmosphere, but the significance of an atmospheric gas or aerosol is not necessarily related to its relative abundance. For example, water vapor, carbon dioxide, and ozone ($O_3$) occur in minute concentrations, yet they are essential for life. By volume, no more than about 4% of the lowest kilometer of the atmosphere is water vapor—even in the warm, humid air over tropical oceans and rainforests. Without water vapor, however, there would be no clouds, and no rain or snow to replenish soil moisture, rivers, lakes, and seas. Although composing only 0.035% of the homosphere, carbon dioxide is essential for photosynthesis. Furthermore, water vapor and carbon dioxide act as a blanket over the Earth's surface, causing the lower atmosphere to retain heat and making the planet warmer and more amenable to life. Although the volume percentage of ozone is minute, this vital gas shields organisms including ourselves from exposure to potentially lethal intensities of ultraviolet radiation from the sun.

The aerosol concentration of the atmosphere is also relatively small, yet these suspended particles participate in important processes. Some aerosols act as nuclei for the development of clouds and precipitation, and some influence air temperature by interacting with solar radiation.

## AIR POLLUTANTS

Human activity also plays a role in the evolution of the atmosphere primarily through our contribution to

# The Martian Atmosphere

Sensors aboard NASA's unmanned Mariner and Viking spacecraft confirmed suspicions that the Martian atmosphere differs considerably from the Earth's atmosphere. The Martian atmosphere is 95% carbon dioxide, 2% to 3% nitrogen, 1% to 2% argon, and 0.1% to 0.4% oxygen. In addition, the Martian atmosphere is much thinner; its surface pressure is only 0.7% of Earth's average sea-level air pressure. In the beginning, more than 4 billion years ago, however, the atmospheres of both planets probably were quite similar, but for several reasons followed distinctly different evolutionary paths.

*Outgassing* produced the primeval atmospheres of both Earth and Mars. Gases were released from ancient planetary rock through volcanic eruptions and through impact when meteorites collided with the rocky surfaces of the planets. The gases were probably the same because the source rocks on both planets were chemically the same. Rock chemistry depends on the temperature at which crystallization takes place, which in turn depends on the planet's distance from the sun. Mars' greater distance from the sun (about 50% farther than Earth) would not produce significantly different crystallization (rock-forming) conditions. Hence, through outgassing, the primeval atmospheres of both planets were mostly carbon dioxide along with some nitrogen and water vapor—the principal gaseous emissions of volcanoes, both ancient and modern.

As noted elsewhere in this chapter, formation of oceans, establishment of the global water and carbon cycles, and the arrival of photosynthetic plants gradually altered the Earth's primeval atmosphere so that eventually nitrogen and oxygen became the primary ingredients and carbon dioxide and water vapor became minor components. On Mars, however, carbon dioxide remained the principal gas.

Within the scientific community currently there is no consensus on why the atmospheres of Earth and Mars evolved differently. One hypothesis attributes the difference to contrasts in the volcanic histories of the two planets. On Mars, the bulk of volcanic activity apparently took place during the planet's first 2 billion years, whereas, on Earth, volcanism has been more or less continuous throughout the planet's history. A decline in Martian volcanism meant less $CO_2$ released to the atmosphere. At the same time, other geological processes began to reduce the density of the Martian atmosphere by removing carbon dioxide. Some $CO_2$ adhered to the fine sediment that blankets the planet's surface, and some carbon dioxide was locked up in carbonate rocks. Consequently, $CO_2$ in the Martian atmosphere gradually fell to a very small fraction of its original value.

As the amount of atmospheric $CO_2$ on Mars declined, so too did the planet's surface temperature. Carbon dioxide slows the loss of heat to space so that as $CO_2$ thins, more heat escapes and air temperatures drop. Three to four billion years ago Martian surface temperatures were so high that water occurred in liquid form. Seas, lakes, and rivers formed; in fact, channels cut by these ancient rivers were photographed by the Mariner spacecraft. Today the mean temperature on the Martian surface is about −53 °C (−63 °F), much too low for running water. Water in unknown quantities likely remains on the planet mixed with solid carbon dioxide (dry ice) in the polar ice caps and as scattered patches of permafrost. Severe cold also meant no life on Mars, no photosynthesis, and hence, no free oxygen in the atmosphere.

The decline in Martian volcanism also cut the supply of nitrogen, and much of the original nitrogen gradually escaped the planet's relatively weak gravitational field. Gravity, the force that holds the atmosphere to a planet, is about 38% weaker on Mars than on Earth. This is because Mars is smaller (53% of the Earth's equatorial diameter) and less massive (10% of the Earth's mass).

In summary, the difference in evolutionary paths taken by the atmospheres of Earth and Mars may be due largely to physical contrasts between the two planets. The Earth has been more volcanically active (greater outgassing) and is more massive (stronger gravitational field) than Mars.

Future spacecraft missions to Mars promise to provide more data that will add to our understanding of the evolution of the Martian atmosphere. The most recent mission to Mars, NASA's Mars Observer spacecraft, was launched in September 1992. However, communications with the spacecraft failed on 21 August 1993, only three days before it was scheduled to orbit the planet. The next missions to Mars are Russian and European ventures scheduled for launch in 1994 and 1996.

air pollution. An **air pollutant** is a gas or aerosol that occurs at a concentration that threatens the well-being of living organisms (especially humans) or disrupts the orderly functioning of the environment. Many of these substances occur naturally in the atmosphere. Sulfur dioxide ($SO_2$) and carbon monoxide (CO), for example, are normal minor components of the atmosphere that are considered pollutants only when their concentrations approach or exceed the tolerance limits of organisms. In sufficiently high concentrations, sulfur dioxide damages the human respiratory system and carbon monoxide is an asphyxiating agent (reduces the blood's oxygen-carrying ability). Certain air pollutants, however, do not occur naturally in the atmosphere, and some of these are hazardous at very low concentrations. An example is asbestos fibers, which are known to cause cancer.

Air pollutants are products of both natural events and human activities. Natural sources of air pollutants include forest fires, dispersal of pollen, wind erosion of soil, decay of dead plants and animals, and volcanic eruptions. The single most important human-related source of air pollutants is the internal combustion engine that propels most motor vehicles. According to the U.S. Environmental Protection Agency (EPA), transportation vehicles yearly emit almost 60 million metric tons (63 million tons) of the major air pollutants (Table 1.3). Many industrial sources also contribute to air quality problems. Unless emissions are controlled, pulp and paper mills, zinc and lead smelters, oil refineries, and chemical plants can be prodigious emitters of air pollutants. Additional pollutants come from fuel combustion for space heating and generation of electricity, from refuse burning, and from various agricultural activities such as crop

dusting. In the United States, almost 130 million metric tons (145 million tons) of the chief air contaminants are emitted to the atmosphere each year as the result of human activity—almost 0.6 metric ton per person.

A substance that is harmful immediately upon emission into the atmosphere is designated a **primary air pollutant.** Carbon monoxide in automobile exhaust is an example. In addition, within the atmosphere, chemical reactions involving primary air pollutants, both gases and aerosols, produce **secondary air pollutants.** An example is photochemical smog, generated by the action of sunlight on automobile exhaust and some industrial emissions (Figure 1.6). We have much more to say about air pollution in Chapter 17.

# Probing the Atmosphere

Much of what we know about the properties of the atmosphere is derived from direct sampling and measurement. At first, scientists explored the atmosphere from the ground, scaling rugged mountain peaks to sample the rarefied air. Through the years, exploration of the atmosphere progressed from tentative probing with primitive instruments to sophisticated remote sensing with space-age technology.

Early efforts at remote sensing of the atmosphere employed kites. In July 1749 in Glasgow, Scotland, Alexander Wilson attached several thermometers to six paper kites and flew them in tandem. Wilson designed the apparatus so that the thermometers would fall to the ground unbroken at predetermined intervals. In this

**TABLE 1.3**
**Estimated Emissions of Principal Air Pollutants in the United States for 1988 (in millions of metric tons)**

| Source | Carbon Monoxide | Sulfur Oxides | Volatile Organics | Particulates | Nitrogen Oxides | Total |
|---|---|---|---|---|---|---|
| Transportation | 41.2 | 0.9 | 6.1 | 1.4 | 8.1 | 57.7 |
| Fuel combustion | 7.6 | 16.4 | 0.9 | 1.7 | 10.8 | 37.4 |
| Industrial processes | 4.7 | 3.4 | 8.5 | 2.6 | 0.6 | 19.8 |
| Solid waste disposal | 1.7 | — | 0.6 | 0.3 | 0.1 | 2.7 |
| Others | 6.0 | — | 2.4 | 0.9 | 0.2 | 9.5 |
| Totals | 61.2 | 20.7 | 18.5 | 6.9 | 19.8 | 127.1 |

Source: U.S. Environmental Protection Agency, *National Air Pollutant Emissions Estimates, 1940–1988,* 1990.

**FIGURE 1.6**
Photochemical smog over Los Angeles. [Photograph by J. M. Moran]

way, he was the first to obtain a free-air temperature profile of the lower atmosphere (up to an altitude of perhaps 60 m or 197 ft).

Benjamin Franklin, an inventive genius, is credited with designing an experiment with a kite to demonstrate the electrical nature of lightning. He proposed flying a kite during a thunderstorm in an effort to attract an electrical discharge to a brass key that was attached to the kite string. Contrary to popular belief, Franklin probably never conducted the experiment himself—a wise decision because a single bolt would have been fatal. But such experiments were conducted

in France during the summer of 1752, and the findings prompted Franklin the next year to extol the effectiveness of lightning rods in his *Poor Richard's Almanac*.

In 1804, the French scientist J. L. Gay-Lussac ushered in the age of manned balloon exploration of the atmosphere. He took air samples and measured temperature and humidity, and on one ascent reached an altitude of 7000 m (23,000 ft). In 1862, the British scientist James Glaisher and his fellow aeronaut Henry Coxwell took weather instruments aloft in a series of balloon ascents from Wolverhampton, England. They almost perished from severe cold and oxygen deprivation when they set a manned balloon altitude record of 9000 m (29,500 ft).

Through the early part of the twentieth century, weather instruments borne by kites, aircraft, and balloons provided data chiefly on the lowest 5000 m (16,000 ft) of the atmosphere. In August 1894 at Harvard's Blue Hill Observatory near Boston, a kite was the first vehicle to carry aloft a self-recording thermometer. The instrument provided an air temperature profile up to an altitude of about 430 m (1400 ft). After World War I, the U.S. Weather Bureau operated a network of weather kites at several locations, mostly in the Central states. Kites were equipped with a recording instrument (meteorgraph) that profiled wind speed, air pressure, temperature, and humidity up to an altitude of about 3000 m (9,800 ft). The longest operating of those stations, at Ellendale, North Dakota, was closed in 1933. For several years afterward, the U.S. Weather Bureau relied mostly on regularly scheduled aircraft to probe the upper atmosphere but this practice proved too costly.

A leap forward in monitoring higher altitudes came in the late 1920s when the first **radiosonde,** a small instrument package equipped with a radio transmitter, was carried aloft by a helium-filled balloon (Figure 1.7). This device transmits to a ground station continuous altitude readings, called a **sounding,** of temperature, air pressure, and humidity. With a radiosonde, data are received immediately; no recovery of a recording instrument is needed. By World War II, radiosonde movements were being tracked from ground stations by radio direction-finding antennas, thus monitoring variations in wind direction and wind speed with altitude. A radiosonde used in this way is called a **rawinsonde.**

Today, radiosondes are launched simultaneously at 12-hour intervals from hundreds of ground stations around the world. The balloon bursts at an altitude of about 30 km (19 mi), and then the instrument package descends to the surface under a parachute. In the United States some radiosondes (about 20%) are recovered, refurbished, and reused. Each radiosonde contains a prepaid mailbag and instructions to the finder for its return to the National Weather Service (NWS).

A **dropwindsonde** is similar to a rawinsonde except that instead of being launched by a balloon from a sur-

**FIGURE 1.7**
Launch of a radiosonde, a balloon-borne instrument package that measures vertical profiles of air temperature, pressure, and relative humidity. [Photograph by Bruce Smith, AMS Project Atmosphere]

face station, it is dropped from an aircraft. The instrument package descends on a parachute at about 18 km (11 mi) per hour and radios data back to the aircraft every few seconds. The dropwindsonde was developed at the National Center for Atmospheric Research (NCAR) in Boulder, Colorado, to obtain soundings over oceans where conventional rawinsonde stations are virtually absent. Dropwindsondes provide vertical profiles of air temperature, pressure, humidity, and wind.

Robert H. Goddard is credited with conducting the first rocket probe of the atmosphere in 1929. The payload of Goddard's primitive rocket included a thermometer and a barometer (for measuring air pressure). World War II spurred advances in rocketry, and, by the late 1940s, rockets were used to investigate the middle and upper atmosphere. In March 1947, a vertically fired V2 rocket took the first successful photographs of the

Earth's cloud cover from altitudes of 110 to 165 km (68 to 102 mi). This and subsequent rocket probes of the atmosphere convinced scientists of the value of cloud pattern photographs in monitoring weather systems and inspired the first serious proposals for orbiting a weather satellite.

In the mid- to late 1950s, the United States' fledgling space program was directed at developing a launch vehicle (rocket) capable of putting a satellite in orbit. Those attempts to be the first in space were thwarted by the Soviet Union's successful orbiting of Sputnik I on 4 October 1957. The age of remote sensing by satellite had begun. On 1 April 1960 the United States orbited the world's first weather satellite, TIROS-I (Television and Infrared Observation Satellite). Since then, a series of increasingly sophisticated weather satellites have been orbited (Figure 1.8).

Weather satellites have proved to be invaluable tools in weather observation. These *eyes in the sky* offer distinct advantages over the network of surface weather stations in providing a broad and continuous field of view; surface weather stations are discrete and often widely spaced data sources. In fact, weather observations are sparse or absent over vast areas of the Earth's surface—especially the oceans. Today's weather satellites carry sophisticated sensors capable of monitoring patterns of temperature and water vapor concentration, upper-air winds, and the life cycles of severe storms. By the early 1990s, vertical profiling of atmospheric temperature and humidity by satellite became a routine practice of the National Weather Service. We have more to say about meteorology by satellite in Chapter 16.

Also since World War II, radar has become an increasingly important tool for surveillance of severe weather systems such as hurricanes and intense thunderstorms. Radar signals locate and track the movement of areas of rainfall and can even determine the rate of rainfall. As we will see in Chapter 14, a new generation of radar can also detect the detailed circulation of air within a severe thunderstorm and thereby provide advance warning of tornado development.

A variety of technologies have thus provided detailed information on the properties of the atmosphere. In the remainder of this chapter we consider two important properties of the atmosphere: its vertical temperature profile and electrical characteristics of the upper atmosphere.

**FIGURE 1.8**
Artist's concept of a geosynchronous meterological satellite positioned over the equator at about 37,000 km (23,000 mi) altitude. This satellite orbits at the same rate as the Earth rotates, so the satellite always surveys the same portion of the planet. From the satellite's relatively high orbit, onboard sensors *see* almost one-third of the Earth's surface. [NASA photograph]

## Temperature Profile of the Atmosphere

For convenience of study, the atmosphere is subdivided into concentric layers based upon the vertical profile of the average air temperature, as shown in Figure 1.9. Almost all weather occurs within the lowest layer, the **troposphere,** which extends from the Earth's surface to an average altitude ranging from about 6 km (3.7 mi) at the poles to about 16 km (10 mi) at the equator. Normally, but not always, the temperature within the troposphere decreases with increasing altitude. Hence, the air temperature is usually lower on mountaintops than in surrounding lowlands (Figure 1.10). On average, within the troposphere, the temperature falls 6.5 Celsius degrees per 1000 meters (3.5 Fahrenheit degrees per 1000 feet). The upper boundary of the troposphere, called the **tropopause,** is a transition zone between the troposphere and the next higher layer, the stratosphere.

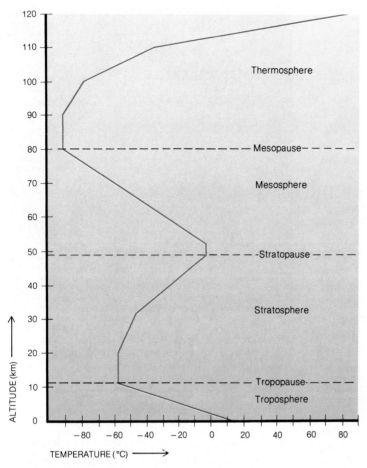

**FIGURE 1.9**
The average variation of temperature with altitude within the atmosphere. Based on this temperature profile, the atmosphere is subdivided vertically into the troposphere, stratosphere, mesosphere, and thermosphere.

The **stratosphere** extends from the tropopause up to about 50 km (31 mi). On average, in the lower portion of the stratosphere, the temperature does not change with altitude. When temperature is constant, the condition is described as **isothermal.** Above about 20 km (12 mi), the temperature rises with increasing altitude up to the top of the stratosphere, the **stratopause.** At the stratopause, the temperature is not much lower than at sea level.

The stratosphere is ideal for jet aircraft travel because it is above the weather. Therefore, it offers excellent visibility and features generally smooth flying conditions. Since the early 1970s, however, scientists have been concerned about possible detrimental effects of certain air pollutants that enter the stratosphere. Because little exchange of air takes place between the troposphere and the stratosphere, pollutants that

reach the lower stratosphere may remain there for long periods. Gases thrown into the stratosphere during violent volcanic eruptions, for example, can persist there for many months to years and perhaps trigger changes in climate. Other pollutants are eroding the protective ozone layer within the stratosphere. We have more to say about these problems in Chapters 2, 17, and 20.

The stratopause is the transition zone between the stratosphere and the next higher layer, the **mesosphere.** Within this layer, the temperature once again falls with increasing altitude. The mesosphere extends up to the **mesopause,** which is about 80 km (50 mi) above the Earth's surface, and features the lowest average temperature in the atmosphere (about $-95\,°C$ or $-139\,°F$). Above this is the **thermosphere,** where temperatures at first are isothermal and then rise rapidly with alti-

**FIGURE 1.10**
Within the troposphere, the temperature falls with increasing altitude. Hence, it is colder on mountain peaks than in lowlands. [Photograph by J. M. Moran]

tude. Within the thermosphere, temperature is more variable with time than in any other region of the atmosphere.

## The Ionosphere and the Aurora

The **ionosphere** is located primarily within the thermosphere, between altitudes of 80 and 400 km (50 and 248 mi). The region is named for its relatively high concentration of ions. An **ion** is an atomic-scale particle that carries an electrical charge. High-energy solar radiation entering the upper atmosphere strips electrons from oxygen and nitrogen atoms and molecules and leaves them as positively charged ions. The highest concentration of ions is in the lower portion of the thermosphere.

Although conditions in the upper atmosphere do not greatly influence day-to-day weather, the ionosphere is important for long-distance radio transmission. The ionosphere reflects radio signals. Radio signals travel in straight lines and bounce back and forth between the Earth's surface and the ionosphere (Figure 1.11). By repeated reflections, a radio signal may travel completely around the globe. We have more to say on this subject in the Special Topic "The Ionosphere and Radio Transmission."

The ionosphere is also the site of the spectacular **aurora borealis** (northern lights) in the northern hemisphere and the **aurora australis** (southern lights) in the Southern Hemisphere. Auroras appear in the night sky as overlapping curtains of greenish-white light, occasionally fringed with pink (Figure 1.12). The bottom of the curtains is at an altitude of about 100 km (62 mi) and the top is at 400 km (248 mi) or higher.

An aurora is triggered by the **solar wind,** a stream of electrically charged subatomic particles (protons and electrons) that continually emanates from the sun and

## The Ionosphere and Radio Transmission

Reception of distant radio signals (waves) at night is not at all unusual. Late-night rado listeners in Illinois and Wisconsin, for example, routinely pick up WBZ, a Boston radio station (1030 on the AM dial) even though the station's transmitter is more than 1500 km (930 mi) away.

Arrival of distant radio waves at night and their subsequent disappearance during sunlit hours are due to interactions of those waves with the ionosphere. A radio wave is a form of radiation (Chapter 2) that travels in straight paths in all directions away from its source transmitter. Earth is a sphere so that its surface gradually curves under and away from direct radio waves. A quiet zone, where direct radio waves are not received, begins about 160 km (100 mi) from the transmitter. At night, beyond the quiet zone, reception resumes because radio waves are reflected back to the Earth's surface from the upper ionosphere.

Recall that the ionosphere is a region of ions and free electrons. Highly energetic radiation from the sun causes the atmosphere's molecular nitrogen ($N_2$) and molecular oxygen ($O_2$) to split into atoms, positively charged ions, and free electrons. The production rate of ions and electrons depends on two factors, both of which vary with altitude: (1) the density of atoms and molecules available for ionization, which decreases rapidly with altitude, and (2) the intensity of solar radiation, which increases with altitude. Combined, these two factors maximize the concentration of ions and free electrons in the ionosphere.

By convention, the ionosphere is subdivided vertically into several layers. From lowest to highest, layers are designated D (60 to 90 km), E (90 to 140 km), and F (above 140 km). The original basis for this subdivision was the belief that each layer is a distinct zone of maximum electron density. Measurements by rock-

travels into space at speeds of 400 to 500 km (250 to 300 mi) per second. The Earth's magnetic field deflects the solar wind and, as a consequence, is deformed into a teardrop-shaped cavity surrounding the planet known as the **magnetosphere** (Figure 1.13). A complex interaction between the solar wind and the magnetosphere generates beams of electrons that collide with atoms and molecules within the ionosphere. Collisions rip apart molecules, excite atoms, and increase ion and

electron densities. As atoms shift down from their excited (energized) states and as ions combine with free electrons they emit radiation, part of which is visible as the aurora. Excited nitrogen molecules emit pinkish or magenta light whereas excited oxygen atoms emit greenish light.

The Earth's magnetic field channels some solar wind particles into two doughnut-shaped belts centered on the planet's north and south geomagnetic poles.

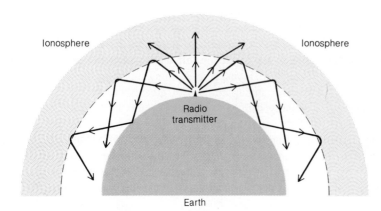

**FIGURE 1.11**
Within the ionosphere, at altitudes above 80 km (50 mi), regions of charged subatomic particles reflect outgoing radio waves. Multiple reflections involving both the ionosphere and the Earth's surface greatly extend the range of radio transmissions.

ets and satellites, however, show that the ionosphere is not made up of discrete layers; rather, electron density increases nearly continuously with altitude to a maximum at an average level close to 300 km (186 mi). Hence, the D, E, and F labels used in this discussion refer to specific subregions within the ionosphere.

Radio waves that enter the ionosphere interact with free electrons in the D, E, and F regions and are either absorbed or reflected back toward the Earth's surface. At night, when there is no ionizing radiation, the D region virtually disappears as ions and electrons recombine into neutral particles. The recombination rate depends on air density; that is, the denser the air, the greater is the likelihood of collision of particles and the capture of electrons by positive ions. In the E and F regions, the air is so rarefied that collisions are infrequent, and although the E region weakens, both regions persist through the night. Radio waves that reach the F region are reflected back toward the Earth's surface. (Exceptions are radio waves that enter the F region at nearly a right angle; these waves pass on into space.) At night, because of F-region reflection, radio waves propagate many hundreds of kilometers from their point of origin.

With the return of the sun's ionizing radiation during the day, the D region redevelops. Most radio waves that reach the D region are absorbed rather than reflected. Waves that do penetrate the D region are reflected by the overlying F region back to the D region where they are absorbed. Consequently, during sunlit hours, radio-wave propagation is not aided by F-region reflection. In summary, then, radio waves travel greater distances at night because they are reflected by the upper ionosphere back to the Earth's surface.

Because the ionosphere is generated by ionizing radiation from the sun, any solar activity that disturbs the flow of this radiation may affect ion density and, consequently, radio communication on Earth. *Sudden ionospheric disturbances (SIDs),* typically lasting 15 to 30 minutes, are caused by bursts of ultraviolet radiation from the sun. Ionization temporarily increases, D-region absorption strengthens, and radio transmissions fade. The same solar activity responsible for the aurora also increases ionization and causes radio fadeout.

**FIGURE 1.12**
The aurora borealis (northern lights). [National Center for Atmospheric Research/ University Corporation for Atmospheric Research/National Science Foundation; Robert Bumpas, photographer]

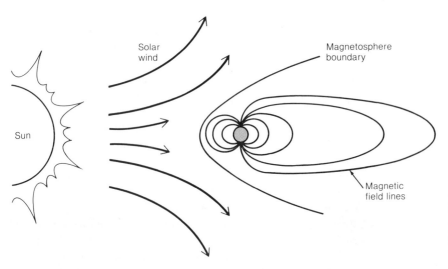

**FIGURE 1.13**
The solar winds deflects the Earth's magnetic field into the magnetosphere. Solar wind particles trigger auroral displays.

These belts of more or less continuous auroral activity, known as **auroral ovals,** are situated between 20 and 30 degrees of latitude from the geomagnetic poles. The northern hemisphere auroral oval is centered on the northwest tip of Greenland at latitude 78.5 degrees N and longitude 69 degrees W.* Consequently, the aurora is usually visible only at high latitudes.

Auroral activity varies with the sun's activity. When the sun is quiet, the auroral oval shrinks, but when the sun is active, the auroral oval expands equatorward, and the aurora may be visible across southern Canada and the northern United States or, rarely, even further south. An active sun features gigantic explosions called **solar flares** (Figure 1.14). A solar flare is a brief event (lasting perhaps an hour) that produces a shock wave that propagates rapidly (500 to 1000 km per second) through the solar wind. Collision of the shock wave with the magnetosphere causes the auroral oval to expand equatorward.

---

*Latitude* is distance on the Earth's surface measured in degrees and minutes north or south of the equator. The latitude of the equator is 0 degrees, and the geographical poles are at 90 degrees N(orth) and 90 degrees S(outh); 1 degree equals 60 minutes. *Longitude* is measured in degrees east and west of the meridian that passes through Greenwich, England. The Greenwich or "prime" meridian is assigned a longitude of 0 degrees. Longitude is measured east and west of the prime meridian to 180 degrees E(ast) and 180 degrees W(est). Note that the Earth's geomagnetic poles do not coincide with either its geographical or compass poles. The compass north is at 76 degrees N, 102 degrees W.

# Conclusions

In this chapter we covered the origin, evolution, composition, and structure of the atmosphere. We emphasized the importance of minor components in the functioning of the atmosphere, and we surveyed the various techniques that are used to monitor the properties of the atmosphere. We also saw how the vertical profile of average air temperature enables us to subdivide the atmosphere into four layers. Because the primary focus of this book is weather and climate, we will be concerned primarily with atmospheric processes operating within the troposphere.

Our next major objective is to examine the driving force behind weather. To do so, we require an understanding of energy input and energy conversions within the atmosphere. In the next chapter we learn how the sun supplies the energy that drives the atmosphere's circulation. As we will see, circulation of the atmosphere ultimately is responsible for the variation of weather from one place to another and with time.

# *Key Terms*

| | |
|---|---|
| **weather** | **climate** |
| **meteorology** | **climatology** |

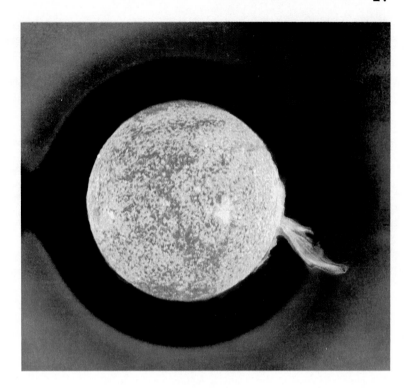

**FIGURE 1.14**
Solar flares emit highly energetic streams of electrically charged subatomic particles into space. [NASA photograph]

atmosphere
scientific method
hypothesis
scientific model
conceptual model
graphical model
physical model
numerical model
photosysnthesis
homosphere
heterosphere
photodissociation
aerosols
air pollutant
primary air pollutant
secondary air pollutant
radiosonde
sounding

rawinsonde
dropwindsonde
troposphere
tropopause
stratosphere
isothermal
stratopause
mesosphere
mesopause
thermosphere
ionosphere
ion
aurora borealis
aurora australis
solar wind
magnetosphere
auroral ovals
solar flares

## *Summary Statements*

☐ Weather varies from one location to another and with time. Weather is defined as the state of the atmosphere at a specified place and time. Climate encompasses average weather plus extremes in weather at some location over a specified time period.

☐ The atmosphere encircles the planet as a thin envelope of gases and suspended particles (aerosols).

☐ The scientific method is a systematic form of inquiry that requires the formulation and testing of hypotheses. A hypothesis is a working assumption that may be accepted, modified, or rejected. Debate and disagreement among scientists on some controversial issue are usual components of the scientific method.

☐ A scientific model is an approximate representation of a real system. Examples in meteorology/climatology include weather maps and numerical models of the atmosphere.

☐ The modern atmosphere is the product of a lengthy evolutionary process that began about 4.6 billion years ago. Volcanic activity played a key role in this evolution by contributing water vapor, carbon dioxide, and nitrogen. Photosynthesis was the primary source of free oxygen.

☐ Within the homosphere, the lowest 80 km (50 mi) of the atmosphere, the principal atmospheric gases, nitrogen and oxygen, occur everywhere in the same relative proportion (about 4 to 1).

☐ The significance of an atmospheric gas or aerosol is not necessarily related to its relative concentration. Water vapor, carbon dioxide, and ozone are *minor* in concentration but extremely important in the roles they play in the atmosphere.

☐ Air pollutants are gases or aerosols occurring in concentrations that adversely affect the well-being of organisms (especially humans) or disrupt the orderly functioning of the environment. Human activities and natural processes generate air pollutants.

☐ Through the years, tools for investigating the atmosphere have progressed from instrumented kites and manned balloons to rockets, radiosondes, satellites, and radar.

☐ A radiosonde consists of an instrument package and radio transmitter carried by balloon to altitudes of about 30 km (19 mi). It provides vertical profiles (soundings) of air temperature, pressure, and relative humidity.

☐ The atmosphere is subdivided into four concentric layers (troposphere, stratosphere, mesosphere, and thermosphere) based on the average vertical temperature profile. Almost all weather takes place in the troposphere, the lowest subdivision of the atmosphere.

☐ The ionosphere is situated primarily within the thermosphere and consists of a relatively high concentration of charged particles (ions and electrons).

☐ An aurora (northern or southern lights) develops in the ionosphere when the solar wind interacts with the Earth's magnetic field. The result is a spectacular display of curtains of color in the night sky of high latitudes.

7. In the geologic past, carbon dioxide was the chief component gas of the Earth's atmosphere. What caused its concentration to decline?

8. How did the formation of oceans and the coming of life on the planet affect the evolution of the atmosphere?

9. Distinguish between the homosphere and the heterosphere.

10. Most aerosols in the atmosphere are products of activities at the Earth's surface. List several natural sources of aerosols. List some sources related to human activities.

11. Present several examples of how some *minor* constituents of the atmosphere are essential for continuation of life on Earth.

12. Under what conditions is a natural constituent (gas or aerosol) of the atmosphere considered to be an air pollutant?

13. In what way was invention of the radiosonde a major step forward in monitoring the atmosphere?

14. List some of the advantages of satellite observations of the atmosphere as contrasted with ground-based techniques of atmospheric monitoring.

15. In what ways does the troposphere differ from the stratosphere?

16. Why do air pollutants that enter the stratosphere tend to persist there for lengthy periods?

17. What is the ratio of nitrogen to oxygen in the troposphere? Does that ratio change within the stratosphere?

18. What is the source of ions within the ionosphere?

19. Why is the aurora visible only at high latitudes?

20. How and why does auroral activity vary with activity on the sun?

## Review Questions

1. Distinguish between weather and climate. Explain why a description of climate only in terms of average weather is incomplete and potentially misleading.

2. List some of the ways whereby the atmosphere sustains life on Earth.

3. In your own words, define the scientific method. What role is played by hypotheses in the scientific method?

4. What is a scientific model and what is its basic purpose? Identify some models that are useful in studying weather and climate.

5. How are numerical models used to predict future climate?

6. What role did volcanic activity play in the evolution of the Earth's atmosphere?

## Questions for Critical Thinking

1. Photosynthesis occurs chiefly during the growing season. Speculate on how variations in the rate of photosynthesis through the course of a year might influence the concentration of carbon dioxide in the atmosphere.

2. Why does a radiosonde balloon burst when it ascends to altitudes above about 30 km (19 mi)?

3. Speculate on how changes in the average temperature of the troposphere might affect the altitude of the tropopause.

4. Mountaintops are closer to the sun than are lowlands, and yet mountaintops are colder than lowlands. Why?

5. What does the usual temperature profile of the troposphere imply about the source of heat for the troposphere?

## Selected Readings

BEVERIDGE, W. I. B. *The Art of Scientific Investigation.* New York: Vintage Books, 1957, 239 pp. Considers the nature of scientific inquiry as a creative art.

CONOVER, J. H. "The Blue Hill Observatory," *Weatherwise* 37, No. 6 (1984):296–303. Presents a historical account of a long-term weather station located near Boston; site of the world's first continuous recording of air temperature variation with altitude.

FARRAND, J., JR. "From Gods to Satellites," *Weatherwise* 44, No.2 (1991):30–36. Summarizes the historical roots of modern meteorology.

HABERLE, R. M. "The Climate of Mars," *Scientific American* 254, No. 5 (1986):54–62. Summarizes what is known about the Martian atmosphere's origin and circulation.

HILL, J. *Weather from Above, America's Meteorological Satellites.* Washington, D.C.: Smithsonian Institution Press, 1991, 89 pp. Chronicles the challenges of developing and launching weather satellites.

KASTING, J. F., ET AL. "How Climate Evolved on the Terrestrial Planets," *Scientific American* 258, No. 2 (1988):90–97. Proposes that differences in the carbon cycle operating on Venus, Earth, and Mars resulted in drastically different climates on these planets.

KERR, R. A. "An 'Outrageous Hypothesis' For Mars: Episodic Oceans," *Science* 259 (1993):910–911. Considers the possibility that Mars was glaciated and had an ocean.

RIEBSAME, W. E., ET AL. "The Social Burden of Weather and Climate Hazards," *Bulletin of the American Meteorological Society* 67 (1986):1378–1388. Includes a summary of trends in mortality and property damage caused by weather extremes.

WICHE, S. A. "Weather on a String." *Weatherwise* 45, No. 3 (1992):10–16. Describes the history of the use of kites in probing the upper atmosphere.

WILLIAMS, J. "The Making of the Weather Page," *Weatherwise* 45, No. 4 (1992):12–18. Describes the origins of the colorful and informative weather page in *USA Today.*

# 2 Radiation

*Splendid with splendor hid you
    come, from your Arab abode,
a fiery topaz smothered in the
    hand of a great prince who rode
before you, Sun—whom you out-
    ran,
piercing his caravan.*

MARIANNE MOORE
*Sun*

The sun supplies the energy that drives the atmosphere's circulation. The most intense portion of the solar energy that reaches the Earth's surface is visible as sunlight. [Photograph by J. M. Moran]

T HE SUN DRIVES THE ATMOSPHERE; that is, the sun is the source of energy that drives the circulation of the atmosphere and powers winds and storms. The circulation of the atmosphere ultimately is responsible for weather and its temporal and spatial variability.

The sun ceaselessly emits energy to space in the form of electromagnetic radiation. A very small portion of that energy is intercepted by the Earth–atmosphere system* and is converted into other forms of energy including, for example, heat and the kinetic energy of the atmosphere's circulation. In this regard, it is important to note that, although energy can be converted from one form to another, it cannot be created nor destroyed. This is the **law of energy conservation,** also known as the **first law of thermodynamics.**

In this chapter, we examine the basic properties of electromagnetic radiation, how solar radiation interacts with the components of the Earth–atmosphere system, and its conversion to heat. We learn how the Earth–atmosphere system responds to solar heating by emitting infrared radiation to space. We also come to appreciate the role of the greenhouse effect in elevating the surface temperature of the planet and making conditions more amenable to life. First we consider the nature of electromagnetic radiation in general, and then we describe some of the specific properties of its various forms.

## Electromagnetic Radiation

Planet Earth is bathed continuously in **electromagnetic radiation,** so named because this form of energy exhibits both electrical and magnetic properties. All known objects emit electromagnetic radiation.* Forms of electromagnetic radiation include radio waves, microwaves, infrared radiation, visible light, ultraviolet radiation, X-rays, and gamma radiation. Together, they make up the **electromagnetic spectrum,** illustrated in Figure 2.1.

Electromagnetic radiation travels as waves, which are usually described in terms of wavelength or frequency. **Wavelength** is the distance between successive wave crests (or, equivalently, wave troughs), as shown in Figure 2.2. **Wave frequency** is defined as the number of crests (or troughs) that pass a given point in a specified period of time, usually 1 second. Passage of one complete wave is called a *cycle,* and a frequency of one cycle per second equals 1.0 hertz (Hz). Wave frequency is inversely proportional to wavelength; that is, the higher the frequency, the shorter the wavelength. Radio waves have frequencies of millions of hertz and wavelengths of up to hundreds of kilometers. At the other end of the electromagnetic spectrum, in contrast, gamma rays have frequencies as high as $10^{24}$ (a trillion trillion) Hz and wavelengths as short as $10^{-14}$ (a hundred trillionth) m.

Electromagnetic waves can travel through space as well as through gases, liquids, and solids. In a

---

*The Earth–atmosphere system is the Earth's surface plus its atmosphere considered together.

*As we will see in Chapter 3, objects emit no electromagnetic radiation at a temperature of absolute zero.

31

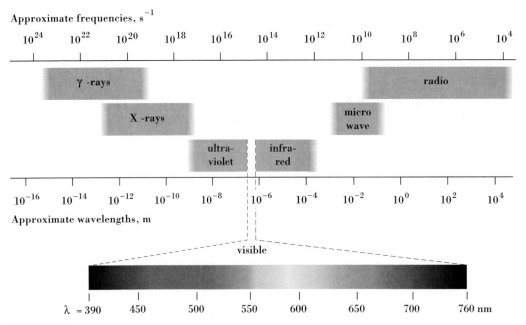

**Approximate frequencies, s$^{-1}$**

**Approximate wavelengths, m**

visible

$\lambda$ = 390   450   500   550   600   650   700   760 nm

**FIGURE 2.1**
The electromagnetic spectrum consists of many types of radiation that are distinguished on the basis of wavelength, frequency, and energy level.

vacuum, all electromagnetic waves travel at maximum speed, 300,000 km (186,000 mi) per second. All forms of electromagnetic radiation slow down when passing through materials, their speed varying with wavelength and type of material. As electromagnetic radiation passes from one medium to another, it may be reflected or refracted (that is, bent) at the interface. This happens, for example, when solar radiation strikes the ocean surface: some is reflected and some bends as it penetrates the water. Electromagnetic radiation may also be absorbed, that is, converted to heat.

Although the electromagnetic spectrum is continuous, it is convenient to assign different names to different segments because we detect, measure, generate, and use those segments in different ways. The different types of electromagnetic radiation do not begin or end at precise points along the spectrum. For example, red light shades into invisible infrared radiation (infrared, meaning *below red*). At the other end of the visible portion of the electromagnetic spectrum, violet light shades into invisible ultraviolet radiation (ultraviolet, meaning *beyond violet*).

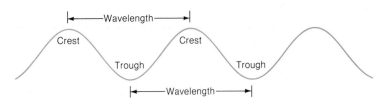

**FIGURE 2.2**
The wavelength of an electromagnetic wave is the distance between successive crests or, equivalently, the distance between successive troughs.

We now briefly consider each segment of the electromagnetic spectrum, beginning at the high-energy, high-frequency, and short-wavelength end.

Beyond visible light on the electromagnetic spectrum and in order of increasing energy level, increasing frequency, and decreasing wavelength, are **ultraviolet radiation (UV), X-rays,** and **gamma radiation.** All three types of radiation occur naturally, and all can be produced artificially. All have medical uses: ultraviolet radiation is a potent germicide; X-rays are a powerful diagnostic tool; and both X-rays and gamma radiation are used to treat cancer patients.

These three highly energetic types of radiation are dangerous as well as useful. Ultraviolet radiation can cause irreparable damage to the light-sensitive cells of the eye. One can be permanently blinded by staring at the sun (say, during a partial solar eclipse), unless a filter is used to block out ultraviolet radiation. Also, overexposure to UV, X-rays, or gamma radiation can cause sterilization, cancer, mutations, or damage to a fetus. Fortunately for us, the Earth's atmosphere blocks out most incoming ultraviolet radiation and virtually all X-rays and gamma radiation. Without this protective atmospheric shield, all life on Earth would be destroyed.

At lower frequencies and longer wavelengths, UV radiation shades into **visible radiation,** that is, visible light. Wavelengths of visible light range from about 0.40 micrometer* at the violet end to approximately 0.70 micrometer at the red end. Visible light is essential for many activities of plants and animals. In plants, light provides the energy needed for photosynthesis; it also coordinates the opening of buds and flowers and the dropping of leaves. For animals, light regulates the timing of reproduction, hibernation, and migration and makes vision possible.

Between visible light and microwaves is **infrared radiation (IR).** IR is not sufficiently energetic to be visible, but we can feel the heat it generates when it is intense—as it is, for example, when emitted by a hot stove. Actually, small amounts of infrared radiation are emitted by every known object, including you and this book. And, as we will see later in this chapter, absorption of IR by certain atmospheric gases is responsible for significant warming of the lower atmosphere.

*One micrometer is a millionth of a meter.

Next comes the **microwave** portion of the electromagnetic spectrum, which includes wavelengths that range from about 0.1 to 300 millimeters. Some microwave frequencies are used for radio communication, for microwave ovens, and for tracking weather systems (radar).

At the low-energy, low-frequency, long-wavelength end of the electromagnetic spectrum are **radio waves.** Their wavelengths range from a fraction of a centimeter up to hundreds of kilometers, and their frequencies can extend to a billion Hz. FM (frequency modulation) radio waves, for example, span 88 million to 108 million Hz; hence, the familiar 88 and 108 at opposite ends of the FM radio dial.

## Radiation Laws

Several physical laws describe the properties of electromagnetic radiation that is emitted by a perfect radiator, a so-called blackbody. (These laws are summarized in the Mathematical Note "Blackbody Radiation Laws" at the end of this chapter.) By definition, at a given temperature, a **blackbody**\* absorbs all radiation incident on it at every wavelength and emits all radiation at every wavelength; no radiation is reflected. A blackbody is therefore a perfect absorber and a perfect emitter. Although neither the sun nor the Earth is a precise blackbody, they so closely approximate perfect radiators that we can apply blackbody radiation laws to them with very useful results. Here we apply two blackbody radiation laws: Wien's displacement law and the Stefan–Boltzmann law.

Although all known objects emit all forms of electromagnetic radiation, the wavelength of most intense radiation is inversely proportional to the temperature of the object. This is a statement of **Wien's displacement law.** Hence, relatively warm objects (such as the sun) emit peak radiation at relatively short wavelengths, whereas colder objects (such as the Earth-atmosphere system) emit peak radiation at longer wavelengths.

*Blackbody* can be misleading because the term does not refer to color. Objects that are not black may be blackbodies, that is, perfect radiators. For example, bright white snow is very nearly a blackbody for infrared radiation.

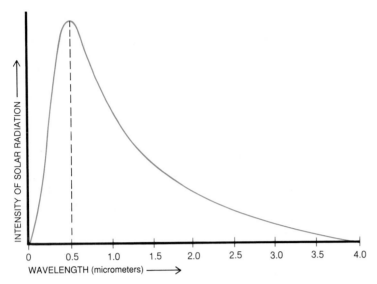

**FIGURE 2.3**
Blackbody radiation emission for the sun. Peak energy intensity is emitted by waves of about 0.50 micrometer in length (in the green portion of the visible spectrum).

A familiar illustration of Wien's displacement law is provided by the top burners of an electric stove. After switching on a burner, the coils warm up and we can readily feel the heat. The coils are still dark and what we are feeling is infrared radiation. But as the coil temperature continues to rise, the coils eventually begin to glow red. At the higher temperature, some of the radiation the coils emit is in the visible range.

Figure 2.3 shows the variation of radiation intensity with wavelength for a blackbody at the equivalent ra-

diating temperature of the sun, about 6000 °C (11,000 °F). Thus, the sun emits a band of radiation (at wavelengths mostly between 0.25 and 2.5 micrometers) that is most intense at a wavelength of about 0.5 micrometer (in the green of visible light). Figure 2.4 shows how the intensity of radiation varies with wavelength for a blackbody at the average radiating temperature of the Earth's surface, about 15 °C (59 °F). The Earth's surface therefore emits a band of infrared radiation (at wavelengths mostly between 4 and 24 micrometers)

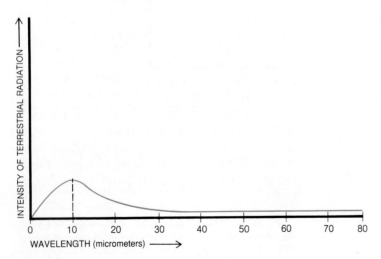

**FIGURE 2.4**
Blackbody radiation emission for the Earth's surface. Peak energy intensity is emitted by waves of about 10 micrometers in length (in the infrared region of the electromagnetic spectrum).

that has peak intensity at a wavelength of about 10 micrometers.

The curves in Figures 2.3 (for the sun) and 2.4 (for the Earth's surface) represent the total radiational energy emitted per unit surface area at all wavelengths. Note, however, that the vertical scales in these two figures are not the same because the sun emits considerably more total radiational energy than does the Earth–atmosphere system. This contrast in energy emission is described by the **Stefan–Boltzmann law:** The total energy radiated by an object across all wavelengths is proportional to the fourth power of its absolute temperature ($T^4$).* The sun radiates at a much higher temperature than does the Earth–atmosphere system, so that the Stefan–Boltzmann law predicts that the sun's energy output per square meter is about 160,000 times that of the Earth–atmosphere system. As solar radiation diverges outward from the sun into space, its intensity diminishes rapidly (as the inverse square of the distance traveled). Hence, when solar radiation reaches Earth, its intensity is reduced considerably.

The total energy (in the form of solar radiation) absorbed by planet Earth is equal to the total energy (in the form of infrared radiation) emitted by the Earth–atmosphere system back to space. This balance between energy input and energy output is known as **global radiative equilibrium** and is an example of the law of energy conservation.

# Input of Solar Radiation

The sun, our closest star, is a huge gaseous body composed almost entirely of hydrogen (about 80% by mass) and helium and featuring internal temperatures that may exceed 20 million °C. The ultimate source of solar energy is a continuous nuclear fusion reaction in the sun's interior. Simply put, in this reaction four hydrogen nuclei (protons) fuse to produce one helium nucleus (alpha particle). However, the mass of the four hydrogen nuclei is about 0.7% greater than the mass of one helium nucleus. This excess mass is converted to energy as described by Albert Einstein's equation

$$E = mc^2$$

whereby mass, $m$, is related to energy, $E$, and $c$ is the speed of light (300,000 km per second). Note that $c^2$ is such a huge number that even a very small mass is converted to an enormous quantity of energy. Some of the energy produced by nuclear fusion is used to bind the helium nucleus together. The rest of the energy is radiated and convected to the sun's surface, and from there, energy is radiated off to space.

At radiating temperatures near 6000 °C (11,000 °F), the visible surface of the sun, known as the **photosphere,** is much cooler than the sun's interior. The photosphere has a honeycomb appearance that is due to a network of huge, irregularly shaped convection cells, called **granules.** A typical granule is about the size of Texas and consists of a broad central area of rising hot gas surrounded by a thin layer of cooler gas sinking back into the sun. Relatively hot spots, called **faculae,** and relatively cool spots, called **sunspots,** dot the photosphere. As we will see in Chapter 20, changes in the number of faculae and sunspots may influence the climate on Earth.

Outward from the photosphere is the **chromosphere** consisting of ions of hydrogen and helium at 4000 to 40,000 °C. Beyond this zone is the outermost portion of the sun's atmosphere, the **corona** (Figure 2.5), a region of hot (1 to 4 million °C) and highly rarefied ionized gases that extends millions of kilometers into space, to the outer limits of the solar system. The solar wind originates in the corona and, as noted in Chapter 1, solar flares that erupt from the photosphere into the corona intensify the solar wind.

Of the enormous quantity of energy radiated by the sun to space, planet Earth intercepts only about one two-billionth of the total amount. Solar radiation that reaches Earth is known as **insolation** (for *in*coming *sol*ar radi*ation*). About 45% is visible as sunlight; the remainder consists of infrared (46%) and ultraviolet (9%) radiation.

*The absolute or Kelvin (K) temperature scale is the number of degrees (kelvins) above absolute zero. The various temperature scales are described in Chapter 3.

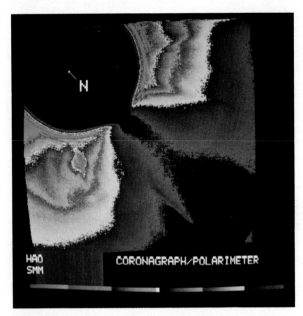

**FIGURE 2.5**
The sun's corona, color-coded to levels of brightness. Photograph was taken by sensors aboard the Solar Maximum Mission satellite. [National Center for Atmospheric Research/University Corporation for Atmospheric Research/National Science Foundation]

**FIGURE 2.6**
On any day of the year, the solar altitude (the sun's angle above the horizon) varies with latitude because the Earth's surface is curved. In this example, the solar altitude is 90 degrees at the equator and decreases with latitude (toward the poles). Hence, solar radiation is most intense at the equator and least intense at the poles.

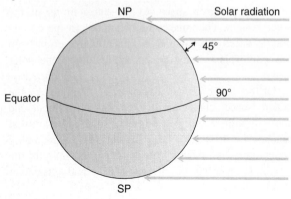

## SOLAR ALTITUDE

We know from experience that the intensity of solar radiation striking the Earth's surface varies significantly over the course of a year. In winter, the sun is lower in the sky, and at the Earth's surface the sun's rays are weaker than in summer. Also in winter, the days are shorter than they are in summer. Even over the course of a single day, noticeable changes occur in insolation: the noon sun is more intense than the rising or setting sun.

It is evident, then, that the angle of the sun above the horizon, called the **solar altitude,** influences the intensity of solar radiation. By intensity we mean the amount of energy striking or passing through a unit area in a unit time.* The sun is so far away from Earth, about 150 million km (93 million mi) on average, that solar radiation reaches the planet as a beam that in cross section is uniform in intensity. The planet is very nearly a sphere and presents a curved surface to incoming solar radiation (Figure 2.6). Hence, solar altitude and the intensity of solar radiation received at the Earth's surface always vary with latitude. As shown in Figure 2.7, where the noon sun is directly overhead (solar altitude of 90 degrees), the sun's rays reaching the Earth's surface are most concentrated and therefore most intense. As the sun moves lower in the sky (that is, as the solar altitude decreases), solar radiation spreads over a greater area of the Earth's surface and thus becomes less intense.

Solar altitude also influences the interaction between insolation and the atmosphere. With decreasing solar altitude, the path of the sun's rays through the atmosphere lengthens (Figure 2.8). As the path lengthens, solar radiation interacts more with the gases and aerosols of the atmosphere and its intensity diminishes. Thus, even with clear skies, the longer the path of solar radiation through the atmosphere, the less intense is the radiation that strikes the Earth's surface.†

Although solar altitude has an important influence on the intensity of solar radiation striking the Earth's surface, the length of day affects the total amount of

---

*This is also called the *flux density.*
†The nature of the interaction (absorption, reflection, and scattering) between solar radiation and the atmosphere is discussed later in this chapter.

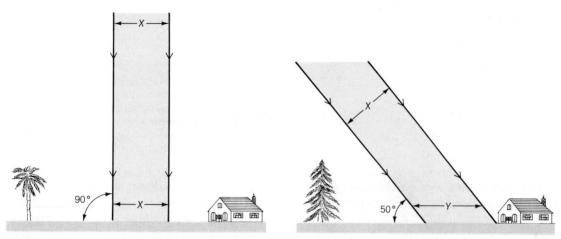

**FIGURE 2.7**

A                                                                 B

The intensity of solar radiation that strikes the Earth's surface varies with changes in solar altitude. (A) Solar radiation is most intense when the sun is directly overhead (solar altitude of 90 degrees). (B) With decreasing solar altitude, the solar radiation received at the Earth's surface spreads over an increasingly larger area ($Y$ is greater than $X$) so that the radiation becomes less intense.

radiational energy that is received. For the same intensity of solar radiation, those areas of the Earth with more daylight hours will receive more total energy. Variations both in solar altitude and in day length accompany the annual march of the seasons. Before examining these relationships, let us first consider the fundamental motions of Earth in space: rotation of the planet on its axis and the planet's orbit about the sun.

## EARTH'S MOTIONS IN SPACE

Rotation of the Earth on its axis accounts for day and night. Approximately once every 24 hours, the Earth makes one complete rotation. Consequently, at any point in time, half the planet is in darkness (night) and the other half is illuminated by solar radiation (day).

In one year, which is actually 365.25 days, the Earth makes one complete revolution about the sun in a slightly elliptical orbit (Figure 2.9). The Earth's orbital eccentricity, that is, its departure from a circular orbit, is so slight that the Earth-to-sun distance varies by only

about 3.3% through the year. Earth is closest to the sun (147 million km or 91 million mi) on about 3 January and farthest from the sun (152 million km or 94 million mi) on 4 July. These are the dates of **perihelion** and **aphelion,** respectively. In the Northern Hemisphere, Earth is closest to the sun in winter and farthest from the sun in summer. The eccentricity of

**FIGURE 2.8**

The path of solar radiation through the atmosphere lengthens as the solar altitude decreases, that is, as the sun moves lower in the sky. $X$ is the path length at high solar altitude, and $Y$ is the path length at low solar altitude.

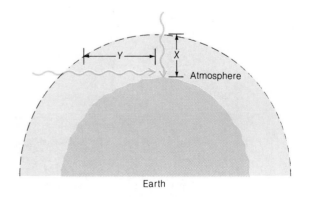

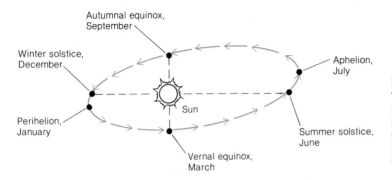

**FIGURE 2.9**
The Earth's orbit is an ellipse, with the sun located at one focus. Earth is closest to the sun at perihelion (about January 3) and farthest from the sun at aphelion (about July 4). Note that in this drawing the eccentricity of the Earth's orbit is greatly exaggerated.

the Earth's orbit about the sun does not, therefore, explain the seasons. What, then, does account for them?

## THE SEASONS

Seasons on Earth are attributed to the 23 degree 27 minute tilt of the Earth's rotational axis to the plane defined by the Earth's orbit (Figure 2.10). This tilt causes the Earth's orientation to the sun to change continually as the planet revolves about the sun. The Northern Hemisphere thus leans away from the sun during winter and leans toward the sun in summer. At the same time that the Northern Hemisphere leans away from the sun, the Southern Hemisphere leans toward the sun. When it is winter in the Northern Hemisphere, it is therefore summer in the Southern Hemisphere, and vice versa.

As Earth's orientation to the sun changes, so too do the solar altitude and length of day. Hence, the intensity and total amount of solar radiation received at the Earth's surface vary seasonally. In the winter hemi-

sphere, solar altitudes are lower, days are shorter, and less solar radiation strikes the Earth's surface. In the summer hemisphere, solar altitudes are higher, days are longer, and more solar radiation strikes the Earth's surface. Less solar radiation in winter than in summer means that winters are colder than summers.

If the Earth's rotational axis were perpendicular to its orbital plane (no tilt), Earth would always have the same orientation to the sun. Without an axial tilt, changes in Earth-to-sun distance between aphelion and perihelion would cause only a slight seasonal contrast.

How does the Earth's orientation with respect to the sun change over the course of a year? Viewed from the Earth's surface, the location of the sun's most intense radiation (solar altitude of 90 degrees) shifts from 23 degrees 27 minutes south of the equator to 23 degrees 27 minutes north of the equator, and then back to 23 degrees 27 minutes south. On 21 March, and again on 23 September, the sun's noon position is directly over the equator. Day and night are of equal length (12 hours) everywhere (Figure 2.11). For this reason,

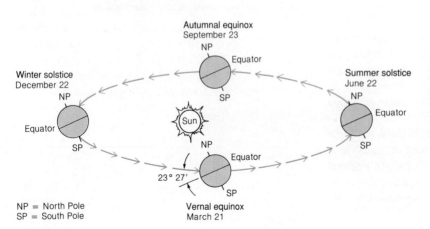

**FIGURE 2.10**
The seasons change because the Earth's equatorial plane is inclined (at 23 degrees, 27 minutes) to its orbital plane. The seasons given are for the Northern Hemisphere. Note that the eccentricity of the Earth's orbit is greatly exaggerated in this drawing.

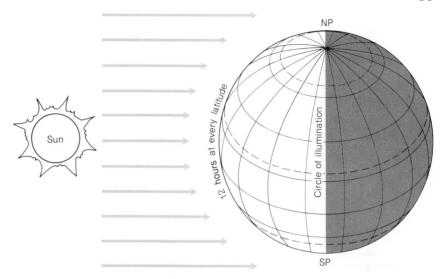

**FIGURE 2.11**
At the autumnal and vernal equinoxes, insolation is maximum at the equator, and day and night are of equal length everywhere.

these dates* are labeled **equinoxes,** from the Latin for *equal nights.*

Following the equinoxes, the sun continues its apparent journey toward maximum poleward locations, its **solstice** latitudes. On 22 June, the sun's noon rays are vertical at 23 degrees 27 minutes N, the latitude belt known as the **Tropic of Cancer.** As shown in Figure 2.12, daylight is continuous north of the **Arctic Circle** (66 degrees 33 minutes N) and absent south of the **Antarctic Circle** (66 degrees 33 minutes S). Elsewhere, days are longer than nights in the Northern Hemisphere, where it is summer, and days are shorter than nights in the Southern Hemisphere, where it is winter.

On 22 December, the noon sun is directly over 23 degrees 27 minutes S latitude, the **Tropic of Capricorn,** and the situation is reversed (Figure 2.13). Daylight is continuous south of the Antarctic Circle and absent north of the Arctic Circle. Elsewhere, nights are longer than days in the Northern Hemisphere, where it is winter, and days are longer than nights in the Southern Hemisphere, where it is summer.

Solar radiation is at its maximum intensity where the noon sun is directly overhead. North and south of this latitude, the intensity of solar radiation diminishes because the solar altitude decreases. For example, at the equinoxes, insolation is most intense at the equa-

tor at noon and decreases with latitude toward the poles. At the Northern Hemisphere summer solstice, insolation is most intense along the Tropic of Cancer and decreases to zero at the Antarctic Circle. At the Northern Hemisphere winter solstice, insolation is most intense along the Tropic of Capricorn and decreases to zero at the Arctic Circle. Furthermore, as shown in Figure 2.14, the seasonal (winter-to-summer) contrast in length of day increases with latitude.

### THE SOLAR CONSTANT

For convenience of study, the solar energy input into the Earth–atmosphere system is often expressed as the solar constant. The **solar constant** is defined as the rate at which solar radiation falls on a surface located at the top of the atmosphere and positioned perpendicular to the sun's rays when Earth is at its mean distance from the sun. The *constant* designation is actually misleading because solar energy output fluctuates by a very small fraction of a percent over a year and exhibits longer term variations (discussed in Chapter 20). Nonetheless, we can approximate the solar constant in energy units as 1.97 calories per square centimeter ($cal/cm^2$) per minute, or in power units as 1372 watts per square meter ($W/m^2$). For the distinction between energy units and power units, see the inside front cover.

---

*Actual dates of equinoxes and solstices vary because of leap year.

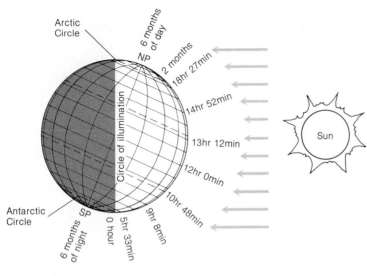

Summer solstice
Northern Hemisphere, June 22

**FIGURE 2.12**
At the Northern Hemisphere summer solstice (June 22), maximum insolation is at 23 degrees, 27 minutes N, and days are longer than nights everywhere north of the equator. Duration of daylight is given for every 20 degrees of latitude.

The rate of solar energy input varies through the course of a year from a maximum when Earth is closest to the sun (perihelion) to a minimum when Earth is farthest from the sun (aphelion). At perihelion, Earth is about 3.3% closer to the sun than at aphelion. The intensity of solar radiation traveling through space diminishes as the inverse square of the distance traveled. (Hence, for example, doubling the distance traveled reduces the radiation intensity to one-quarter of its initial value). Consequently, the planet receives about 6.7% more radiation at perihelion (2.04 cal/cm$^2$ per minute) than at aphelion (1.91 cal/cm$^2$ per minute).

The perihelion/aphelion contrast in solar energy coupled with the seasonal variation in radiation has some implications for global climate. The Southern

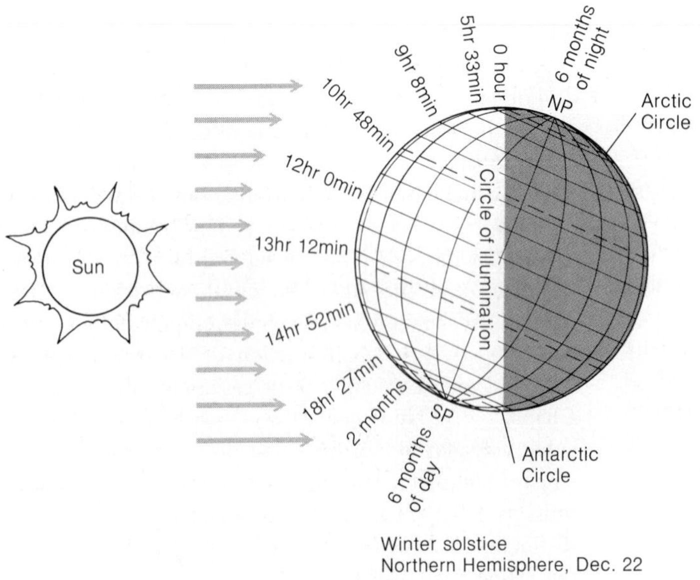

Winter solstice
Northern Hemisphere, Dec. 22

**FIGURE 2.13**
At the Northern Hemisphere winter solstice (December 22), maximum insolation is at 23 degrees, 27 minute S, and days are shorter than nights everywhere north of the equator. Duration of daylight is given for every 20 degrees of latitude.

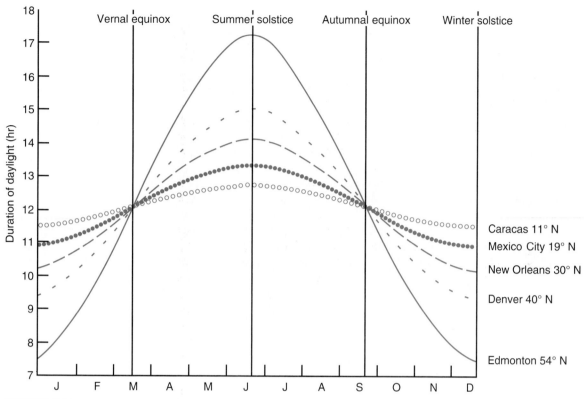

**FIGURE 2.14**
Variation in the length of daylight by latitude.

Hemisphere receives more radiation during summer and less radiation during winter than does the Northern Hemisphere. Consequently, all other factors being equal, we might expect the winter-to-summer temperature contrast to be greater in the Southern Hemisphere than in the Northern Hemisphere. As we will see in Chapter 18, however, in the Southern Hemisphere the larger percentage of ocean surface area plus the great thermal stability of ocean water moderates seasonal temperature differences and largely offsets the greater seasonal contrast in insolation.

## Solar Radiation and the Atmosphere

As solar radiation travels through the atmosphere, it interacts with gases and aerosols. These interactions consist of reflection, scattering, and absorption. Solar ra-

diation that is not reflected or scattered back to space, or absorbed by gases and aerosols, reaches the Earth's surface, where further interactions take place. Within the atmosphere, the sum of the percentage of insolation that is absorbed (the *absorptivity*) plus the percentage reflected or scattered (the *albedo*) plus the percentage transmitted to the Earth's surface (the *transmissivity*) must equal 100%. This relationship is another example of the law of energy conservation.

**Reflection** takes place at the interface between two different media such as air and cloud when some of the radiation striking that interface is thrown back. At such an interface (Figure 2.15), the angle of incident radiation *(i)* equals the angle of reflected radiation *(r)*; this is known as the **law of reflection.** The fraction of incident radiation that is reflected by some surface (or interface) is the **albedo** of that surface, that is,

$$\text{albedo} = \frac{\text{reflected radiation}}{\text{incident radiation}}$$

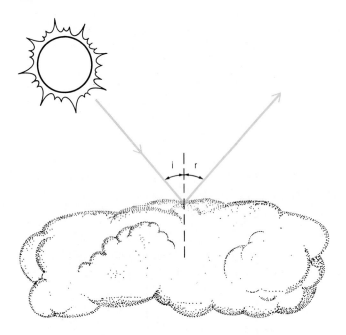

**FIGURE 2.15**
An illustration of the law of reflection whereby the angle of incident radiation *(i)* equals the angle of reflected radiation *(r)*.

where albedo is expressed either as a fraction or a percentage. As a rule, surfaces that have a high albedo (reflect a large fraction of incident solar radiation) appear light-colored and surfaces with a low albedo are perceived as dark-colored.

Within the atmosphere, the tops of clouds are the most important reflectors of insolation (Figure 2.16). The albedo of cloud tops depends primarily on cloud thickness and varies from under 40% for thin clouds (less than 50 m, or 164 ft, thick) to 80% or more for thick clouds (more than 5000 m, or 16,400 ft, thick). The average albedo for all cloud types and thicknesses is about 55%, and at any given time, clouds cover about 60% of the planet.

**Scattering** is the dispersal of radiation in all directions—up, down, and sideways. Actually, reflection is a special case of scattering. Both gas molecules and aerosols in the atmosphere scatter solar radiation but with some important differences. Scattering by molecules is wavelength dependent. As described in the Special Topic "Why Is the Sky Blue?," the preferential scattering of blue-violet light by oxygen and nitrogen molecules is the principal reason for the color of the daytime sky. On the other hand, water droplets and ice crystals that compose clouds scatter visible solar radiation equally at all wavelengths so that clouds appear white.

Reflection and and scattering within the atmosphere merely change the direction of insolation. However, through **absorption,** radiation is converted to heat. Oxygen, ozone, water vapor, and various aerosols are the principal absorbers of solar radiation within the atmosphere. Absorption by an atmospheric gas is selective by wavelength, that is, each gas absorbs strongly in some wavelengths and weakly or not at all in other wavelengths.

The clear atmosphere is essentially transparent to solar radiation in the wavelength range between about 0.3 and 0.8 micrometer (visible radiation). In the stratosphere, oxygen and ozone are strong absorbers of solar ultraviolet radiation at wavelengths shorter than 0.3 micrometer. Oxygen absorbs very short UV (less than 0.2 micrometer), and ozone absorbs longer UV (0.22 to 0.29 micrometer). The net effect of this absorption is twofold: (1) a significant reduction in the intensity of UV that reaches the Earth's surface and (2) a marked warming of the upper stratosphere (see Figure 1.9).

Water vapor absorbs solar infrared radiation in bands of wavelength greater than 0.8 micrometer. Clouds, on the other hand, are poor absorbers of solar radiation. Typically, clouds absorb less than 10% of the radiation that strikes the cloud top, although exceptionally thick clouds such as thunderclouds (cumulonimbus) absorb somewhat more.

**FIGURE 2.16**
Cloud tops strongly reflect solar radiation. Note the brightness of this cloud layer viewed from an airplane window. [Photograph by J. M. Moran]

## The Ozone Shield

UV absorption in the stratosphere prevents potentially lethal levels of UV from reaching the Earth's surface. UV is absorbed during both the formation and destruction of ozone within the stratosphere. Without this **ozone shield,** life as we know it could not exist on Earth.

At altitudes between about 10 and 50 km (6 and 31 mi), UV dissociates oxygen molecules ($O_2$) into oxygen atoms (O) that subsequently combine with other oxygen molecules to form ozone molecules ($O_3$). While all this is going on, other photochemical reactions—some quite complex—destroy ozone. The consequence of these competing chemical reactions is a peak ozone concentration of only about 10 ppm (parts per million) at altitudes of 20 to 25 km (12 to 16 mi). Some UV penetrates the ozone shield and reaches the Earth's surface where overexposure may cause serious human health problems. For more on this subject, refer to the Special Topic "The Hazards of Sunbathing."

One of today's major environmental concerns involves the role of a group of chemicals, known as CFCs (for *chlorofluorocarbons*), which may be eroding the ozone shield. CFCs have been used for decades as refrigerants, propellants in aerosol sprays, and blowing agents for foam insulation (Table 2.1). Within the troposphere, CFCs are virtually inert (chemically nonreactive), but eventually some CFCs drift into the stratosphere where intense UV radiation breaks them down and liberates chlorine atoms. Chlorine (Cl) acts as a catalyst* in chemical reactions that convert ozone to oxygen. In this way, each chlorine atom can destroy tens of thousands of ozone molecules.

A thinner ozone shield would likely mean more intense UV radiation at the Earth's surface and, for humans, a greater risk of skin cancer, eye damage including cataracts, and immune deficiencies. As a rule of thumb, every 1% decline in stratospheric ozone concentration translates into a 2% increase in the intensity

---

*A catalyst accelerates a chemical reaction without itself being altered by that reaction,

## SPECIAL TOPIC

# Why Is the Sky Blue?

Visible light is composed of the entire spectrum of colors—red, orange, yellow, green, blue, and violet. If a beam of sunlight passes through a glass prism and onto a screen (Figure 1), we see a spectrum of colors because the light beam is refracted (bent) by the glass.

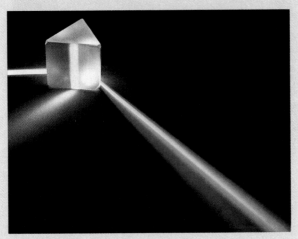

**FIGURE 1**
A glass prism disperses visible sunlight into its component colors. [Photograph © David Parker, Science Source/Photo Researchers]

The more energetic violet wavelengths of the visible spectrum are bent the most, and the less energetic red wavelengths are bent the least.

Refraction is not the only way that visible light is dispersed into its component colors. Scattering of sunlight can produce a similar effect and is responsible for the color of the daytime sky. Scattering occurs when tiny particles in the atmosphere interact with light waves and send those light waves in random directions. If the radii of the scattering particles are much smaller than the wavelength of the scattered light, then the amount of scattering varies with wavelength. This is the case, for example, when visible light is scattered by the gas molecules composing the atmosphere. In a now classic experiment performed in 1881, Lord Rayleigh demonstrated that this scattering of light is inversely proportional to the fourth power of the wavelength. This means that violet light, at the short-wavelength end of the visible spectrum, is scattered much more than red light, which is at the long-wavelength end. As sunlight travels through the atmosphere, component colors are scattered selectively out of the solar beam: violet is scattered more than blue, blue more than green, green more than yellow, and so forth.

**TABLE 2.1**
**The Most Commonly Used Chlorofluorocarbons (CFCs)**

| Type | Formula | Primary use | Residence time* (in years) |
|------|---------|-------------|------------------|
| CFC-11 | $CCl_3F$ | Foam blowing agent; aerosol propellant | 76 |
| CFC-12 | $CCl_2F_2$ | Refrigeration and air conditioning | 139 |
| CFC-113 | $C_2Cl_3F_3$ | Solvent for cleaning electronic microcircuits | 92 |

*Time required for 63% of the CFC to be washed from the atmosphere.

of biologically active UV* that passes through the ozone shield. The actual amount that reaches the Earth's surface hinges on cloudiness and air quality. Various studies suggest that a 2.5% thinning of the ozone shield could boost the incidence of human skin cancers by 10%.

The first indications of a thinning of the ozone shield came from Antarctica. During the Southern Hemisphere spring (mainly September and October), the ozone layer in the Antarctic stratosphere thins drasti-

*The portion of UV radiation that is responsible for sunburn and causes or contributes to skin cancer spans wavelengths from 0.28 to 0.32 micrometer and is known as UVB.

The dependence of scattering on wavelength predicts that scattered sunlight should be mostly violet in color. Why, then, does the sky appear blue rather than violet? The principal reason is that the human eye is more sensitive to blue light than to violet light, so that the sky appears bluer than it really is. Another factor contributing to the sky's blueness may be the dilution of violet light by all the other scattered colors. Although the other colors are scattered less than violet, they tend to wash violet into blue.

We can now understand why on a clear evening the setting sun is red (Figure 2). At very low solar altitudes, the sun's rays traverse the maximum thickness of the atmosphere. Hence, at sunset (or sunrise) there is considerable interaction between incoming solar radiation and the atmosphere's component gas molecules. Consequently, nearly all the blue-violet light is scattered out of the solar beam, leaving it rich in the red end of the visible spectrum.

If we were on the moon, which has a highly rarefied atmosphere, the sun would appear as a white disk in a black sky and stars would be visible even in bright sunshine; within a rarefied atmosphere, scattering is inconsequential.

Different results occur when radiation is scattered by particles that have radii approaching or exceeding the wavelength of the radiation being scattered. In such instances, scattering is not wavelength dependent; instead, radiation is scattered equally at all wavelengths. This is known as *Mie scattering*. The particles composing clouds (tiny ice crystals and/or water droplets) and most other atmospheric aerosols are sufficiently large to scatter sunlight in this way. For this reason, clouds appear white, and when the atmosphere contains a considerable concentration of aerosols, the entire sky is a hazy white.

**FIGURE 2**
The red of the setting sun. [Photograph by Mike Brisson]

cally. By November (December at the latest) the ozone level recovers. Although the British Antarctic Survey team first reported this phenomenon in 1985, massive ozone depletion had been measured during the prior eight Antarctic springs, but was erroneously attributed to instrument error. The area of ozone depletion, dubbed the **Antarctic ozone hole,** is about the size of the continental United States (Figure 2.17).

Satellite monitoring revealed that the Antarctic ozone hole deepened from the late 1970s into the 1990s. A field investigation conducted during the Antarctic spring of 1987 not only measured a record ozone loss of about 50% but also detected exceptionally high concentrations of chlorine monoxide (ClO), a product of chemical reactions known to destroy ozone. For many scientists, ClO was the "smoking gun" linking CFCs to the Antarctic ozone hole. Record or near-record ozone depletion was measured in the Antarctic stratosphere during the springs of 1989 through 1992.

Discovery of the Antarctic ozone hole and its possible link to CFCs spurred questions regarding trends in stratospheric ozone elsewhere around the globe. In the spring of 1992, Richard Stolarski of the Goddard Space Flight Center reported a downward trend in stratospheric ozone at midlatitude ground stations for all seasons since 1970. Shorter duration satellite measurements (about 12 years) show a negative trend in both hemispheres except near the equator where no significant change was indicated. In midlatitudes, the greatest rate of depletion occurs in late win-

## The Hazards of Sunbathing

Many people believe that a dark tan is attractive and a sign of good health (Figure 1), but mounting evidence indicates that too much sun can lead to health problems. One of the most noticeable effects is premature aging of the skin, that is, the skin develops a leathery texture and wrinkles and dark spots appear. The sun also contributes to certain types of cataracts, opaque regions in the eye that interfere with vision and, if untreated, may cause blindness. Most worrisome is the role of the sun in skin cancer, the most common form of cancer in the United States. More than 400,000 new cases of skin cancer are diagnosed each year; about one in seven Americans will develop this disease during their lifetime. The culprit is overexposure to the ultraviolet portion of solar radiation, specifically the long-wavelength end of ultraviolet, known as *UVB*. How can we protect ourselves from this dangerous radiation?

The body has natural defenses against UVB radiation. At the base of the epidermis, the outer layer of skin, are cells that regularly divide and produce new cells. Some of these new cells migrate toward the skin surface, while other cells remain behind and continue to divide. The migrating cells die on their way to the surface. Eventually these dead cells reach the surface where they form a layer of tightly packed dead cells, usually about 20 cells thick. This outer layer of dead cells shields the underlying layers of skin by absorbing potentially damaging UVB radiation.

The outer layer of dead cells gradually erodes as these cells are shed from the skin. You can observe this by vigorously rubbing your arm in a bright light: tiny flakes of skin resemble fine dust particles. Cells migrating from the base of the epidermis replace cells that are shed and thus maintain the protective layer of dead cells. The sequence from cell formation to shedding typically takes three to four weeks. UVB radiation

**FIGURE 1**

Sunbathers run the risk of excessive exposure to solar ultraviolet (UV) radiation that could cause health problems. [Photograph by J. M. Moran]

speeds up cell formation, so that a thicker layer of dead cells forms that better protects the underlying layers of skin.

The body's second line of defense consists of specialized cells called *melanocytes,* which are scattered among the dividing cells at the base of the epidermis. Melanocytes respond to UVB by producing *melanin,* a dark pigment that absorbs UVB and is responsible for darkening the skin. Melanin spreads to surrounding cells, which then migrate upward, die, and become part of the outer, dense layer of dead cells. Migration and death of melanin-containing cells explain why a tan gradually fades: melanin is lost during shedding.

Melanin does not provide total protection against UVB, however. Typically, several days elapse between the skin's initial exposure to UVB and the development of a protective tan. In the meantime, considerable damage can occur, especially if the initial exposure causes sunburn. Also, melanin does not absorb all the incident UVB so that no one is immune to the hazards of overexposure. Although the risk of contracting skin cancer is greater for fair-skinned individuals whose melanocytes produce less melanin, people with dark complexions, whose melanocytes produce more melanin, can also develop skin cancer.

Like other forms of cancer, skin cancer usually takes 20 or more years to develop. Thus, the average age for the first diagnosis of skin cancer is about 50. Unfortunately, that age is declining as skin cancer becomes increasingly more common among younger people. If our natural defenses are insufficient to protect us from developing skin cancer, what more can we do?

We must first recognize that we are exposed to much more UVB than we realize. Ultraviolet radiation penetrates clouds more readily than does visible radiation so that even when the sky is completely overcast, some protection from UVB is necessary. At the beach a person may believe that the best way to avoid too much sun is to sit under a beach umbrella, jump in the water, or wear a T-shirt. Such strategies, however, provide inadequate protection. Sand reflects up to 50% of the incident UVB, thereby exposing a person to dangerous radiation even in the shade of a beach umbrella. Water transmits UVB to a depth of a meter or so, and a wet T-shirt allows 20% to 30% of incident UVB to reach the skin. The beach is not the only place where a person is likely to be exposed to high levels of UVB. Skiing at high mountain elevations, where UVB is more intense than at sea level, results in significant exposure. In addition, snow is even more reflective of UVB than is beach sand.

Common sense can help us to avoid overexposure to UVB. Studies indicate that one or more severe sunburns greatly increases a person's chances of contracting skin cancer later in life. Thus, initial exposure to the sun should be brief until a protective tan develops. Whenever possible, we should avoid the sun at its greatest intensity, which is between 10 A.M. and 3 P.M.. A good rule of thumb is that if your shadow is shorter than you are tall, then you should apply a sunscreen; if your shadow is longer than you are tall, there is no problem. Because most skin cancers appear on commonly exposed surfaces such as the forearms, face, and neck, wearing protective clothing (e.g., a wide-brimmed hat) is a wise strategy.

Applying a sunscreen to all exposed skin is always prudent. Sunscreens have ingredients that selectively block UV wavelengths. Because of recent concerns about overexposure to UVA (a portion of the ultraviolet spectrum that can damage deeper layers of the skin), a broad-spectrum sunscreen that provides protection from both *UVB* and *UVA* is recommended. When purchasing a sun-screen, be aware of the degree of protection offered: the sun-screen's *sun protection factor (SPF).* The SPF is a measure of the time that the skin can safely be exposed to the sun. The higher the SPF value, the longer the protection lasts. Experts recommend a minimum SPF of 15. Unfortunately many people fail to apply sufficient amounts of sunscreen to their skin. To compensate, many medical personnel recommend products with higher SPF values. Note that suntan lotions are not sunscreens. Suntan lotions merely help to keep the skin moist; they do not provide any protection from UV.

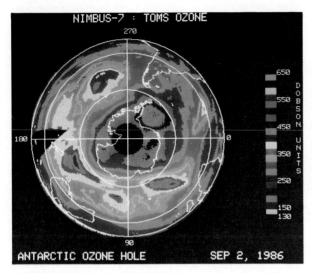
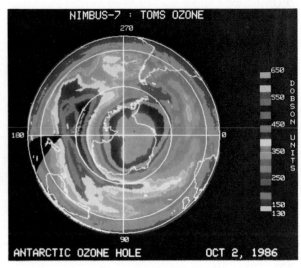

**FIGURE 2.17**
Maps of total ozone in the Southern Hemisphere illustrating the Antarctic ozone hole. These images were produced at NASA's Goddard Space Flight Center using data obtained by the TOMS (Total Ozone Mapping Spectrometer) aboard the Nimbus-7 satellite. The ozone hole is portrayed in gray and violet colors and is surrounded by a ring of relatively high total ozone (yellow, green, and brown) at middle latitudes. As illustrated in the color bar at the right, ozone concentrations are expressed in Dobson units. (The depth of ozone produced if all the ozone in a column of the atmosphere is brought to sea-level temperature and pressure is expressed in Dobson units where one Dobson unit is a hundredth of a millimeter.) Between 2 September 1986 (A) and 2 October 1986 (B), the region of very low total ozone expanded and covered an area about the size of the continental United States. [Courtesy of NASA]

ter and early spring and peaks at slightly more than 6% per decade.

At present, scientists are unable to say much about trends in UV levels at the Earth's surface because of limited data. Furthermore, ozone is a component of photochemical smog and some scientists argue that ozone pollution of the lower troposphere may at least partially offset thinning of the ozone shield. We have more to say on threats to the ozone shield and what is being done about this problem in Chapter 17.

## Solar Radiation and the Earth's Surface

Solar radiation that passes directly through the atmosphere to the Earth's surface constitutes **direct insolation.** This is augmented by **diffuse insolation,** solar radiation that is scattered and/or reflected to the Earth's surface. The direct plus diffuse insolation that strikes the Earth's surface is either reflected or absorbed by it depending on the surface albedo. The fraction that is not reflected is absorbed—that is, converted to heat—and the fraction that is not absorbed is reflected.

As noted earlier, light surfaces are more reflective of solar radiation than are dark surfaces. For example, skiers who have been sunburned on the slopes know full well that a snow cover is very reflective of sunlight. Fresh-fallen snow typically has an albedo of between 75% and 95%; that is, 75% to 95% of the solar radiation striking the surface of a snow cover is reflected, and the rest (5% to 25%) is absorbed. At the other extreme, the albedo of a dark surface, such as a blacktopped road or a spruce forest, may be as low as 5%. From these values, we can understand why light-colored clothing is usually a more comfortable choice than dark-colored clothing during hot weather. The albedos of some common surfaces are listed in Table 2.2.

The albedo of some but not all surfaces also varies

**TABLE 2.2**
**Albedo (Reflectivity) of Some Common Surface Types for Visible Solar Radiation**

| Surface | Albedo (% reflected) |
|---|---|
| Grassland | 16–20 |
| Deciduous forest | 15–20 |
| Coniferous forest | 5–15 |
| Green crops | 15–25 |
| Tundra | 15–20 |
| Desert | 25–30 |
| Sand | 18–28 |
| Blacktopped road | 5–10 |
| Sea ice | 30–40 |
| Fresh snow | 75–95 |
| Old snow | 40–60 |
| Glacier ice | 20–40 |
| Water (high solar altitude) | 3–10 |
| Water (low solar altitude) | 10–100 |
| Cities | 14–18 |

**FIGURE 2.18**
Variation of albedo with solar altitude for a flat and undisturbed water surface. The albedo increases sharply for solar altitudes less than 30 degrees. A wave-covered surface has a slightly higher albedo at high solar altitudes and slightly lower albedo at low solar altitudes.

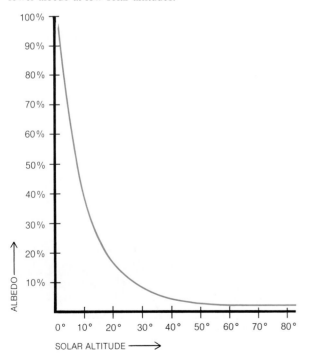

with the angle of incidence of solar radiation, that is, the angle of the sun above the horizon (solar altitude). The variation of albedo with solar altitude is especially pronounced for the surfaces of oceans and lakes. As shown in Figure 2.18, under clear skies, the albedo of a water surface increases with decreasing solar altitude. The increase in albedo is especially sharp for solar altitudes less than 30 degrees and approaches a mirrorlike 100% near sunrise and sunset. On the other hand, when the sky is completely cloud covered, only diffuse insolation is incident on a water surface and the albedo varies little with solar altitude and is uniformly very low (less than 10%). On a global basis, the average albedo of the ocean surface is only about 8%; that is, the ocean is a strong absorber of solar radiation.

Solar radiation is selectively absorbed by wavelength as it penetrates ocean or lake waters. Absorption increases for increasing wavelength so that within clear, clean water, red light is totally absorbed within about 15 m (49 ft) of the surface, whereas blue-violet light may penetrate to depths of 250 m (820 ft). Usually, however, suspended sediments significantly boost the rate of absorption so that sunlight is completely absorbed within shallower depths. In fact, most water bodies are so turbid (cloudy) that little if any sunlight reaches below 10 m (33 ft).

Significant changes in surface albedo occur seasonally and affect the portion of solar radiation that is converted to heat. Over land, a thick winter snow cover elevates the surface albedo considerably. In autumn in forested areas, the loss of leaves from deciduous trees raises the surface albedo. In middle and high latitudes, significant increases in surface albedo accompany the winter freeze-over of lakes and the formation of sea ice.

## Solar Radiation Budget

Satellite measurements indicate that the Earth–atmosphere system reflects or scatters off to space about 31% of incoming solar radiation. This is known as the Earth's **planetary albedo.** By contrast, the moon's albedo is only about 7% primarily because of the absence of clouds in the highly rarefied lunar atmosphere. This

**TABLE 2.3**
**Solar Radiation Budget**

| | |
|---|---|
| Reflected by the Earth–atmosphere system | 31% |
| Absorbed by the atmosphere | 23% |
| Absorbed by the Earth's surface | 46% |
| Total | 100% |

means that Earth viewed from the moon (by U.S. astronauts) is more than four times brighter than the moon viewed from Earth on a clear night.

Of the total solar radiation that is intercepted by the Earth–atmosphere system, only about 23% is absorbed by the atmosphere. In other words, the atmosphere is relatively transparent to solar radiation. The remaining 46% of solar radiation (31% was reflected or scattered to space) is absorbed by the Earth's surface, chiefly because of the low average albedo of ocean waters covering about 71% of the globe. The solar radiation budget is summarized in Table 2.3.

The Earth's surface is the principal recipient of solar heating, and heat continuously flows from Earth's surface to the atmosphere, which eventually radiates it back into space. The Earth's surface is thus the main source of heat for the atmosphere. This is evident in the vertical temperature profile of the troposphere: Normally, air is warmest close to the Earth's surface, and air temperature drops with increasing altitude, that is, away from the main source of heat.

# Infrared Response and the Greenhouse Effect

If solar radiation were continually absorbed by the Earth–atmosphere system without any compensating flow of heat out of the system, the temperature of the Earth's surface and atmosphere would rise steadily. In reality, the average global air temperature changes little from one year to the next. Global radiative equilibrium keeps the planet's temperature in check. That is, solar heating is balanced by the escape of heat (via infrared radiation) from the Earth–atmosphere system to space. Although solar radiation is supplied only to the illuminated portion of the planet, infrared radiation is emitted ceaselessly, day and night, by the entire Earth–atmosphere system.

*GREENHOUSE WARMING*

Because insolation and terrestrial infrared radiation peak in different portions of the electromagnetic spectrum, they have different properties, and they interact differently with the atmosphere. As noted earlier, atmospheric gases and aerosols absorb only about 23% of incoming solar radiation. In contrast, the atmosphere absorbs greater amounts of infrared radiation emitted by the Earth's surface. The atmosphere, in turn, reradiates some infrared radiation back to the Earth's surface (Figure 2.19). This reradiation by the atmosphere slows the escape of heat to space and causes the lower troposphere to have a higher average temperature than the upper troposphere. The lower atmosphere is thus hospitable to life. Thanks to this reradiation, the Earth's average surface temperature is considerably higher than it would otherwise be. Viewed from space, the planet radiates at about −18 °C (0 °F), whereas the average temperature at the Earth's surface is about 15 °C (59 °F). Hence, infrared reradiation elevates the Earth's

**FIGURE 2.19**
Schematic illustration of the greenhouse effect. (A) With no atmosphere, infrared radiation escapes readily to space. (B) With an atmosphere, certain gases (mainly water vapor) absorb and reradiate infrared radiation both downward and upward. The net effect is warming of the lower atmosphere.

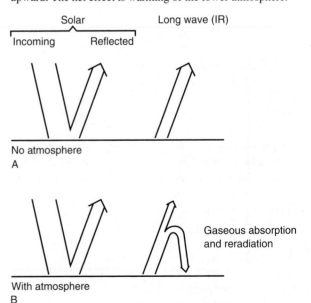

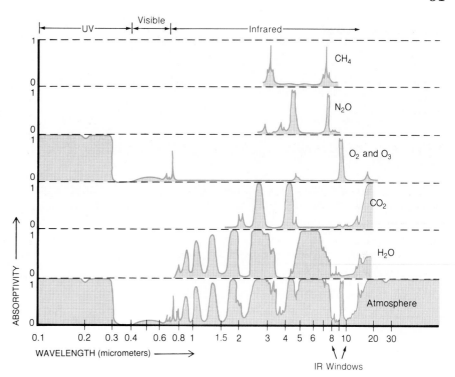

**FIGURE 2.20**
Absorption of radiation by selected components of the atmosphere as a function of wavelength. Absorptivity is the fraction of the radiation absorbed and ranges from 0 to 1 (0% to 100% absorption). Absorptivity is very low or near zero in atmospheric windows. Note the infrared windows near 8 and 10 micrometers. [From R. G. Fleagle and J. Businger, *An Introduction to Atmospheric Physics.* New York: Academic Press, 1980, p. 232.]

surface temperature by

$$[15\ °C - (-18\ °C)] = 33\ \text{Celsius degrees}$$
or
$$[59\ °F - 0\ °F] = 59\ \text{Fahrenheit degrees}$$

Certain atmospheric gases, primarily water vapor and to a lesser extent carbon dioxide, ozone, methane ($CH_4$), and nitrous oxide ($N_2O$), absorb terrestrial infrared radiation and thus slow its escape to space. The percentage of infrared radiation absorbed varies with wavelength, as shown in Figure 2.20. Interestingly, the percentage absorbed is very low at wavelengths near the peak infrared intensity. It is through these so-called **atmospheric windows,** wavelength bands in which there is little or no absorption of radiation, that most heat from the Earth's surface eventually escapes to space as infrared radiation.

Like the Earth's atmosphere, window glass is relatively transparent to visible solar radiation but slows the transmission of infrared radiation. Greenhouses, where plants are grown, are designed to take advantage of this property of glass by being constructed almost entirely of glass panes (Figure 2.21). Sunlight

**FIGURE 2.21**
The glass of a greenhouse behaves in an analogous manner to certain gases in the atmosphere that strongly absorb infrared radiation. Greenhouse gases include water vapor and carbon dioxide. [Photograph by J. M. Moran]

## Greenhouse Effect on Mars and Venus

On Earth, water vapor is the principal greenhouse gas. Water vapor plus the other greenhouse gases (e.g., carbon dioxide, ozone, methane) is responsible for elevating the average temperature of the lower atmosphere by 33 Celsius degrees (59 Fahrenheit degrees). A greenhouse effect also operates on our neighboring planets, Mars and Venus. On both planets, the principal atmospheric gas, carbon dioxide, is also the main green-

house gas. The Martian atmosphere is considerably thinner than the Earth's atmosphere; its greenhouse effect raises the average surface temperature by about 10 Celsius degrees (18 Farenheit degrees). On the other hand, the Venusian atmosphere is considerably denser than the Earth's atmosphere; its greenhouse warming is estimated at 523 Celsius degrees (941 Fahrenheit degrees).

readily enters a greenhouse and is absorbed (that is, converted to heat). Some of the heat energy is radiated as IR, which is absorbed and not readily transmitted by the glass. Because of similarities in radiative properties, the behavior of IR-absorbing atmospheric gases is often referred to as the **greenhouse effect** and the gases are known as **greenhouse gases.**

The greenhouse analogy is not always correct, however. Trapping of IR radiation by glass is only part of the reason why some greenhouses retain internal heat. Those greenhouses cut heat loss principally by acting as a shelter from the wind, thereby reducing conductive and convective heat loss (discussed in Chapter 3). As a rule, the thinner the greenhouse glass and the greater the external wind speed, the more important is the shelter effect compared to radiation trapping. Nonetheless, because reference to the greenhouse effect is so common in discussions of the Earth–atmosphere radiation balance, we use the term here.

To illustrate the greenhouse effect in the atmosphere, compare the typical summer weather in the American Southwest with that of the Gulf of Mexico coast. Both areas are at about the same latitude and hence receive about the same intensity of solar radiation. In both regions, afternoon temperatures typically top 30 °C (86 °F). At night, however, air temperatures in the two areas often differ markedly. In the Southwest, there is usually less water vapor in the air to impede the escape of infrared radiation, and heat is readily lost to space. Surface air temperatures may fall below 15 °C (59 °F) by sunrise. On the other hand, along the Gulf Coast, the air often is more humid and thus absorbs more infrared radia-

tion. Because a portion of this heat is reradiated back toward the Earth's surface, nighttime temperatures may fall only into the 20s Celsius (the 70s Fahrenheit).

Clouds, which are composed of water droplets and/or ice crystals, also produce a greenhouse effect. Hence, all other factors being equal, nights usually are colder when the sky is clear than when the sky is cloud covered. Even a high, thin overcast through which the moon is visible can elevate temperatures at the Earth's surface by several Celsius degrees. Recall, however, that clouds also have a high albedo for solar radiation. Hence, clouds affect climate in opposing ways: By absorbing and reradiating IR, they warm the Earth's surface and by reflecting solar radiation, they cool the Earth's surface. On a global scale, which one of these two effects is more important? Recent analyses of satellite measurements of radiation indicate that clouds have a net cooling effect on global climate. That is, all other factors being equal, a greater cloud cover would tend to cool the planet.

### GREENHOUSE GASES AND GLOBAL WARMING

Today, considerable concern is directed at the potential global climatic consequences of increasing concentrations of greenhouse gases, especially carbon dioxide. Atmospheric $CO_2$ has been on the rise since the Industrial Revolution because of the burning of coal and oil and, to a lesser extent, because of the clearing of forests (Figure 2.22). More $CO_2$ in the atmosphere enhances the greenhouse effect and, unless compen-

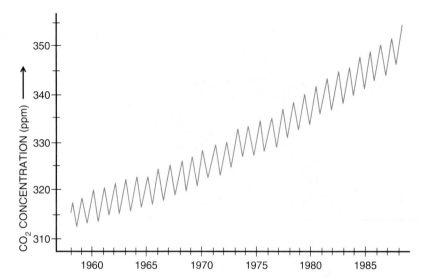

**FIGURE 2.22**
Upward trend in atmospheric carbon dioxide level as measured at Mauna Loa Observatory, Hawaii. Annual cycles of photosynthesis and respiration account for the yearly oscillation in $CO_2$. [Data from C. D. Keeling et al., Scripps Institution of Oceanography, as reported in *Geophysical Monitoring for Climate Change*, NOAA]

sated for by other changes in the Earth–atmosphere system, may trigger significant global warming.*

As noted in Chapter 1, atmospheric scientists have employed numerical models of the atmosphere in an effort to predict the magnitude of global warming that could attend a continued increase in $CO_2$. Depending on the specific model, these experiments predict that the Earth's average surface temperature could rise between 1.5 and 4.5 Celsius degrees (2.7 and 8.1 Fahrenheit degrees) with a doubling of $CO_2$—possible by the middle of the next century. As Stephen Schneider of the National Center for Atmospheric Research (NCAR) pointed out, the warming could match the total post-Ice Age temperature rise, albeit at a rate 10 to 100 times faster. The agricultural, socioeconomic, and political implications of such a climate change are likely to be far reaching and disruptive. Already some atmospheric scientists are attributing the global warming trend of the 1980s to an enhanced greenhouse effect.

In addition to $CO_2$, upward trends in other atmospheric IR-absorbing trace gases may also enhance the natural greenhouse effect. Specifically, concentrations of methane ($CH_4$), nitrous oxide ($N_2O$), and CFCs (chlorofluorocarbons) are rising (Table 2.4). Scientists propose a number of causes for the buildup of these greenhouse gases. Methane is the product of anaerobic decay (in the absence of oxygen) of organic matter and its increase may be linked to biological activity in

termites and the stomachs of cattle and sheep, or as effluvium of marshes, rice paddies, and landfills. The upward trend in nitrous oxide is likely the product of industrial air pollution, and, as noted in Table 2.1, CFCs have numerous sources.

Although atmospheric concentrations of methane, nitrous oxide, and CFCs are considerably less than carbon dioxide, they are more efficient absorbers of IR because they strongly absorb within the atmospheric windows. In fact, absorption by gases is directly proportional to concentration, so that doubling the concentration of a gas doubles the absorption. The combined climatic impact of rising levels of these gases could equal $CO_2$-induced global warming. We have more to say about prospects for an enhanced greenhouse effect in Chapter 20.

**TABLE 2.4**
**Trends in Greenhouse Gases**

| Greenhouse gas | Present concentration (ppb)[a] | Pre-Industrial concentration (ppb) | Annual rate of increase (%) |
|---|---|---|---|
| $CO_2$ | 353,000 | 280,000 | 0.5 |
| $CH_4$ | 1,738 | 790 | 0.9 |
| $N_2O$ | 310 | 288 | 0.8 |
| $O_3$[b] | 20–40 | 10 | 0.5–2.0 |
| CFCs | 0.28–0.48 | 0 | 4.0 |

*In the popular media, the term greenhouse effect often refers to $CO_2$-induced global warming; the more appropriate term is an enhanced greenhouse effect.

[a]Parts per billion by volume at the Earth's surface.

[b]Northern Hemisphere tropospheric ozone only.

**FIGURE 2.23**
The pyranometer is the standard instrument used to measure solar radiation. [Courtesy of Qualimetrics, Inc.]

## Radiation Measurement

The **pyranometer** is the standard instrument for measuring the intensity of solar radiation striking a horizontal surface. The instrument consists of a sensor enclosed in a transparent hemisphere (Figure 2.23) that transmits total (direct plus diffuse) short-wave (less than 3.5-micrometer wavelength) insolation. The sensor is a disk consisting of alternating black and white wedge-shaped segments that form a starlike pattern. The black wedges are highly absorptive and the white wedges are highly reflective of solar radiation. Differences in absorptivity and albedo mean that the temperatures of the black and white portions of the sensor respond differently to the same intensity of solar radiation. The temperature contrast between the black and white segments is calibrated in terms of radiation flux (W/m$^2$, for example). A pyranometer may be linked electronically to a pen recorder that traces a continuous record of insolation (Figure 2.24), or the instrument's output signal may be recorded on a magnetic tape cassette for processing and storage.

Special care must be taken in mounting and maintaining a pyranometer. The instrument should be situated where it will not be affected by shadows, by any highly reflective surfaces nearby, or by other sources of radiation. The glass bulb must also be kept clean and dry.

## Conclusions

Earth intercepts a tiny fraction of the total radiational energy output of the sun. Movements of the planet in space distributes this energy unequally within the Earth–atmosphere system. As solar radiation passes through the atmosphere, a portion is scattered and reflected back to space and a portion is absorbed (that is, converted to heat). Overall, however, the atmosphere is relatively transparent to solar radiation and that which reaches the Earth's surface is either reflected or absorbed. The Earth–atmosphere system responds to solar heating by emitting infrared

**FIGURE 2.24**
A recording pyranometer produces a continuous record of the incoming solar radiation. Scattered clouds cast shadows on the radiation sensor and produce spikes in the curve. This record was obtained at Green Bay, Wisconsin, in early July.

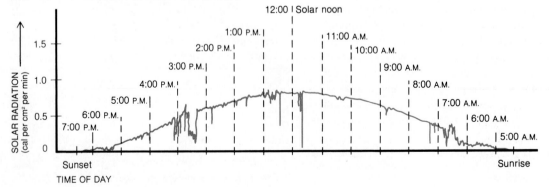

MATHEMATICAL NOTE

## Blackbody Radiation Laws

Several laws describe properties of the electromagnetic radiation emitted by a *perfect radiator.* A perfect radiatior, usually called a *blackbody,* is a hypothetical object that absorbs all the radiation that strikes it; that is, a perfect radiator neither reflects nor transmits any radiation. In reality, no perfect radiators exist, but the sun and the Earth's surface radiate *approximately* as blackbodies. We can therefore apply blackbody radiation laws to solar and terrestrial radiation, with some qualifications. Here, in brief, are the basic blackbody radiation laws.

*Kirchhoff's law* holds that a perfect absorber of radiation of a given wavelength is also a perfect emitter of radiation at that same wavelength. A blackbody thus emits, as well as absorbs, all incident radiation at all wavelengths. In general, for all objects the efficiency of radiation absorption, called *absorptivity,* equals the efficiency of radiation emission, called *emissivity.* A good absorber is therefore a good emitter, and a poor absorber is a poor emitter. The emissivity and absorptivity of a blackbody are both 100%.

As noted elsewhere in this chapter, in a transparent substance such as the atmosphere, the law of energy conservation requires the sum of the absorptivity plus albedo plus transmissivity to equal 100%. From Kirchhoff's law, because absorptivity equals emissivity, it follows that the sum of emissivity plus albedo plus

transmissivity equals 100%. For an opaque substance, such as the ground, transmissivity equals zero, and the sum of emissivity plus albedo equals 100%.

The *Stefan–Boltzmann law* states that the rate at which a blackbody radiates energy across all wavelengths (called *emittance, E*) is directly proportional to the fourth power of the absolute temperature, $T$ (measured in kelvins), of the radiating body. The mathematical statement of this law is

$$E = \sigma T^4$$

where $\sigma$ is the Stefan–Boltzmann constant equal to $5.67 \times 10^{-8}$ W m$^{-2}$ K$^{-4}$.

*Planck's law* states that the rate at which radiation is emitted by a blackbody depends on the absolute temperature of the body and the specific wavelength of the radiation. This law enables one to compute the amount of radiation emitted at some wavelength and at a specified temperature.

*Wien's displacement law* holds that the wavelength at which a blackbody emits the maximum intensity of radiation, $\lambda_{max}$, is inversely proportional to the absolute temperature, $T$, of the blackbody, that is,

$$\lambda_{max} = C/T$$

where $C$ is the constant of proportionality (equal to 2897 if $\lambda_{max}$ is expressed in micrometers).

radiation to space. Greenhouse gases and clouds absorb and reradiate some IR back to the Earth's surface thereby significantly elevating the average temperature of the lower atmosphere.

Net incoming solar radiation is balanced by the infrared radiation emitted to space by the Earth–atmosphere system. Absorption of solar radiation causes warming, and emission of infrared radiation to space causes cooling. Within the Earth–atmosphere system, however, the rates of radiational heating and radiational cooling are not the same everywhere. Before we examine the reasons for these energy imbalances and their implications for atmospheric circulation, we first need to distinguish between heat and temperature and de-

scribe heat transfer processes. This we do in the next chapter.

## *Key Terms*

| | |
|---|---|
| law of energy conservation | wave frequency |
| first law of thermodynamics | ultraviolet radiation (UV) |
| | X-rays |
| electromagnetic radiation | gamma radiation |
| electromagnetic spectrum | visible radiation |
| wavelength | infrared radiation (IR) |
| | microwave |

radio waves
blackbody
Wien's displacement law
Stefan–Boltzmann law
global radiative
  equilibrium
photosphere
granules
faculae
sunspots
chromosphere
corona
insolation
solar altitude
perihelion
aphelion
equinoxes
solstice
Tropic of Cancer

Arctic Circle
Antarctic Circle
Tropic of Capricorn
solar constant
reflection
law of reflection
albedo
scattering
absorption
ozone shield
Antarctic ozone hole
direct insolation
diffuse insolation
planetary albedo
atmospheric windows
greenhouse effect
greenhouse gases
pyranometer

## Summary Statements

☐ All objects emit energy as electromagnetic radiation. The many forms of electromagnetic radiation make up the electromagnetic spectrum and are distinguished on the basis of wavelength, frequency, and energy level.

☐ Earth and sun closely approximate perfect radiators (blackbodies) so that blackbody radiation laws may be applied to them with useful results.

☐ Wien's displacement law predicts that the wavelength of most intense radiation is inversely proportional to the absolute temperature of the radiating object. Hence, solar radiation peaks in the visible, and Earth–atmosphere (terrestrial) radiation peaks in the infrared (IR).

☐ According to the Stefan–Boltzmann law, the total energy radiated by an object at all wavelengths is directly proportional to the fourth power of its absolute temperature. Hence, the sun emits considerably more radiational energy per unit area than does the Earth–atmosphere system.

☐ The total energy (in the form of solar radiation) that is absorbed by the Earth–atmosphere system equals the total energy (in the form of infrared radiation) emitted by the Earth–atmosphere system to space. This is known as global radiative equilibrium.

☐ Solar altitude, the angle of the sun above the horizon, influences the intensity of solar radiation that strikes the Earth's surface. All other factors being equal, as the solar altitude increases, the intensity of solar radiation received at the Earth's surface also increases.

☐ As a consequence of the Earth's elliptical orbit, nearly spherical shape, and tilted rotational axis, solar radiation is distributed unevenly over the Earth's surface and changes through the course of a year.

☐ In the summer hemisphere, solar altitudes are higher, days are longer, there is more solar radiation, and air temperatures are higher. In the winter hemisphere, solar altitudes are lower, days are shorter, there is less solar radiation, and air temperatures are lower.

☐ Solar radiation that is not absorbed by the atmosphere (converted to heat) or reflected or scattered off to space, reaches the Earth's surface.

☐ Absorption of ultraviolet radiation during the natural formation and destruction of ozone within the stratosphere shields organisms from potentially lethal levels of UV.

☐ Solar radiation that strikes the Earth's surface is either reflected or absorbed depending on the surface albedo. Surfaces that appear light-colored have a relatively high albedo for visible radiation and surfaces that appear dark-colored have a relatively low albedo.

☐ The atmosphere is heated from below, that is, heat flows from the Earth's surface to the overlying air. This is evident in the average temperature profile of the troposphere, that is, the temperature drops with altitude.

☐ Water vapor, carbon dioxide, and several other atmospheric gases absorb and reradiate IR back toward the Earth's surface, thereby significantly elevating the average temperature of the lower atmosphere. Clouds also contribute to this so-called greenhouse effect.

☐ Rising levels of atmospheric $CO_2$ and several other IR-absorbing gases may enhance the greenhouse effect and cause global warming. Combustion of fossil fuels and, to a lesser extent, clearing of forests are responsible for an upward trend in atmospheric carbon dioxide.

☐ The pyranometer is the standard instrument for monitoring the flux of solar radiation.

## Review Questions

1. What is the relationship between the wavelength and frequency of electromagnetic radiation?
2. Describe how energy level varies within the electromagnetic spectrum.
3. What is a *blackbody*?
4. In your own words define *Wien's displacement law* and *Stefan–Boltzmann law*. Apply these laws to the sun and the Earth–atmosphere system.

5. In the Northern Hemisphere, we are closer to the sun during winter than during summer. Why then is winter colder than summer?

6. Distinguish between aphelion and perihelion. Also distinguish between the equinox and solstice.

7. What is the significance of the Tropic of Cancer and the Tropic of Capricorn?

8. What is the significance of the Arctic Circle and the Antarctic Circle?

9. How and why does solar altitude affect the intensity of solar radiation received at the Earth's surface?

10. Define the *solar constant.*

11. What is the difference between scattering and reflection?

12. When an object absorbs radiation, what happens to its temperature?

13. Compare the average albedo of the Earth's surface with the average albedo of clouds.

14. Although insolation reaches its maximum intensity around noon, air temperatures in the lower troposphere typically do not reach a maximum until several hours later. Explain why.

15. How does the high albedo of a snow cover influence air temperature?

16. The atmosphere is relatively transparent to solar radiation. Explain this statement.

17. What is the significance of the greenhouse effect for life on Earth?

18. How does cloudiness influence the rate of fall of air temperatures at night?

19. What is an *atmospheric window* for infrared radiation?

20. An enhanced greenhouse effect may be the consequence of rising trends in certain greenhouse gases. Identify those gases.

## Quantitative Questions

1. Suppose that the solar radiation intercepted by Earth were uniformly distributed over the entire planet. Compute the number of watts incident on each square meter. *Hint*: The ratio of the area of a circle to the surface area of a sphere having the same diameter as the circle is 1 to 4.

2. The intensity of radiation decreases as the inverse square of the distance traveled. If the mean Earth-to-sun distance were three times what it is today, the solar constant would be reduced to what fraction of its present value?

3. Compute the albedo if the solar radiation incident on some surface is 200 W/m$^2$ and the solar radiation reflected by that surface is 160 W/m$^2$. Does the surface appear relatively light or dark in color?

4. According to the law of reflection, if solar radiation strikes the top of a cloud at an angle of 45 degrees with a line perpendicular to the cloud surface, at what angle does the reflected solar radiation leave the cloud surface?

## Questions for Critical Thinking

1. For the entire globe, why must net insolation balance infrared radiation emitted to space? What would be the implications for global climate if this energy balance did not prevail?

2. If Earth had no atmosphere, the total solar radiation received per 24-hour day at the Earth's surface would reach a maximum in summer at the South Pole. Explain why the maximum is at the pole and not at the Tropic of Capricorn.

3. Why is the albedo of the moon considerably less than the planetary albedo of the Earth?

4. Speculate on what natural or human-related activities could raise or lower the Earth's planetary albedo. What are the implications of such changes for global climate?

5. What basic assumptions are made when blackbody radiation laws are applied to the sun and the Earth–atmosphere system?

## Selected Readings

BRUNE, W. H. "Ozone Crisis: The Case Against Chlorofluorocarbons," *Weatherwise* 43, No. 3 (1990):136–143. Reviews what is considered to be the chief threat to the stratospheric ozone shield.

FOUKAL, P. V. "The Variable Sun," *Scientific American* 262, No. 2 (1990):34–41. Describes how the sun's output of radiation and particles varies with time.

LINDZEN, R. S. "Some Coolness Concerning Global Warming," *Bulletin of the American Meteorological Society* 71 (1990):288–299. Questions the widely held assumption that rising levels of atmospheric $CO_2$ will cause atmospheric warming.

STOLARSKI, R., ET AL. "Measured Trends in Stratospheric Ozone," *Science 256* (1992):342–349. Describes trends in stratospheric ozone derived from both ground-based and satellite monitors.

WALSH, J. E. "Snow Cover and Atmospheric Variability," *American Scientist* 72, No. 1 (1984):50–57. Discusses the influence of a regional snow cover on the radiation balance.

WHITE, R. M. "The Great Climate Debate," *Scientific American* 263, No. 1 (1990):36–43. Examines prospects for global warming due to a buildup of greenhouse gases.

# 3 Heat and Temperature

*Maycomb was an old town, but it was a tired old town when I first knew it. . . . Somehow, it was hotter then: a black dog suffered on a summer's day; bony mules hitched to Hoover carts flicked flies in the sweltering shade of the live oaks on the square. Men's stiff collars wilted by nine in the morning. Ladies bathed before noon, after their three-o'clock naps, and by nightfall were like soft tea-cakes with frostings of sweat and sweet talcum.*

HARPER LEE
*To Kill A Mockingbird*

Heat and temperature are two distinct concepts. The temperatures of different substances respond differently to the addition (or loss) of the same quantity of heat. This is one reason why, at the beach in summer, the sand feels warmer than the water. [Photograph by J. M. Moran]

T EMPERATURE IS ONE of the most important and common variables used to describe the state of the atmosphere; it is a usual component of weather reports and forecasts. From everyday experience, we know that air temperature varies with time: from one season to another, between day and night, and even from one hour to the next. Air temperature also varies from one place to another: highlands and higher latitudes are usually colder than lowlands and lower latitudes.

We also know that temperature and heat are closely related concepts. When we heat a pan of soup on the stove, the temperature of the soup rises. When we drop an ice cube into a beverage, the temperature of the beverage falls. Granted that the two concepts are related, what then is the precise distinction between heat and temperature? This is one of the questions we deal with in this chapter. We also contrast the major temperature scales, consider how temperature is measured, describe how heat is transported in response to temperature gradients, and explain how the temperature of a substance responds to an addition or loss of heat.

## Distinguishing Heat and Temperature

All substances are made up of a multitude of minute particles (atoms or molecules) that are continually in rapid, random motion. Hence, atoms and molecules possess energy of motion, called **kinetic energy. Heat** is defined as the *total* kinetic energy of the atoms or molecules composing a substance.* The atoms or molecules in a substance do not all move at the same velocity so that there's actually a range of kinetic energy among the atoms or molecules. **Temperature** is directly proportional to the *average* kinetic energy of the individual atoms or molecules.

The distinction between heat and temperature is made clearer by the following illustration. A cup of water at 80 °C (176 °F) is much hotter than a bathtub of water at 30 °C (86 °F). That is, the average kinetic energy of individual water molecules at 80 °C is greater than at 30 °C. However, the greater volume of water in the bathtub means that it contains more total kinetic molecular energy (heat) than does the cup of water. Consequently, the cup of water will cool down to room temperature much more rapidly than will the bathtub of water. Much more heat energy must be removed from the bathtub water than from the cup of water in order for both to cool to the same temperature.

From the above discussion it would appear that air temperature always changes whenever air gains or loses heat. However, this is not necessarily the case. Water is a component of air and occurs in all three phases (ice crystals, droplets, and vapor), and, as we will see in Chapter 4, heat is either required or released when water changes phase. Furthermore, air is a highly compressible mixture of gases; that is, an air sample can change volume. As we will discuss in Chapter 6, heat energy is required for the work of expansion or compression of air. Hence, as a sample of air gains or loses heat, that heat may be involved in some combination of temperature change, phase change of water, or volume change.

*Note that heat is a characteristic of a substance and not a separate entity. Strictly speaking, therefore, it is imprecise to refer to heat apart from the particular substance that possesses the heat. For example, heat does not rise, but heated air does. In our discussions of heat in this book, we refer either explicitly or implicitly to the heat energy of some substance.

## Temperature Scales

For most scientific purposes, temperature is described in terms of the Celsius scale. First proposed by the Swedish astronomer Anders Celsius in 1736, the Celsius temperature scale has the numerical convenience of a 100-degree interval between the melting point of ice and the boiling point of pure water. The United States is virtually the only nation that still uses the numerically more cumbersome Fahrenheit scale for everyday temperature measurements, including weather reports. This scale was introduced in 1714 by a German physicist, Gabriel Fahrenheit. If a thermometer graduated in both scales is immersed in a glass containing a mixture of ice and water, the Celsius scale will read 0 °C and the Fahrenheit scale will read 32 °F. In boiling water at sea level, the readings will be 100 °C and 212 °F.

The average kinetic energy of individual molecules is less in cold substances than in hot substances. There is, theoretically at least, a temperature at which all molecular motion ceases. It is called **absolute zero** and corresponds to −273.15 °C (−459.67 °F). Actually, some atomic-level activity likely occurs at absolute zero, but as noted in Chapter 2, an object at that temperature emits no electromagnetic radiation.

On the Kelvin scale, temperature is the number of kelvins above absolute zero; hence, the Kelvin scale is a more direct measure of average kinetic molecular activity than either the Celsius or Fahrenheit temperature scales. Whereas units of temperature are expressed as *degrees Celsius* on the Celsius scale and *degrees Fahrenheit* on the Fahrenheit scale, on the Kelvin scale they are expressed simply as *kelvins (K)*. Because nothing can be colder than absolute zero, there are no negative temperatures on the Kelvin scale. A 1-kelvin increment corresponds precisely to a 1-degree increment on the Celsius scale. The three scales are contrasted in Figure 3.1, and conversion formulas are

$$°F = \tfrac{9}{5}\,°C + 32°$$
$$°C = \tfrac{5}{9}\,(°F - 32°)$$
$$K = °F + 459.67$$
$$K = °C + 273.15$$

## Temperature Measurement

A **thermometer** is the usual instrument for monitoring variations in air temperature. Galileo is credited with its invention in 1593. Perhaps the most common type of thermometer consists of a liquid-in-glass tube attached to a graduated scale (Figure 3.2). Typically, the liquid is either mercury (which freezes at −39 °C or −38 °F) or alcohol (which freezes at −117 °C or −179 °F). Both the glass and the liquid (mercury or alcohol) expand when heated but the liquid much more so than

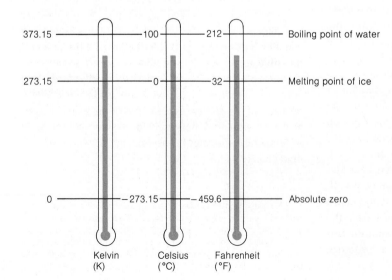

**FIGURE 3.1**
A comparison of the three temperature scales: Kelvin, Celsius, and Fahrenheit.

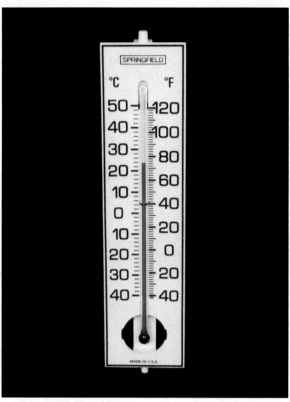

**FIGURE 3.2**
A liquid-in-glass thermometer graduated in both the Celsius (°C) and Fahrenheit (°F) scales. In this case, the liquid is alcohol. [Photograph by J. M. Moran]

the glass. As the air temperature rises and heat is transferred to the thermometer, the liquid expands and rises in the glass tube; as the air and thermometer cool, the liquid contracts and drops in the tube.

Some liquid-in-glass thermometers are designed to register the maximum and minimum temperatures over a specified period. In one common design, maximum and minimum thermometers are mounted side by side (Figure 3.3). The maximum thermometer has a small constriction in the tube just above the bulb. As heat is supplied to the bulb, the fluid (usually mercury) expands upward and beyond the constriction. Then, when the temperature falls, the fluid thread breaks at the constriction so that the end of the mercury column is positioned at the highest (maximum) temperature. The maximum thermometer is reset by whirling the tube which drives the fluid past the constriction and back into the bulb. In the minimum thermometer, one end of

a dumbbell-shaped glass index floats just below the surface of the fluid column (usually alcohol). As the temperature falls, the index is drawn downward; as the temperature rises again, the fluid expands and the index is left behind at the lowest (minimum) temperature. The minimum thermometer is reset by tilting the thermometer, bulb end upward. Usually, maximum and minimum thermometers are read and reset once every 24 hours.

A second type of thermometer uses a bimetallic sensing element to take advantage of the expansion and contraction that accompany the heating and cooling of metals. A bimetallic sensing element consists of two different metal strips that are welded together side by side. The two metals have different rates of thermal expansion; that is, one metal expands more than the other in response to the same heating. Because the two metals are bonded together, heating causes the bimetallic strip to bend; the greater the heating, the greater the bending. For example, the rate of thermal expansion of brass is about twice that of iron so that a bimetallic strip composed of those two metals will bend in the direction of the iron when heated. A series of gears or levers translates the response of the bimetallic strip to a pointer and a dial calibrated to read in °C or °F. Alternatively, this device may be rigged to a pen and a clock-driven drum to give a continuous trace of temperature with time (Figure 3.4). This instrument is called a **thermograph.**

Another type of thermometer employs a thermistor, a type of electrical conductor whose resistance changes as the temperature fluctuates. Variations in electrical resistance are calibrated in terms of temperature. Radiosondes are equipped with this type of thermometer. In addition, a thermistor thermometer may be designed to give remote temperature readings by mounting the sensor at the end of a long cable joined to the instrument. This type of system is currently being installed at National Weather Service facilities in place of the standard liquid-in-glass thermometers shown in Figure 3.3.

Regardless of the type of thermometer used, two important considerations in selecting an instrument are accuracy and response time. For most meteorological purposes, a thermometer that is reliable to within 0.3 Celsius degree (0.5 Fahrenheit degree) is sufficient. Response time refers to the instrument's capability of resolving oscillations in temperature. Most liquid-in-glass and electrical resistance thermometers have rapid

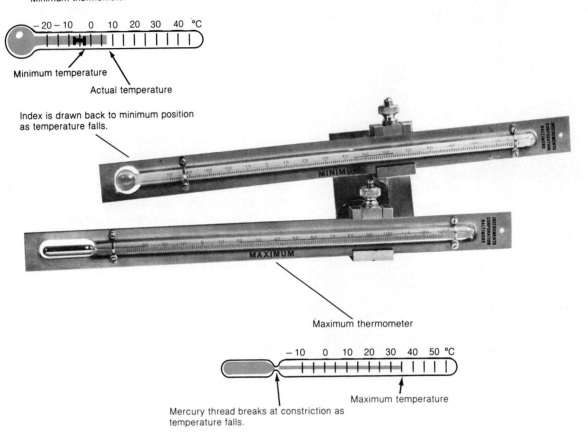

Minimum thermometer

Minimum temperature

Actual temperature

Index is drawn back to minimum position as temperature falls.

Maximum thermometer

Mercury thread breaks at constriction as temperature falls.

Maximum temperature

**FIGURE 3.3**
Liquid-in-glass thermometers mounted side by side and designed to register maximum and minimum temperatures, usually over a 24-hr period. [Courtesy of Belfort Instrument Company]

response times, whereas bimetallic thermometers tend to be more sluggish.

For accurate measurements of air temperature, ideally a thermometer should be adequately ventilated and shielded from precipitation, direct sunlight, and the night sky. Enclosing thermometers (and other weather instruments) in a louvered wooden shelter painted white (such as shown in Figure 3.5A) has been standard practice for official temperature measurements.* The sensor for the new National Weather Service thermistor system is mounted inside a shield made of

louvered plastic (Figure 3.5B), and the digital read-out box is indoors. So that temperature readings are representative, an instrument shelter should be located in an open grassy area well away from trees, buildings, or other obstacles. As a rule of thumb, the shelter should be no closer than four times the height of the nearest obstacle. Where a shelter is not available, mounting a thermometer outside a window on the shady north side of a building is usually sufficient for general purposes.

In addition to using standard thermometers, air temperature can sometimes be deduced in unconventional ways. One interesting and surprisingly accurate approach is to count cricket chirps. For air temperatures above about 12 °C (54 °F), the number of cricket chirps heard in an 8-second (sec) period plus 4 approximates the air temperature in degrees Celsius.

*Widespread use of instrument shelters dates only to the 1870s. In North America, the earliest report of a sheltered thermometer was at the Toronto Magnetic and Meteorological Observatory in 1841. Previously, official thermometers were usually suspended unprotected just outside a window (not always north-facing).

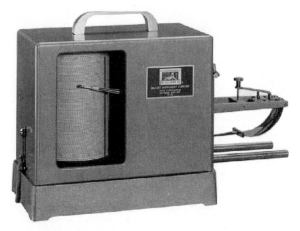

**FIGURE 3.4**
A thermograph provides a continuous trace of fluctuations in air temperature. [Courtesy of Belfort Instrument Company]

# Heat Units

Although temperature is a convenient way to describe the degree of hotness or coldness, we can quantify heat energy directly. Until recently, meteorologists commonly measured heat energy in units called calories. A **calorie (cal)** is defined as the quantity of heat needed to raise the temperature of 1 gram of water 1 Celsius degree (technically, from 14.5 to 15.5 °C). (The *calorie* used to measure the energy content of food is actually 1000 heat calories or 1.0 kilocalories.) Today, the more usual unit for energy of any form, including heat, is the joule (J). One calorie equals 4.1868 J. In this book we use both units of heat measurement.

In the English system, heat is quantified as British thermal units. A **British thermal unit (Btu)** is defined as the amount of heat required to raise the temperature of 1 pound (lb) of water 1 Fahrenheit degree (technically, from 62 to 63 °F). One Btu is equivalent to 252 cal and to 1055 J.

**FIGURE 3.5**
National Weather Service (NWS) thermometers are mounted inside a shelter that provides ventilation and protection from precipitation and direct sunlight. (A) Standard white wooden louvered instrument shelter. (B) Shelter for the new NWS thermistor system. [Photographs by J. M. Moran]

A

B

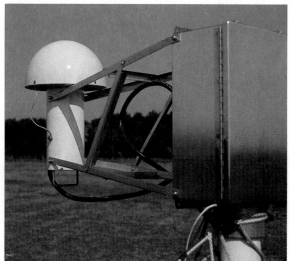

# Transport of Heat

In response to unequal rates of radiational heating and radiational cooling within the Earth–atmosphere system, air temperature varies from one place to another. (In Chapter 4, we examine the nature and implications of these imbalances in radiational heating and cooling rates.) A change in temperature with distance is known as a **temperature gradient.** A familiar temperature gradient is between the hot equator and the cold poles (a horizontal temperature gradient). Another is the temperature gradient between the relatively mild Earth's surface and the relatively cold tropopause (a vertical temperature gradient).

In response to a temperature gradient, heat flows in accordance with the **second law of thermodynamics.** Simply put, this law states that all systems tend toward disorder. You probably have personal experience with some implications of the second law. For example, if you avoid cleaning up your home or room, it rapidly becomes more and more disorganized. The presence of a gradient of any kind within a system signals order within that system. Hence, as a system tends toward disorder, gradients are eliminated. The second law predicts that where a temperature gradient exists, heat flows in a direction so as to erase the gradient, that is, heat flows from locations of higher temperature toward locations of lower temperature. In addition, the greater the temperature difference (that is, the steeper the temperature gradient), the more rapid is the rate of heat flow.

In response to temperature gradients within the Earth–atmosphere system, heat is transferred via conduction, convection, and radiation.

## *CONDUCTION*

Conduction occurs within a substance or between substances that are in direct physical contact. With **conduction,** the kinetic energy of atoms or molecules (that is, heat) is transferred by collisions between neighboring atoms or molecules. This is why a metal spoon heats up when placed in a steaming cup of coffee. As the more energetic molecules of the hot coffee collide with the less energetic atoms of the cooler spoon, some kinetic energy is transferred to the

**TABLE 3.1**
**Heat Conductivities of Some Familiar Substances**

| Substance | Heat Conductivity[a] |
|---|---|
| Copper | 0.92 |
| Aluminum | 0.50 |
| Iron | 0.16 |
| Ice (at 0 °C) | 0.0054 |
| Limestone | 0.0048 |
| Concrete | 0.0022 |
| Water (at 10 °C) | 0.0014 |
| Dry sand | 0.0013 |
| Air (at 20 °C) | 0.000061 |
| Air (at 0 °C) | 0.000058 |

[a]Heat conductivity is defined as the quantity of heat (cal) that would flow through a unit area of a substance (cm$^2$) in one second in response to a temperature gradient of one Celsius degree per centimeter. Hence, heat conductivity is in units of calories per cm$^2$ per sec per C° per cm.

atoms of the spoon. These atoms then transmit some of their heat energy, via collisions, to their neighboring atoms, so heat is eventually conducted up the handle of the spoon and the handle becomes hot to the touch.

Some substances conduct heat much more readily than others. As a rule, solids are better conductors than liquids, and liquids are better conductors than gases. At one extreme, metals are excellent conductors of heat, and at the other extreme, air is a very poor conductor of heat. Heat conductivities of some common substances are listed in Table 3.1.

Differences in heat conductivity can cause one substance to feel colder than another, even though both substances have the same temperature. For example, at the same relatively low temperature, a metallic object feels colder than a wooden object. The heat conductivity of metals is much greater than that of wood so that when you grasp the two objects, your hand more rapidly conducts heat to the metallic object than to the wooden object. Consequently, you have the sensation that the metal is colder than the wood.

The relatively low heat conductivity of air makes it useful as a heat insulating medium. Heat conductivity is lower for still air than for air in motion. Thus, to take maximum advantage of air as a heat insulator, air must be confined. For example, when a fiberglass blanket is used as attic insulation, it is primarily the still air trapped between individual fiberglass fibers that inhib-

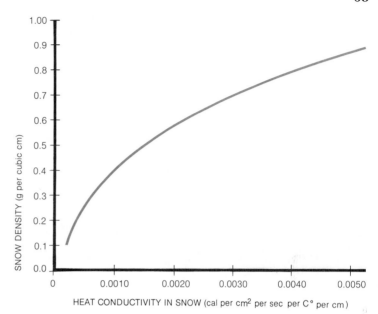

**FIGURE 3.6**
A thick fresh cover of snow is a good heat insulator primarily because of the air trapped between the snowflakes. (Air is a poor conductor of heat.) But, in time, the snow settles, air escapes, snow density increases, and the snow cover's insulating property diminishes.

its heat loss. In time, as the fibers settle, air is excluded, and the blanket loses much of its insulating value.

A fresh snow cover has extremely low heat conductivity because of the air trapped between individual snowflakes. A thick snow cover (20 to 30 cm or 8 to 12 in.) can thus inhibit or prevent freezing of the underlying soil, even though the temperature of the overlying air may drop well below freezing. In time, however, like the fiberglass, the snow cover loses some of its insulating property as the snow settles and air escapes (Figure 3.6).

Heat is conducted from warm ground to cooler overlying air, but because air has a low heat conductivity, conduction is significant only in a very thin layer of air that is in immediate contact with the Earth's surface. Convection is much more important than conduction in transporting heat vertically within the troposphere.

## CONVECTION

Although conduction takes place in solids, liquids, or gases, convection generally occurs only in liquids or gases.* **Convection** is the transport of heat within a fluid via motions of the fluid itself.

---
*Geologists point out an important exception: convection currents likely occur in the Earth's solid interior under conditions of great confining pressures.

Convection occurs within the atmosphere as a consequence of differences in air density. As heat is conducted from the relatively warm ground to cooler overlying air, the air becomes warmer than the surrounding air. Warm air is less dense than cold air so that the warm air rises and cooler, denser air sinks. The cooler air is then heated by the ground and the process is repeated. In this way, as shown in Figure 3.7, a convective circulation of air transports heat vertically from the Earth's surface into the troposphere. As we will see in Chapter 4, heat is also transported horizontally within the atmosphere; such a process is known as **advection.**

We can readily see convection currents by placing a pan of water on a hot stove and then adding a drop or two of food coloring to the water. We actually see the circulating water redistributing heat that is conducted from the bottom of the pan into the water. In this instance and in the troposphere, conduction and convection work together; heat transported by the combined processes of conduction and convection is known as **sensible heating.**

## RADIATION

As we saw in Chapter 2, **radiation** is a form of energy transport consisting of electromagnetic waves traveling at the speed of light. Unlike conduction and

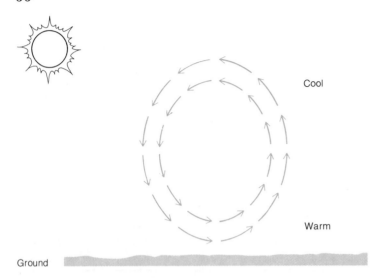

Cool

Warm

Ground

**FIGURE 3.7**
Convection currents transport heat from near the Earth's surface into the troposphere.

convection, radiation does not require an intervening physical medium; it can take place in a vacuum. Although not precisely a vacuum, interplanetary space is so highly rarefied that conduction and convection play no significant role in the transport of heat from the sun to Earth (or any of the other planets). Rather, radiation is the principal means whereby the Earth–atmospheresystem gains heat from the sun. Radiation is also the principal means whereby heat escapes from the planet to space.

As we also saw in Chapter 2, absorption of radiation involves a conversion of electromagnetic energy to heat. In contrast, emission of electromagnetic energy represents a loss of heat. All objects both absorb and emit electromagnetic radiation. If an object absorbs more than it emits, then its temperature rises or if an object emits more than it absorbs, then its temperature falls. At equilibrium, when absorption and emission of radiation are equal, the object's temperature is constant.

Radiative equilibrium does not necessarily mean that the temperature stays constant among all the components of a system. The Earth–atmosphere system is in radiative equilibrium with its surroundings. Nonetheless, heat may be redistributed among the components of the Earth–atmosphere system (for example, oceans, land, glaciers) so that air temperature at a specified location may undergo significant short- and long-term variations. Hence, global radiative equilibrium does not preclude changes in climate.

## Specific Heat

Whether by conduction, convection, or radiation, transport of heat from one place to another within the Earth–atmosphere system is accompanied by changes in air temperature. But, as pointed out at the beginning of this chapter, some of that heat may, instead, be used to change either the phase of water or the volume of the air. In circumstances where heat gain and heat loss affect the temperature, net heat gain causes air temperature to rise, whereas net heat loss causes air temperature to drop.

In a more general sense, the temperature response to an input (or output) of some specified quantity of heat varies from one substance to another. The amount of heat required to change the temperature of 1 gram of a substance by 1 Celsius degree is defined as the **specific heat** of that substance. Joseph Black, a Scottish chemist, first proposed the concept of specific heat back in 1760. Two different materials registering the same temperature do not necessarily possess the same amount of heat energy. Different quantities of heat may also be required to raise or lower the temperature of equal amounts of two different substances by one degree. The specific heat of all substances is measured relative to that of water, which is 1 cal per gram per Celsius degree (at 15 °C). Table 3.2 lists the specific heats of some familiar materials.

**TABLE 3.2**
**Specific Heats of Some Familiar Substances**

| Substance | Specific Heat (cal per g per C°) |
|---|---|
| Water | 1.000 |
| Ice (at 0 °C) | 0.478 |
| Wood | 0.420 |
| Aluminum | 0.214 |
| Brick | 0.200 |
| Granite | 0.192 |
| Sand | 0.188 |
| Dry air[a] | 0.171 |
| Copper | 0.093 |
| Silver | 0.056 |
| Gold | 0.031 |

[a]At constant volume.

On exposure to the same heat source, a substance with low specific heat warms up more than a substance with a higher specific heat. Water has the greatest specific heat of any naturally occurring substance. For example, its specific heat is about five times that of dry sand. Hence, 1 cal of heat will raise the temperature of 1 gram of water 1 Celsius degree, whereas 1 cal will raise the temperature of 1 gram of sand about 5 Celsius degrees. This is one reason why in summer the sand at the beach feels hot relative to the water (Figure 3.8).

The difference in specific heats is one of the principal reasons why the surface temperature of land is more variable with time than that of a body of water-such as a lake or the ocean. Compared to an adjacent body of water, land heats up more during the day and in summer and cools down more at night and during the winter. Hence, a body of water exhibits greater resistance to temperature change, called **thermal stability,** than does a land mass.

Another factor that contributes to the greater thermal stability of water bodies compared to land is a difference in heat transport. Solar radiation penetrates water to significant depths, but not the opaque land surface. Furthermore, ocean and lake waters circulate so that heat is readily transported to great volumes of water, whereas heat is conducted only very slowly through soil. Thus, an identical input of heat causes a land surface to warm up more than an equivalent surface area of a water body.

The contrast in thermal stability between land and water has important implications for climate because air temperatures are regulated to a considerable extent by the temperature of the surface over which the air resides or travels. Localities that are immediately downwind of an ocean or large lake (*maritime* localities) exhibit smaller seasonal temperature variations than do localities that are situated well inland (*continental* localities). An illustration of this effect is the comparison of the average monthly temperatures of

**FIGURE 3.8**
The contrast in specific heat is one reason why the sand is hotter than the water. [Photograph by J. M. Moran]

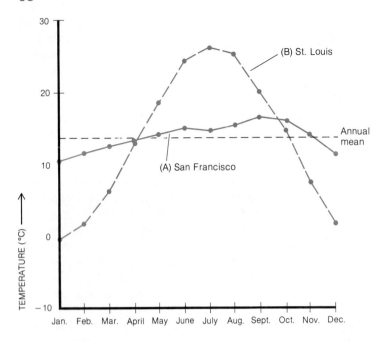

**FIGURE 3.9**
Variation of average monthly temperatures for (A) maritime San Francisco and (B) continental St. Louis. [NOAA data]

San Francisco and St. Louis (Figure 3.9). Although the two cities are at about the same latitude, summers are cooler and winters are milder in maritime San Francisco than in continental St. Louis.

Climatologists use an **index of continentality** to describe the degree of maritime influence on average air temperatures. Several different indexes are available, but most are based on the difference between average winter and average summer temperatures. A generalized index of continentality for North America is shown in Figure 3.10. Note that because prevailing winds blow from west to east, western North America is more maritime (less continental) than eastern North America.

## Heating and Cooling Degree-Days

With the contemporary concern for energy conservation, television and newspaper weather summaries routinely report heating or cooling degree-day totals in addition to daily maximum and minimum temperature. Heating and cooling degree-days are indicators of household energy consumption for space heating and cooling, respectively.

In the United States, **heating degree-days** are based on the Fahrenheit temperature scale and are computed only for days when the mean outdoor air temperature is lower than 65 °F (18 °C). Heating engineers who formulated this index early in this century found that when the mean outdoor temperature drops below 65 °F, space heating is required in most buildings to maintain an indoor air temperature of 70 °F (21 °C). The mean daily temperature is the simple arithmetic average of the 24-hour maximum and minimum air temperatures. Subtracting the mean daily temperature from 65 °F yields the number of heating degree-days for that day. For example, suppose that this morning's low temperature was 36 °F (2 °C), and this afternoon's high temperature was 52 °F (11 °C). Today's mean temperature would then be 44 °F (7 °C), giving a total of 65 − 44 = 21 heating degree-days. It is usual to keep a running total of heating degree-days, that is, to add degree-days for successive days through the heating season (actually from July of one year through June of the next).

Fuel distributors and power companies closely monitor heating degree-days. Fuel oil dealers base fuel use rates on cumulative degree-days and schedule home deliveries accordingly. Natural gas and electrical utilities anticipate power demands on the basis of degree-

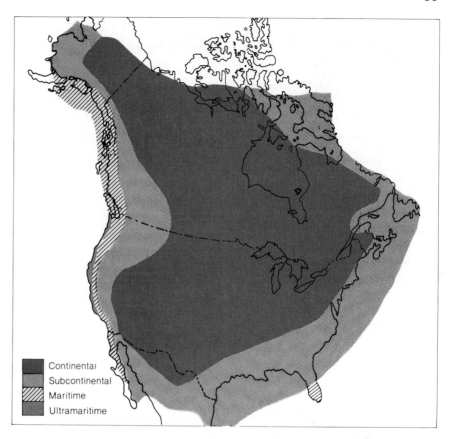

**FIGURE 3.10**
An index of continentality gauges the influence of oceans on air temperature over continents. In this scheme, North America is divided into zones of increasing maritime influence: continental, subcontinental, maritime, and ultramaritime. The greater the maritime influence, the less is the contrast between average summer and average winter temperatures. [Modified after D. R. Currey, "Continentality of Extratropical Climates," *Annals of the Association of American Geographers 64*, No. 2 (1974):274]

Continental
Subcontinental
Maritime
Ultramaritime

day totals, and implement priority use policies on the same basis when capacity fails to keep pace with demand.

Figure 3.11 shows the average annual heating degree-day totals over the United States and Canada. Note that in Canada, heating degree-days are based on the Celsius temperature scale and are computed for days when the mean temperature is below 18 °C. Outside of mountainous areas, regions of equal heating degree-day totals tend to parallel latitude circles with degree-day totals increasing poleward. As an example, the average annual space heating requirement in Chicago (6100 heating degree-days) is about four times that of New Orleans (1500 heating degree-days). If per unit fuel costs are the same in both cities, then, in an average winter, Chicago homeowners can expect to pay four times as much for space heating as homeowners in New Orleans.

**Cooling degree-days** are computed only for days when the mean outdoor air temperature is higher than 65 °F.* Supplemental air conditioning may be needed on such days. Again, a cumulative total is maintained through the cooling (summer) season. Generally, however, air conditioning is needed mostly in those localities where cumulative cooling degree-days exceed 700.

Note that indexes of heating and cooling requirements are based on outside air temperatures and do not take into account other weather elements, such as air circulation and humidity, which influence human comfort and demands for space heating and cooling. Heating and cooling degree-days are therefore only approximations of our residential fuel demands for heating and cooling.

*Higher base temperatures are sometimes used to take into account the human comfort range of 20 to 25 °C (68 to 77 °F).

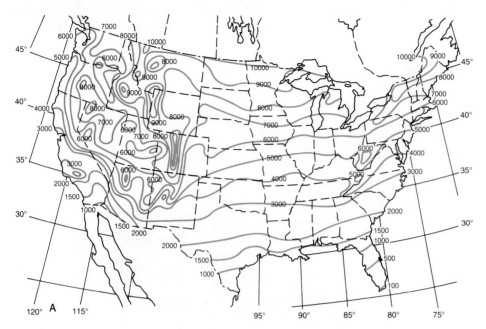

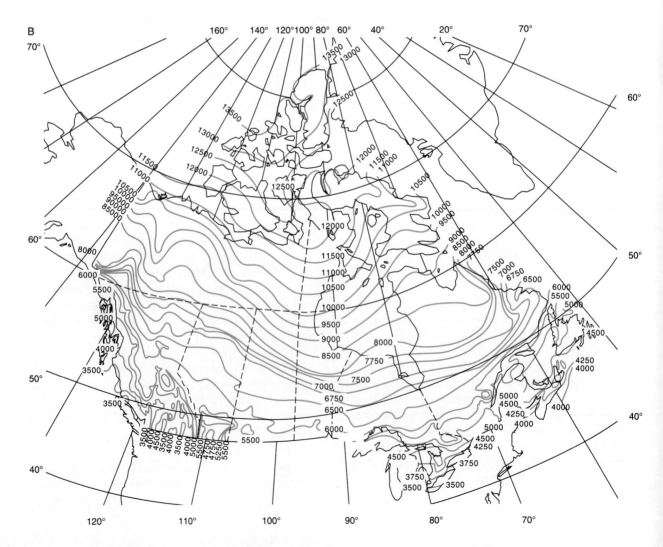

**FIGURE 3.11**
Average annual heating degree-day totals over 48 states of the United States (A) and Canada (B). Note that the base temperature for computing heating degree-days is 65 °F in the United States and 18 °C in Canada. [Data from NOAA and Atmospheric Environment Service of Canada]

# Windchill

At low air temperatures, the wind increases human discomfort outdoors and heightens the danger of *frostbite,* the freezing of body tissue. Air in motion increases the rate of sensible heat loss (the combined effect of conduction and convection) from the body. Immediately adjacent to the body is a very thin (measured in millimeters) layer of still air, called the **boundary layer,** that helps to insulate the body from heat loss. Within the boundary layer, heat loss is by the very slow process of conduction. (Recall that still air is a poor conductor of heat.) However, as wind speed increases, the thickness of the boundary layer diminishes, and the rate of sensible heat loss from the body increases.

Because of the danger of frostbite, weather reports during winter in northern localities and in mountainous regions include the **windchill equivalent temperature (WET),** sometimes referred to simply as the *windchill index.* This index, first introduced in the early 1940s by the polar scientist P. A. Siple and his colleague C. E. Passel, and later refined by A. Court, is presented in Table 3.3. As an illustration, suppose that the actual air temperature is $-1$ °C (30 °F) and the air is calm; then the windchill equivalent temperature is the same as the air temperature. If, however, the wind speed increases to 32 km (20 mi) per hour, the WET drops to $-15$ °C (5 °F). Contrary to popular opinion, this does not mean that skin temperature actually drops to $-15$ °C. Through sensible heat transfer, skin temperature can drop no lower than the temperature of the surrounding (ambient) air, which in this example is $-1$ °C (30 °F). What it does mean is that any exposed body parts lose heat at a *rate* equivalent to conditions induced by calm winds at $-15$ °C. Hence, air in motion (wind) is more effective in removing heat from the body than air temperature alone would imply. As a

**TABLE 3.3A**
**Windchill Equivalent Temperature (°C)**

| Wind Speed (m/sec) | Air Temperature (°C) | | | | | | | | | | | | | | | |
|---|---|---|---|---|---|---|---|---|---|---|---|---|---|---|---|---|
| | *6* | *3* | *0* | *−3* | *−6* | *−9* | *−12* | *−15* | *−18* | *−21* | *−24* | *−27* | *−30* | *−33* | *−36* | *−39* |
| 3 | 3 | −1 | −4 | −7 | −11 | −14 | −18 | −21 | −24 | −28 | −31 | −34 | −38 | −41 | −45 | −48 |
| 6 | −2 | −6 | −10 | −14 | −18 | −22 | −26 | −30 | −34 | −38 | −42 | −46 | −50 | −54 | −58 | −62 |
| 9 | −6 | −10 | −14 | −18 | −23 | −27 | −31 | −35 | −40 | −44 | −48 | −53 | −57 | −61 | −65 | −70 |
| 12 | −8 | −12 | −17 | −21 | −26 | −30 | −35 | −39 | −44 | −48 | −53 | −57 | −62 | −66 | −71 | −75 |
| 15 | −9 | −14 | −18 | −23 | −27 | −32 | −37 | −41 | −46 | −51 | −55 | −60 | −65 | −69 | −74 | −79 |
| 18 | −10 | −14 | −19 | −24 | −29 | −33 | −38 | −43 | −48 | −52 | −57 | −62 | −67 | −71 | −76 | −81 |
| 21 | −10 | −15 | −20 | −25 | −29 | −34 | −39 | −44 | −49 | −53 | −58 | −63 | −68 | −73 | −77 | −82 |
| 24 | −10 | −15 | −20 | −25 | −30 | −35 | −39 | −44 | −49 | −54 | −59 | −63 | −68 | −73 | −78 | −83 |

**TABLE 3.3B**
**Windchill Equivalent Temperature (°F)**

| Wind Speed (mi/hr) | Air Temperature (°F) | | | | | | | | | | | | | | | | | |
|---|---|---|---|---|---|---|---|---|---|---|---|---|---|---|---|---|---|---|
| | *45* | *40* | *35* | *30* | *25* | *20* | *15* | *10* | *5* | *0* | *−5* | *−10* | *−15* | *−20* | *−25* | *−30* | *−35* | *−40* | *−45* |
| 5 | 43 | 37 | 32 | 27 | 22 | 16 | 11 | 6 | 1 | −5 | −10 | −15 | −20 | −26 | −31 | −36 | −41 | −47 | −52 |
| 10 | 34 | 28 | 22 | 16 | 10 | 4 | −3 | −9 | −15 | −21 | −27 | −33 | −40 | −46 | −52 | −58 | −64 | −70 | −76 |
| 15 | 29 | 22 | 16 | 9 | 2 | −5 | −11 | −18 | −25 | −32 | −38 | −45 | −52 | −58 | −65 | −72 | −79 | −85 | −92 |
| 20 | 25 | 18 | 11 | 4 | −3 | −10 | −17 | −25 | −32 | −39 | −46 | −53 | −60 | −67 | −74 | −82 | −89 | −96 | −103 |
| 25 | 23 | 15 | 8 | 0 | −7 | −15 | −22 | −29 | −37 | −44 | −52 | −59 | −66 | −74 | −81 | −89 | −96 | −104 | −111 |
| 30 | 21 | 13 | 5 | −2 | −10 | −18 | −25 | −33 | −41 | −48 | −56 | −63 | −71 | −79 | −86 | −94 | −102 | −109 | −117 |
| 35 | 19 | 11 | 3 | −4 | −12 | −20 | −28 | −35 | −43 | −51 | −59 | −67 | −74 | −82 | −90 | −98 | −106 | −113 | −121 |
| 40 | 18 | 10 | 2 | −6 | −14 | −22 | −29 | −37 | −45 | −53 | −61 | −69 | −77 | −85 | −93 | −101 | −108 | −116 | −124 |
| 45 | 17 | 9 | 1 | −7 | −15 | −23 | −31 | −39 | −47 | −55 | −62 | −70 | −78 | −86 | −94 | −102 | −110 | −118 | −126 |

# Temperature and Human Comfort

To appreciate how we respond to temperature stress, we must first understand that humans are *homeothermic*. That is, we regulate our internal or core temperature within 2 Celsius degrees (3.6 Fahrenheit degrees) of 37 °C (98.6 °F), despite much greater variations in the temperature of the *ambient* (surrounding) air. The core refers to those regions of the body that contain vital organs such as the brain, heart, lungs, and digestive tract. If vital organs are not maintained at a nearly constant temperature, they do not function properly. Other parts of the body, such as legs and arms, however, may undergo much greater temperature changes without adverse effects.

At ambient air temperatures of 20 to 25 °C (68 to 77 °F), someone who is fully clothed and indoors usually feels comfortable at rest. Within this temperature range, the body readily maintains a core temperature of 37 °C without having to resort to special temperature-regulating mechanisms.

When we are exposed to ambient air temperatures above or below the 20 to 25 °C range, the body must initiate processes that maintain the core temperature at 37 °C. For example, if we stand in the sun on a day when the air temperature hits 30 °C (86 °F), our core temperature will begin to rise. In response, we perspire. The heat required to evaporate the perspiration is provided by our skin, and as a consequence, our skin cools. We experience the same cooling effects of evaporation as we step out of a shower or climb out of a swimming pool. *Evaporative cooling* reduces skin temperature, which in turn, helps restore the body's core temperature to a normal level.

If we are exposed to air temperatures below 20 °C (68 °F), the core temperature may begin to fall, and we start to shiver. This increased muscular activity produces additional heat, which helps raise the core temperature back to 37 °C. Perspiring and shivering are examples of *thermoregulation,* natural physiological responses that assist in maintaining a nearly constant core temperature regardless of the ambient air temperature.

Another thermoregulation process involves blood flow. Heat transfer from the body core to the skin occurs via the circulatory system. When we are exposed to air temperatures below 20 °C, our body can limit heat loss by restricting blood flow to the skin. By direction of the nervous system, many of the tiny blood vessels in the skin constrict. As the ambient air temperature drops, additional blood vessels constrict, further reducing blood flow to the body surface. You can observe this phenomenon by immersing your hand in a container of ice water; the skin becomes paler. Reduced blood flow to the skin lowers skin temperature. As a consequence, the body-to-air temperature gradient declines and cooling by radiation, conduction, and convection slows.

In contrast, exposure to ambient air temperatures above 25 °C (77 °F) triggers increased blood flow to the body surface. Blood vessels near the skin surface dilate and give the skin a flushed or reddish appearance. Greater flow of blood to the body surface raises skin temperature. Hence, the body-to-air temperature gradient increases, and cooling by radiation, conduction, and convection is enhanced. Greater blood flow to the skin also increases evaporative cooling by supplying more water for perspiring.

In addition to physiological changes, behavioral responses assist in thermoregulation. For example, if we

consequence, the body may be unable to supply heat to the body surface quickly enough to prevent the temperature of exposed skin from dropping to subfreezing levels.

Many factors besides wind speed affect human comfort during cold weather. The windchill equivalent temperature does not take into account variations in rates of body heat production due to changes in physical activity or metabolic processes. Nor does it consider the effects of humidity or radiational heating and cooling. Furthermore, we may argue that comfort is very subjective and perceived differently by differ-

feel hot, we shed clothing, seek shelter from the sun, or turn on a fan or air conditioner.

Under some circumstances, thermoregulation is insufficient to maintain a 37 °C core temperature. For example, if a person hiking in the woods is drenched by a cold rain and then overexerts himself and becomes exhausted, thermoregulation may not be able to compensate for heat loss. Consequently, the core temperature drops, and hypothermia may ensue.

*Hypothermia* refers to those responses that occur when the human core temperature drops below 35 °C (95 °F). Initially, shivering becomes more violent and uncontrollable. In addition, the victim begins to have difficulty speaking and becomes apathetic and lethargic. If the core temperature falls below 32 °C (90 °F), shivering is replaced by muscular rigidity, and coordination deteriorates. Mental abilities are impaired, and the victim is generally unable to help himself. At a core temperature of 30 °C (86 °F), the person may drift into unconsciousness. Death may occur at core temperatures below about 24 °C (75 °F) because the heart rhythm becomes uncontrollably irregular (ventricular fibrillation) or uncontrollably halted (cardiac arrest).

Once hypothermia begins, the victim is in serious trouble. With only a 3-Celsius-degree (5.4-Fahrenheit degree) drop in core temperature, the body's ability to regulate its core temperature is already greatly impaired. If the core temperature drops to 29 °C (84 °F), thermoregulation is essentially ineffective.

The first signs of hypothermia should never be ignored; one should take action immediately. Treatment takes two forms: prevention of further heat loss and addition of heat. Further heat loss can be prevented by replacing wet clothing with dry clothing, finding shelter, and insulating the person from the ground, so that body heat is not conducted to the colder ground surface. The body can be heated by an external source, such as a space heater or other human bodies. Administering a hot, nonalcoholic beverage, if the victim is conscious, helps warm the core from the inside. In any event, medical attention should be sought as soon as possible.

In some situations, thermoregulation may be unable to prevent a rise in core temperature. For instance, a person exposed to hot desert conditions with an inadequate supply of water will eventually experience a rise in core temperature. If the core temperature continues to climb, hyperthermia may occur. *Hyperthermia* refers to those responses that take place when the core temperature tops 39 °C (102 °F). As the core temperature climbs to 41 °C (106 °F), thermoregulation breaks down, and a person may suddenly and quite unexpectedly collapse. The victim also experiences muscle cramps or spasms and slips into unconsciousness. Perspiring ceases, although it is not known if this is a cause or a result of hyperthermia. With serious heat stress, the individual may die within a few hours unless the core temperature is lowered artificially. These responses are collectively identified by various names, including heatstroke, sunstroke, and heat apoplexy.

A victim of hyperthermia must be treated promptly because, once thermoregulation fails, the core temperature rises rapidly. To save the victim, the core temperature must be lowered from outside the body. The victim should be moved to a cooler environment, and, if possible, the body should be placed in cold water. Alternatively, sponging the body with alcohol enhances cooling as the alcohol evaporates. Again, one should seek medical attention as soon as possible.

The human body has a remarkable capability to adjust to changing air temperatures and remain comfortable. This capacity is limited, however. If our core temperature begins to deviate from normal, we or our companions must take corrective action promptly. Failure to do so may prove fatal.

ent individuals. Nonetheless, the WET has proven to be a very useful guide in regions prone to harsh winter weather.

When the WET is significantly lower than the air temperature, we are well advised to dress more warmly than the actual air temperature might suggest.

Especially at low windchill equivalent temperatures, all body parts should be protected if a person expects to be exposed to the wind for more than a few minutes at a time. Low air temperatures and high winds are especially hazardous to those body parts that are usually exposed and have a high surface-to-

## Temperature and Crop Yields

In the Midwest, there's an old saying that corn should not be planted before the leaves of an oak tree reach the size of a squirrel's ear. Although this adage is quaint, farmers today use more sophisticated and reliable indexes of crop–climate relationships to help optimize crop yields. Two important indexes are length of growing season and growing-degree units.

The period during the year when air temperatures remain sufficiently mild to permit plant growth is a critical determinant of agricultural productivity. For many years, the *growing season* was defined as the number of days between the date of last killing frost in spring and the date of first killing frost in fall. However, "killing frost" is an ambiguous term. Whether or not a frost kills a plant depends on the plant species and the stage of its life cycle, the duration of freezing temperatures, and the rate of freezing. Hence, no single air temperature criterion can fully convey the actual impact of subfreezing temperatures on agriculture, particularly where a variety of crops are grown. Currently, the period most commonly used to delineate the growing season is the freeze-free period, that is, the time between the last day of 0 °C (32 °F) or lower in the spring and the first

date of 0 °C (32 °F) or lower in the autumn. (See Appendix V for a listing of average freeze-free periods for selected locations in the United States and Canada.)

As a rule, the average length of the growing season shortens with increasing latitude so that the types of crops that can be grown successfully also change with latitude. The growing season may be lengthened locally by the moderating influence of nearby large bodies of water. A good example comes from Wisconsin, where localities bordering Lake Michigan have a growing season that is 20 to 40 days longer than sites at the same latitude, but 100 km (62 mi) inland. Topography can also influence the length of the growing season. Because cold air is relatively dense, it tends to drain downhill so that the growing season typically is several weeks shorter in valleys than on surrounding hillsides. For that reason, vineyards and orchards are usually situated on hillsides rather than in valley bottoms.

One problem with relying on growing season length as a predictor of agricultural yields is that temperatures above 0 °C (32 °F) are not equally effective in promoting crop growth. For example, corn grows very little at temperatures below 10 °C (50 °F) or above 30 °C (86 °F). Hence, the actual air temperatures during the

volume ratio, such as the ears, nose, and fingers. These body parts are especially susceptible to frostbite. For more information on human responses to heat stress, see the Special Topic "Temperature and Human Comfort."

Temperature is also a critical factor in agricultural success. Of all the climatic variables, air temperature is the single most important consideration in determining where specific crops can be grown. For information on this subject, see the Special Topic "Temperature and Crop Yields."

## Conclusions

Heat and temperature are distinct and yet closely related quantities. In response to temperature gradients, conduction, convection, and radiation redistribute heat from one place to another within the Earth–atmosphere system. Because air is a poor conductor of heat, convection is much more important than conduction in distributing heat within the troposphere. The temperature response of a substance that gains or loses heat depends on the specific heat of that substance. The temperature of a body of water is much less variable than that of land surfaces. This contrast in thermal stability influences the overlying air so that regions downwind of the ocean (or large lakes) exhibit less seasonal temperature contrast.

growing season can either accelerate or retard plant growth and often are as significant as the length of the growing season in governing crop yields.

To better gauge the influence of air temperature on plant growth during the growing season, scientists developed an index involving *growing-degree units (GDUs)*. Because one of its most successful applications is in the cultivation of corn, we will use corn as an example of how the index works. To determine the GDUs for a particular day, the average daily temperature is computed by adding the maximum and minimum temperatures in °F for the day (24 hours) and dividing by two. Because most corn growth occurs when air temperatures are between 50 and 86 °F (10 and 30 °C), any daily maximum temperature higher than 86 °F is counted as 86 °F and any daily minimum temperature below 50 °F is counted as 50 °F. The number 50 (the lower threshold for growth of corn) is then subtracted from the daily average temperature to obtain the number of GDUs for that day. GDUs are then added from one day to the next to give a cumulative total.

There are several important applications of GDUs. One is to match the GDU requirement of a particular corn hybrid with the long-term average number of GDUs for a region. For example, a farmer living in central Illinois or Indiana typically selects a corn hybrid that requires 3000 to 3400 GDUs to mature. If the planting season is delayed (perhaps by early spring rains) or if fields have to be replanted (perhaps because of poor germination owing to low soil moisture after planting), the farmer may have to switch to a shorter season hybrid (that is, a hybrid that requires fewer GDUs). By computing the number of GDUs that have already accumulated during the growing season and subtracting this total from the number of GDUs for an average growing season, the farmer obtains a good estimate of what hybrid to use so that the corn reaches maturity before the first freeze of the fall.

For grains such as corn, oats, and wheat, GDUs provide a useful guide in selecting the best varieties to plant in specific areas and can be used in planning planting and harvesting times. For fruit crops such as apples, pears, and peaches, GDUs have been used to successfully predict bloom dates and other stages of development, including ripening. Thus, calculation of GDUs aids in choosing varieties that bloom after the danger of freezing temperatures has passed but are still capable of producing fruit before the end of the growing season. GDUs have also been used to predict the time of insect infestation, which aids in planning pesticide spraying and other pest-control measures.

Although GDUs are valuable in modeling the growth and development of plants, often specific weather factors affect quality, color, and size of fruits. For example, day-to-night temperature shifts are critical for growing wine grapes.

Imbalances in rates of radiational heating and radiational cooling ultimately are responsible for producing temperature gradients within the Earth–atmosphere system. We examine the reasons for these imbalances in the next chapter.

## Key Terms

| | |
|---|---|
| kinetic energy | British thermal unit (Btu) |
| heat | temperature gradient |
| temperature | second law of |
| absolute zero | thermodynamics |
| thermometer | conduction |
| thermograph | convection |
| calorie (cal) | advection |

| | |
|---|---|
| sensible heating | heating degree-days |
| radiation | cooling degree-days |
| specific heat | boundary layer |
| thermal stability | windchill equivalent |
| index of continentality | temperature (WET) |

## Summary Statements

☐ Heat is the total kinetic energy of atoms or molecules that compose a substance. Temperature, on the other hand, is directly proportional to the average kinetic energy of the individual atoms or molecules.

☐ As a sample of air gains or loses heat, that heat may be used for some combination of changes in temperature, phase of water, or volume of the air sample.

☐ The Fahrenheit temperature scale is still commonly used in the United States. The Celsius temperature scale is more convenient in that a 100-degree interval separates the freezing and boiling points of pure water. The Kelvin scale is based on absolute zero and is a more direct measure of average kinetic molecular activity than either the Fahrenheit or Celsius scale.

☐ An object at absolute zero (0 kelvin) emits no electromagnetic radiation.

☐ Three common types of thermometers are liquid-in-glass, bimetallic, and thermistor. Thermometers should be mounted where they are sheltered from precipitation and direct sunshine.

☐ Heat energy is quantified as calories, joules, or British thermal units.

☐ In response to a temperature gradient, heat flows from locations of higher temperature toward locations of lower temperature. This is a consequence of the second law of thermodynamics. Heat flows via conduction, convection, and radiation.

☐ As a rule, solids (especially metals) are better conductors of heat than are liquids, and liquids are better conductors than are gases. Still air is a very poor conductor of heat.

☐ Convection is the transport of heat within a fluid via motions of the fluid itself. Convection is much more important than conduction in transporting heat within the troposphere.

☐ All objects absorb and emit radiation. At radiative equilibrium, absorption balances emission and the temperature of the object is constant. If absorption exceeds emission, the temperature rises and if emission exceeds absorption, the temperature falls.

☐ The temperature response to an input or output of heat differs from one substance to another depending on the specific heat of each substance.

☐ Water bodies such as lakes or seas exhibit less temperature variability than do land areas because the specific heat of water is higher than that of land, solar radiation readily penetrates water, and water circulates.

☐ The winter-to-summer temperature contrast is greater in continental climates than in maritime climates.

☐ Heating and cooling degree-days are indicators of household energy demand for space heating and cooling, respectively.

☐ Windchill equivalent temperature (WET) is a measure of the combined effect of low air temperature and wind on human comfort.

## Review Questions

1. Distinguish between heat and temperature.
2. Why is temperature a misleading way of comparing the heat content of different substances?
3. Why is the Celsius temperature scale more convenient than the Fahrenheit temperature scale for most scientific purposes?
4. What is the significance of absolute zero (0 K)?
5. In localities where winter temperatures commonly dip below −40 °C (−40 °F), what types of thermometers must be used?
6. What is meant by a thermometer's response time? Why is it an important consideration in selecting a thermometer?
7. Why is it a good idea to shelter thermometers?
8. Identify two temperature gradients within the Earth–atmosphere system.
9. Heat always flows from a warmer object to a colder object. How is this an example of the second law of thermodynamics?
10. Identify and describe three processes whereby heat flows in response to a temperature gradient. Provide a common example of each process.
11. Air has a very low heat conductivity. How is this property of air useful in insulating a home?
12. Why is convection much more important than conduction in the transport of heat within the troposphere?
13. Contrast the heat conductivities of solids, liquids, and gases.
14. How does a deep snow cover inhibit the freezing of the underlying soil, even though the air temperature drops well below freezing?
15. Define specific heat. How does specific heat relate to thermal stability?
16. How and why do maritime climates differ from continental climates?
17. Why is the west coast of North America more maritime than the east coast?
18. Identify three factors that prevent dramatic changes in the temperature of large bodies of water.
19. Explain how heating degree-days are computed. What is the significance of cumulative heating degree-days?
20. How does the wind affect the rate of heat transport from the human body?

## Quantitative Questions

1. Convert the following.

    0 °C = _____°F = _____K

    20 °C = _____°F = _____K

    65 °F = _____°C = _____K

    0 °F = _____°C = _____K

    −5 °C = _____°F = _____K

2. What is the boiling point of pure water at sea level on the Kelvin scale?

3. An increment of 20 Celsius degrees corresponds to an increment of how many Fahrenheit degrees?

4. How much heat must be added to raise the temperature of 100 g of water from 20 °C to 90 °C?

5. Calculate the temperature change in Celsius degrees when

    a. 5 cal of heat is added to 5 g of water.

    b. 5 cal of heat is added to 5 g of ice at −5 °C.

    c. 1000 cal of heat is added to a brick that weighs 1 kg.

    d. 100 cal of heat is added to a layer of dry sand weighing 100 g.

6. Compute the number of heating degree-days when

    a. the day's mean outdoor temperature is 45 °F.

    b. the day's maximum temperature is 27 °F and the day's minimum temperature is 2 °F.

    c. the day's maximum temperature is 10 °C and the day's minimum temperature is −2 °C.

7. What is the windchill equivalent temperature if the actual air temperature is 25 °F and the wind speed is 25 miles per hour?

## Questions for Critical Thinking

1. Explain how a thermos bottle keeps hot beverages hot and cold beverages cold.

2. Because air temperature is directly proportional to the mean kinetic energy of molecules, would you expect the speed of the wind to affect air temperature? Explain your response.

3. A Pennsylvania highway sign warns that a bridge surface freezes before the road surface. Why?

4. Air is a very poor conductor of heat. Why, then, do some people go to the trouble of wrapping insulating materials around hot water pipes?

5. Design an experiment to determine the temperature response of samples of dry and wet sand to the same input of heat. Predict the outcome of such an experiment.

6. In winter, double-glazed windows or storm windows are used to cut heat loss from buildings. Speculate on how such windows cut conductive and convective heat flow.

## Selected Readings

Asimov, I. *Understanding Physics: Motion, Sound, and Heat.* New York: New American Library Mentor Book, 1969, 248 pp. Includes well-written explanations of basic physical principles with chapters on heat and temperature.

Driscoll, D. M. "Windchill: The 'Brr' Index," *Weatherwise* 40, No. 6 (1987):321–326. Presents a critique of the windchill concept along with its historical background.

Middleton, W. E. K. *A History of the Thermometer and Its Use in Meteorology.* Baltimore, MD: Johns Hopkins University Press, 1966, 249 pp. Provides an authoritative summary of the history of development of an instrument essential in weather observation.

Quayle, R., and F. Doehring. "Heat Stress: A Comparison of Four Different Indices," *Weatherwise* 34, No. 3 (1981):120–124. Contrasts four means of measuring heat stress.

Snow, J. T., and S. B. Harley. "Basic Meteorological Observations for Schools: Temperature," *Bulletin of the American Meteorological Society* 68 (1987):468–496. Describes methods of temperature measurement using economical instruments.

# 4 Heat Imbalances and Weather

*Then the sea*
*And heaven rolled as one and from*
*the two*
*Came fresh transfigurations of*
*freshest blue.*

WALLACE STEVENS
*Sea Surface Full of Clouds*

Imbalances in radiational heating and cooling trigger processes that redistribute heat within the Earth-atmosphere system. Evaporation of water at the Earth's surface and its subsequent condensation as clouds form an important heat-transfer process. [Photograph by J. M. Moran]

W EATHER IS NOT a capricious act of nature. Rather, weather is a response to unequal rates of radiational heating and radiational cooling within the Earth–atmosphere system. As we saw in Chapter 2, absorption of solar radiation causes heating whereas emission of infrared radiation causes cooling. Imbalances in rates of radiational heating and cooling from one place to another within the Earth–atmosphere system produce temperature gradients. In response to temperature gradients, the atmosphere circulates and thereby redistributes heat.

In this chapter we examine the heat imbalances that develop within the Earth–atmosphere system. We then consider the various means whereby heat is redistributed in response to temperature gradients. All this enables us to better understand atmospheric circulation, energy conversions within the Earth–atmosphere system, and the processes that govern air temperature within the troposphere.

# Heat Imbalance: Atmosphere Versus Earth's Surface

In Figure 4.1, we show how solar radiation intercepted by the planet interacts with the Earth–atmosphere system. For convenience of study, we assume that 100 units of solar radiation strike the upper atmosphere. As noted in Chapter 2, 31% is reflected or scattered to space, 23% is absorbed by the atmosphere, and 46% is absorbed by the Earth's surface. Also shown in Figure 4.1 are infrared (longwave) emissions by the Earth's surface and atmosphere. Of the 115 units of infrared radiation emitted by the Earth's surface, 106 units are absorbed by clouds and greenhouse gases. Of the 106 units absorbed, 100 are reradiated back to the Earth's surface (the greenhouse effect).

The global distribution of incoming solar radiation and outgoing infrared radiation implies net warming of the Earth's surface and net cooling of the atmosphere (Table 4.1). In the atmosphere, the rate of cooling due to infrared emission is greater than the rate of warming due to absorption of solar radiation. At the Earth's surface, however, it's just the opposite. The rate of

**TABLE 4.1**
**Global Radiation Balance**

| | |
|---|---|
| Solar radiation intercepted by Earth | 100 units |
| Solar radiation budget | |
|    Scattered and reflected to space (8 + 17 + 6) | 31 |
|    Absorbed by the atmosphere (19 + 4) | 23 |
|    Absorbed by the Earth's surface | 46 |
|    Total | 100 units |
| Radiation budget at the Earth's surface | |
|    Infrared coolong (100 − 115) | −15 |
|    Solar heating | +46 |
|    Net heating | +31 units |
| Radiation budget of the atmosphere | |
|    Infrared cooling (−40 − 20 + 6) | −54 |
|    Solar heating | +23 |
|    Net cooling | −31 units |
| Heat transfer: Earth's surface to atmosphere | |
|    Sensible heating (conduction plus convection) | 7 |
|    Latent heating (phase changes of water) | 24 |
|    Net transfer | 31 units |

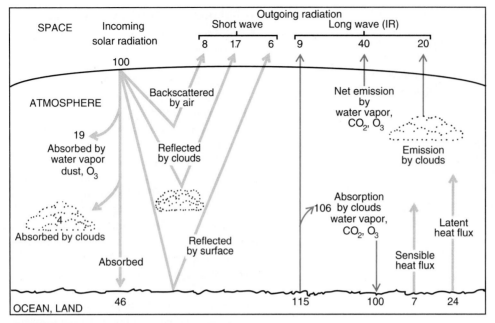

**FIGURE 4.1**

The distribution of 100 units of incoming solar radiation and outgoing infrared radiation on a global scale indicates excess heating at the Earth's surface. This excess heat is transferred to the atmosphere via sensible and latent heating. [Adapted from M. C. MacCracken and F. M. Luther, eds. *Detecting The Climatic Effects of Increasing Carbon Dioxide.* Report DOE/ER-0235, 1985 U.S. Department of Energy, Washington, DC]

warming due to absorption of solar radiation exceeds the rate of cooling due to infrared emission.

In reality, the atmosphere is *not* cooling relative to the Earth's surface. Hence, a compensating transfer of heat must be taking place from the Earth's surface to the atmosphere via processes other than radiation. How is this heat transfer accomplished? A combination of sensible heating (conduction plus convection) and latent heating (phase changes of water) is responsible. In Figure 4.1, 31 units of heat energy are transferred from the Earth's surface to the atmosphere by these two processes: 7 units by sensible heating (7/31 or about 23%) and 24 units by latent heating (24/31 or about 77%).

### SENSIBLE HEATING

The term *sensible* is used to describe this heat-transfer process because heat redistribution brought about by sensible heating can be monitored, or "sensed," as temperature changes. **Sensible heating** involves both conduction and convection. Heat is con-

ducted from the relatively warm surface of the Earth to the cooler overlying air. Heated air rises and is replaced by cooler, denser air subsiding from above. The consequence is a convective transport of heat from the Earth's surface into the troposphere. Because air is a poor conductor of heat (Chapter 3), convection is much more important than conduction in heat transfer within the atmosphere.

### LATENT HEATING

**Latent heating** refers to the movement of heat from one place to another as a consequence of changes in the phase of water. Depending on the specific type of phase change, water either absorbs heat from its environment or releases heat to its environment. The quantity of heat that is involved in phase changes is known as *latent heat*. Heat is required for phase changes because of differences in molecular activity represented by the three physical phases of water. In the solid phase (ice), water molecules are relatively inactive and vi-

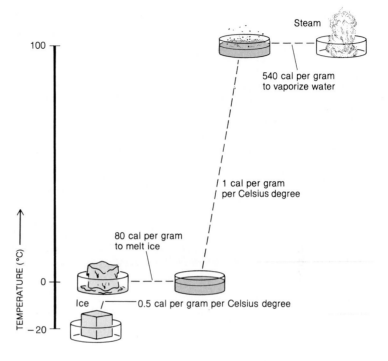

**FIGURE 4.2**
Heat is added to raise the temperature of ice and water and to change the phase of water. Note, however, that during a phase change (ice to water, water to vapor, for example), the temperature of the water undergoing the phase change is constant. [From J. M. Moran, M. D. Morgan, and J. H. Wiersma, *Introduction to Environmental Science.* New York: W. H. Freeman and Company, p. 273. © 1980]

brate about fixed locations. Hence, at subfreezing temperatures, an ice cube maintains its shape. In the liquid phase, molecules move about with greater freedom, so that liquid water takes the shape of its container. In the vapor phase, molecules exhibit maximum activity and diffuse readily throughout the entire volume of a container. A change of phase is thus linked to a change in level of molecular activity, which is brought about by either a gain or loss of heat.

During any phase change, heat is exchanged between water and its environment. Although the temperature of the environment changes in response, the temperature of the water undergoing the phase change remains constant until the phase change is complete. That is, the available heat, latent heat, is involved exclusively in changing the phase of water and not in changing its temperature. Interestingly, unusually great quantities of heat are required to bring about phase changes of water as compared to phase changes of other naturally occurring substances. The basic reasons for this contrast are given in the Special Topic "The Unique Thermal Properties of Water."

To illustrate further the concept of latent heat, consider the fate of a 1-g ice cube in Figure 4.2 as it is heated from an initial temperature of −20 °C (−4 °F).

The specific heat of ice is about 0.5 cal per gram per Celsius degree, which means that 0.5 cal of heat must be supplied for every 1 Celsius degree of temperature rise. Hence, warming our ice cube from −20 to 0 °C (−4 to 32 °F) requires an input of 10 cal of heat. Once the freezing (or melting) point is reached, an additional 80 cal of heat (called the **latent heat of fusion,** or the **latent heat of melting**) must be supplied per gram to break the forces that bind water molecules in the ice phase. The temperature of the water and ice remains at 0 °C until all the ice melts.

The specific heat of liquid water is 1 cal per gram per Celsius degree. Thus, once our ice cube melts, 1 cal is needed for every 1 Celsius degree rise in water temperature. A phase change from liquid water to water vapor (called **evaporation**) can occur at any temperature and requires the addition of much more heat than does a phase change from ice to liquid water. The **latent heat of vaporization** varies from about 600 cal per gram at 0 °C to 540 cal per gram at 100 °C (212 °F). For our 1-g ice cube to vaporize directly without melting, a process called **sublimation,** the latent heat of fusion plus the latent heat of vaporization must be supplied to the ice cube. This amounts to 680 cal per gram at 0 °C.

# The Unique Thermal Properties of Water

Water exhibits some unusual thermal properties considering its low molecular weight of 18 atomic mass units. Normally the boiling point of related substances increases as the molecular weight increases. However, chemically related substances of greater molecular weight have boiling points 100 to 150 Celsius degrees lower than water. In fact, water's freezing and boiling temperatures are so high that within the range of temperatures observed on Earth, water occurs in all three phases—as solid, liquid, and vapor. (Interestingly, water is the only substance on Earth that exists naturally in all three phases.) Water's latent heat of fusion and latent heat of vaporization are also unusually high, and its specific heat is the highest among commonly occurring substances. These thermal properties of water have important implications for weather and climate as is evident in this and other chapters.

Why does water behave in such an unusual manner? The answer is found in its molecular structure. As shown in Figure 1, each water molecule consists of two hydrogen atoms bonded to one atom of oxygen. Bonding involves the sharing of two electrons, one from the hydrogen atom and the other from the oxygen atom. Because the oxygen atom has a stronger attraction for the shared electrons than do the hydrogen atoms, it acquires a small negative charge and the hydrogen is left with a small positive charge. Because the hydrogen–oxygen–hydrogen atoms describe a 105 degree angle, a charge separation develops on water molecules. Molecules having such a separation of positive and negative charges are said to be *dipolar.*

Opposite charges attract so that, like tiny magnets, the dipolar water molecules link together. The negative (oxygen) pole of one is attracted to a positive (hydrogen) pole of a neighboring molecule. This special attractive force, *hydrogen bonding,* is particularly strong among water molecules. Water molecules form hydrogen bonds in three directions because each molecule has three potential sites for hydrogen bond formation.

Hydrogen bonding inhibits changes in the kinetic activity of individual water molecules. Hence, as heat is added to water, the resulting increase in kinetic molecular activity and corresponding rise in tempera-

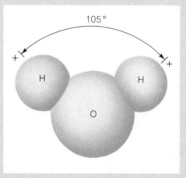

**FIGURE 1**
The dipolar structure of the water molecule.

ture are unusually small. That is, water exhibits an anomalously high specific heat. In addition, greater amounts of heat, and thus unusually high tempertures, are required for water to reach its melting and boiling points.

The stabilizing effect of hydrogen bonding also means that an exceptionally great amount of heat is required to change the phase of water. For water to change from solid to liquid, heat energy must be supplied to break some of the hydrogen bonds that maintain water in the crystalline (solid) phase. The relative strength of the hydrogen bond dictates an unusually high latent heat of fusion for water. Not all hydrogen bonds are broken as water changes from ice to liquid; that is, numerous small clusters of bonded molecules persist into the liquid phase. Considerably greater amounts of heat are required for water to vaporize (either evaporate or boil) because all hydrogen bonds must be broken. For this reason, water's latent heat of vaporization is considerably greater than its latent heat of fusion. When the phase-change direction is reversed, anomalously large quantities of heat are released to the environment.

In summary, in the solid and liquid phases, dipolar water molecules are linked by hydrogen bonds. Hydrogen bonding causes water temperature to respond sluggishly to an addition or loss of heat. Hydrogen bonding also means that water requires or releases unusually great quantities of heat when it changes phase.

If the sequence just described is reversed—that is, if the water vapor is cooled until it becomes liquid and then ice—the water temperature drops and phase changes take place as equivalent amounts of heat are released to the environment. When water vapor becomes liquid, a process called **condensation,** latent heat of vaporization is released to the environment, and when water freezes, latent heat of fusion is released. If water vapor changes to ice without first becoming liquid, a process known as **deposition,** the latent heats of vaporization plus fusion are released to the environment.

Applying this phase-change concept to the Earth–atmosphere system illustrates the mechanism of heat transfer by latent heating. As the Earth's surface absorbs radiation (both solar and infrared), some of the heat produced thereby is used to vaporize water from oceans, lakes, rivers, soil, and vegetation. Within the troposphere, some of the water vapor condenses to tiny liquid water droplets or deposits as ice crystals that are visible as clouds. During cloud formation, then, water changes phase and latent heat is released to the atmosphere. The heat required for vaporization is supplied at the Earth's surface, and that same heat is subsequently released to the atmosphere during cloud development. Through latent heating, then, heat is transferred from the Earth's surface into the troposphere.

Formation of familiar convective clouds combines

**FIGURE 4.3**
Cloud development transports excess heat at the Earth's surface into the troposphere via conduction, convection, and latent heat transfer. [Photograph by J. M. Moran]

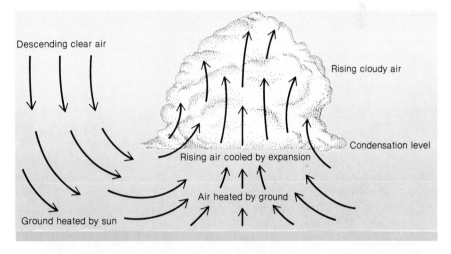

sensible heating with latent heating to channel heat from the Earth's surface into the troposphere. Updrafts (the ascending branch) in convective currents often produce **cumulus clouds,** which resemble puffs of cotton floating in the sky (Figure 4.3). These clouds are sometimes referred to as "fair weather" cumulus because they are seldom accompanied by rain or snow. On the other hand, when certain atmospheric conditions develop (described in Chapter 13), convective currents surge to great altitudes, and cumulus clouds billow upward to form towering cumulonimbus clouds, also known as thunderstorm clouds (Figure 4.4). In retrospect, then, two important heat transfer processes took place last summer when that thunderstorm washed out your ball game or sent you scurrying for shelter at the beach.

Ocean waters cover about 71% of the Earth's surface, so it is not surprising that latent heating is more significant than sensible heating on a global scale. The **Bowen ratio** describes how net radiational heating (solar plus infrared) is partitioned between sensible heating and latent heating. That is,

$$\text{Bowen ratio} = \frac{\text{sensible heating}}{\text{latent heating}}$$

At the global scale, the Bowen ratio is 7/24 or 0.29. As shown in Table 4.2, the Bowen ratio varies from one locality to another depending on the amount of surface moisture. The ratio of sensible heating to latent heating is about 1:10 for oceans, but about 2:1 for a

**TABLE 4.2**
**Bowen Ratio for Various Geographical Areas**

| Geographical Area | Bowen Ratio[a] |
|---|---|
| Europe | 0.62 |
| Asia | 1.14 |
| North America | 0.74 |
| South America | 0.56 |
| Africa | 1.61 |
| Australia | 2.18 |
| Atlantic Ocean | 0.11 |
| Indian Ocean | 0.09 |
| Pacific Ocean | 0.10 |
| All land | 0.96 |
| All oceans | 0.11 |

Source: W. D. Sellers, *Physical Climatology.* Chicago, IL: The University of Chicago Press, 1965, p. 105.

[a]The Bowen ratio is the ratio of heat used for conduction and convection (sensible heating) to heat used for vaporization of water (latent heating).

relatively dry region like the desert interior of Australia. The drier the land surface is, the less important is latent heating and the more important is sensible heating.

In some places at some times, heat transport is directed from the troposphere to the Earth's surface, which is the reverse of the average global situation that we just described. This occurs, for example, when mild air flows over cold, snow-covered ground or when warm air blows over a relatively cool ocean or lake sur-

**FIGURE 4.4**
When convection currents surge to great altitudes within the troposphere, cumulus clouds may billow upward and form thunderclouds. [Photograph by J. M. Moran]

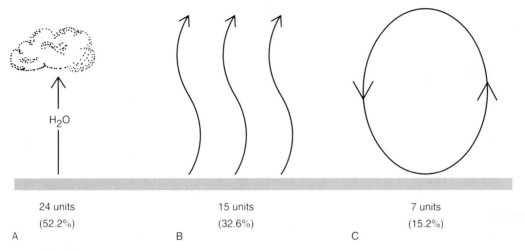

**FIGURE 4.5**
The Earth's surface is cooled via (A) vaporization of water, (B) net emission of infrared radiation, and (C) conduction plus convection. Units represent global averages; see Figure 4.1 for an explanation.

face. Heat transport from atmosphere to Earth's surface frequently occurs at night when radiational cooling causes the Earth's surface to become colder than the overlying air.

We noted in Chapter 1 that nearly all weather is confined to the troposphere, the lowest subdivision of the atmosphere. This implies that sensible and latent heat transfer operate primarily within the troposphere. Heat and temperature distributions above the troposphere are determined primarily by radiational processes.

By way of summary, it is important to note that the Earth's surface is cooled via three processes: (1) vaporization of water, (2) conduction plus convection, and (3) emission of infrared radiation (Figure 4.5). From Figure 4.1, we conclude that vaporization of water (24 units) is more important than radiational cooling (15 units) which, in turn, is more important than conduction plus convection (7 units).

## Heat Imbalance: Variation by Latitude

On a global scale, imbalances in radiational heating and radiational cooling occur not only vertically, but also horizontally, with latitude. Because Earth is nearly a sphere, parallel beams of incoming solar radiation strike lower latitudes more directly than higher latitudes. At higher latitudes, solar radiation spreads over a greater area and is less intense per unit surface area than at lower latitudes. Emission of infrared radiation (IR) by the Earth–atmosphere system also varies with latitude but less so than solar radiation. IR emission declines with increasing latitude in response to the drop in temperature with latitude. (Recall from Chapter 2 that radiation emission is temperature dependent.) Consequently, over the course of a year at higher latitudes, the rate of infrared cooling exceeds the rate of warming caused by absorption of solar radiation. At lower latitudes the reverse is true: the rate of solar radiational warming is greater than the rate of infrared radiational cooling (Figure 4.6).

Averaged over all latitudes, incoming energy (absorbed solar radiation) must equal outgoing energy (infrared radiation emitted to space). That is, the areas under the two curves in Figure 4.6 are equal. This global radiative equilibrium is an illustration of the law of energy conservation.

Satellite measurements indicate that the division between regions of *net* radiational cooling and regions of *net* radiational warming is close to the 30 degree lati-

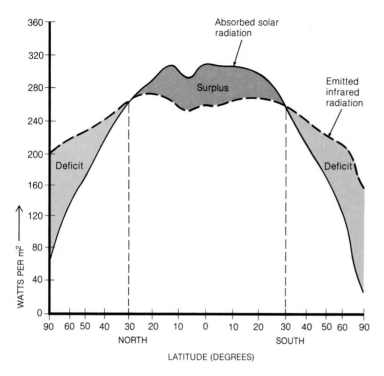

**FIGURE 4.6**
Variation by latitude of absorbed solar radiation (solid line) and outgoing infrared radiation (dashed line) as obtained by satellite measurements made between June 1974 and February 1978. Radiative heating and cooling rates are equal at about 30 degrees N and S. [From J. S. Winston et al., *Earth-Atmosphere Radiation Budget Analyses Derived from NOAA Satellite Data*, June 1974–February 1978. Washington, DC: NOAA Meteorological Satellite Laboratory, 1979.]

tude circle in both hemispheres. The implication is that latitudes poleward of about 30 degrees N and 30 degrees S experience net cooling over the course of a year, whereas tropical latitudes are sites of net warming. In fact, lower latitudes do not become progressively warmer relative to higher latitudes. Heat must therefore be transported from the tropics into middle and high latitudes.

**Poleward heat transport** is brought about primarily by north–south exchange of air masses. An **air mass**\* is a huge volume of air that covers thousands of square kilometers and is relatively uniform in temperature and water vapor concentration. The properties of an air mass largely depend on the characteristics of the surface over which the air mass resides and travels. Air masses that form at high latitudes over cold, often snow- or ice-covered surfaces are relatively cold. Those air masses that form at low latitudes are relatively warm. Air masses that develop over oceans are humid and those that form over land are dry. Hence, there are four basic types of air masses: cold

\*We have more to say about air masses and air mass modification in Chapter 11.

and humid, cold and dry, warm and humid, and warm and dry. Warm air masses from lower latitudes flow poleward and are replaced by cold air masses that flow toward the equator from source regions at higher latitudes. In this way, sensible heat is transported poleward.

Air mass exchange accounts for about half the total poleward heat transport. The remaining poleward heat transport is due to release of latent heat in storms (about 30%) and to ocean currents (about 20%). In low latitudes, water that evaporates from the warm ocean surface is drawn into the circulation of developing storms. As storms travel poleward, some of that water vapor condenses as clouds, thereby releasing latent heat. The latent heat of vaporization acquired in lower latitudes is thus transported into higher latitudes. In addition, cold ocean currents drift to-ward the tropics, whereas warm ocean currents drift poleward. In the tropics, a relatively cool surface ocean current is a *heat sink,* that is, heat is conducted from the warm air to the cool ocean water. In middle and high latitudes, on the other hand, a relatively warm surface ocean current is a *heat source* for the cooler atmosphere; that is, heat is conducted from sea to air.

# Weather: Response to Heat Imbalances

As we have seen, imbalances in rates of radiational heating and cooling give rise to temperature gradients between (1) the Earth's surface and troposphere and (2) low and high latitudes. In response, heat is transported within the Earth–atmosphere system via conduction, convection, cloud development, air mass exchange, and storms. That is, the atmosphere circulates and brings about changes in weather. A cause-and-effect chain thus operates in the Earth–atmosphere system starting with the sun as the prime energy source and resulting in weather.

## *ENERGY CONVERSIONS*

We have also seen that within the Earth–atmosphere system, some solar radiation is converted to heat through absorption, and eventually all of this heat is emitted to space as infrared radiation. Some solar energy is also converted to **kinetic energy,** the energy of motion, in the circulation of the atmosphere. Kinetic energy is manifested in winds, in convection currents,

and in the north–south exchange of air masses. Circulation (weather) systems do not last indefinitely, however. The kinetic energy of atmospheric circulation ultimately is dissipated as frictional heat as winds blow against the Earth's surface. This heat, in turn, is emitted to space as infrared radiation. Figure 4.7 is a schematic diagram of the major energy transformations operating within the Earth–atmosphere system. Technology also taps solar energy and transforms it into heat and electricity. For more on such conversions, see the Special Topic "Solar Power." Conversion of solar radiation into heat, kinetic energy, or electricity follows the law of energy conservation.

In summary then, the sun drives the atmosphere: imbalances in solar heating spur atmospheric circulation (weather), which redistributes heat. Hence, solar energy is the ultimate source of kinetic energy, which is manifest in the circulation of the atmosphere.

## *SEASONAL CONTRASTS*

The rate of heat redistribution within the Earth–atmosphere system varies with season, and, hence, atmospheric circulation and weather also change through the

**FIGURE 4.7**
A series of energy transformations operate within the Earth–atmosphere system. [Reprinted with permission of Merrill, an imprint of Macmillan Publishing Company, from Meteorology, 5th ed., by Albert Miller and Richard A. Anthes, Copyright © 1985 Merrill Publishing Company, Columbus, Ohio.]

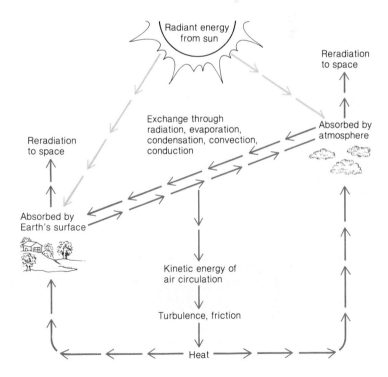

**SPECIAL TOPIC**

## Solar Power

Solar power is a renewable energy source that will last as long as the sun itself—billions of years. Furthermore, solar power is an environmentally attractive alternative to conventional energy sources, especially coal, oil, and nuclear fuels.

If all the solar radiation that penetrates the atmosphere were uniformly distributed over the entire Earth's surface, about 178 watts would illuminate every square meter. (By comparison, recall from Chapter 2 that the solar constant is about 1372 W per m$^2$). This amounts to about 4 kilowatt-hours per square meter per day, sufficient to meet nearly half of the heating and cooling requirements of an average home in the United States. The total amount of solar energy that falls yearly on U.S. lands is about 600 times greater than the nation's total annual energy demand. In fact, if solar energy could be collected and sold at average electricity rates, more than $9000 worth of energy shines on the roof of a small house (93 m$^2$ or 1000 ft$^2$) over the course of a year.

To tap solar power, we use *solar collectors,* panels that collect and concentrate the sun's rays. At present, solar collectors are used for space or water heating in small buildings such as homes, schools, and apartment houses. In the near future, solar-powered air conditioners may cool these same dwellings. Solar collectors do not produce temperatures high enough to turn water to steam, so that these devices are not adequate for most industrial purposes.

Basically, a solar collector is a framed panel of glass. Sunlight passes through two layers of glass before it is absorbed by a blackened (low albedo) metal plate. Heat is then conducted from the absorbing plate to either air or a liquid, which is conveyed by fans or pumps to wherever the heat is needed. Figure 1 is an example of the type of solar panels currently in use. Solar collectors typically capture 30% to 50% of the solar energy that reaches them.

All technologies based on the collection of solar radiation are limited by the fact that the energy source is not continuous. There's no insolation at night; during

**FIGURE 1**
Solar panels on the roof of an apartment building in Quincy, Massachusetts. [Photograph by J. M. Moran]

the day, its intensity varies seasonally, geographically, and with cloud cover. In the United States, the desert Southwest has the greatest potential for solar power because cloudiness is minimal and insolation is intense.

Seasonal fluctuations in solar radiation are especially troublesome in middle and high latitudes. For example, in eastern Washington State, a relatively sunny locality, the average annual insolation is 194 W per m$^2$, but monthly mean values vary from a low of 50 to a high of 343 W per m$^2$, a sevenfold difference through the course of a year. This seasonal variation underscores the need for long-term (summer-to-winter) heat storage systems for such localities.

Solar panels often collect more heat during the day than can be used at that time. Excess heat is usually stored in insulated water tanks or in compartments filled with rocks. In middle and high latitudes, conventional heating systems are needed as a backup to solar panels, particularly during extended periods of cloudy or very cold weather.

To reduce the variability of insolation due to changes in solar altitude, solar collectors are tilted, and some are designed to track the sun so that collectors are always oriented perpendicular to the solar beam. The advantage of tilted and tracking collectors over ones

**FIGURE 2**
A power tower stands in the middle of a field of mirrors. Computer-controlled mirrors (called *heliostats*) track the sun and focus solar radiation on a heat exchanger in the tower. This radiation is converted to heat that produces steam in a boiler. The steam then drives a turbine that generates electricity. This facility near Barstow, California, is operated by Southern California Edison. [U.S. Department of Energy photograph]

that are fixed and horizontal depends on the average cloud cover and latitude of the site. An optimal situation occurs in winter at a midlatitude locality favored by clear skies. There, tilted and tracking solar collectors can double the amount of absorbed insolation.

Scientists and engineers are currently studying ways to convert solar energy to electricity on a large scale. In one conversion method, a *power tower system,* computer-controlled mirrors, called *heliostats,* track the sun and focus its energy on a single heat-collection point at the top of a tower. Concentrated sunlight in these systems can produce temperatures up to 480 °C (900 °F), high enough to convert water into high-pressure steam for driving turbines that generate electricity. A solar-powered system of this sort requires a large land area covered with tracking mirrors. In fact, generation of 50 megawatts of electricity (enough for 15,000 homes) requires about 2.6 km² (1 mi²) of

land, making that land unsuited for other purposes. A federally sponsored experimental power tower system is located outside Barstow, California (Figure 2). Southern California Edison operates the Solar One facility, which consists of an array of 1818 heliostats and can generate 10 megawatts of electricity.

An alternative to solar-driven turbines for generating electricity is the *photovoltaic cell,* also called *solar cell.* Solar cells convert insolation directly into electricity and routinely power a number of devices ranging from handheld calculators to parking lot lights to space vehicles. Photovoltaic cells use sunlight to create a *voltage,* that is, a difference in electrical potential, in a device called a *diode* (Figure 3). When sunlight falls on the diode, an electric current flows through the circuit to which it is connected. Only certain materials develop the necessary voltage to produce a direct current when they are illuminated. These materials, known as *semiconductors,* are composed of highly purified silicon to which tiny amounts of specific impurities have been added. Semiconductors are manufactured in the form of thin wafers or sheets that are then placed perpendicular to incoming sunlight. Electricity moves through metal contact wires on the front and back sides of the wafer. Groups of wafers are wired together to form photovoltaic modules, and these are interconnected to form a photovoltaic panel (Figure 4). About 40 wafers must be linked to match the power output of a single automobile battery.

One serious drawback of today's solar cells is low efficiency. In this context, *efficiency* is defined as the percentage of insolation striking a photovoltaic cell that is converted to electric energy. A handheld solar-powered calculator, for example, has an efficiency of only 3% or less. Some mass-produced photovoltaic panels have conversion efficiencies of 10% to 12%. In 1988, researchers at Sandia National Laboratories were able to achieve an all-time high efficiency of 31% by stacking two solar cells (a gallium arsenide cell on top and a silicon cell on the bottom). Among the many factors that contribute to the low efficiency of solar cells are cell reflectivity, conversion of radiation to heat, and the sensitivity of cells to only a portion of the solar spectrum.

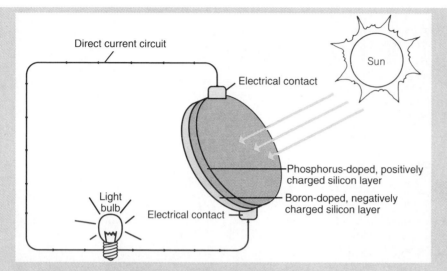

**FIGURE 3**
This silicon photovoltaic cell converts sunlight directly into electricity. Sunlight frees electrons from silicon atoms, producing a direct current that travels through the circuit.

The principal obstacle to greater use of solar cells for generating electricity is cost, although future prospects are encouraging. The cost of solar-cell-generated electricity has dropped by a factor of three since 1980. Future reductions in cost are unlikely to be achieved by boosting the efficiency of traditional solar cells because, as a general rule, production costs spiral with increasing efficiency. A more promising alternative is further development of thin-film solar

cells which are less efficient but also much less expensive to manufacture than traditional solar cells. A *thin-film solar cell* consists of a film of silicon or other light-sensitive substance that is deposited on a base material, whereas traditional solar cells consist of individual crystals.

Commercial-scale testing of the world's largest array of thin-film photovoltaic modules took place at Davis, California, in late 1992. By early the next year, the array was feeding electricity to the grid operated by the Pacific Gas and Electric Company. The photovoltaic system supplies up to 479 kilowatts, sufficient for the electrical needs of about 125 households.

About 50 megawatts of photovoltaic capacity are installed annually around the world. Before the end of the century, multimegawatt solar-cell power plants are expected to be feeding electricity into regional grids at costs that are competitive with conventional power plants. This outlook is based on current technological trends in developing more efficient thin-film solar cells (up to 11.2% efficient in 1988), plus declining manufacturing costs made possible by mass production and economy of scale. In addition, environmental concerns and the inevitable decline in supplies of fossil fuels should spur demand for solar power in the future.

**FIGURE 4**
Solar electricity-generating panel composed of interconnected solar cells. [U.S. Department of Energy photograph]

year. For example, when steep temperature gradients prevail across North America, the weather tends to be energetic. Storm systems are large and intense, winds are strong, and the weather is changeable. Such weather is typical of winter, when it is not unusual for daily temperatures in the southern United States to be more than 30 Celsius degrees (54 Fahrenheit degrees) higher than temperatures across southern Canada.

In contrast, when air temperature varies little across the continent, as in summer, the weather tends to be more tranquil, and large-scale weather systems are generally weak and not well-defined. Nevertheless, summer weather is sometimes very active. Intense heating of the ground by the summer sun often triggers strong convection and the development of thunderstorms. Some of these weather systems spawn destructive hail, strong and gusty winds, and heavy rains. However, these systems are usually shorter lived and more localized than winter storms.

# Variation of Air Temperature

Air temperature is variable, fluctuating from hour to hour, from one day to the next, with the seasons, and from one place to another. Our discussion of the basic reasons for weather provides some insight as to why air temperature is variable. The radiation balance plus the movement (advection) of air masses regulate air temperature locally. Although these two factors actually work in concert, for the purpose of study we first consider them separately.

## RADIATIONAL CONTROLS

Conditions that influence the local radiation balance and, hence, the local air temperature include (1) the time of day and day of the year, which determine the solar altitude and the intensity and duration of incoming solar radiation, (2) cloud cover because cloudiness affects the flux of both solar and terrestrial radiation, and (3) the nature of the surface cover because surface characteristics determine the albedo and the percentage of absorbed radiation (heat) used for sensible heating and latent heating. Hence, air temperature is generally higher in June than in January (in the Northern Hemisphere), during the day than at night, under clear rather than cloudy afternoon skies, when the ground is bare instead of snow covered, and when the ground is dry rather than wet.

The annual temperature cycle, also called the march of mean monthly temperature, clearly reflects the systematic variation in incoming solar radiation over the

**FIGURE 4.8**
The march of mean monthly temperature at Clevelandia, Amazon Basin (4 degrees N, 52 degrees W). At this near-equator location, little temperature change occurs through the course of a year because of minimal variation in solar radiation and length of daylight. [World Meteorological Organization data]

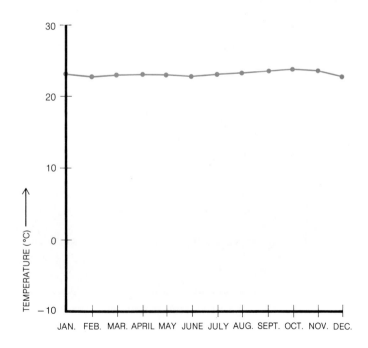

course of a year. In the latitude belt between the Tropics of Cancer and Capricorn, solar radiation varies little over the course of a year. Thus, the variation in average monthly air temperature during the year exhibits little if any seasonal contrast (Figure 4.8). In fact, in the tropics, the temperature difference between night and day often is greater than the winter-to-summer temperature contrast.

In middle latitudes, solar radiation features a pronounced annual maximum and minimum. At high latitudes, poleward of the Arctic and Antarctic circles, the seasonal difference in solar radiation is extreme, varying from zero in winter to a maximum in summer. This marked periodicity of insolation outside of the tropics accounts for the distinct winter-to-summer temperature contrasts observed in middle and high latitudes (Figure 4.9).

In middle and high latitudes, the march of mean monthly temperature lags behind the monthly variation in insolation, so that the warmest and coldest months of the year typically do not coincide with the times of maximum and minimum solar radiation, respectively.

This is because the troposphere's temperature profile takes time to adjust to the changing solar energy input. Typically, the warmest portion of the year is about a month after the summer solstice, and the coldest part of the year usually occurs about a month after the winter solstice. In the United States, the temperature cycle lags the solar cycle by an average of 27 days. However, in coastal localities with a strong maritime influence (Florida, the shoreline of New England, and coastal California, for example), the average lag time is up to 36 days. In addition, as we saw in Chapter 3, the maritime influence reduces the amplitude of the annual march of monthly mean temperature; that is, the winter-to-summer temperature contrast is less.

To some extent, the variation of air temperature over the course of a 24-hour day reflects the day-to-night (diurnal) variation in radiation. Typically, the day's lowest temperature occurs near sunrise as the culmination of a night of radiational cooling. The day's highest temperature is usually recorded in early or midafternoon, even though insolation peaks around noon. The

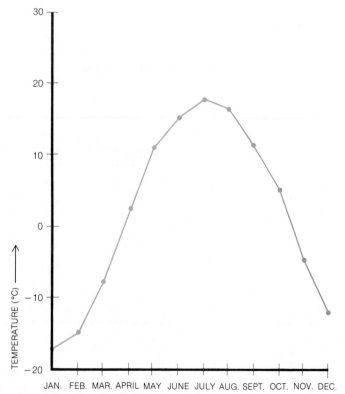

**FIGURE 4.9**
The march of mean monthly temperature at Regina, Saskatchewan, Canada (50 degrees 26 minutes N, 104 degrees 40 minutes W). The temperature regime at this extremely continental midlatitude locality is strongly influenced by the seasonal variation of incoming solar radiation. [Atmospheric Environment Service data]

# Why Mountaintops Are Cold

Even though mountaintops are closer to the sun, air temperatures are lower than in lowlands. There are two basic reasons for this phenomenon. As noted in Chapter 2, the Earth's surface is the principal recipient of solar radiational heating. Heat, in turn, flows from the Earth's surface to the atmosphere via latent heating and sensible heating. Hence, the troposphere is heated from below and its temperature profile mirrors this, that is, air temperature drops with altitude. Furthermore, the concentration of water vapor (the main greenhouse gas) is greatest at the Earth's surface, declines with altitude, and is virtually absent above 10,000 m (32,500 ft). Very little water vapor in the atmosphere above mountaintops means that infrared radiation readily escapes to space. Hence, air temperatures on mountaintops are relatively low in spite of intense sunshine.

time required for the temperature of the lower troposphere to adjust to the day-to-night variation in radiation accounts for the lag of several hours between radiational forcing and air temperature response.

The diurnal lag between insolation and air temperature explains why in summer the greatest risk of sunburn is about noon and not during the warmest time of day. Incoming solar ultraviolet radiation, the cause of sunburn, is most intense near noon, but the air temperature usually reaches a maximum several hours later.

As a further illustration of local radiational controls, consider the influence of ground characteristics on air temperature. All other factors being equal, in response to the same insolation, the air over a dry surface warms up more than it would if that surface were moist. When the surface is dry, absorbed radiation is used primarily for sensible heating of the air (conduction and convection of heat from the surface into the overlying air). Hence, the Bowen ratio is relatively high, and so is the air temperature. On the other hand, when the surface is moist, much of the absorbed radiation is used to vaporize water, the Bowen ratio is lower, and so is the air temperature. This relationship suggests a simple means of reducing summer air conditioning needs: where water is in plentiful supply, shallow pools of water placed on flat rooftops reduce direct solar heating of a building.

A relatively high Bowen ratio also helps explain why unusually high air temperatures often accompany **drought,** a lengthy period of extreme moisture deficit. Soils dry out, crops wither and die, and lakes and other reservoirs shrink. Because less surface moisture is available for vaporization, more of the available heat is channeled into raising the air temperature through conduction and convection. Consider as an example the severe drought that gripped a ten–state area of southeastern United States between December 1985 and July 1986. In most places, rainfall was less than 70% of the long-term average, and in the hardest hit areas, portions of the Carolinas, it was less than 40%. By July, many weather stations in the drought-stricken region were setting new high-temperature records. Columbia, South Carolina, Savannah, Georgia, and Raleigh–Durham, North Carolina, all reported the warmest July on record. Also contributing to record heat was more intense solar radiation reaching the ground, a consequence of less than the usual daytime cloud cover.

The same association between exceptionally dry surface conditions and unusually high air temperatures was observed during the severe drought that afflicted the Midwest and Great Plains during the summer of 1988. At many long-term weather stations, the summer of 1988 was one of the driest and hottest on record. Record high temperatures triggered much speculation (especially in the media) that a major climatic change was underway and that the much-heralded $CO_2$-induced global warming had at last begun. But, as we will see in Chapter 20, one long, hot summer does not necessarily signal a climatic change. Actually, researchers at the National Center for Atmospheric Research (NCAR) reported that the drought of 1988 was caused by an atmospheric circulation pattern that was linked to an unusual sea-surface temperature pattern over the tropical Pacific.

Because of the heat required for vaporization or melting of snow, a snow cover reduces sensible heating of the overlying air. In addition, the relatively high albedo of snow substantially decreases the amount of solar radiation that is absorbed at the surface of the snow cover and converted to heat. Consequently, a snow cover lowers the day's maximum air temperature. Because snow is also an excellent radiator of infrared, nocturnal radiational cooling is extreme where the ground is snow-covered—especially when skies are clear and winds are very light or calm. On such nights, the air temperature near the surface may be 10 Celsius degrees (18 Fahrenheit degrees), or more, lower than it would if the ground were bare of snow. The net effect of a snow cover, then, is to reduce significantly the 24-hour average temperature.

## AIR MASS CONTROLS

**Air mass advection** refers to the movement of an air mass from one locality to another. **Cold air advection** occurs when the wind blows in a direction that crosses regional isotherms from a colder area to a warmer area (arrow A in Figure 4.10), and **warm air advection** takes place when the wind blows in a direction that crosses regional isotherms from a warmer area to a colder area (arrow B in Figure 4.10). **Isotherms** are lines drawn on a map through localities having the same air temperature.* Air mass advection thus occurs when one air mass replaces another air mass having different temperature characteristics. Recall that air mass exchange is the most important process in poleward heat transport.

In terms of temperature variations at a given locality, the significance of air mass advection depends on the initial temperature characteristics of the new air mass, as well as on the degree of modification the air mass undergoes as it travels over the Earth's surface. For example, a surge of bitterly cold arctic air loses much of its punch when it travels over ground that has no snow cover because the arctic air is warmed from below by sensible heating (conduction and convection).

*Isotherms are interpolated between weather stations. Selection of the interval between successive isotherms hinges on temperature range and the desired resolution of the temperature field.

**FIGURE 4.10**
Cold air advection occurs when (A) the horizontal wind blows across isotherms from cold areas toward warmer areas, and warm advection occurs when (B) the horizontal wind blows across isotherms from warm areas toward colder areas. Solid lines are isotherms

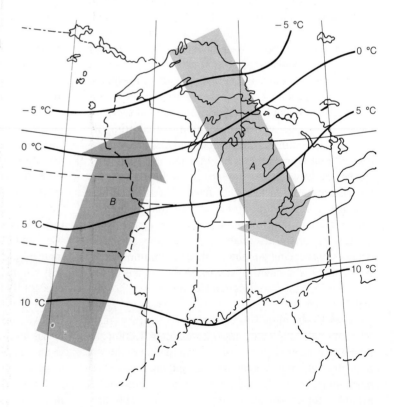

In contrast, modification of an arctic air mass by sensible heating is minimized when the air mass travels over a cold, snow-covered surface.

So far, we have been describing how horizontal movement of air (advection) influences air temperature at some locality. However, as we saw in our discussion of convection currents, air also moves vertically. As air moves up and down, its temperature changes: air cools as it rises and warms as it descends. We discuss the reasons for these temperature changes in Chapter 6.

Although we have considered local radiational controls and air mass advection separately, the two actually regulate air temperature together. For example, air mass advection may compensate for, or even overwhelm, local radiational influences on air temperature. As noted earlier, the local radiation balance usually causes the air temperature to rise from a minimum around sunrise to a maximum in early or midafternoon. This typical pattern can change, however, if an influx of cold air occurs during the same period. Depending on how cold the incoming air is, air temperatures may climb more slowly than usual, remain steady, or even fall during daylight hours. If cold air advection is extreme, air temperatures may drop precipitously throughout the day, in spite of bright, sunny skies. In another example, air temperatures may climb through the evening hours as a consequence of strong warm air advection, so the day's high temperature occurs just before midnight.

# Conclusions

Imbalances in radiational heating and radiational cooling are ultimately responsible for the circulation of the atmosphere. These imbalances occur both vertically (between the Earth's surface and the atmosphere) and horizontally (between tropical and higher latitudes). Through circulation of the atmosphere, heat is redistributed within the Earth–atmosphere system. This enables us to understand how air temperature is regulated by a combination of local radiational controls and air mass advection.

Another important consequence of atmospheric circulation is the formation of clouds that can produce rain and snow. Before examining cloud- and precipitation-forming processes in detail, we must first consider another variable of the atmosphere, air pressure. We do this in the next chapter.

## *Key Terms*

| | |
|---|---|
| sensible heating | cumulus clouds |
| latent heating | Bowen ratio |
| latent heat of fusion | poleward heat transport |
| latent heat of melting | air mass |
| evaporation | kinetic energy |
| latent heat of | drought |
| vaporization | air mass advection |
| sublimation | cold air advection |
| condensation | warm air advection |
| deposition | isotherms |

## *Summary Statements*

☐ In the atmosphere, the rate of cooling due to infrared emission is greater than the rate of warming due to absorption of solar radiation. At the Earth's surface, on the other hand, the rate of warming due to absorption of solar radiation is greater than the rate of cooling due to emission of infrared radiation. In response, heat is transported from the Earth's surface to the atmosphere via sensible heating (conduction plus convection) and latent heating (vaporization of water followed by cloud development).

☐ Latent heat is either required or released when water changes phase. A phase change means a change in molecular activity. Heat is released to the environment when water vapor condenses or when water freezes, and heat is absorbed from the environment when ice melts or water evaporates.

☐ The ratio of sensible heating to latent heating, the Bowen ratio, depends on the amount of moisture at the Earth's surface. The Bowen ratio is relatively high for dry surfaces and relatively low for wet surfaces. The global Bowen ratio is 0.29.

☐ Poleward of about 30 degrees latitude, over the course of a year, the rate of cooling due to infrared emission to space is greater than the rate of warming due to absorption of solar radiation. In tropical latitudes, on the other hand, the rate of warming due to absorption of solar radiation is greater than the rate of cooling due to emission of infrared radiation. Poleward heat transport is the consequence.

☐ Poleward heat transport is brought about chiefly by north–south air mass exchange and, to a lesser extent, by release of latent heat in storm systems and by the flow of warm and cold ocean currents.

☐ The sun drives the circulation of the atmosphere by causing heat imbalances within the Earth–atmosphere system. Some solar energy is converted to the kinetic energy of atmospheric circulation, which is ultimately dissipated as frictional heat that is radiated to space.

☐ The radiation balance plus air mass advection regulate local air temperature. Radiation balance varies with time of day, day of year, cloud cover, and characteristics of the Earth's surface. Air mass advection occurs when winds blow across regional isotherms. In some cases, cold or warm air advection overwhelms the local radiation balance.

## Review Questions

1. Distinguish between radiational heating and radiational cooling.
2. On a global scale and over the course of a year, compare the radiational heating and cooling of the atmosphere with the radiational heating and cooling of the Earth's surface.
3. How is excess heat at the Earth's surface transported into the troposphere? Which of these processes is most important and why?
4. Describe sensible heating of the atmosphere.
5. Describe latent heating of the atmosphere.
6. How does molecular activity vary with the three physical phases of water? Why is heat either absorbed or released when water changes phase?
7. Explain how sensible heating and latent heating are both involved in the formation of cumulus clouds.
8. Which requires more heat per gram, melting ice or evaporating water? Why the difference?
9. When clouds form, heat is released to the atmosphere. Explain why.
10. Define the Bowen ratio. What is the global Bowen ratio?
11. How does the Bowen ratio change as a drought

progresses? What does this change imply about the relationship between drought and air temperature?
12. Identify the three processes whereby the Earth's surface cools.
13. Under what conditions is the net flow of heat directed from the troposphere to the Earth's surface?
14. Describe how radiational heating (due to absorption of solar radiation) and radiational cooling (due to emission of infrared radiation) vary with latitude.
15. What processes are involved in poleward heat transport? Which one of these processes is most important?
16. Why is air temperature variable?
17. All other factors being equal, how does a snowcover affect air temperature during the day and at night?
18. Why is the warmest time of day usually several hours after the day's peak solar radiation?
19. Distinguish between cold air advection and warm air advection.
20. Under what conditions might the day's high temperature occur at 11 P.M.?

## Quantitative Questions

1. How many calories of heat are required to melt 20 g of ice at 0 °C to 20 g of water at 0 °C?
2. How many calories of heat are required to vaporize 10 g of water at 100 °C?
3. How much heat is released to the environment when 45 g of water freezes at 0 °C?
4. How much heat is required to change 10 g of ice at $-10$ °C to liquid water at 20 °C?
5. How much heat is needed to sublimate 40 g of ice at 0 °C?

## Questions for Critical Thinking

1. In spite of clear skies, the air temperature remains about steady throughout the daylight hours. Explain this unusual observation.
2. Speculate on whether there might be seasonal changes in the magnitude of poleward transport of heat.
3. How might temperature affect the rate of evaporation of water?
4. Distinguish between evaporation and boiling of water.
5. In a northerly climate, how might a deep snow cover in late winter delay the onset of spring-like weather.

## Selected Readings

HAMAKAWA, Y. "Photovoltaic Power," *Scientific American* 256, No. 4 (1987):87–92. Discusses new developments in photovoltaic technology and future prospects for large-scale power plants.

INGERSOLL, A. P. "The Atmosphere," *Scientific American* 249, No.3 (1983):162–174. Concisely summarizes properties of the atmosphere and the global radiation balance.

TRENBERTH, K. E. "What Are the Seasons?" *Bulletin of the American Meteorological Society* 64 (1983):1276–1282. Analyzes the distinction between meteorological seasons and astronomical seasons.

# 5 Air Pressure

*"Yes, what a climb that was!
I was scared to death, I can tell
    you.
Sixteen hundred meters—that is
    over five thousand feet, as I
    reckon it. . . ."*

*And Hans Castrop took in a deep,
experimental breath of the strange
air. It was fresh, and that was all.
It had no perfume, no content, no
humidity; it breathed in easily, and
held for him no association.*

Thomas Mann
*The Magic Mountain*

Air pressure and air density decline
with increasing altitude. At high elevations, the air is so thin that
mountain climbers may require a
supplementary oxygen supply. [Pho-
tograph by J. M. Moran]

I N DESCRIBING the state of the atmosphere, television and radio weathercasts usually include the latest air pressure reading along with air temperature and relative humidity. Although we are physically aware of changes in temperature and humidity, we do not sense changes in air pressure as readily. If we follow air pressure reports over a period of time, however, we quickly learn that important shifts in weather accompany relatively small variations in air pressure.

In this chapter, we examine the properties of air pressure and the reasons for spatial and temporal variations in air pressure. In later chapters, we describe how this variability of air pressure contributes to the circulation of the atmosphere.

# Defining Air Pressure

Air exerts a force on the surfaces of all objects that it contacts. Air pressure is a measure of that force per unit of surface area. Molecules composing air are always in rapid, random motion, and each molecule exerts a force as it collides with the surface of a solid (the ground, for example) or liquid (the ocean, for example). In a millionth of a second, billions upon billions of gas molecules bombard every square centimeter of the Earth's surface. The total air pressure exerted, then, is the cumulative force of a multitude of molecules colliding with the surface of any object in contact with air.

The amount of pressure produced by the gas molecules composing air depends on (1) the mass of the molecules, (2) the pull of gravity,* and (3) the kinetic molecular activity. Usually, **air pressure** at a given location on the Earth's surface is described as the weight per unit area of the column of air above that location. The pressure at any point within the atmosphere is equal to the weight per unit area of the atmosphere above that point. Weight is the force exerted by gravity on a unit mass, that is,

$$\text{weight} = \text{mass} \times \text{acceleration of gravity}$$

# Pressure Balance

The average air pressure at sea level is about 1.0 kg per square centimeter (14.7 lb per square in.). This same pressure is produced by a column of water about 10 m (33 ft) high. Hence, the total weight of the atmosphere on the roof of a typical three-bedroom ranch-style house at sea level is about 2.1 million kg (4.6 million lb), equivalent to the combined weight of 1500 full-size autos. Why doesn't the roof collapse? It doesn't because air pressure at any point is the same in all directions—up, down, and sideways. Thus, the air pressure within the house exactly counterbalances the air pressure outside the house, and the *net* pressure acting on the roof is zero. This pressure balance (or equilibrium) is the prevailing condition in the atmosphere.

*Gravity is the force that holds us and all other objects on the Earth's surface.

## Variation with Altitude

We know from pumping air into a bicycle tire that air is highly compressible; that is, its volume and density are variable. The pull of gravity compresses the atmosphere so that the maximum **air density** (mass of molecules per unit volume) is at the Earth's surface. In other words, the atmosphere's gas molecules are most closely spaced at the Earth's surface, and the spacing between molecules increases with increasing altitude. The number of gas molecules per unit volume thus decreases with altitude. This "thinning" of air is so rapid that at an altitude of 16 km (10 mi), air density is only about 10% of its average sea-level value.

Thinning of air with altitude is accompanied by a decline in air pressure (Figure 5.1). This drop in air pressure with altitude was first verified by Florin Périer in 1658 when he directed mountain climbers to take barometer* readings as they ascended the Puy-de-Dome in France. Because of the compressibility of air, the rate of air pressure drop with altitude is greatest in the lower troposphere and then becomes more gradual aloft. For example, air pressure decreases about 25% in the first 2500 m (8200 ft), but a further ascent of 3000 m (9800 ft) is required for another 25% drop in air pressure.

*A barometer, the instrument used to monitor air pressure, is described later in this chapter.

**FIGURE 5.1**
Variation of air pressure in millibars (mb) with altitude. The average air pressure at sea level is 1013.25 mb.

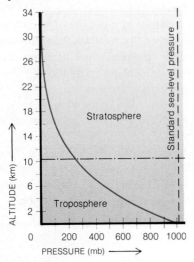

Vertical profiles of air pressure (Figure 5.1) and air temperature (Figure 1.9) are based on the **standard atmosphere,** a model of the real atmosphere. The standard atmosphere is the state of the atmosphere averaged for all latitudes and seasons. It features a fixed sea-level air temperature (15° C or 59° F) and pressure (1013.25 mb) and fixed vertical profiles of temperature and pressure. Appendix II presents the standard atmosphere in tabular format.

Although air pressure and air density drop with altitude, it is not possible to specify an altitude where the Earth's atmosphere definitely ends. That is, no one altitude can be clearly identified as the beginning of interplanetary space. At best, we can describe the vertical extent of the atmosphere in terms of the relative distribution of its mass with altitude. Half the atmosphere's mass lies between the Earth's surface and an altitude of about 5500 m (about 18,000 ft). As passengers in an aircraft flying at that altitude, we can readily identify features on the ground such as highways and towns (as long as clouds do not block our view). Such an experience makes us keenly aware of how thin the atmosphere is.

About 99% of the atmosphere's mass is below 32 km (20 mi). Above 80 km (50 mi), the relative proportions of atmospheric gases change markedly, and by about 1000 km (620 mi), the atmosphere merges with the highly rarefied interplanetary gases, hydrogen and helium. Interestingly, if the Earth's atmosphere had a uniform density throughout, it would have a well-defined top. If we assume a temperature equal to the average value at sea level, the top of a uniform density atmosphere would be at an altitude of only 8 km (5 mi).

From a slightly different perspective, at an altitude of only 32 km (20 mi), air pressure is less than 1% of its average sea-level value. This rapid pressure drop means that appreciable changes in air pressure accompany even relatively minor changes in land elevation. For example, the average air pressure at Denver, the "mile-high" city, is about 83% of the average air pressure at Boston, located just above sea level.

The expansion and thinning of air that accompany the fall in air pressure with altitude can cause discomfort for people who visit high altitudes. For more on this, see the Special Topic "Human Responses to Changes in Air Pressure."

Very low air density at high altitudes also has interesting implications for air temperature and heat transfer. In the thermosphere, the highest thermal subdivision of the atmosphere (Figure 1.9), individual atoms and molecules move about with an average kinetic activity that would indicate very high temperatures. There are so few atoms and molecules per unit volume, however, that the *total* kinetic molecular energy (that is, heat) is relatively low. Hence, in spite of temperatures that approach 1200 °C (2200 °F) in the thermosphere, heat is not readily conducted to cooler bodies. Satellites orbiting at these altitudes, for example, do not acquire such temperatures.

## Horizontal Variations

Air pressure differs from one place to another, and variations are not always due to differences in the eleva-

tion of the land. In fact, meteorologists are more interested in air pressure variations that arise from factors other than land elevation. Hence, weather observers usually determine an equivalent sea-level value; that is, they adjust local air pressure readings upward to what the air pressure would be if the station were actually located at sea level. The simplest method of adjustment is to assume that the imaginary column of air extending from the station down to sea level has the properties of the standard atmosphere. When this **reduction to sea level** is carried out everywhere, air pressure still varies from one place to another and fluctuates from day to day and even from hour to hour (Figure 5.2).

Air pressure readings, reduced to sea level, vary from one place to another, but the magnitude of this variation is much less than the vertical drop in air pressure. In fact, the same pressure change observed in the lowest 30 m (98 ft) of the troposphere may not be

**FIGURE 5.2**

A trace from a barograph showing the variation in air pressure in millibars reduced to sea level at Green Bay, Wisconsin, from 30 March through 4 April 1982. Note that significant changes in air pressure occur from day to day and even from one hour to the next.

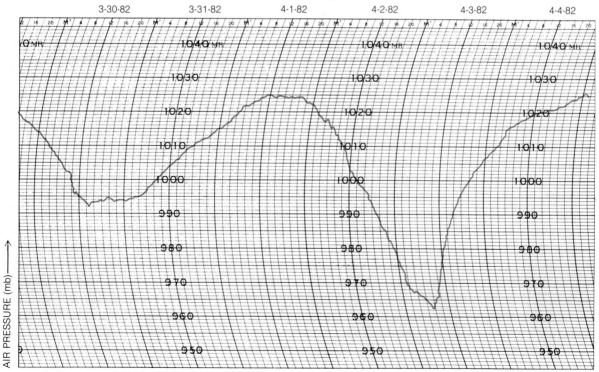

M = midnight on scale at top of record.
Time lines are every 2 hr.

equalled over a horizontal distance of 200 km (124 mi) at sea level. Nonetheless, these relatively small horizontal changes in air pressure may be accompanied by important changes in weather.

In middle latitudes, weather is dominated by a continuous procession of different air masses that bring about changes in air pressure and changes in weather. Recall from Chapter 4 that an air mass is a huge volume of air that is relatively uniform in temperature and water vapor concentration. As air masses move from place to place, surface air pressures fall or rise, and the weather changes. As a general rule, weather becomes stormy when air pressure falls and becomes fair when air pressure rises.

Why do some air masses exert greater pressure than other air masses? One reason is the difference in air density that arises from differences in air temperature, or from differences in water vapor concentration, or from both. As a rule, temperature has a much more important influence on air pressure than does water vapor concentration.

## INFLUENCE OF TEMPERATURE AND WATER VAPOR

Recall from Chapter 3 that temperature is a measure of the average kinetic energy of individual molecules. Rising air temperature means an increase in average kinetic molecular activity. If air is heated within a closed container, such as a rigid metal can, we would expect the air pressure on the internal walls of the container to rise as the increasingly energetic molecules bombard the walls with greater force. The air density inside the container does not change because no air is added to or removed from the container and the air volume is constant. In contrast, the atmosphere is not confined by walls (except the Earth's surface), so the air is free to

**FIGURE 5.3**
A hot air balloon rises within the atmosphere because heated air within the balloon is less dense than the cooler air surrounding the balloon. [Photograph by Mike Brisson]

expand and contract. That is, within the atmosphere, air density is variable.

Within the atmosphere, when air is heated (e.g., by conduction, convection, or radiation), air pressure decreases. This is because the greater activity of the heated molecules increases the spacing between neighboring molecules and thus reduces air density. As the density of a column of air declines, so too does the pressure exerted by the column. Warm air is thus lighter (less dense) than cold air and consequently exerts less pressure (Figure 5.3).

The greater the concentration of water vapor in air, the less dense is the air. This statement is contrary to the popular perception that humid air is "heavier" than dry air. Although hot, muggy air may weigh heavily on a person's disposition, humid air is, in fact, less dense than dry air at the same temperature. Water vapor reduces the density of air because the molecular weight of water is less than the average molecular weight of dry air.*

The atmosphere is a continuous fluid. This implies that when water molecules enter the atmosphere as a gas, they take the place of other gas molecules, principally nitrogen and oxygen. The molecular weight of water vapor is less than that of either $N_2$ or $O_2$. Hence, as the water vapor concentration in air increases (perhaps as the consequence of evaporation of water at the Earth's surface), the net effect is for air density to decrease. At the same temperature, then, a humid air mass exerts less pressure than a relatively dry air mass.

*The molecular weight of water is 18 atomic mass units, and the mass-weighted mean molecular weight of dry air is about 29 atomic mass units.

Cold, dry air masses produce higher surface pressures than do warm, humid air masses. Warm, dry air, in turn, produces higher surface pressures than an equally warm, but more humid, air mass. Hence, a change in surface air pressure usually accompanies the replacement of one air mass by another, that is, **air mass advection.** For example, we would expect the air pressure to rise as a surge of cold air replaces the mild air that has been with us for the past several days. Air mass modification (changes in air mass temperature and/or water vapor concentration) also produces changes in surface air pressure. These modifications may occur when an air mass travels over different surface types (from cold snow cover to mild bare ground, for example) or, if the air mass is stationary, when the air is locally heated or cooled. In Chapter 4, we examined the regulation of air temperature by the local radiation balance plus air mass advection. From the above discussion, we see that local conditions and air mass advection also influence surface air pressure.

### DIVERGENCE AND CONVERGENCE

In addition to variations in temperature and (to a lesser extent) water vapor concentration, divergence or convergence of winds may also change the surface air pressure. Consider some examples. Suppose that at the Earth's surface, horizontal winds blow radially away from a central point, as in Figure 5.4A. This is an example of **divergence** of air. At the center, air descends from above and takes the place of the air diverging at the surface. If more air diverges at the surface than descends from aloft, then the air density and air pressure

**FIGURE 5.4**
Air descends from aloft and then diverges at the Earth's surface (A). Air converges at the Earth's surface and then ascends (B). Such patterns of airflow can cause changes in air density and air pressure.

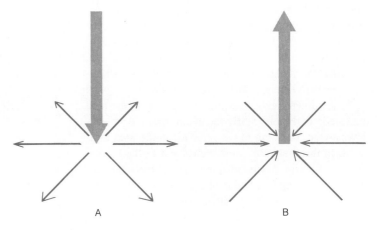

A                    B

## Human Responses to Changes in Air Pressure

While traveling or hiking in mountains above 2440 m (8000 ft), we may develop symptoms of mountain sickness, that is, headache, nausea, fatigue, and shortness of breath. If we do, we are not alone. Nearly one out of every four visitors to mountains experiences these uncomfortable symptoms. What causes them? Mountain sickness is caused by low oxygen levels in the body, particularly in the brain. Low oxygen levels result from breathing the thinning air that accompanies the fall in air pressure with altitude. Thus, the easiest way to alleviate mountain sickness is to descend to a lower altitude.

If we remain at altitude, the body quickly responds to lower oxygen levels by increasing the heartbeat and rate of breathing. These emergency adjustments protect the body while longer term changes take place. Symptoms of mountain sickness gradually subside after several days as the number of red blood cells, which transport oxygen in the bloodstream, increases. The body thus adjusts to low oxygen levels. These adjustments, which enable people to live with levels of oxygen that otherwise would cause serious health problems, constitute *acclimatization*.

Some people are better able to acclimatize than others because of differences in general health and genetic makeup. The ability to adjust to altitude is limited, however. For example, long-term residence at high altitudes cannot fully restore the body's capacity to perform work to the level that is possible at sea level. Furthermore, no humans have acclimatized permanently to altitudes higher than about 5200 m (17,000 ft). At such altitudes acclimatization cannot prevent a continual decline in body weight and deterioration of all bodily functions.

Health problems associated with thinning air mean that an aircraft cabin must be pressurized at altitudes above 4570 m (15,000 ft) unless a supplemental oxygen supply is available to the flight crew and passengers. In actual practice, to avoid mountain sickness, commercial aircraft cabins are pressurized at takeoff and remain pressurized throughout the flight. A cabin is typically pressurized to about 75% of sea-level air pressure.

Although the cabin is pressurized, we commonly feel the effects of changing air pressure in a rapidly ascending or descending aircraft by a popping sensation in our ears. Rapid ascent or descent in an express elevator often produces the same sensation. Ear-popping is symptomatic of a natural response that helps to protect the eardrum from damage.

decrease. On the other hand, suppose that at the Earth's surface, horizontal winds blow radially inward toward a central point, as in Figure 5.4B. This is an example of **convergence** of air. If more air converges at the surface than ascends, then air density and air pressure increase.

In the two examples in Figure 5.4, wind direction induces divergence and convergence of air. In addition, downstream changes in wind speed can cause divergence or convergence. In Figure 5.5A, the wind speed increases downstream, causing divergence in the area indicated. In Figure 5.5B, the wind speed decreases downstream, resulting in convergence in the area indicated. In the first case, air stretches and in the second case, air piles up. In later chapters, we discuss in greater detail how divergence and convergence of air cause air pressure changes within weather systems.

**FIGURE 5.5**
A downstream increase in wind speed causes divergence (A), whereas a downstream decrease in wind speed causes convergence (B).

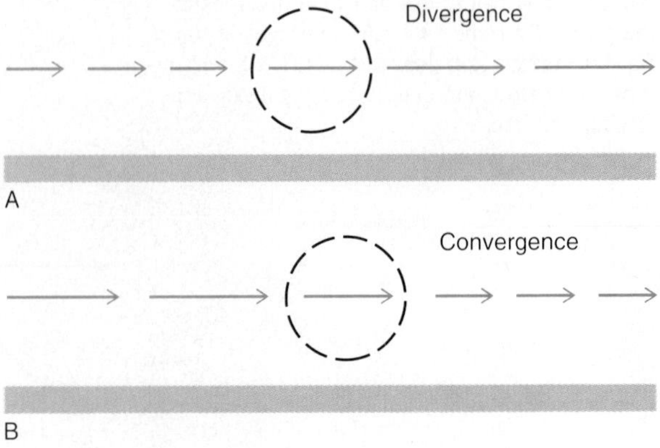

As Figure 1 indicates, the eardrum separates the outer ear from the middle ear chamber. As an aircraft takes off and cabin pressure drops, the air pressure on the outer ear declines. As air pressure on the outer ear changes, the eardrum becomes distorted unless a comparable pressure change takes place in the middle ear. If the pressure does not equalize between the outer and middle ear, the eardrum bulges outward. Such deformation not only causes physical discomfort, but the bulging eardrum does not vibrate efficiently and sounds are muffled. If the air pressure difference between the middle ear and outer ear continues to increase, the eardrum could rupture and perhaps cause a permanent hearing loss.

Fortunately, there is a natural mechanism whereby the air pressure on the middle ear is altered. The *eustachian tube* connects the middle ear with the upper throat region, which, in turn, leads to the outside via the oral and nasal cavities. Normally, the eustachian tube is closed where it enters the throat, but it will open if a sufficient air pressure difference develops between the middle ear and the throat. When the eustachian tube opens, air pressure in the middle ear quickly adjusts to the external air pressure and the eardrum pops back to its normal shape. Vibrations of the eardrum that are associated with its rapid change in shape are what we hear as "ear-popping", the human body's way of preventing a permanent hearing loss when we experience a rapid change in air pressure.

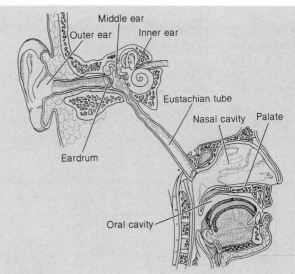

**FIGURE 1**
When differences in air pressure develop between the middle ear and the outer ear, the eardrum is distorted and sounds are muffled. Opening of the eustachian tube equalizes the pressure and causes a popping sensation in the ears.

To reduce discomfort, we can hasten the opening of the eustachian tube by yawning or swallowing. For this reason, air travelers are advised to chew gum because the subsequent swallowing helps to equalize air pressure on both sides of the eardrum.

## Highs and Lows

We now have some insight into the meaning of those *H* and *L* symbols on newspaper and television weather maps (Figure 5.6). We determine the spatial pattern of surface air pressure on a weather map by first reducing air pressure readings at all weather stations to sea level. Then, we draw isobars. An **isobar** joins locations having the same air pressure.* An *H* or *HIGH* symbol is used to designate places where sea-level air pressure is relatively high compared with the air pressure of surrounding areas, and an *L* or *LOW* symbol is used to indicate regions where sea-level air pressure is relatively low by comparison. For

*Isobars must be interpolated between weather stations.

reasons presented in Chapter 9, a *HIGH* usually is a fair weather system, and a *LOW* is a stormy weather system.

## Air Pressure Measurement

A **barometer** is the instrument used to monitor changes in air pressure. There are two basic types of barometer: mercurial and aneroid.

The more accurate, though cumbersome, of the two is the **mercurial barometer,** invented in 1643 by Evangelista Torricelli, an Italian mathematician and student of Galileo. The instrument consists of a glass tube a

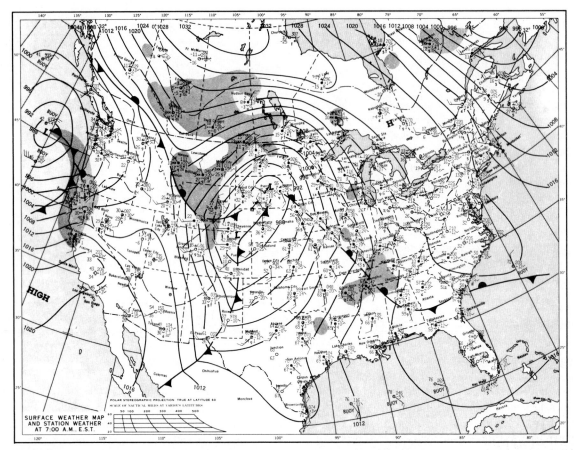

**FIGURE 5.6**
A typical surface weather map showing variations in air pressure (reduced to sea level) from one
place to another. Solid lines are isobars, lines joining locales having the same air pressure.
Shaded areas are where rain or snow is falling. [NOAA]

little less than 1.0 m (39 in.) long, sealed at one end,
open at the other end, and filled with mercury, a very
dense liquid. The open end of the tube is inverted into a
small open container of mercury, as shown in Figure
5.7. Mercury settles down the tube (and into the container) until the weight of the mercury column exactly
balances the weight of the atmosphere acting on the
surface of the mercury in the container. Hence, the
height of the mercury column is directly proportional to
atmospheric pressure.

In addition to a reduction to sea level, air pressure
readings from a mercurial barometer require adjustments for (1) the expansion and contraction of mercury
that accompany changes in temperature, and (2) the
slight variation of gravity with latitude. By convention,

readings are adjusted to standard conditions of 0 °C (32
°F) and 45 degrees latitude.

The average atmospheric pressure at sea level will
support the mercury column in the tube to a height of
760 mm (29.92 in.). When air pressure changes, however, the height of the mercury column changes. Falling air pressure allows the mercury column to drop, and
increasing air pressure forces the mercury column to
rise. This, then, is the origin of the practice of express-
ing air pressure in units of length.

The **aneroid** (nonliquid) **barometer,** pictured in Figure 5.8A, is less precise but more portable than the mercurial barometer. It consists of a flexible chamber from
which much of the air has been evacuated (Figure
5.8B). A spring keeps the chamber from collapsing. As

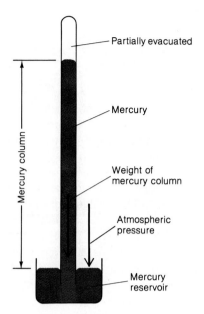

**FIGURE 5.7**
Schematic drawing of a mercury barometer.

air pressure changes, the chamber flexes, compressing when pressure rises and expanding when pressure drops. A series of gears and levers transmits these movements to a pointer on a dial, which is calibrated to read in equivalent millimeters (or inches) of mercury, or to read directly in units of air pressure. Some new aneroid barometers provide direct digital readouts.

Indoor air pressure quickly adjusts to changes in outdoor air pressure. Hence, a barometer is usually mounted indoors. The instrument should be anchored to a sturdy wall that does not receive direct sunlight. Mercurial barometers must be vertical and most aneroid barometers are designed to be read in a vertical position.

Aneroid barometers that are intended for home use typically have dials with legends, such as *fair, changeable,* and *stormy,* corresponding to certain ranges of air pressure. These designations should not be taken literally because a given air pressure reading does not always correspond to a specific type of weather.

Much more useful than these legends for local weather forecasting is **air pressure tendency,** that is,

**FIGURE 5.8**
An aneroid barometer, a portable instrument used to measure variations in air pressure. External view (A) and internal view (B). [Photograph by J. M. Moran; diagram adapted from Snow et al., "Basic Meteorological Observations for Schools: Atmospheric Pressure." *Bulletin of the American Meteorological Society,* 73 (1992): 785]

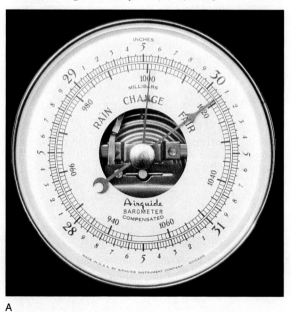

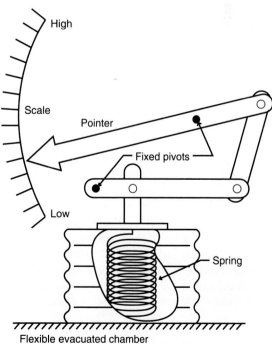

## Altimetry

Altimetry is the determination of altitude above mean sea level based on air pressure. An *altimeter* is an aneroid barometer that is graduated in increments of altitude. The graduation (that is, the calibration of altitude against air pressure) is prescribed by the *standard atmosphere,* which is described elsewhere in this chapter and in Appendix II.

At any time and place the real atmosphere usually differs from the standard atmosphere so that an altimeter typically does not give the true altitude. The indicated altitude (the altimeter reading) is the same as the true altitude only when air pressure and temperature match the standard atmosphere. Unless adjustments are made, the discrepancy between indicated and true altitudes can pose serious problems, especially during the crucial takeoff and landing phases of flight.

Changes in surface air pressure en route are one cause of differences between indicated and true altitudes. For example, as an aircraft travels toward a destination reporting a lower surface air pressure than its departure point, the altimeter will read higher than the true altitude. Hence, aircraft altimeters are equipped with a movable scale that enables the pilot to adjust altimeter readings. The Federal Aviation Administration (FAA) requires all aircraft flying below 5500 m (18,000 ft) to calibrate altimeters to surface air pressure radioed from flight service stations en route. (Above 5500

m, aircraft fly along an isobaric surface with the altimeter zeroed at the standard sea-level pressure of 1013.25 mb.)

In-flight adjustments of altimeters to surface conditions, however, do not correct for pressure variations that arise en route principally from temperature variations within the air column beneath the aircraft. Cold air is denser than warm air so that air pressure drops more rapidly with altitude in cold air than it does in warm air. Hence, within a column of cold air a given air pressure occurs at a lower altitude than does the same air pressure in a column of warm air. This means, for example, that as an aircraft flies into a column of air that is warmer than specified by the standard atmosphere (Figure 1), the altitude indicated by the altimeter will be lower than the true altitude. Conversely, the altimeter aboard an aircraft flying into air colder than specified by the standard atmosphere will read too high.

The danger, of course, is that an erroneous altimeter reading may impede a pilot's ability to clear an obstacle such as a mountain peak. In practice this hazard can be greatly reduced by an onboard computer that measures air temperature at flight level and makes an appropriate adjustment to the altimeter reading. Note that this correction is not based on the mean temperature of the air column so that, although the error is reduced, it is not totally eliminated.

---

the change in air pressure with time. Rising air pressure usually means continued fair or clearing weather, whereas falling air pressure generally signals stormy weather. For determining pressure tendency, some aneroid barometers are equipped with a second pointer that serves as a reference marker. By turning the knob on the barometer face, the user sets the second pointer to correspond to the current air pressure reading. At a later time (perhaps in 2 or 3 hours), the user can observe the new pressure reading and compare it with the earlier set reading to determine the air pressure tendency.* Weather map symbols for air pres-

*In taking a reading from a dial-type aneroid barometer, it is a good idea to first gently tap the barometer because friction in the mechanism may cause the dial to stick.

**FIGURE 5.9** (opposite)
A barograph provides a continuous trace of air pressure variations with time. An example is shown as Figure 5.2. [Photograph by J. M. Moran]

sure tendency are given in Appendix III (page 478).

An aneroid barometer may also be linked to a pen that records on a clock-driven drum chart, as shown in Figure 5.9. This instrument, called a **barograph,** provides a continuous trace of air pressure with time. Because air pressure drops with altitude, an aneroid barometer can be calibrated to monitor altitude. Such an instrument is called an **altimeter.** For more on this, see the Special Topic "Altimetry."

In summary, differences between altimeter readings and true altitude arise from en route changes in surface air pressure and/or average air temperature. Even with adjustments in altimeter readings, pilots are well advised to follow the adage *"cold or low, look out below."* Hence, pilots should always select a flight altitude that will allow for a margin of safety, especially when flying over mountainous terrain or during conditions of restricted visibility.

An alternative to an air pressure-based altimeter is a *radio altimeter,* which is flown aboard most commercial aircraft. This instrument emits radio waves from the plane to the ground, where the signal is reflected back to the aircraft. Altitude is calibrated in terms of the time elapsed between emission and reception of the radio signal. The longer the signal takes, the higher is the indicated altitude of the aircraft. Use of this device requires a thorough knowledge of the underlying terrain.

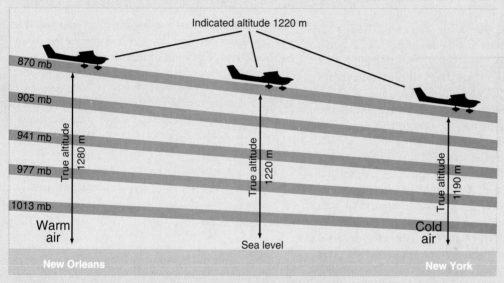

**FIGURE 1**
An aircraft altimeter is initially calibrated to the standard atmosphere. But as the aircraft flies into columns of colder or warmer air, the indicated altitude may differ significantly from the true altitude.

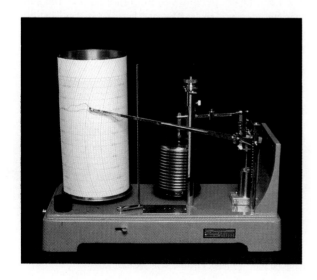

## Pressure Units

On television and radio weather reports, air pressure readings are usually given in units of length (millimeters or inches), based on the type of instrument used to measure the pressure. It is more appropriate, however, to express air pressure in units of pressure. Physicists use the *pascal (Pa)* as the metric unit of pressure and have determined that the average air pressure at sea level is 101,325 Pa, 1013.25 hectopascals (hPa), or 101.325 kilopascals (kPa). Meteorologists, on the other hand, traditionally designate air pressure in *millibar (mb)* units, where 1 mb equals 1 hPa or 100 Pa. In turn, 1 mb is the equivalent of 0.02953 in. of mercury (Table 5.1).

**TABLE 5.1**
**Conversion Factors for Units of Air Pressure**

1 bar = 1000 millibars (mb)
1 mb = 0.02953 in. of mercury
1 inch of mercury = 33.8639 mb
1 kilopascal (kPa) = 1000 pascals (Pa)
1 hectopascal (hPa) = 100 Pa
1 mb = 1 hPa
1 inch of mercury = 33.8639 hPa
1 inch of mercury = 25.4 mm of mercury

The average sea-level air pressure reading is 1013.25 mb (29.92 in.), and the usual worldwide range in sea-level air pressure is about 970 to 1040 mb (28.64 to 30.71 in.). The lowest sea-level air pressure ever recorded was 870 mb (25.69 in.), measured on 12 October 1979 in the eye of Typhoon Tip over the Pacific Ocean northwest of Guam. In some tornadoes, the air pressure is likely lower than 870 mb, but such low readings have never been measured because tornadic winds destroy barometers. The highest sea-level air pressure ever recorded was 1083.8 mb (32.01 in.) at Agata, Siberia, on 31 December 1968 and was associated with an extremely cold air mass.

Meteorologists often express altitudes in terms of an isobaric surface, that is, a surface where the air pressure is the same everywhere. Examples are the 200-mb, 500-mb, and 750-mb surfaces. Air pressure drops more rapidly with altitude in a column of cold (dense) air than in a column of warm (less dense) air. Hence, equivalent pressure surfaces (e.g., the 500-mb level) occur at a lower altitude in a cold column of air than in a warm column of air. Thus, within the troposphere, isobaric surfaces slope downward from the relatively warm tropical latitudes toward the relatively cold polar latitudes.

## The Gas Law

To this point, we have described the state of the atmosphere in terms of variations in temperature, pressure, and density. These important properties, collectively known as **variables of state,** change in magnitude from one place to another across the Earth's surface, with altitude above the Earth's surface, and with time. The three variables of state are also interrelated through the

**gas law.** Although the gas law was derived for a single ideal gas,* the law provides a reasonably accurate description of the behavior of air, which, as noted earlier is a mixture of many different gases.

Simply put, the gas law states that the pressure exerted by air is directly proportional to the product of its density and temperature. Expressed as a word equation, the gas law becomes

air pressure = constant × density × temperature

The constant is an experimentally derived number that changes a proportional relationship to an equation. The physical basis for this law is presented in the Mathematical Note at the end of this chapter.

From the gas law equation, the following is evident.

1. If the density is held constant, a rise in temperature is accompanied by an increase in pressure. Recall that this is what happens when we heat air inside a rigid container.

2. If the temperature is held constant, an increase in pressure is accompanied by an increase in density.

3. At constant pressure, a rise in temperature means a decrease in density.

In the atmosphere, however, the situation is more complicated because all three variables of state may change simultaneously. Hence, as the air temperature rises (perhaps as a consequence of warm air advection), the air density decreases, and the air pressure at the Earth's surface falls. On the other hand, in winter, it is usual for air temperatures to drop (not rise) as the surface air pressure rises. The gas law is satisfied in this case because air density increases at the same time as the temperature drops.

## Conclusions

We now have a working definition of air pressure, and we have examined the causes of spatial and temporal variations in air pressure within the Earth–atmosphere system. Air pressure drops rapidly with altitude in the lower troposphere and then more gradually aloft. At

*An ideal gas follows the kinetic molecular theory precisely. For more on this, see this chapter's Mathematical Note.

great altitudes, the atmosphere gradually merges with the gases of interplanetary space so that the atmosphere has no clearly defined top. Reduction to sea level eliminates the influence of weather station elevation on air pressure. Then, surface air pressure depends on air density, which, in turn, is governed by air temperature and, to a lesser extent, by the concentration of water vapor in air. Divergence and convergence of air may also affect air density and surface air pressure.

The range of air temperature and pressure in the Earth–atmosphere system means that water occurs in all three phases. Earlier we saw how phase changes of water help to redistribute heat within the atmosphere (latent heating). Changes in air temperature also trigger the phase changes of water that cause clouds to form or to dissipate. In the next three chapters, we take a closer look at water within the atmosphere with a special emphasis on cloud- and precipitation-forming processes.

## MATHEMATICAL NOTE

## The Gas Law

Although the atmostphere is a mixture of many gases, it behaves much as if it were a single ideal gas. By definition, an *ideal gas* follows the kinetic molecular theory precisely. That is, an ideal gas is made up of a very large number of minute particles, called *molecules,* that are in rapid and random motion. As they move about, molecules experience perfectly elastic collisions, so they lose no momentum. Molecules are so small that the attractive forces between them are negligible.

An ideal gas follows Charles's law and Boyle's law exactly, whereas a real gas behaves only approximately as these laws dictate. *Charles's law* holds that with constant pressure, $P$, the absolute temperature, $T$ (in kelvins), of an ideal gas is inversely proportional to the density, $\rho$, of the gas. That is,

$$T \propto 1/\rho$$

As a sample of gas is heated, the gas expands and its density decreases. According to Boyle's law, on the other hand, when the temperature is held constant, the pressure and density of an ideal gas are directly proportional. That is,

$$P \propto \rho$$

As the pressure on an ideal gas increases, its volume decreases and its density increases.

Charles's and Boyle's laws are combined as the *ideal gas law,* that is

$$P \propto \rho T$$

The constant of proportionality, $R$, varies depending on the specific gas. The ideal gas law is thus expressed as

an equation relating variables of state,

$$P = \rho R T$$

This so-called *equation of state* describes approximately the behavior of dry air (air minus water vapor) when we assign $R$ a value of 287 J/kg-K.

We can modify the equation of state so that it relates to variables of state for humid air, the more realistic situation. One approach is to keep the same $R$, the gas constant for dry air, and replace $T$ by $T_v$, the virtual temperature. The *virtual temperature* is defined to be the temperature of dry air having the same density and pressure as a given sample of humid air. The virtual temperature is computed from the specific humidity, $q$, that is,

$$T_v = T(1 + 0.61q)$$

*Specific humidity* is the ratio of the mass of water vapor to the mass of humid air containing the water vapor and is usually expressed as grams of water vapor per kilogram of humid air (Chapter 6). The virtual temperature is higher than the actual temperature but rarely by more than 3 Celsius degress.

For humid air, the equation of state thus becomes

$$P = \rho R T_v$$

This equation is valid for humid air in which no condensation takes place. Condensation complicates matters because it is accompanied by release of latent heat of vaporization, which elevates air temperature and lowers density.

## Key Terms

air pressure
air density
standard atmosphere
reduction to sea level
air mass advection
divergence
convergence
isobar

barometer
mercurial barometer
aneroid barometer
air pressure tendency
barograph
altimeter
variables of state
gas law

## Summary Statements

☐ The pressure (force per unit area) exerted by the atmosphere depends on the pull of gravity and the mass and kinetic energy of the gas molecules that compose air.

☐ At any specified point within the atmosphere, air pressure has the same magnitude in all directions.

☐ Air pressure and air density decrease rapidly with altitude in the lower atmosphere and then more gradually aloft.

☐ The atmosphere has no clearly defined upper boundary. Rather, the atmosphere of Earth gradually merges with the highly rarefied hydrogen/helium atmosphere of interplanetary space.

☐ About 99% of the mass of the atmosphere is situated below an altitude of 32 km (20 mi).

☐ Air pressure readings are reduced to sea level in order to remove the influence of station elevation.

☐ Within the atmosphere, air density is inversely proportional to air temperature. Hence, all other factors being equal, cold air masses are more dense and exert higher pressure than do warm air masses.

☐ Within the atmosphere, air density is also inversely proportional to water vapor concentration. Hence, at equivalent temperatures, dry air masses are denser and exert higher pressure than do humid air masses.

☐ As a rule, temperature has a much more significant influence on air pressure than does humidity.

☐ Air pressure may fluctuate in response to divergence or convergence of air, which is produced by changes in wind speed or direction.

☐ Important changes in weather often accompany relatively small changes in air pressure at the Earth's surface. As a rule, high or rising pressure signals fair weather and low or falling pressure means stormy weather.

☐ A barometer monitors changes in air pressure. The mercurial barometer is more accurate but far less portable than the aneroid barometer.

☐ Variables of state of the atmosphere (temperature, pressure, and density) are related through the gas law.

## Review Questions

1. Define air pressure.
2. Why is it that buildings do not have to be constructed to withstand the weight of the atmosphere?
3. Explain why air density is greatest at the Earth's surface.
4. Is there a clearly defined top to the Earth's atmosphere? Explain your response.
5. Why are the cabins of commercial jet aircraft pressurized?
6. Why do weather station personnel routinely adjust air pressure readings to sea-level values?
7. How does air temperature influence the pressure exerted by a column of air?
8. How does the concentration of water vapor influence the pressure exerted by a column of air?
9. Why is it that increasing the concentration of water vapor in a unit volume of air decreases its density?
10. As a general rule, how does the weather change as the air pressure at the Earth's surface rises or falls?
11. Which air mass exerts a greater surface air pressure, a warm and humid air mass or an equally warm but dry air mass? Explain your response.
12. What is the meaning of the *H* and *L* symbols on newspaper and television weather maps?
13. Air pressure readings are often reported in units of length (millimeters or inches) rather than units of pressure. Why?
14. Explain the principle of the mercurial barometer.
15. What are the advantages of an aneroid barometer over a mercurial barometer?
16. Why is a barometer also an altimeter?
17. Explain why air pressure tendency can be useful in forecasting the weather locally.
18. What is the *standard atmosphere* and how is it used?
19. State the gas law in your own words.
20. In winter, cold air masses produce higher surface air pressures than do mild air masses. Explain how the gas law is satisfied.

## Questions for Critical Thinking

1. If we use water instead of mercury in a glass-tube barometer, compute the required height for the tube. The density of mercury (Hg) is 13.6 g/cm$^3$ and the density of water is 1.0 g/cm.$^3$

2. A jet aircraft is cruising at the 400-mb level, that is, at the altitude where the air pressure is 400 mb. What fraction of the atmosphere's mass is below the aircraft?

3. Design an experiment that demonstrates that, at equal temperatures, relatively dry air exerts more pressure than relatively humid air.

4. What is an *ideal gas?* What assumptions are made in applying the gas law to the atmosphere?

5. On a particularly hot and muggy evening, a sportscaster comments that baseballs hit to the outfield will not carry far in the "heavy" air. How valid is this observation?

## Selected Readings

FAA AND NOAA. *Aviation Weather.* Washington, DC: Federal Aviation Administration, 1975, 219 pp. Intended for pilots and flight operations personnel; a well-illustrated handbook that applies the basics of meteorology to flight.

HOUSTON, C. S. "Mountain Sickness," *Scientific American 267,* No. 4 (1992):58-66. Describes human responses to low oxygen levels.

SNOW, J. T., M. E. AKRIDGE, AND S. B. HARLEY. "Basic Meteorological Observations for Schools: Atmospheric Pressure," *Bulletin of the American Meteorological Society* 73(1992):781–794. Reviews some commercially available barometers, describes how to build your own barometer, and suggests some activities using barometers.

# 6 Humidity and Stability

*All the rivers run into the sea, yet the sea is never full; unto the place from whence the rivers come thither they return again.*

*Ecclesiastes* 1:7

Water occurs naturally in three phases: ice, liquid, and vapor. Water changes phase as it circulates between the Earth's surface and the atmosphere. This circulation is part of the global water cycle that supplies us with all of our fresh water. [Photograph by Mike Brisson]

114

W ITHIN THE ATMOSPHERE, water occurs in all three phases: as water vapor (an invisible gas) and as tiny ice crystals and water droplets (visible as clouds). The total amount of water within the atmosphere is very small, and most of that is in the lower portion of the troposphere. Indeed, if at any moment all water were removed from the atmosphere as rain and distributed uniformly over the globe, this water would cover the Earth's surface to a depth of only about 2.5 cm (1.0 in.).

Water continually cycles into the atmosphere as vapor from reservoirs of water at the Earth's surface, and water continually leaves the atmosphere and returns to the Earth's surface as rain, snow, and other forms of precipitation. On average, the residence time of a water molecule in the atmosphere is about 10 days. This cycling is an essential component of a global-scale system known as the hydrologic cycle.

In this chapter, we consider how the hydrologic cycle functions, particularly as it relates to the transfer of water between the Earth's surface and the atmo-sphere. We learn how to quantify the water vapor concentration of air, how air becomes saturated through up-lift, and how atmospheric stability influences ascent of air. All this is important because as air approaches saturation, the development of clouds becomes more and more likely and clouds are required for precipitation.

## The Hydrologic Cycle

The total amount of water on the planet is constant and has been that way for perhaps millions of years. Water is distributed among oceanic, terrestrial, and atmospheric reservoirs (Table 6.1). The ocean, the largest of these reservoirs, contains 97.2% of the planet's water; most of the remainder is tied up as ice in Antarctic and Greenland glaciers. The ceaseless flow of water among the reservoirs, known as the **hydrologic cycle,** is illustrated in Figure 6.1. In brief, water vaporizes from the sea and land to the atmosphere where clouds form. From clouds, rain and snow fall back to the Earth's surface, thus supplying rivers, which flow back to the sea. The endlessness of the hydrologic cycle is expressed in the verse from Ecclesiastes on the opening page: "All the rivers run into the sea, yet the sea is never full; unto the place from whence the rivers come thither they return again." Here we focus on a critical link in the hydrologic cycle, the link that joins the atmosphere to the oceanic and terrestrial reservoirs of water.

Within the usual range of air temperature and pressure on planet Earth, all three phases of water coexist

**TABLE 6.1**
**Water Stored in Global Reservoirs of the Hydrologic Cycle**

| Reservoir | % of Earth's total water |
|---|---|
| World oceans | 97.2 |
| Ice sheets and glaciers | 2.15 |
| Groundwater | 0.62 |
| Lakes (freshwater) | 0.009 |
| Inland seas, saline lakes | 0.008 |
| Soil water | 0.005 |
| Atmosphere | 0.001 |
| Rivers and streams | 0.0001 |

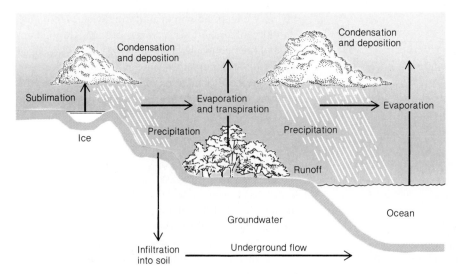

**FIGURE 6.1**
The hydrologic cycle is a continuous transfer of water among terrestrial, oceanic, and atmospheric reservoirs.

naturally. That is, water vapor is in equilibrium with water's liquid and solid phases. Under these conditions, water molecules continually change phase. At the interface between water and air (the surface of the ocean, for example), some water molecules escape from the water surface and enter the air as vapor, whereas other water molecules leave the vapor phase and return to the water surface as liquid. **Evaporation** occurs if more water molecules enter the air than return to liquid water, and **condensation** takes place if more water molecules return to liquid water than enter the atmosphere as vapor.

Water evaporates from the surface of seas, lakes, and rivers as well as from soil and the damp surfaces of plant leaves and stems. Evaporation of ocean water is the principal source of atmospheric water vapor. **Transpiration** is the process by which water absorbed by plant roots eventually escapes as vapor through tiny pores on the surface of green leaves. On land, transpiration is considerable and is often more important than direct evaporation from the surfaces of lakes, streams, and the soil. For example, a single hectare (2.5 acres) of corn typically transpires 34,000 liters (L) (8800 gal) of water per day. Measurements of direct evaporation and transpiration are usually combined as **evapotranspiration.**

A two-way exchange of water molecules also takes place at the interface between ice (or snow) and air, except that water molecules cannot escape from an ice surface as readily as they can from a liquid water sur-

face. Water molecules are more tightly bonded in the solid phase than in the liquid phase. **Sublimation** occurs when more water enters the vapor phase than returns to ice, and **deposition*** takes place when more water returns to the ice than vaporizes.

The gradual shrinkage of snowbanks, even though the air temperature remains well below the freezing point, results from both sublimation and settling. Condensation or deposition on exposed surfaces on the ground is visible as dew or frost, respectively. These same processes operating within the atmosphere produce clouds.

**Precipitation**—rain, drizzle, snow, ice pellets, and hail—returns a major portion of atmospheric water from clouds to the Earth's surface, where most of it vaporizes back into the atmosphere. Evaporation or sublimation, followed by condensation or deposition, purifies water. As water vaporizes from the Earth's surface, suspended and soluble substances like sea salts are left behind. Through these cleansing mechanisms, saline water from the sea eventually falls on land as freshwater precipitation, which replenishes terrestrial reservoirs.

Comparing the movement of water into and out of the terrestrial reservoirs with the movement of water into and out of the ocean is instructive. The balance sheet for the inputs and outputs of water to and from

*Sometimes *sublimation* is applied to the phase change from vapor to solid as well as the phase change from solid to vapor.

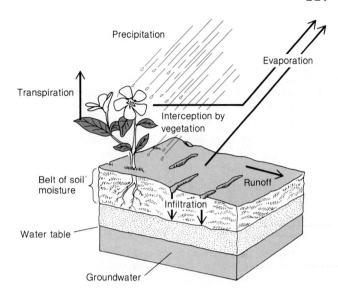

**FIGURE 6.2**
The various pathways taken by precipitation that falls on the land. [From J. M. Moran et al., *Introduction to Environmental Science*. New York: W. H. Freeman, p. 163. Copyright © 1980.]

the various global reservoirs is called the **global water budget** (Table 6.2). Each year on the continents the total precipitation exceeds evapotranspiration by about one-third. At sea, however, the annual evaporation exceeds precipitation. Over the course of a year, the water budget therefore shows a net gain of water on land and a net loss of water from the oceans, with the excess on land equal to the ocean's deficit. The land, however, is not getting any soggier, nor are the world's oceans drying up. This is because the excess precipitation on land drips, seeps, and flows from the land back to the ocean.

Once precipitation reaches the ground, it follows various routes (Figure 6.2). Some water vaporizes, and the remainder either seeps into the ground (as soil moisture or groundwater) or runs off as rivers and streams to the sea. The ratio of the portion that infiltrates the

Earth's surface to the portion that runs off depends on the intensity of rainfall and on the vegetation, topography, and physical properties of the surface. For example, frozen soil is impermeable so that heavy rain falling on frozen soil mostly runs off and may cause flooding.

A river plus its tributaries drains water from a fixed geographical region called a *drainage basin*. Climate, vegetation, topography, and various activities, both natural and human, in the drainage basin affect both the quantity and the quality of surface and ground water.

It is well worth emphasizing that the sun drives the hydrologic cycle. As we saw in Chapter 4, some of the radiation that strikes the Earth's surface is absorbed, that is, converted to heat. Some of this heat is used to vaporize water. If water did not vaporize, there would be no clouds, no precipitation, and no hydrologic cycle.

## How Humid Is It?

"It's not the heat, it's the humidity." This popular statement attributes the discomfort we feel on a hot, muggy day to the water vapor content of the air. Water vapor concentration in air is an important determinant of our physical comfort, as discussed in the Special Topic "Humidity and Human Comfort."

**TABLE 6.2**
**Global Water Budget[a]**

| Source | $m^3$ per year (gal per year) |
|---|---|
| Precipitation on sea | $3.24 \times 10^{14}$ $(85.5 \times 10^{15})$ |
| Evaporation from sea | $3.60 \times 10^{14}$ $(95.2 \times 10^{15})$ |
|    Net loss from sea | $-0.36 \times 10^{14}$ $(-9.7 \times 10^{15})$ |
| Precipitation on land | $0.98 \times 10^{14}$ $(26.1 \times 10^{15})$ |
| Evaporation from land | $0.62 \times 10^{14}$ $(16.4 \times 10^{15})$ |
|    Net gain on land | $+0.36 \times 10^{14}$ $(+9.7 \times 10^{15})$ |

[a]Note that the excess of water on land equals the deficit of water at sea.

# Humidity and Human Comfort

Why are we uncomfortable on a hot, muggy day? Many people ascribe to the old saying "It's not the heat, but the humidity" as the explanation for this discomfort. In fact, our physical comfort does depend, in part, on our body's ability to lose heat via evaporative cooling. This ability depends, in turn, on the vapor pressure gradient between our skin (the evaporative surface) and the surrounding air. The greater the vapor pressure of the ambient air (that is, the more humid the air), the smaller is the vapor pressure gradient. Consequently, less water evaporates from the skin surface. Hence, evaporative cooling decreases as air becomes more humid.

In light of the relationship between ambient vapor pressure and evaporative cooling, we can better understand why people often experience greater discomfort when the weather is both hot and humid. At air temperatures above 25 °C (77 °F), perspiring and consequent evaporative cooling promote heat loss from the body. If the air is also quite humid, the rate of evaporation is reduced, which hampers the body's ability to maintain a nearly constant core temperature (37 °C or 98.6 °F). In contrast, drier air facilitates evaporative cooling. Hence, at air temperatures higher than 25 °C (77 °F), we feel more comfortable when the air is dry than when the air is humid.

Scientists at the National Oceanic and Atmospheric Administration (NOAA) estimate that heat waves accompanied by high humidities kill approximately 150 people each year in the United States. High temperatures and humidities adversely affect everyone to some extent. As the weather causes increasing discomfort, people become more irritable and are less able to perform physical and mental tasks. During hot, humid weather, the efficiency of factory workers declines, and students do not concentrate as well on their studies.

In recent decades, scientists have developed a variety of indexes that gauge the combined effect of temperature and humidity on human comfort and advise people of the potential danger of heat stress. Since the summer of 1984, the National Weather Service has regularly reported the *heat index,* or *apparent temperature index,* developed by R. G. Steadman in 1979. As Table 1 illustrates, high temperature and high humidity combine to produce an apparent temperature that is considerably higher than the actual temperature. For example, a person exposed to an air temperature of 38 °C (100 °F) and a relative humidity of 50% experiences the same discomfort and stress as if the air temperature were 49 °C (120 °F) because the high humidity retards evaporative cooling of the skin.

**TABLE 1**
**Apparent Temperature Index**

| Relative humidity (%) | Air temperature (°F) | | | | | | | | | | |
|---|---|---|---|---|---|---|---|---|---|---|---|
| | 70 | 75 | 80 | 85 | 90 | 95 | 100 | 105 | 110 | 115 | 120 |
| | Apparent temperature (°F) | | | | | | | | | | |
| 0 | 64 | 69 | 73 | 78 | 83 | 87 | 91 | 95 | 99 | 103 | 107 |
| 10 | 65 | 70 | 75 | 80 | 85 | 90 | 95 | 100 | 105 | 111 | 116 |
| 20 | 66 | 72 | 77 | 82 | 87 | 93 | 99 | 105 | 112 | 120 | 130 |
| 30 | 67 | 73 | 78 | 84 | 90 | 96 | 104 | 113 | 123 | 135 | 148 |
| 40 | 68 | 74 | 79 | 86 | 93 | 101 | 110 | 123 | 137 | 151 | |
| 50 | 69 | 75 | 81 | 88 | 96 | 107 | 120 | 135 | 150 | | |
| 60 | 70 | 76 | 82 | 90 | 100 | 114 | 132 | 149 | | | |
| 70 | 70 | 77 | 85 | 93 | 106 | 124 | 144 | | | | |
| 80 | 71 | 78 | 86 | 97 | 113 | 136 | | | | | |
| 90 | 71 | 79 | 88 | 102 | 122 | | | | | | |
| 100 | 72 | 80 | 91 | 108 | | | | | | | |

Source: National Weather Service, NOAA.

The heat index can be divided into four categories based on the severity of potential impact on human well-being; these categories and associated heat stress symptoms are listed in Table 2. Apparent temperatures that fall in Category IV pose the least hazard. When apparent temperatures are between 27 and 32 °C (80 and 90 °F), fatigue is possible with prolonged exposure and physical activity. At the opposite end of the spectrum, Category I apparent temperatures pose the greatest danger. When apparent temperatures exceed 54 °C (130 °F), heatstroke (hyperthermia) is imminent. Keep in mind that the actual degree of stress experienced varies with age, general health, and body characteristics.

Where do people actually experience the most uncomfortable (and potentially life-threatening) summer weather? Given our discussion of the impact of high humidity on evaporative cooling, many of us would probably point to locales near the Gulf of Mexico where high temperature and high humidity are the norm. For example, during a summer heat wave, a typical temperature/relative humidity combination for New Orleans is 32 °C (90 °F)/60%. Referring to Tables 1 and 2, we see that such conditions translate into an apparent temperature of 100 °F, a category III hazard. Certainly, under such weather conditions, most of us would experience some discomfort and would need to closely monitor our exposure and level of physical activity.

Surprisingly, however, the most uncomfortable summer weather does not occur along the Gulf Coast. Rather we need to turn our attention to the deserts of Arizona and California. Here we find that although relative humidity values are considerably lower, much higher temperatures translate into greater heat stress. For example, residents of Phoenix, Arizona, typically experience summer afternoon temperatures approaching 43 °C (110 °F) with a relative humidity of perhaps 20%. This gives an apparent temperature of 44 °C (112 °F), a Category II hazard.

The examples of New Orleans and Phoenix illustrate that our perception of what contributes to heat stress often depends upon where we live. Residents of the eastern half of North America correctly associate greater discomfort with the increasingly muggy days of summer. At the same time, residents of the Southwest correctly ascribe their days of greatest discomfort to the relatively dry, but very hot conditions of the desert in summer.

Let us now turn our attention to the impact of very low humidity on human comfort. At very low humidity, we experience the apparent temperature as lower than the actual temperature (Table 1), another indication of the role of evaporative cooling. At low humidities, the ambient air feels cooler than it actually is. It is, therefore, desirable to raise indoor humidity during the winter heating season. Increasing the humidity makes us feel comfortable at a lower room temperature, and we can turn down the thermostat and save on fuel bills.

Thus far in our discussion of the combined impact of air temperature and humidity on human comfort, we have ignored the effects of radiative exchange between the body and its surroundings. Although the apparent temperature suggests comfortable conditions, discomfort might result from radiative heat loss to cold walls or windows or through radiative heat gain from hot surfaces. Our discussion has also neglected the role played by wind in promoting human comfort when conditions are hot and humid. Air in motion accelerates evaporative cooling and helps to transport heat away from the body. Hence, our sense of physical comfort depends on a variety of environmental factors.

**TABLE 2**
**Hazards Posed by Heat Stress by Range of Apparent Temperature[a]**

| Category | Apparent temperature | Heat syndrome |
|---|---|---|
| I | 54 °C or higher (130 °F or higher) | Heatstroke or sunstroke *imminent*. |
| II | 41 to 54 °C (105 to 130 °F) | Sunstroke, heat cramps, or heat exhaustion *likely*. Heatstroke *possible* with prolonged exposure and physical activity. |
| III | 32 to 41 °C (90 to 105 °F) | Sunstroke, heat cramps, or heat exhaustion *possible* with prolonged exposure and physical activity. |
| IV | 27 to 32 °C (80 to 90 °F) | Fatigue *possible* with prolonged exposure and physical activity. |

Source: National Weather Service, NOAA.
[a]Apparent temperature combines the effects of heat and humidity on human comfort.

We know from experience that humidity is quite variable. In this section, we examine some ways to quantify the water vapor concentration of air.

### VAPOR PRESSURE AND MIXING RATIO

Water vapor, just like all other gaseous components of the atmosphere, obeys Dalton's law. **Dalton's law** states that the total pressure of a mixture of gases equals the sum of the pressures exerted by each constituent gas. When water enters the atmosphere as vapor, water molecules readily disperse and mix with the other gases composing air and contribute to the total pressure exerted by the atmosphere. **Vapor pressure** is simply the pressure contributed by water vapor alone and is directly proportional to the concentration of water vapor in air.

From a practical perspective, the vapor pressure is very unlikely anywhere to be greater than about 40 mb. As noted in Chapter 1, no more than about 4% of the atmosphere's lowest kilometer is water vapor even in the warm, humid air over tropical oceans and rainforests. Hence, we can estimate the maximum possible vapor pressure to be 4% of the mean sea-level air pressure (1013.25 mb) or about 40 mb.

Alternatively, we can describe the water vapor concentration of air as the mass of water vapor per mass of dry air—usually expressed as grams of water vapor per kilogram of dry air. This measure of water vapor in air is called the **mixing ratio** because it specifies how much water vapor is mixed with the other atmospheric gases. Vapor pressure and mixing ratio are two of the more common ways of specifying the water vapor concentration of air and we use both in this book.

### SPECIFIC AND ABSOLUTE HUMIDITY

Other, less frequently used measures of the water vapor content of air are specific humidity and absolute humidity. **Specific humidity** is defined as the ratio of the mass of water vapor (in grams) to the mass (in kilograms) of the air containing the water vapor (that is, the combined mass of dry air plus water vapor). Specific humidity usually differs from the mixing ratio by no more than about 2%. **Absolute humidity** is defined as the mass of water vapor (in grams) per unit volume

of dry air (in cubic meters), equivalently, the vapor density of water. However, absolute humidity is not very useful for most meteorological purposes because the volume of a unit mass of air may change. A change in volume means a change in absolute humidity even though there is no gain or loss of water vapor.

## The Saturation Concept

Earlier in this chapter we noted that a two-way exchange of water molecules takes place at the interface between water and air (or between ice and air). Water molecules are in a continual state of flux between the liquid and vapor phases. During evaporation, more water molecules enter the vapor phase than return to the liquid phase, and during condensation, more water molecules return to the liquid phase than enter the vapor phase. Eventually, a state of dynamic equilibrium may develop such that liquid water becomes vapor at the same rate that water vapor becomes liquid. That is, the flux of water molecules is the same in both directions. At equilibrium, the vapor pressure is constant and is known as the **saturation vapor pressure.**

If we elevate the temperature, we upset this dynamic equilibrium at least temporarily. The higher the temperature, the greater is the average kinetic energy of individual molecules, and the more readily do water molecules escape the water surface as vapor. Initially, evaporation prevails. If there's a sufficient supply of water and if the water vapor is not continually carried away by winds, eventually a new equilibrium is established. That is, a new balance develops between the flux of water molecules becoming vapor and the flux of water molecules becoming liquid. Now, however, at a higher temperature than before, the water vapor concentration is greater so that the saturation vapor pressure is higher.

Ultimately then, the saturation vapor pressure depends on the rate of vaporization of liquid water, which, in turn, is regulated by temperature. The same argument applies to the **saturation mixing ratio.** Hence, as indicated in Figures 6.3 and 6.4 and Tables 6.3 and 6.4, the saturation vapor pressure and saturation mixing ratio increase as the temperature rises. As a rule of thumb, the saturation vapor pressure and saturation mixing

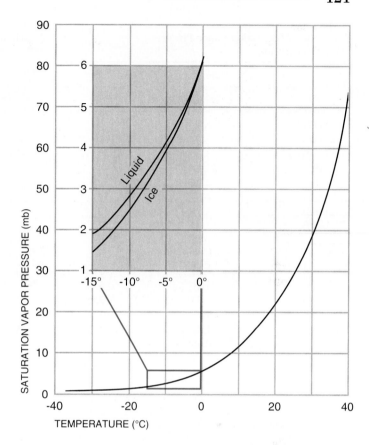

**FIGURE 6.3**
Variation of the saturation vapor pressure with air temperature.

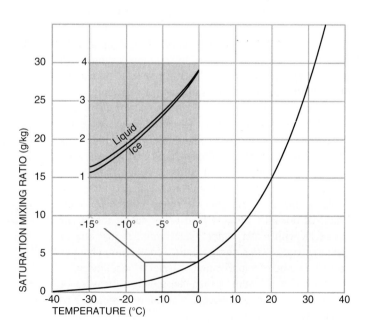

**FIGURE 6.4**
Variation of the saturation mixing ratio with air temperature.

**TABLE 6.3**
**Variation of Saturation Vapor Presure with Temperature**

| Temperature °C (°F) | Saturation vapor pressure (mb) | |
|---|---|---|
| | Over water | Over ice |
| 50 (122) | 123.40 | |
| 45 (113) | 95.86 | |
| 40 (104) | 73.78 | |
| 35 (95) | 56.24 | |
| 30 (86) | 42.43 | |
| 25 (77) | 31.67 | |
| 20 (68) | 23.37 | |
| 15 (59) | 17.04 | |
| 10 (50) | 12.27 | |
| 5 (41) | 8.72 | |
| 0 (32) | 6.11 | 6.11 |
| −5 (23) | 4.21[a] | 4.02[a] |
| −10 (14) | 2.86 | 2.60 |
| −15 (5) | 1.91 | 1.65 |
| −20 (−4) | 1.25 | 1.03 |
| −25 (−13) | 0.80 | 0.63 |
| −30 (−22) | 0.51 | 0.38 |
| −35 (−31) | 0.31 | 0.22 |
| −40 (−40) | 0.19 | 0.13 |
| −45 (−49) | 0.11 | 0.07 |

[a]Note that for temperatures below freezing, two different values are given: one over supercooled water and the other over ice. Supercooled water remains liquid at subfreezing temperatures.

**TABLE 6.4**
**Variation of Saturation Mixing Ratio with Temperature**

| Temperature °C (°F) | Saturation mixing ratio (g/kg) | |
|---|---|---|
| | Over water | Over ice |
| 50 (122) | 88.12 | |
| 45 (113) | 66.33 | |
| 40 (104) | 49.81 | |
| 35 (95) | 37.25 | |
| 30 (86) | 27.69 | |
| 25 (77) | 20.44 | |
| 20 (68) | 14.95 | |
| 15 (59) | 10.83 | |
| 10 (50) | 7.76 | |
| 5 (41) | 5.50 | |
| 0 (32) | 3.84 | 3.84 |
| −5 (23) | 2.64[a] | 2.52[a] |
| −10 (14) | 1.79 | 1.63 |
| −15 (5) | 1.20 | 1.03 |
| −20 (−4) | 0.78 | 0.65 |
| −25 (−13) | 0.50 | 0.40 |
| −30 (−22) | 0.32 | 0.24 |
| −35 (−31) | 0.20 | 0.14 |
| −40 (−40) | 0.12 | 0.08 |
| −45 (−49) | 0.07 | 0.05 |

[a]Note that for temperatures below freezing, two different values are given: one over supercooled water and the other over ice. Supercooled water remains liquid at subfreezing temperatures.

ratio double for about every 11-Celsius-degree (20-Fahrenheit-degree) rise in air temperature. At water's boiling point (100 °C at sea level), the saturation vapor pressure is the same as the surrounding air pressure.

The relationship between temperature and the saturation vapor pressure (or saturation mixing ratio) is popularly interpreted to imply that warm air can *hold* more water vapor than can cold air. That is, the atmosphere is likened to a sponge that can accommodate only so much water. This viewpoint can be misleading. Air does not literally hold water vapor; rather, air coexists with water vapor. Water vapor is just one of the many gases that compose air. Temperature changes affect the rate of vaporization of water. Hence, the saturation vapor pressure, rather than being a measure of air's *water-holding capacity,* is actually a measure of water's vaporization rate.

# Relative Humidity

Relative humidity is perhaps the most familiar way of describing the water vapor content of air. It is the water vapor measurement most often cited by television and radio weathercasters. **Relative humidity** compares the actual concentration of water vapor in the air with the concentration of water vapor in that same air at saturation. Relative humidity (RH) is expressed as a percentage and can be computed approximately from either the vapor pressure or the mixing ratio. That is,

$$RH = \frac{vapor\ pressure}{saturation\ vapor\ pressure} \times 100\%$$

or

$$RH = \frac{mixing\ ratio}{saturation\ mixing\ ratio} \times 100\%$$

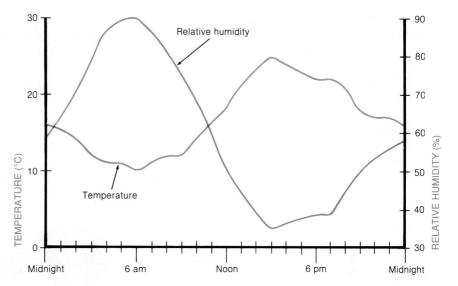

**FIGURE 6.5**
The variation of relative humidity on a day with no air mass advection (calm winds). The relative humidity varies inversely with the air temperature.

Note that when the actual concentration of water vapor in air is equal to the water vapor concentration at saturation, the relative humidity is 100%, that is, the air is saturated with respect to water vapor.

Consider an example of how relative humidity is computed. Suppose that the air temperature is 10 °C (50 °F), and the vapor pressure is 6.1 mb. From Table 6.3, we determine that at 10 °C the saturation vapor pressure of air is 12.27 mb. Using the formula above, we compute the relative humidity to be 49.7%, that is,

$$ RH = \frac{6.1 \text{ mb}}{12.27 \text{ mb}} \times 100\% = 49.7\% $$

Note that the relative humidity is directly proportional to the vapor pressure and inversely proportional to the saturation vapor pressure. Because the saturation vapor pressure is directly proportional to air temperature, the relative humidity is also inversely proportional to air temperature. The dependence of relative humidity on air temperature can be confusing because even if the actual concentration of water vapor in air does not change, the relative humidity rises or falls depending on how the air temperature changes.

On a clear and calm day, the air temperature usually rises from a minimum near sunrise to a maximum during early to midafternoon and falls off thereafter. If the vapor pressure (or mixing ratio) does not change through the day, then the relative humidity varies inversely with air temperature: the relative humidity is highest when the temperature is lowest and vice versa

(Figure 6.5). After sunrise, as the air warms, the relative humidity drops, but not because the air is drying out. The relative humidity decreases because rising air temperature means that the saturation vapor pressure increases.

In the previous example, the relative humidity responded only to variations in local air temperature; the wind was calm, and there was no air mass advection. The situation is more complicated when air mass advection occurs. Advection can influence both air temperature and vapor pressure. For example, when a warm and humid air mass replaces a cool and dry air mass, the relative humidity is affected by both higher temperatures and increasing vapor pressure.

Perhaps surprisingly, the mean annual water vapor content of the troposphere is about the same over the desert area of the southwestern United States as it is over the Great Lakes region. However, the relative humidity is significantly different in the two localities. Higher average air temperatures in the Southwest mean lower relative humidities and reduced likelihood of cloudiness and precipitation.

## Humidification

In public buildings as well as in individual homes, it is sometimes desirable to alter the relative humidity in the interest of human comfort. When indoor air is extremely dry, a humidifier adds water vapor to the air, and when indoor air is excessively muggy, a dehumidifier removes water vapor.

In winter, as the air that is drawn in from outdoors is heated by a furnace, its relative humidity declines—sometimes to uncomfortably low levels. For example, when outdoor air at −20 °C (−4 °F) and relative humidity of 50% is drawn indoors and heated to 20 °C (68 °F), the relative humidity drops to about 3%. At such low relative humidities, people may experience discomfort caused by irritation of mucous membranes in the nose and throat. House plants require more frequent watering, and wood furniture may crack and become unjointed.

One remedy for excessively low relative humidity indoors is a humidifier. A humidifier raises the relative humidity of indoor air to more comfortable levels by vaporizing water into the air and thereby increasing the vapor pressure (or mixing ratio). A common type of humidifier consists of a wheel that continually rotates a porous belt into and out of a reservoir of water. As a fan blows air through the wetted belt, water evaporates into the air stream.

On the other hand, in summer when the weather is hot and humid, a dehumidifier may be desirable. A dehumidifier lowers the relative humidity of indoor air to more comfortable levels by lowering the vapor pressure (or mixing ratio) by inducing condensation. In a standard dehumidifier, a fan draws humid air past cold refrigerated coils. The air is cooled to saturation, water vapor condenses on the cold coils, and liquid water drips from the coils into a collection reservoir.

## Humidity Measurement

Leonardo da Vinci, a creative genius of the fifteenth century, was probably the first to conceive of an instrument to gauge the water vapor concentration of air. His design was a simple balance. A small wad of dry cotton on one side of the scale was exactly balanced by a weight on the other side of the scale. As the cotton absorbed water vapor from the air, an imbalance would develop, the amount of imbalance being a measure of humidity.

Today, an instrument commonly used to measure atmospheric humidity is the **psychrometer.** It consists of two identical mercury-in-glass thermometers mounted side by side, as shown in Figure 6.6. The bulb of one thermometer is wrapped in a muslin wick. Readings are taken by first soaking the wick in distilled water, and then whirling both thermometers (or aerating them with a fan) until the wet-bulb thermometer reading steadies. The dry-bulb thermometer measures the actual air temperature. Water vaporizes from the muslin wick into air streaming past the wet-bulb thermometer, and evaporative cooling lowers the wet-bulb temperature. The

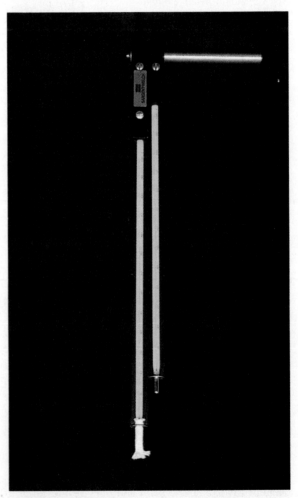

**FIGURE 6.6**
A sling psychrometer measures atmospheric humidity. Using a psychrometric table (inside back cover), we can determine the relative humidity from the air temperature and the difference between the air temperature and the wet-bulb temperature (wet-bulb depression). [Photograph by J. M. Moran]

**FIGURE 6.7**
A hydrograph provides a continuous
trace of relative humidity variations.
[Courtesy of Qualimetrics]

drier the air, the greater the evaporation, and the lower the reading will be on the wet-bulb thermometer compared to the dry-bulb reading. The temperature difference between the two thermometers, known as the **wet-bulb depression,** is calibrated in terms of percent relative humidity on a psychrometric table (see inside back cover).

For example, if the air temperature (dry-bulb reading) is 20 °C (68 °F), and the wet-bulb depression (dry-bulb reading minus wet-bulb reading) is 5 Celsius degrees (9 Fahrenheit degrees), the relative humidity is 58%. If there is no wet-bulb depression, the air is saturated; that is, the relative humidity is 100%.

An instrument that is less accurate than a psychrometer, but measures atmospheric humidity directly and conveniently without the use of tables, is the **hair hygrometer.** As the name implies, this device uses hair as the humidity sensor. Hair lengthens slightly as the relative humidity increases and shrinks slightly as the relative humidity drops. Hair typically changes length by about 2.5% over the full range of relative humidity from 0 to 100%. Usually, a sheaf of blond human hairs is linked mechanically to a pointer on a dial that is calibrated to read in percent relative humidity.

A hair hygrometer may be designed to drive a pen on a clock-driven drum, as shown in Figure 6.7. This instrument, called a **hygrograph,** traces a continuous record of fluctuations in relative humidity with time.

The type of hygrometer used in radiosondes (described in Chapter 1) is based on changes in the electrical resistance of certain chemicals as they adsorb* water vapor from the air. The adsorbing element may be a thin carbon coating on a glass or plastic strip. The more humid the air, the more moisture that is adsorbed, and the greater is the change in the resistance to an electric current passing through the sensing element. Variations in electrical resistance are calibrated in terms of percent relative humidity.

## Achieving Saturation

The relative humidity is 100% at saturation, but it is possible for air to be *supersaturated,* that is, for the relative humidity to exceed 100%. As the relative humidity approaches saturation, however, condensation or deposition of water vapor becomes more and more

---

*Adsorption refers to the adhesion of a gas or soluble substance to the *surface* of some object.

likely. Clouds are products of condensation or deposition within the atmosphere. Hence, the probability of cloud development increases as the relative humidity approaches saturation.

What, then, causes the relative humidity to increase? Relative humidity increases

1. When air is cooled; the saturation vapor pressure declines while the vapor pressure remains constant.
2. When water vapor is added to the air; the vapor pressure increases while the saturation vapor pressure remains constant.

In this chapter we focus on the first process, and in the next chapter we give some examples of the second process.

## EXPANSIONAL COOLING AND COMPRESSIONAL WARMING

One way whereby air cools and increases in relative humidity is by expansion. In fact, expansional cooling is the principal means of cloud formation in the atmosphere. According to the gas law (Chapter 5), as the pressure on an air parcel* decreases, both the temperature and density decrease. That is, air cools as it expands, a process known as **expansional cooling.** We experience expansional cooling when we let the air out of a tire; the air is cool to the touch. Within the tire, air is under considerably greater pressure than when it flows through the open tire valve and into the atmosphere. In expanding, the air does work as it pushes aside the air that formerly occupied the volume into which it expands. Work requires energy and the energy used in the work of expansion is drawn from the kinetic molecular energy (heat) of the air. Hence, the temperature of the expanding volume of air drops.

We saw in Chapter 5 that air pressure falls with altitude. Hence, as an air parcel ascends in the atmosphere, it expands in the same way as a helium-filled balloon drifting skyward. As an air parcel rises and expands, its temperature drops; if it is unsaturated, the air parcel's relative humidity increases.

Conversely, the gas law predicts that as the pressure

*An *air parcel* is a unit mass (say, 1 g) of air.

on an air parcel increases, both the temperature and density increase. That is, as air is compressed, it warms, a process known as **compressional warming.** A familiar example is the warming of the cylinder wall of a bicycle tire pump as air is pumped (compressed) into a tire. The work of compressing air is converted into heat. Hence, as an air parcel descends within the atmosphere, it is compressed, its temperature rises, and its relative humidity drops.

In a more general sense, we can imagine that upward and downward currents of air are composed of continuous streams of air parcels. The behavior of an air current is the same as that of a component air parcel. In summary, then, ascending unsaturated air cools and its relative humidity increases, and descending air warms and its relative humidity decreases.

## ADIABATIC PROCESSES

During an **adiabatic process,** no heat is exchanged between a mass and its environment. Within the atmosphere, expansional cooling and compressional warming of unsaturated air are adiabatic processes. That is, during its ascent or descent within the atmosphere, an air parcel is neither heated nor cooled by radiation, conduction, phase changes of water, or mixing with its surroundings.

Cooling of ascending unsaturated air is due to the work of expansion only and amounts to 10 Celsius degrees per 1000 m (5.5 Fahrenheit degrees per 1000 ft) of ascent. This is the **dry adiabatic lapse rate** (Figure 6.8).* On the other hand, the atmosphere compresses descending air, and the air temperature rises 10 Celsius degrees for every 1000 m (5.5 Fahrenheit degrees per 1000 ft) of descent. Thus, the dry adiabatic lapse rate enables us to accurately predict the temperature change of unsaturated air as it moves vertically within the atmosphere. For more on energy conservation and the adiabatic process, refer to the Mathematical Note at the end of this chapter.

Should rising air parcels cool to the point that the relative humidity approaches 100% and condensation or deposition takes place, the ascending air then no

*The *dry* designation in *dry adiabatic lapse rate* can be misleading. Ascending and descending air parcels contain some water vapor.

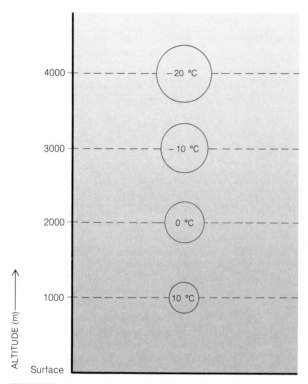

**FIGURE 6.8**

As an unsaturated parcel of air ascends within the atmosphere, it expands and cools at the dry adiabatic lapse rate (10 Celsius degrees per 1000 m of ascent).

longer cools at the dry adiabatic rate (Figure 6.9). This is because latent heat that is liberated during condensation or deposition (Chapter 4) *partially* counters expansional cooling. Consequently, an ascending saturated (cloudy) air parcel cools more slowly than an ascending unsaturated (clear) air parcel. Rising saturated air parcels cool at the **moist adiabatic lapse rate.**

Although the dry adiabatic lapse rate is fixed in value, the moist adiabatic lapse rate varies with temperature. As noted earlier, temperature governs the rate of vaporization of water so that the saturation vapor pressure (or saturation mixing ratio) increases with rising temperature. That is, warm saturated air has a higher vapor pressure (or mixing ratio) than cool saturated air. Thus, condensation or deposition in warm saturated air releases more latent heat than condensation or deposition in cool saturated air. The greater the quantity of latent heat released, the more

the expansional cooling is offset, and the smaller is the moist adiabatic lapse rate. The moist adiabatic lapse rate ranges from about 4 Celsius degrees per 1000 m (2.2 Fahrenheit degrees per 1000 ft) for very warm saturated air to almost 9 Celsius degrees per 1000 m (5 Fahrenheit degrees per 1000 ft) for very cold saturated air. For convenience, we use an average value of 6 Celsius degrees per 1000 m (3.3 Fahrenheit degrees per 1000 ft) for the moist adiabatic lapse rate.

Regardless of temperature, the relative humidity of ascending saturated (cloudy) air remains constant at about 100%. As the air expands and cools, however, its saturation mixing ratio declines and so does its mixing ratio. That is, some water vapor is converted to water droplets or ice crystals. Consider an illustration. Saturated air at 10 °C (50 °F) has a mixing ratio of about 7.8 g/kg (Table 6.4). If that air is lifted 1000 m, it cools moist adiabatically to 4 °C (39 °F). At that temperature, the saturation mixing ratio is about 5.1 g/kg. Because the relative humidity is still 100%, the mixing ratio is also about 5.1 g/kg. This means that 2.7 g/kg of water vapor (7.8 − 5.1 g/kg) was converted to cloud droplets via condensation.

As long as an unsaturated air parcel continues to ascend, its temperature will drop and its relative humidity will approach saturation. Forces arising from density differences within the atmosphere may either enhance or suppress this vertical motion of air. The net effect depends on the stability of the atmosphere.

## Atmospheric Stability

An air parcel is subject to buoyant forces that arise from density differences between the parcel and the surrounding *(ambient)* air. The higher the temperature of an air parcel, the lower is its density. Parcels that are warmer (lighter) than the ambient air tend to rise, and parcels that are cooler (denser) than the ambient air tend to sink. An air parcel continues to rise or sink until it reaches air of equivalent temperature (and density).

We determine **atmospheric stability** by comparing the temperature change of an ascending or descending

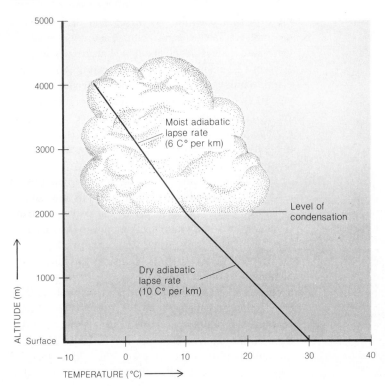

**FIGURE 6.9**
Rising parcels of unsaturated air cool at the dry adiabatic lapse rate, but as the rising parcels cool, their relative humidity increases. At the level of condensation, the relative humidity is 100%; that is, the parcels are saturated. Continued ascent of the saturated (cloudy) parcels triggers condensation (or deposition) and release of latent heat. The latent heat partially offsets the adiabatic cooling so that rising saturated air parcels do not cool as rapidly as rising unsaturated air parcels.

air parcel with the temperature profile, or sounding, of the ambient air layer in which the parcel ascends or descends. As we have seen, the cooling rate of a rising air parcel depends on whether the parcel is saturated (moist adiabatic lapse rate) or unsaturated (dry adiabatic lapse rate), and the warming rate of a descending air parcel is 10 Celsius degrees per 1000 m. Recall from Chapter 1 that a balloon-borne radiosonde monitors the vertical temperature profile (or sounding).

Within a **stable air layer,** an ascending air parcel becomes cooler (denser) than the ambient air, and a descending air parcel becomes warmer (less dense) than the ambient air (Figure 6.10). Hence, any upward or downward displacement of an air parcel in stable air gives rise to forces that will tend to return the parcel to its original altitude. Within an **unstable air layer,** on the other hand, an ascending air parcel remains warmer (less dense) than the ambient air and continues to ascend, or a descending air parcel remains cooler (denser) than the ambient air and continues to descend (Figure 6.11).

Although lapse rates for both unsaturated and saturated air parcels are essentially fixed, the temperature profile—and hence atmospheric stability—changes sig-

**FIGURE 6.10**
Upward and downward displacements of an unsaturated air parcel within stable air. The parcel is subject to a force that restores it to its original altitude. Hence, stable air inhibits vertical motion.

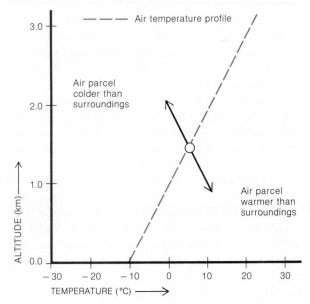

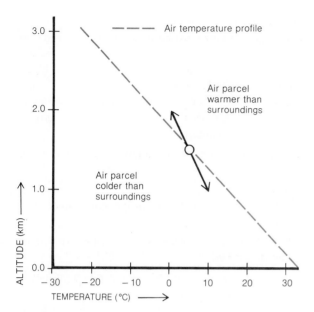

**FIGURE 6.11**
Upward and downward displacements of an unsaturated air parcel within unstable air. The parcel accelerates away from its original altitude. Hence, unstable air enhances vertical motion.

nificantly from season to season, from day to day, and even from one hour to the next. Stability can change as a consequence of (1) local radiational heating or cooling, (2) air mass advection, or (3) large-scale ascent or descent of air.

On a clear and calm night, radiational cooling of the ground stabilizes the overlying air, whereas during the day, intense solar heating of the ground destabilizes the overlying air. This situation is illustrated in Figure 6.12. An air mass is stabilized as it travels over a colder surface (snow-covered ground, for example), and it is destabilized as it flows over a warmer surface. Whether by radiation or by advection, stability changes because the temperature profile (sounding) changes. Air is stabilized when cooling from below makes the sounding steeper, and air is destabilized when heating from below makes the sounding less steep.

Generally, an air layer becomes more stable when it descends (subsides) and less stable when it ascends. When air subsides, the upper portion undergoes more compressional warming than does the lower portion so that the sounding becomes steeper and air layer

**FIGURE 6.12**
(A) Nocturnal radiational cooling stabilizes the overlying air by dawn. (B) By noon, however, bright sunshine has warmed the ground and destabilized the air.

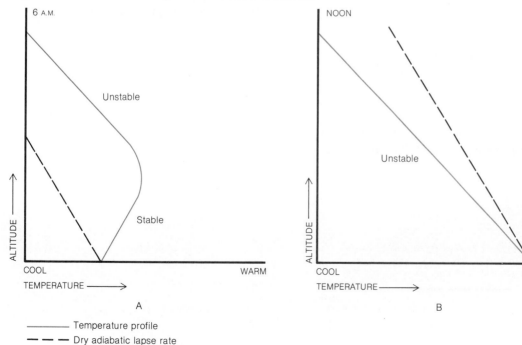

stability increases. The lower portion of an ascending air layer is usually more humid than the upper portion and so achieves saturation sooner. Consequent release of latent heat in the lower portion of the air layer makes the sounding less steep and reduces the stability.

A further complication is that stability can change with altitude. For example, as shown in Figure 6.13, a stable air layer may be sandwiched between unstable air layers. Different types of air mass advection at different altitudes within the troposphere are often responsible for such variations in stability.

Figure 6.14 summarizes a number of possible atmospheric stability conditions, along with the dry adiabatic and average moist adiabatic lapse rates. If the sounding indicates that the temperature of the ambient air is dropping more rapidly with altitude than the dry adiabatic lapse rate (that is, more than 10 Celsius degrees per 1000 m), then the ambient air is unstable for both saturated and unsaturated air parcels. This situation is called **absolute instability.**

If the sounding lies between the dry adiabatic and moist adiabatic lapse rates, **conditional stability** prevails. That is, the air layer is stable for unsaturated air parcels and unstable for saturated air parcels. This situation is relatively common. Conditional stability means that unsaturated air must be forced upward in order to reach saturation. But once saturation is achieved, the cloudy air will be buoyed upward, that is, the cloud will build vertically.

An air layer is stable for both saturated and unsaturated air parcels when the sounding indicates any of the following conditions.

**1.** The temperature of the ambient air drops more slowly with altitude than the moist adiabatic lapse rate.

**2.** The temperature does not change with altitude (*isothermal*).

**3.** The temperature increases with altitude (*temperature inversion*).

Any one of these three types of temperature profiles is a case of **absolute stability.**

What happens when a sounding coincides with either the dry or moist adiabatic lapse rate? A sounding that equals the dry adiabatic lapse rate is neutral for unsaturated air parcels and unstable for saturated air parcels. A sounding that is the same as the moist adiabatic lapse rate is neutral for saturated air parcels and

**FIGURE 6.13**
Air stability may change with altitude. In this example, a stable air layer (a temperature inversion) is sandwiched between unstable air layers.

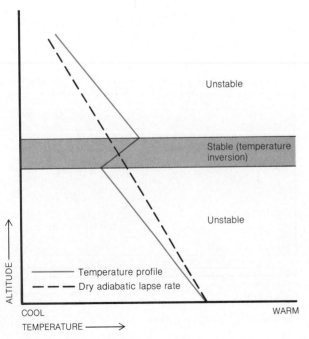

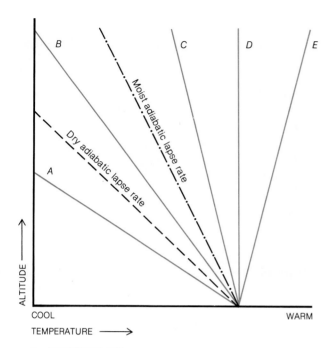

A    Absolute instability
B    Conditional stability
C    Absolute stability (lapse)
D    Absolute stability (isothermal)
E    Absolute stability (inversion)

**FIGURE 6.14**
Air stability is determined by comparing vertical temperature profiles (red lines) with the dry adiabatic lapse rate in the case of unsaturated (clear) air parcels and with the moist adiabatic lapse rate in the case of saturated (cloudy) air parcels. Here a number of possible stability cases are illustrated.

stable for unsaturated air parcels. Within a **neutral air layer,** a rising or descending air parcel always has the same temperature (and density) as its surroundings. Hence, a neutral air layer neither impedes nor spurs upward or downward motion of air parcels.

It is evident that atmospheric stability influences weather by affecting the vertical motion of air. Stable air suppresses vertical motion, and unstable air enhances vertical motion, convection, expansional cooling, and cloud development. Because stability also affects the rate at which polluted air mixes with clean air, stability must be considered in assessing air pollution potential. We examine this point in Chapter 17.

# Lifting Processes

Ascending air cools and its relative humidity increases. With sufficient ascent and expansional cooling, the relative humidity approaches 100% and condensation or deposition takes place, that is, clouds form. What, then, causes air to rise? Air rises (1) as the ascending branch of a convection current, (2) along the surface of a warm or cold front, (3) up the slopes of a hill or mountain, or (4) as the consequence of convergence. As described in the Special Topic "Clouds by Mixing," saturation and cloud development are also possible when unsaturated air masses (or air layers) mix.

As we saw in Chapter 4, convection is an important means of heat transfer that develops when the sun heats the Earth's surface, which then heats the overlying air. Heated air rises, expands, and cools. Eventually, the air becomes so cool and dense that it sinks back toward the surface where it is heated again. Cumulus clouds may form where convection currents ascend, and the sky is cloud-free where convection currents descend (Figure 6.15). In general, the higher the altitude reached by convection currents, the greater the expansional

**FIGURE 6.15**
The sky is partly covered with cumulus clouds. Convection currents are upward where there are clouds, and convection currents are downward where the sky is cloud free. [Photograph by J. M. Moran]

SPECIAL TOPIC

# Clouds by Mixing

Clouds (or fog) develop as the consequence of cooling caused by uplift of air or, as we will see in Chapter 7, by contact with a surface that has been chilled by emission of IR radiation. A third process of cloud formation involves mixing of air masses (or air layers within the same air mass) that differ in temperature and vapor pressure. Surprisingly, a cloud may form even though the two air masses (or air layers) are unsaturated to begin with.

As emphasized earlier, saturation vapor pressure increases rapidly with rising temperature. This relationship is displayed schematically in Figure 1 where temperature is plotted on the horizontal axis and vapor pressure is plotted on the vertical axis. The average temperature and average vapor pressure of a specific air mass (or air layer) plot as a single point on that diagram. An air mass that plots below the saturation vapor pressure curve is unsaturated, on the curve is saturated, and above the curve is supersaturated.

Suppose that two different unsaturated air masses plot as points $A$ and $B$ on the diagram. It is reasonable to assume that mixing the two air masses would produce a new air mass with properties (average temperature and vapor pressure) that plot somewhere along a straight line connecting points $A$ and $B$. The precise location of that point (the mixture) along the line depends on the relative volume of the two air masses. Note that, in this case, the mixture of the two air masses is unsaturated so that no cloud forms.

Consider, however, two other unsaturated air masses, $C$ and $D$, one cold and dry, the other warm and humid. In this case, the straight line linking $C$ and $D$ intersects the saturation vapor pressure curve. This means that the new air mass resulting from the mixing of air masses $C$ and $D$ is saturated and a cloud forms.

This description of cloud formation by mixing provides insight as to the cause of aircraft contrails (Figure 2). Jet engine exhaust mixes with the ambient (surrounding) air and may form a *contrail*, that is, a condensation trail, behind the aircraft. Heat and water are combustion products of jet engines so that the exhaust is hot and humid. Turbulence in the wake of a jet engine facilitates the mixing of exhaust with the ambient air. If the ambient air has the right combination

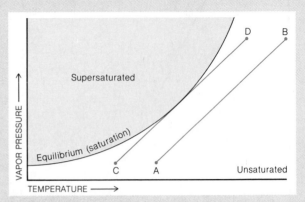

**FIGURE 1**
Variation of saturation vapor pressure with temperature. Air masses that plot as points above the curve are supersaturated, on the curve are saturated, and below the curve are unsaturated.

of low vapor pressure and low temperature, the mixture is saturated and a contrail forms. Such conditions are most likely in the upper troposphere where commercial jetliners travel.

All of us probably have experienced another example of cloud formation by mixing on a cold winter day. On such a day, as we exhale, we can "see our breath." What we actually see is a small cloud formed by the mixing of our warm, humid breath with the colder and drier ambient air.

**FIGURE 2**
Jet contrails form when hot, humid exhaust gases mix with cold, dry air. [Photograph by J. M. Moran]

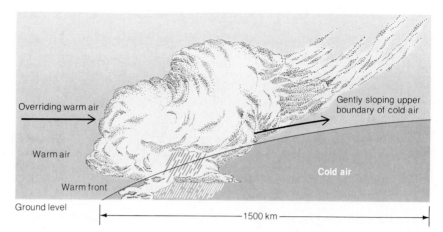

**FIGURE 6.16**
Warm, light air displaces cooler, denser air by overriding the cool air along a gently sloping frontal surface (a warm front). The rising warm air expands and cools to saturation, and clouds develop.

cooling, and the more likely it is that clouds and precipitation will form. Convection currents that soar to great altitudes within the troposphere typically spawn thunderstorms (cumulonimbus clouds).

Clouds and precipitation are often triggered by frontal uplift, which occurs when contrasting air masses meet. A **front** is a narrow zone of transition between two air masses that differ in temperature and/or humidity. A warm and humid air mass is less dense than a cold and dry air mass and thus, displaces cold air by riding up and over it (Figure 6.16). The leading edge of the advancing warm air is known as a **warm front.** In contrast, cold and dry air displaces warm and humid air by sliding under it and forcing it upward (Figure 6.17). The leading edge of the advancing cold air is known as a **cold front.** The net effect of the replace-

ment of one air mass by another air mass is the lifting of air, which, in turn, leads to expansional cooling, cloud development, and perhaps rainfall or snowfall. Hence, clouds and precipitation are associated with fronts. We have much more to say about fronts and associated weather in Chapter 11.

**Orographic lifting** occurs when air is forced upward by topography, the physical relief of the land. As winds sweep across the landscape, hills and valleys cause the moving air to alternately ascend and descend. If relief is sufficiently great, the resulting expansional cooling and compressional warming of air affects the development of clouds and precipitation. For example, a mountain range that is oriented perpendicular to prevailing winds forms a natural barrier that results in a cloudier and wetter climate on one side of the range

**FIGURE 6.17**
Cold, dense air displaces warmer, lighter warm air by sliding under the warm air, thereby forcing the warm air to rise along a steeply sloping frontal surface (a cold front). The rising warm air expands and cools to saturation, and clouds develop.

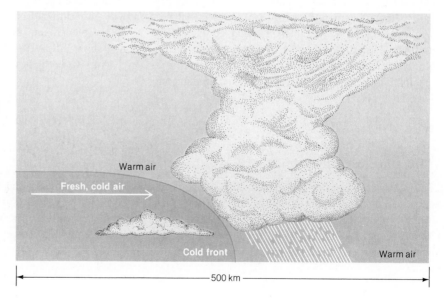

Moist windward slope

Dry leeward slope

**FIGURE 6.18**
A mountain range that intercepts a flow of humid air may induce a cloudy, rainy climate on its windward slopes and a dry climate on its leeward slopes and beyond.

than on the other side (Figure 6.18). As air is forced to rise along *windward* slopes (facing the wind), it expands and cools, which increases its relative humidity. With sufficient cooling, clouds and precipitation develop. Meanwhile, on the mountain's *leeward* slopes (downwind side), air descends and warms, which reduces its relative humidity, so that clouds and precipitation are less likely. In this way, mountain ranges induce two contrasting climatic zones: a moist climate on the windward slopes and a dry climate on the leeward slopes. Dry conditions often extend many hundreds of kilometers to the lee of a prominent mountain range; this region is known as a **rain shadow.**

An orographically induced contrast in precipitation is especially apparent from west to east across Washington and Oregon, where the north–south Cascade Range intercepts the prevailing flow of humid air from the Pacific Ocean. Exceptionally rainy conditions prevail in the western portions of those states, whereas semiarid conditions characterize much of the eastern portions. Figure 6.19 shows two contrasting Pacific Northwest landscapes that underscore the disparity in precipitation. Markedly different plant and animal communities live on the windward versus leeward slopes of a mountain barrier. This climatic contrast also affects the domestic water supply, the crops that can be grown, and the type of human shelter required. For example, Denver and nearby communities on the leeward side of the Colorado Rocky Mountains must import water from the wetter, western slopes via tunnels through the mountains.

As we saw in Chapter 5, when winds near the Earth's surface converge, upward motion of air is the consequence. Upward motion means expansional cooling, increasing relative humidity, and eventually cloud and precipitation formation. In later chapters, we describe several weather systems in which convergence plays an important role in triggering stormy weather.

Note that fronts, topography, and converging winds can force stable unsaturated air upward to the point that clouds develop. If the air is conditionally stable, clouds will surge upward and precipitation may be heavy. If, on the other hand, the air is absolutely stable, clouds will show little vertical development and precipitation is likely to be light if it ocurrs at all.

**WEATHER FACT**

## The Rainiest Place on Earth

The windward slopes of Mount Waialeale on the northeast coast of the island of Kauai in the Hawaiian chain is the rainiest location on Earth. A weather station at an elevation of 1569 m (5142 ft) receives an average annual rainfall of 1199 cm (39.3 ft). Heavy rainfall is triggered by orographic lifting of warm, humid air borne by persistent northeast trade winds. By contrast, on the leeward slopes of the same volcanic mountain, annual rainfall averages less than 50 cm (20 in.).

A

B

**FIGURE 6.19**
The luxuriant vegetation of the Hoh Rain Forest in the Olympic National Park of northwestern Washington (A) is in marked contrast to the cold desert vegetation of eastern Oregon (B). Mountain ranges between the two areas account for the striking difference in climate. [Photographs by Jack W. Dykinga (A) and Chris Migdol (B)]

# Conclusions

The atmosphere is one of many reservoirs of water in the global hydrologic cycle. Water vapor enters the atmosphere via evaporation of liquid water, transpiration by plants, and sublimation of ice and snow. We express the concentration of water vapor in air as vapor pressure, mixing ratio, or relative humidity. The saturation vapor pressure and saturation mixing ratio increase with rising air temperature so that the water vapor concentration is greater in saturated warm air than in saturated cold air. On the other hand, relative humidity is inversely proportional to air temperature.

Expansional cooling is the principal means whereby air approaches saturation and clouds form. Ascending air expands and cools, its relative humidity approaches 100%, and water vapor condenses (or deposits) as tiny water droplets (or ice crystals) that are visible as clouds. Conversely, descending air is compressed and warms, its relative humidity drops, and clouds vaporize or fail to develop.

Ascending air cools at the dry adiabatic lapse rate if unsaturated and at the moist adiabatic lapse rate if saturated. Atmospheric stability influences vertical motion in the atmosphere such that stable air suppresses ascent of air and cloud development whereas unstable air enhances ascent of air and cloud development. Atmospheric stability is determined by comparing the cooling rate of ascending air parcels with the temperature profile of the ambient air through which the air parcels ascend. Air ascends along fronts, in

convection currents, and up the slopes of mountain ranges.

Radiational cooling and local increases in water vapor concentration also raise the relative humidity to saturation and cause the formation of dew, frost, or fog. We explore these processes in Chapter 7, along with the mechanics of cloud development and the classification of clouds.

## MATHEMATICAL NOTE

# Energy Conservation and the Dry Adiabatic Process

Energy cannot be created or destroyed, but it can change from one form to another. Another way to express this *law of energy conservation* is to observe that in any physical or biological system we can account for all the original energy, regardless of the energy transformations that take place. Application of this law to atmospheric processes provides us with valuable information about the workings of weather.

Heat gained by an air parcel is either added to the parcel's store of internal energy or is used to do work on the parcel. Conversely, heat released by an air parcel is either subtracted from the internal energy of the parcel or is the consequence of the parcel's work on its surroundings. In order to understand this concept, let us first examine separately (1) internal energy and (2) work done on or by an air parcel. We will then combine the two in one expression of energy conservation.

*Internal energy* refers to the total kinetic molecular energy of the mixture of gases composing an air parcel, that is, *heat*. A change in the internal energy, or heat energy ($\Delta Q$), of an air parcel is therefore proportional to a change in the temperature ($\Delta T$) of the parcel. That is,

$$\Delta Q \propto \Delta T$$

As we saw in Chapter 3, however, internal energy and temperature are related through the specific heat $(C)$. So,

$$\Delta Q = C\Delta T$$

Actually, the specific heat of dry air can be evaluated at constant pressure ($C_p = 1005$ J per kg per K) or, as in this case, at constant volume ($C_v = 718$ J per kg per K). Hence,

$$\Delta Q = C_v \Delta T$$

Work done on or by an air parcel consists of compression or expansion of the air parcel. An illustration will demonstrate this concept. Suppose we have a sealed cylindrical container of air, as in Figure 1, equipped at one end with a piston that can compress the air sample or allow the air sample to

**FIGURE 1**
A piston is used to compress a sample of air or to allow the air sample to expand. Pressure *(P)* is applied over a distance ($\Delta D$). The piston has cross-sectional area A.

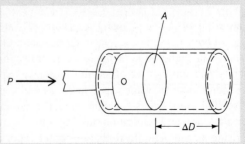

expand. If we compress the air, we use energy to do work on the air sample. If we then release the piston, the air sample expands and works against (pushes) the piston. When the air is compressed, work is therefore done on it, but when the air expands, the air does work on its surroundings. Energy is either supplied (during compression) or released (during expansion).

We can calculate the work (or energy) involved in the compression or expansion of air by referring to the piston example. The energy required to compress or expand the air sample is simply the product of the required force *(F)* times the distance over which the force is applied. Recall that pressure is a force per unit area. If we apply a pressure *(P)* on an air sample by moving a piston of fixed cross-sectional area *(A)* a variable distance *(ΔD)*, the volume *(ΔV)* of the air sample changes.

$$\Delta V = A\Delta D$$

The energy required *(ΔQ)* for the volume change would be

$$\Delta Q = F\Delta D$$

or

$$\Delta Q = PA\Delta D$$

or

$$\Delta Q = P\Delta V$$

To this point, we have considered internal energy and the energy involved in the compression and expansion of air separately. We now combine the two components in one equation.

$$\Delta Q = C_v \Delta T + P\Delta V$$

The heat flow into and out of an air parcel is thus accounted for by changes in internal energy, by the work of expansion or compression of air, or by both.

With this statement of the law of energy conservation, we can now explore the adiabatic process. According to the *adiabatic assumption,* there is no net heat flow through the imaginary walls of an air parcel. That is, in the previous equation,

$$\Delta Q = 0$$

Hence,

$$C_v \Delta T + P\Delta V = 0$$

This means that when an air parcel undergoes an adiabatic process, a temperature change *(ΔT)* accompanies a volume change *(ΔV)*. The temperature of an air parcel drops when its volume increases as when the air parcel ascends within the atmosphere. Conversely, the temperature of an air parcel rises when its volume decreases as when the air parcel descends within the atmosphere. Ascending unsaturated air expands and cools adiabatically at a constant rate of 10 Celsius degrees per 1000 m of ascent. This is called the *dry adiabatic lapse rate.* Descending air is compressed and warms at the rate of 10 Celsius degrees per 1000 m of descent.

Adiabatic processes can be displayed graphically on a *Stüve thermodynamic diagram* like the one in Figure 2. The Stüve diagram presents the relationships among several atmospheric variables. Horizontal lines are *isobars,* lines of constant air pressure, and vertical lines are *isotherms,* lines of constant temperature. Straight red sloping lines are *dry adiabats,* the temperature change of an unsaturated air parcel that is subjected to adiabatic expansion or compression. Dry adiabats are labeled in kelvins (K). Curved dashed lines are *moist adiabats,* the temperature change of a saturated air parcel that is subjected to expansional cooling. Black sloping lines are lines of equal saturation mixing ratio in grams of water vapor per kilogram of dry air.

As an illustration of the usefulness of a Stüve diagram, consider an unsaturated air parcel that undergoes dry adiabatic expansion as a consequence of uplift (along a front, for example). Initially, the air parcel has a temperature of 20 °C (68 °F), a pressure of 1000 mb, and a relative humidity of 50%. These conditions are plotted as point *A* in Figure 2. On the diagram, we see that the saturation mixing ratio is about 16 g/kg, so that a relative humidity of 50% means that the mixing ratio of the parcel is 8 g per kg. The parcel is then lifted so that it cools dry adiabatically until it becomes saturated. The air parcel temperature thus follows the 293 K dry adiabat to point *B,* where the actual mixing ratio equals the saturation mixing ratio (8 g/kg) and the relative hu-

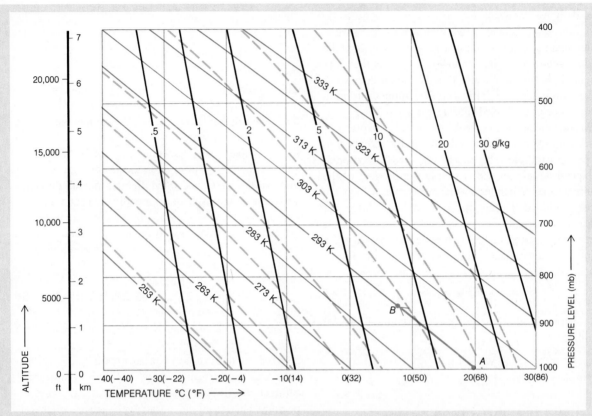

**FIGURE 2**
On a Stüve thermodynamic diagram, an air parcel at point *A* is subjected to a dry adiabatic expansion to point *B*.

midity is 100%. Hence, saturation is achieved when the air parcel reaches an altitude of about 1500 m (4900 ft), where the parcel temperature is 8 °C (46 °F). At this altitude, water vapor in the air parcel begins to con-

dense and a cloud forms. With continued ascent above this altitude, the saturated (cloudy) air parcel follows amoist adiabat.

## Key Terms

| | | | |
|---|---|---|---|
| hydrologic cycle | global water budget | relative humidity | stable air layer |
| evaporation | Dalton's law | psychrometer | unstable air layer |
| condensation | vapor pressure | wet-bulb depression | absolute instability |
| transpiration | mixing ratio | hair hygrometer | conditional stability |
| evapotranspiration | specific humidity | hygrograph | absolute stability |
| sublimation | absolute humidity | expansional cooling | neutral air layer |
| deposition | saturation vapor pressure | compressional warming | front |
| precipitation | saturation mixing ratio | adiabatic process | warm front |
| | | dry adiabatic lapse rate | cold front |
| | | moist adiabatic lapse rate | orographic lifting |
| | | atmospheric stability | rain shadow |

## Summary Statements

☐ The hydrologic cycle is the ceaseless circulation of a fixed quantity of water among the Earth's oceanic, atmospheric, and terrestrial reservoirs.

☐ Transfer of water between the Earth's surface and the atmosphere involves the processes of evaporation, condensation, transpiration, sublimation, deposition, and precipitation.

☐ The global water budget implies that there is a net flow of water from land to sea. Precipitation falling on land vaporizes, infiltrates the ground, or runs off as rivers or streams.

☐ Vapor pressure and mixing ratio are two direct measures of the water vapor concentration of air. Relative humidity indicates how close air is to saturation and is temperature dependent.

☐ Saturation vapor pressure and saturation mixing ratio represent an equilibrium in the flux of water molecules entering and leaving the vapor phase. Both parameters increase with rising temperature because temperature regulates the rate of vaporization of water.

☐ When the water vapor concentration of air equals the water vapor concentration at saturation, the relative humidity is 100%. For unsaturated air, the relative humidity varies inversely with air temperature. Hence, on a day with no air mass advection, the relative humidity is highest when the air temperature is lowest.

☐ The psychrometer is a standard instrument for measuring humidity. The greater the wet-bulb depression, the lower is the relative humidity. When the wet-bulb and dry-bulb temperatures are equal, the relative humidity is 100%.

☐ The relative humidity of unsaturated air increases when the air is cooled (lowering the saturation vapor pressure) or when water vapor is added to the air (increasing the vapor pressure). As air approaches a relative humidity of 100%, cloud development becomes more and more likely.

☐ Expansional cooling and compressional warming are adiabatic processes, that is, there is no net heat exchange between an air parcel and its surroundings as that parcel ascends or descends within the atmosphere.

☐ We determine atmospheric stability by comparing the temperature (or density) of an air parcel moving vertically (up or down) within the atmosphere with the temperature (or density) of the surrounding (ambient) air. Stable air inhibits vertical motion; unstable air enhances vertical motion.

☐ Local radiational heating or cooling, advection, or large-scale ascent or subsidence can change the stability of an air mass.

☐ Expansional cooling and cloud development occur through uplift of air in convection currents, along fronts or mountain slopes, or as the consequence of convergence of surface winds.

## Review Questions

1. Distinguish among evaporation, transpiration, and sublimation.
2. Provide some common examples of sublimation and depositon.
3. What is the difference between condensation and deposition?
4. Explain how heat energy is involved in the phase changes of water.
5. How does the global hydrologic cycle purify water?
6. The global water budget implies that there must be a net flow of water from the continents to the ocean. Explain why.
7. Distinguish between air pressure and vapor pressure. How do they compare in magnitude?
8. How do air temperature changes influence (a) the saturation vapor pressure, (b) the saturation mixing ratio, and (c) the relative humidity?
9. Under what condition is the mixing ratio equal to the saturation mixing ratio?
10. Why does the relative humidity usually fall between sunrise and early afternoon on a clear and calm day?
11. In localities where winters are cold, some central home heating systems are equipped with humidifiers. Why?
12. Describe how we use a sling psychrometer to measure relative humidity.
13. Identify two ways whereby the relative humidity of air is increased.
14. Describe the temperature change of an unsaturated parcel of air as it alternately ascends and descends within the atmosphere.
15. In what sense are expansional cooling and compressional warming of air adiabatic processes?
16. Explain why rising parcels of saturated (cloudy) air cool more slowly than rising parcels of unsaturated (clear) air.
17. How does the stability of ambient air influence the upward and downward movement of air parcels?
18. Both clear air and clouds are associated with convection within the atmosphere. Explain why.

**19.** Why are clouds and precipitation often associated with fronts?

**20.** How does topography influence the development of clouds.

## Quantitative Questions

**1.** Determine the relative humidity if
   a. the mixing ratio is 5 g/kg and the saturation mixing ratio is 15 g/kg.
   b. the vapor pressure is 3 mb and the saturation vapor pressure is 12 mb.
   c. the vapor pressure equals the saturation vapor pressure.

**2.** Determine the mixing ratio if
   a. the relative humidity is 25% and the saturation mixing ratio is 24 g/kg.
   b. the relative humidity is 100% and the air temperature is 30 °C (86 °F).
   c. the relative humidity is 80% and the saturation mixing ratio is 15 g/kg.

**3.** Outdoor air at 0 °C (32 °F) and relative humidity of 90% is brought indoors and heated to 20 °C (68 °F). Compute the relative humidity of indoor air.

**4.** Determine the relative humidity if
   a. the air temperature is 10 °C (50 °F) and the wet-bulb temperature is 7 °C (45 °F).
   b. the air temperature is 15 °C (59 °F) and the wet-bulb temperature is 15 °C (59 °F).
   c. the air temperature is 25 °C (77 °F) and the wet-bulb depression is 10 Celsius degrees (18 Fahrenheit degrees).

**5.** An unsaturated air parcel initially has a temperature of 0 °C (32 °F). Assuming that the relative humidity remains less than 100%, predict the temperature of the parcel after it is lifted 2500 m.

**6.** Predict the altitude (in m and ft) of the base of a cumulus cloud if the air near the surface of the Earth has a temperature of 15 °C (48 °F) and a relative humidity of 30%.

**7.** Determine whether the following soundings are stable, unstable, or neutral for both saturated and unsaturated air parcels.
   a. +10 °C/1000 m
   b. −5 °C/1000 m
   c. −8 °C/1000 m
   d. −10 °F/1000 ft
   e. +1 °F/1000 ft
   f. −13 °C/1000 m

**8.** If the air is 2% water vapor, the air pressure is 990 mb, and the temperature is 25 °C (77 °F), determine the vapor pressure and the relative humidity.

**9.** How much more water vapor is in saturated air at 30 °C (86 °F) than in saturated air at 5 °C (41 °F)?

## Questions for Critical Thinking

**1.** Does a relative humidity of 25% measured on a cold day in January mean the same as a relative humidity of 25% measured on a hot day in July?

**2.** With intense radiational cooling during the early morning hours, the air temperature is observed to fall rapidly until frost forms. Thereafter, until sunrise, the air temperatureeither remains steady or falls slowly. Why the change in air temperature behavior?

**3.** In early winter, cold air flows from snow-covered land out over the ice-free waters of Lake Ontario. Describe how and why the stability of the air mass changes.

**4.** Is it possible for thunderstorms to form when the lower atmosphere is conditionally stable? Explain your answer.

**5.** Explain why less humid air must be lifted higher than more humid air before condensation or deposition of water vapor begins.

## Selected Readings

BOHREN, C. F., AND G. M. BROWN. "Genies in Jars, Clouds in Bottles, and a Bucket with a Hole in It," *Weatherwise* 35, No. 2 (1982):86–89. Includes a lucid explanation of the concept of saturation vapor pressure.

FORRESTER, F. H. "An Inventory of the World's Water," *Weatherwise* 38, No. 2 (1985):82–105. Describes the various reservoirs in the global hydrologic cycle.

KALKSTEIN, L. S., AND K. M. VALIMONT. "An Evaluation of Summer Discomfort in the United States Using a Relative Climatological Index," *Bulletin of the American Meteorological Society* 67 (1986):842–848. Uses a regional approach to the climatology of stress caused by a combination of high temperature and high humidity.

LEOPOLD, L. B. *Water: A Primer.* New York: W. H. Freeman, 1974, 166 pp. Presents a concise and well-illustrated summary of the functioning of the global water cycle.

PEPI, J. W. "The Summer Simmer Index," *Weatherwise* 40, No. 3 (1987):143–145. Provides a critique of existing discomfort indexes and a proposal for a new one.

SISKIND, D. E. "How Humid Is Humid?" *Weatherwise* 44, No. 3 (1991):24–26. Considers the likelihood of a 90 °F temperature with a 90% relative humidity.

# 7 Dew, Frost, Fog, and Clouds

*I am the daughter of earth
  and water,
And the nursling of the sky:
I pass through the pores of the
  ocean and shores;
I change, but I cannot die.*

PERCY BYSSHE SHELLY
*The Cloud*

Clouds appear in a variety of forms. They are composed of tiny water droplets or ice crystals or a combination of the two. [Photograph by J. M. Moran]

C LOUDS WHISK across the sky in ever-changing patterns of white and gray; fog lends an eerie silence to a dreary day; dew and frost glisten in the morning sun. Clouds, fog, dew, and frost are all products of condensation or deposition of atmospheric water vapor. Most clouds are the consequence of saturation brought about by uplift and expansional cooling of air, but dew, frost, and most fogs develop when the lowest layers of air are chilled to saturation via other processes. In this chapter, we describe the formation of dew, frost, and fog and the development and classification of clouds.

## Low-Level Saturation Processes

Air that is in contact with the Earth's surface becomes saturated if its temperature is lowered sufficiently. Such cooling decreases the saturation vapor pressure (or saturation mixing ratio) and thus increases the relative humidity. As the relative humidity approaches 100%, dew, frost, or fog may form.

### DEW AND FROST

Dew and frost are primarily the consequence of nocturnal radiational cooling. At night, an object on the Earth's surface (a plant leaf or automobile windshield, for example) emits infrared radiation to the atmosphere and eventually off to space, and thereby the object cools. At the same time, the atmosphere emits infrared radiation back to the Earth's surface, where some is absorbed by the object, and thereby the object warms. On a clear night (minimal greenhouse effect), the object emits more radiation than it receives from the atmosphere. Consequently, the surface of the object becomes cooler than the air adjacent to it and heat is conducted from the air to the object. With sufficient cooling, the air in immediate contact with the object becomes saturated. If the air temperature remains above the freezing point, water vapor may condense on the object as **dew** (Figure 7.1); if the air temperature drops below freezing, water vapor may deposit as **frost** (Figure 7.2).

Note that dew and frost are not forms of precipitation, because they do not *fall* from clouds but, rather, develop in place on exposed surfaces. A similar phenomenon occurs when drops of water appear on the outside surface of a cold can of soda on a hot summer day. The *sweat* on the can is actually dew.

The temperature to which air must be cooled, at constant pressure, to reach saturation (relative to liquid water) is called the **dew point.** In order for water vapor to condense as dew on the surface of an object, the temperature of that surface must drop *below* the dew point. The dew point is also a measure of humidity. The higher the dew point, the greater is the water vapor concentration. It also follows that when the difference between the actual air temperature and the dew point is small, the relative humidity is high and vice versa. Furthermore, dew point is an index of human comfort. Although the relative humidity is an important consideration, as a rule, most people experience discomfort when the dew point rises above about 20 °C (68 °F). In summer, dew points are highest in the states bordering the Gulf of Mexico. There, July mean dew points are between 21 and 24 °C (70 and 75 °F). The lowest summer dew points are in the Rocky Mountain states

**FIGURE 7.1**
If winds are calm, nocturnal radiational cooling may bring a shallow layer of air to saturation. At temperatures above freezing, water vapor condenses as dew on exposed surfaces, such as on the strands of this spider web. [Photograph by Tranquality/Philip Chaudoir]

and desert Southwest. From New Mexico northwestward into western Montana, for example, July mean dew points are between −1 and 7 °C (30 and 45 °F).

Note that the dew point should not be confused with the wet-bulb temperature (described in Chapter 6); they are not the same. Recall that the wet-bulb temperature is determined by inducing evaporative cooling. Adding water vapor to the air raises the temperature at which dew will form. Hence, except at saturation, the wet-bulb temperature is higher than the dew point. At saturation, the dew-point, wet-bulb, and ambient air temperatures are the same.

The dew point can be obtained from measurements of the dry-bulb (ambient air) temperature plus the wet-bulb depression (see the psychrometric tables inside the back cover). For example, if the dry-bulb temperature is 20 °C (68 °F) and the wet-bulb depression is 5 Celsius degrees (9 Fahrenheit degrees), then the dew point is 11.6 °C (53 °F). Note that in this case, the wet-bulb temperature is 15 °C (59 °F), some 3.4 Celsius degrees (6 Fahrenheit degrees) higher than the dew point.

When cooling at constant pressure produces saturation at temperatures at or below 0 °C (32 °F), the temperature is called the **frost point.** Water vapor deposits as frost on the surface of an object if the temperature of that surface falls *below* the frost point. Frost occurs in a variety of forms: Delicate feathery patterns of crystals may develop on a windowpane during a cold winter night, or fernlike crystals of **hoarfrost** may grow to a length of several centimeters on the twigs and branches of trees and shrubs (Figure 7.3).

When neither cold air advection nor warm air advection is expected, the dew point (or frost point) sometimes can be used to predict the next morning's minimum air temperature. Suppose, for example, it is autumn in New England and weather forecasters are calling for clear nighttime skies and calm winds, conditions ideal for extreme nocturnal radiational cooling. Gardeners know that all the ingredients are present for chilly temperatures by early morning, and they want to know whether they should go to the trouble of protecting their freeze-sensitive plants. As a general rule, under the conditions described, the late afternoon's dew point is a reasonable estimate of the next morning's low temperature: If the dew point is near or below 0 °C (32 °F), freeze protection is advisable.

What is the physical basis for this rule? In response to nocturnal radiational cooling, the air temperature falls continually until the relative humidity nears 100% and condensation or deposition occurs (producing dew, frost, or perhaps fog). Latent heat released during condensation or deposition offsets radiational cooling to some extent so that the air temperature tends to stabilize near the dew point or frost point. Many other factors, however, may complicate this simple rule. For one, the length of the night is an important control of the amount of radiational cooling. Summer nights may be too short for radiational cooling sufficient to lower the air temperature to the dew point, particularly if the air is relatively dry (low dew point).

It is a popular misconception that Florida and California citrus growers fear a winter frost. What they actually fear is a solid freeze, that is, a cold snap that

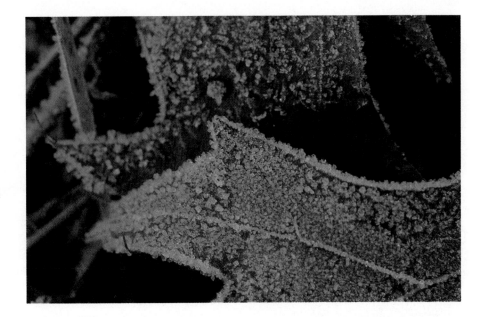

**FIGURE 7.2**
If winds are calm, nocturnal radiational cooling may bring a shallow layer of air to saturation. At temperatures below freezing, water vapor deposits as frost on exposed surfaces. [Photograph by Tranquality/Philip Chaudoir]

freezes the water in plant tissues and causes potentially lethal damage to the plant. As a general rule of thumb, the air temperature must fall to −5 °C (23 °F) or lower for at least 4 hours for serious damage to citrus trees. For more information on freeze protection for plants, see the Special Topic "Freeze Prevention." It is also a popular misconception that frost causes the leaves of deciduous trees to change color in autumn. For more on this subject, refer to the Special Topic "Jack Frost and Autumn Color."

**FIGURE 7.3**
Deposition of water vapor from a layer of humid, cold air causes crystals of hoarfrost to grow to several centimeters in length on the twigs of trees and bushes. [Photograph by J. M. Moran]

SPECIAL TOPIC

# Freeze Prevention

One of the most effective strategies to protect crops against late spring or early fall freezes is to avoid cultivating sites on the receiving end of cold air drainage. In such freeze-prone areas as valley bottoms, the growing season may be several weeks shorter than it is on surrounding highlands. For this reason, orchards and vineyards are often situated on hillslopes rather than on valley floors. Even on hillslopes site selection must be made carefully. Obstructions such as hedgerows, roads, or railroad embankments can impede the downslope drainage of cold air, and thus crops immediately upslope from the barrier may be damaged. To avoid this problem, growers construct channels through the barrier so that cold air can continue to drain downslope.

Because cranberries are grown in freeze-prone, low-lying bogs, their cultivation requires special precautions to avoid the threat of cold air drainage. Prior to harvest in late September and early October, the berries require cool autumn nights to develop their rich, red color. Killing freezes, however, also typically occur during this ripening period. When freezing temperatures threaten, growers flood the bogs with water that effectively insulates the berries. Continuous circulation keeps the water from freezing.

Based on their understanding of conditions that favor extreme nocturnal radiational cooling, scientists and growers have developed several other strategies to reduce the incidence of radiation freezes. One factor that contributes to such freezes is the absence of clouds. To protect crops on clear nights, growers can create their own clouds. Clouds absorb infrared radiation emitted by the ground and vegetation, and subsequently reemit a portion of that radiation back to the crop. It is widely believed that a smoke cloud emanating from smudge pots inhibits freezing temperatures by enhancing the local greenhouse effect. However, smoke particles are actually nearly transparent to infrared radiation so that their influence on the local greenhouse effect is negligible. (Furthermore, smudge pots have

FIGURE 1
Young tomato seedlings in the San Joaquin Valley, California, are protected from wind and nocturnal radiational cooling by "hot caps" until the roots are established. [U.S. Bureau of Reclamation photograph]

been banned in many locales because of their deleterious impact on air quality.) On the other hand, a mist cloud formed by a fine water spray does enhance the local greenhouse effect and can provide crops with some protection.

For small plants, other types of radiation screens are effective. For example, plastic "hot caps" placed over plants create a protective microclimate around them (Figure 1). During the day, the sun warms the soil and plants. If hot caps are placed over plants in the late afternoon, the heat gained during the day is better conserved at night, reducing the chances of frost formation on the plants. For somewhat larger plants, other radiation screens, such as wooden slats or cheesecloth, can restrict exposure to clear night skies without significantly blocking the daytime solar radiation needed for crop growth.

Nocturnal radiational cooling often produces a temperature inversion within the lowest air layer. Whereas air temperatures at ground level may drop below freezing, air temperatures near the top of the inversion (perhaps 15 m or 49 ft above the ground) may be several

**FIGURE 2**
Ten-meter-high wind machines circulate warmer air down among the orange trees during cold nights in the San Joaquin Valley, California. [U.S. Department of Agriculture photograph]

degrees above freezing. The extreme stability of the inversion layer prevents the warmer air aloft from mixing with the cooler air at ground level. A logical approach, then, is to use large motor-driven fans or propellers mounted on towers to circulate the warmer air aloft down to ground level. This is standard practice in citrus groves (Figure 2), where the economic value of the crop justifies the high capital investment in freeze-prevention equipment.

Crops are sometimes protected from freezing by spraying them with a fine water mist when the temperature of plant tissues drops to 0 °C (32 °F). Our initial reaction may be to question how a coating of ice can help plants survive. Although water on the plant surface is at 0 °C, the latent heat of fusion released during the phase change from liquid water to ice helps stabilize plant temperatures. Stabilization near 0 °C often prevents plant damage because most actively growing tissues are not injured until their temperature drops to −1 to −5 °C (23 to 30 °F). Nevertheless, this strategy requires careful monitoring. As long as sprinkling and freezing continue, the temperature of the ice remains at about 0 °C. If, however, sprinkling is discontinued before ambient air temperatures rise high enough to melt the ice, then heat is conducted from the leaves to the ice, and the leaf temperature drops to potentially lethal levels. In addition, care must be taken that the ice burden does not become so great that plants are damaged by excess weight. For this reason, the sprinkling method is most suitable for low-growing vegetable crops such as cucumbers and strawberries.

Spraying reduces the threat of freeze damage by adding latent heat. Sensible heat can also be added directly through fuel combustion in heaters. In addition, heaters emit infrared radiation. Heaters are most beneficial for those plants directly exposed to the warm plume of air, so using many small heaters is considerably more effective than using a few large ones. Such an arrangement can raise both air and plant temperatures by several degrees.

These freeze-prevention strategies are not always sufficient. For three days around Christmas 1983, an arctic air mass surged into Florida dropping temperatures to well below freezing through much of the state. Damage to the citrus crop exceeded $1 billion. Again, in mid-January 1985, another cold snap of comparable severity caused further damage to surviving citrus trees. Losses from this double blow reduced citrus-producing acreage by almost 90% in Lake County, Florida, formerly the state's second largest citrus-producing county. Subsequently, a number of growers replanted, and the region was recovering when another deep freeze over Christmas weekend 1989 again ruined much of central Florida's citrus crop. This most recent blow may well mean an end to the citrus industry in the northern reaches of Florida's citrus belt.

## Jack Frost and Autumn Color

As autumn approaches, leaves on many plants change from their familiar green of spring and summer to brilliant hues of purple, yellow, orange, and red (Figure 1). Autumn colors lure millions of people to the countryside to enjoy one of nature's most spectacular shows. How does nature produce autumn color? Contrary to popular opinion, frosts and freezing temperatures have very little to do with the color of autumn leaves. Weather does play a role, but in order to understand that role, we must first consider the chemical changes that take place within leaves during the growing season.

In spring and summer, leaves are green because large quantities of green pigments called *chlorophylls* are present in leaf cells. Although chlorophyll molecules break down during the normal metabolic functions of a leaf, they also are continually manufactured. Consequently, the chlorophyll content of leaves remains high and leaves stay green. With the coming of autumn, however, chlorophyll production slows and leaf chlorophyll content declines, thereby gradually unmasking the yellow, brown, and orange pigments that are also in the leaves throughout the growing season. These pigments, which are called *carotenoids,* are common in many living things such as carrots, bananas, and canaries. In autumn, carotenoids tint the leaves of such trees as maples, ashes, birches, and cottonwoods.

The leaves of some plants turn red or purple, hues

**FIGURE 1**
Deciduous tree in brilliant autumn colors. [Photograph by J. M. Moran]

### *FOG*

**Fog** is a visibility-restricting suspension of tiny water droplets or ice crystals in an air layer next to the Earth's surface. Simply put, fog is a cloud in contact with the ground. By international convention,* fog is defined as restricting visibility to 1000 m (3250 ft) or less; otherwise the suspension is called **mist.** (The popular definition of mist is a light drizzle.) Fog may

*For aviation purposes, the criterion for fog is a visibility restriction of 10 km (6.2 mi) or less.

develop when air becomes saturated through radiational cooling, advective cooling, the addition of water vapor, or expansional cooling.

With a clear night sky, light winds, and an air mass that is humid near the ground and relatively dry aloft, radiational cooling may cause the air near the ground to approach saturation. When these conditions occur, a cloud develops. A ground-level cloud formed in this way is called **radiation fog** (Figure 7.4). High humidity at low levels within the air mass is usually due to evaporation of water from a moist surface. Hence, radiation fogs are most common in marshy areas or where

caused by another group of pigments called *anthocyanins*. Unlike the carotenoids, these pigments are not normally manufactured by leaves until late in the growing season. Anthocyanins are responsible for the color of such fruits as cranberries, purple grapes, and strawberries. In autumn, anthocyanins are also produced in the leaves of such trees as oaks, maples, dogwoods, and persimmon. Often the colors of carotenoids and anthocyanins combine to deepen oranges and produce the fiery reds and bronzes that are typical of the autumn leaves of many trees.

What then is the role of weather in autumn color? The leaves of deciduous plants cannot survive the dryness of cold winter air because they would lose much more water to the dry air than the roots could absorb from the frozen soil. Hence, the shorter daylengths that accompany the coming of autumn signal deciduous plants to drop their leaves as they become dormant for winter. As part of this preparatory process, the rate of chlorophyll production declines, the formerly hidden carotenoids appear, and, in some species, anthocyanins are manufactured. Thus, leaves turn color.

Although these processes occur every fall, personal experience tells us that colors are more brilliant in some years than in others. Why the difference? The most brilliant leaf color usually develops when days are sunny and cool and nights are chilly but not subfreezing. For many years, people assumed that bright sunlight triggered the synthesis of sugars in leaves and that frosty nights slowed the transport of these sugars out of the leaf. The leaf uses sugars to manufacture anthocyanins. One problem with this explanation is that as the aging leaf loses its chlorophyll, it also loses its ability to produce sugars. Scientists now know that sunlight promotes the synthesis of anthocyanins. As for frosty nights, they probably have little to do with the brilliance of leaf color. Rather, sunny days are often followed by clear nights which enhance radiational cooling and lead to subfreezing air temperatures.

the soil has been saturated by a recent period of heavy rainfall or rapid snowmelt.

Note that light winds rather than calm conditions favor development of radiation fog. Light winds produce a slight mixing that more effectively transfers heat from throughout the layer of humid air to the relatively cold ground. Consequently, the entire air layer is chilled to below the dew point. If winds are calm, however, there is no mixing and heat transfer is by conduction alone. Because still air is a poor conductor of heat, only a very thin layer of air immediately in contact with the ground is cooled to saturation. Hence, calm winds favor dew or frost rather than radiation fog. On the other hand, if winds become too strong, the humid air at low levels mixes with the drier air aloft, the relative humidity drops, and radiation fog disperses or fails to develop.

In hilly regions, air chilled by radiational cooling drains downslope and settles in low-lying areas such as river valleys. Hilltops may thus be clear of fog, while in deep valleys, the fog is thick and persistent.

At night, once a fog bank forms, radiational cooling becomes a little more complicated. The top surface

**FIGURE 7.4**

Radiation fog develops as a consequence of extreme radiational cooling of a layer of humid air on a night when there is some air movement. [Photograph by Arjen and Jerrine Verkaik/SKYART]

**FIGURE 7.5**
Radiation fog dispersing at Boston's Logan International Airport as viewed from an approaching
aircraft. [Photograph by J. M. Moran]

of the fog bank very efficiently emits IR radiation sky-ward. Meanwhile, some of the IR radiation emitted by the ground is absorbed by the fog droplets and reradi-ated back to the ground. Hence, the ground cools more slowly and the maximum cooling shifts to the top of the fog bank, the fog thickens, and visibility continues to drop.

Typically, radiation fog persists for only a few hours after sunrise. Then, the fog gradually thins and dis-perses as saturated air at low levels mixes with drier air above the fog bank (Figure 7.5); the relative humid-ity drops and fog droplets vaporize. Mixing may be caused by convection triggered by solar heating of the ground or by a strengthening of regional winds. In win-ter, however, when the weak rays of the sun are readily reflected by the top of the fog layer, radiation fog may

be quite persistent. For example, in the valleys of the Great Basin and in California's San Joaquin Valley, winter radiation fogs may persist for many days to weeks at a time.

Air mass advection is sometimes accompanied by fog. The temperature and the water vapor concentra-tion of an air mass depend on the nature of the surface over which the air mass forms and travels. As an air mass moves from one place to another, termed **air mass advection,** those characteristics change, partly as a re-sult of the modifying influence of the surfaces over which the air mass travels. When the advecting air passes over a relatively cold surface, the air mass may be chilled to saturation in its lower layers. This type of cooling is known as *advective cooling* and occurs, for example, in spring when mild, humid air flows over

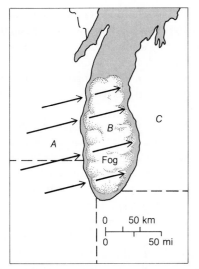

**FIGURE 7.6**
In summer, as warm and humid air (A) flows over the relatively cool surface of Lake Michigan, the air is chilled to saturation and dense fog (B) forms. As the air flows over the warmer land surface of lower Michigan (C), air temperatures rise, the relative humidity drops, and the fog dissipates. [From *Weather and Climate of the Great Lakes Region,* Figure 31, p. 109, by Val L. Eichenlaub, © 1979 by the University of Notre Dame Press.]

relatively cold, snow-covered ground. Snow on the ground may chill the air to the point that fog develops. Fog formed by advective cooling is known as **advection fog.**

Advection fog also develops when warm and humid air streams over relatively cold ocean or lake water. Persistent, dense fogs develop in this way over Lake Michigan in summer as shown in Figure 7.6. Thick fogs over the frigid waters of the Grand Banks of Newfoundland are a consequence of mild maritime air flowing northward from over the warm Gulf Stream waters.

Nocturnal radiational cooling sometimes combines with air mass advection to produce fog. An initially dry air mass becomes more humid in its lower levels after a long trajectory over the open waters of a lake or the ocean. Over land areas, downwind from the water body, if the night sky is clear, the modified air mass is subjected to extreme radiational cooling and fog develops. San Francisco's famous fog forms in this way as onshore winds transport cool, humid air into the city and nocturnal radiational cooling chills the air to saturation.

**Arctic sea smoke** is fog that develops in winter when extremely cold, dry air flows over a large unfrozen body of water. The lower portion of the air mass modifies through contact with the relatively warm water. Evaporation and sensible heating cause the lower portion of the air mass to become more humid and warmer than the air above. Heating from below destabilizes the air and the consequent mixing of mild, humid air with cold, dry air brings the air to saturation and fog forms.* Because the air is destabilized, fog appears as rising filaments or streamers that resemble smoke or steam (Figure 7.7). For this

*Recall the Special Topic "Clouds by Mixing" in Chapter 6.

**FIGURE 7.7**
Steam fog forms when relatively cold air comes in contact with relatively warm water. [Photograph by Mike Brisson]

reason, a more general name for fog produced when cold air comes in contact with warm water is **steam fog.** Steam fog also develops on a cold day over a heated outdoor swimming pool or hot tub and sometimes over a wet highway or field when the sun comes out after a rain.

Another way of explaining the physics of steam fog is to consider the distribution of vapor pressure over the water surface. Within about 1 to 2 m (3 to 7 ft) of the water surface, the vapor pressure is relatively high because of evaporation into the heated air. Above this shallow layer the air is colder and has a much lower vapor pressure. Hence, a vertical vapor pressure gradient develops over the water body and, in response, water molecules stream (diffuse) upward. (As we will see in Chapter 9, gases move from high toward low pressure in response to a pressure gradient.) As the humid air streams upward, it cools and the water vapor condenses into tiny droplets that we see as steam. About 5 to 10 m (16 to 33 ft) above the water surface, the rising streamers encounter drier air and evaporate.

Fog also develops on hillsides or mountain slopes as a consequence of the upslope movement of humid air (Figure 7.8). Ascending humid air undergoes expansional cooling and eventually reaches saturation. Any further ascent of the saturated air produces fog. Fog formed in this way is called **upslope fog.**

## Cloud Development

Water vapor is an invisible gas, but the condensation and deposition products of water vapor are visible. Clouds are the visible manifestations of the condensation and deposition of water vapor within the atmosphere. They are composed of tiny water droplets or ice crystals or a mixture of both. In this section, we consider the process of cloud development.

Laboratory studies have demonstrated that in clean air, air that is free of dust and other aerosols, condensation (or deposition) of water vapor requires supersaturated conditions (that is, a relative humidity greater than 100%). In clean air, the degree of supersaturation

needed for cloud development increases rapidly as the radius of the droplets decreases. For example, formation of relatively small droplets with radii of 0.001 micrometer requires a supersaturation of nearly 340%. In contrast, relatively large droplets, with radii greater than 1.0 micrometer, need only slight supersaturation to form.

Why does the degree of supersaturation hinge on droplet size? The values of saturation vapor pressure listed in Table 6.3 apply only to the situation where air overlies a *flat* surface of pure water.* At equivalent temperatures, the saturation vapor pressure is higher in the air surrounding a spherical water droplet than in the air over a flat surface of pure water. As a water surface exhibits increasing curvature, it becomes easier for water molecules to escape the liquid and become vapor. Water molecules that form a curved surface have fewer neighboring molecules and hence, are more weakly held (by hydrogen bonding) than water molecules that form a flat surface. In the case of spherical water droplets, the curvature increases with decreasing radius. Thus, at the same temperature, water molecules more readily escape small droplets than large droplets. This implies that the saturation vapor pressure (and degree of required supersaturation) must increase as droplet size decreases.

We might reasonably assume that, within the atmosphere, cloud droplets grow from much smaller water droplets. However, the great supersaturations required for condensation of very small droplets do not develop in the real atmosphere. How, then, do cloud droplets form? The simple answer is that they have a head start.

In the atmosphere, at most, only slightly supersaturated conditions are necessary for cloud development. This is because the atmosphere contains abundant **nuclei,** tiny solid and liquid particles that provide relatively large surface areas on which condensation or deposition can take place. Nuclei are products of both natural and human activity. Forest fires, volcanic eruptions, wind erosion of the soil, saltwater spray, and the discharge from domestic and industrial chimneys are all sources of nuclei. Typically, a cubic centimeter of air contains about 10,000 nuclei.

---

*Note that the saturation vapor pressure (or saturation mixing ratio) for air over a flat surface of pure water is the basis for computing relative humidity and the degree of supersaturation.

Some nuclei have radii greater than 1.0 micrometer, large enough to facilitate droplet condensation at relative humidities that rarely exceed 100% by more than a fraction of 1%. More important, however, is the presence in air of a plentiful supply of relatively small nuclei that are hygroscopic. **Hygroscopic nuclei** have a special chemical affinity (attraction) for water molecules. Condensation begins on these nuclei at relative humidities under 100%. In fact, magnesium chloride, a salt in sea spray, is a hygroscopic substance that can initiate condensation at relative humidities as low as 70%. Because some nuclei are relatively large and many are hygroscopic, we can expect cloud development when the relative humidity nears 100%.

Depending on their specific function, nuclei are classified as one of two types: **cloud condensation nuclei (CCN)** and **ice-forming nuclei.** CCNs are active (that is, they promote condensation) at temperatures both

**FIGURE 7.8**
Upslope fog forms when humid air ascends a mountainside. Expansional cooling brings the air to saturation. [Photograph by Arjen and Jerrine Verkaik/SKYART]

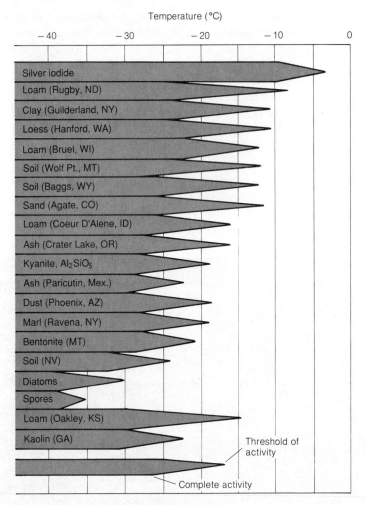

Temperature (°C)

| | −40 | −30 | −20 | −10 | 0 |

Silver iodide
Loam (Rugby, ND)
Clay (Guilderland, NY)
Loess (Hanford, WA)
Loam (Bruel, WI)
Soil (Wolf Pt., MT)
Soil (Baggs, WY)
Sand (Agate, CO)
Loam (Coeur D'Alene, ID)
Ash (Crater Lake, OR)
Kyanite, $Al_2SiO_5$
Ash (Paricutin, Mex.)
Dust (Phoenix, AZ)
Marl (Ravena, NY)
Bentonite (MT)
Soil (NV)
Diatoms
Spores
Loam (Oakley, KS)
Kaolin (GA)

Threshold of
activity

Complete activity

**FIGURE 7.9**
Some deposition (sublimation) nuclei and their activation temperatures. Note that most nuclei are soil particles. [From V. J. Schaefer, "The Formation of Ice Crystals by Sublimation," *Weatherwise* 32 (1979):256]

above and below freezing because water vapor can condense into droplets that remain liquid even when the cloud temperature is well below 0 °C (32 °F). These are *supercooled* water droplets. Ice-forming nuclei are much less abundant than cloud condensation nuclei and become active only at temperatures well below freezing. There are two types of ice-forming nuclei: (1) *freezing nuclei,* which cause liquid droplets to freeze, and (2) *deposition nuclei* on which water vapor deposits directly as ice. Most freezing nuclei are active only at temperatures below −9 °C (16 °F). Deposition nuclei do not become fully active until temperatures fall below −20 °C (−4 °F), as shown in Figure 7.9.

## Classification of Clouds

Even a casual observer of the sky notices that clouds occur in a wide variety of forms. These forms are not arbitrary. Clouds are shaped by many processes operating in the atmosphere. In fact, keeping track of changes in clouds and cloud cover often provides clues about future weather.

The British naturalist Luke Howard was among the first to devise a classification of cloud types. Formulated in 1802-1803, the essentials of Howard's scheme are still in use today. Cloud forms are given special Latin names and are organized by appearance and by altitude of occurrence.

Based on *appearance,* the simplest distinction is

among cirrus, stratus, and cumulus clouds. Cirrus clouds are fibrous, stratus clouds are layered, and cumulus clouds occur as heaps or puffs. On the basis of *altitude,* the most common clouds in the troposphere are grouped into four families (Table 7.1): high clouds, middle clouds, low clouds, and clouds exhibiting vertical development. The adjectives listed in Table 7.2 may be applied to cloud names in order to more fully specify their characteristics.

High, middle, and low clouds are produced by gentle uplift of air (typically less than 5 cm/sec or 1.0 mi/hr) over broad areas. These clouds spread laterally to form layers and are called **stratiform clouds.** Clouds with vertical development generally cover smaller areas and are associated with much more vigorous uplift (sometimes in excess of 30 m/sec or 70 mi/hr). Consequently, these clouds are heaped or puffy in appearance and are called **cumuliform clouds.** In later chapters, we de-

scribe the various weather systems that trigger development of stratiform and cumuliform clouds. Here, we are concerned primarily with brief descriptions of the most common clouds in each family.

## HIGH CLOUDS

The base* of high clouds is at altitudes above 7000 m (23,000 ft). Temperatures are so low in this region of the atmosphere (below $-25$ °C or $-13$ °F) that clouds are composed almost exclusively of ice crystals, a composition that gives them a fibrous or filamentous

*The actual altitude of the cloud base varies seasonally and with latitude. For example, in polar regions, the base of high clouds may be as low as 3000 m (9800 ft). Cloud base altitudes quoted here are intended as general guidelines.

**TABLE 7.1**
**Cloud Classification**

| Genus | Altitude of cloud base above ground (km) | Shape and appearance |
|---|---|---|
| High clouds | | |
| Cirrus (Ci) | 7–18 | Delicate streaks or patches |
| Cirrostratus (Cs) | 7–18 | Transparent thin white sheet or veil |
| Cirrocumulus (Cc) | 7–18 | Layer of small white puffs or ripples |
| Middle clouds | | |
| Altostratus (As) | 2–7 | Uniform white or gray sheet or layer |
| Altocumulus (Ac) | 2–7 | White or gray puffs or waves in patches or layers |
| Low clouds | | |
| Stratocumulus (Sc) | 0–2 | Patches or layers of large rolls or merged puffs |
| Stratus (St) | 0–2 | Uniform gray layer |
| Nimbostratus (Ns) | 0–4 | Uniform gray layer from which precipitation is falling |
| Clouds with vertical development | | |
| Cumulus (Cu) | 0–3 | Detached heaps or puffs with sharp outlines and flat bases, and slight or moderate vertical extent |
| Cumulonimbus (Cb) | 0–3 | Large puffy clouds of great vertical extent with smooth or flattened tops, frequently anvil shaped, from which showers fall, with thunder. |

Source: M. Neiburger, J. G. Edinger, and W. D. Bonner. *Understanding our Atmospheric Environment.* New York: W. H. Freeman, 1973, p. 11. Copyright © 1973. All rights reserved.

**TABLE 7.2**
**Characteristics of Various Cloud Types**

| Characteristic | Meaning | Applied to |
|---|---|---|
| Castellanus | Towerlike vertical development | Cirrocumulus, altocumulus |
| Congestus | Crowded in heaps | Cumulus |
| Fractus | Broken | Stratus, cumulus |
| Humilis | Little vertical development | Cumulus |
| Lenticularis | Lens-shaped | Cirrostratus, altocumulus, stratocumulus |
| Mammatus | Hanging protuberances | Cumulonimbus |
| Uncinus | Hook-shaped | Cirrus |

appearance.* Their names include the prefix cirro from the Latin meaning "a curl of hair."

**Cirrus** (Ci) clouds are nearly transparent and occur as delicate silky strands, sometimes called mares' tails (Figure 7.10). Strands are actually streaks of falling ice crystals blown laterally by strong winds. Like cirrus clouds, **cirrostratus** (Cs) clouds are also nearly transparent, so the sun or moon readily shines through them.

---

*Kenneth Sassen of the University of Utah reports that supercooled water droplets were detected in cirrus clouds at −40 to −50 °C (−40 to −58 °F) over Kansas on 5–6 December 1991. Sassen attributes this rare occurrence to an influx of volcanic nuclei from the June 1991 eruption of Mount Pinatubo.

They form a thin, white veil or sheet that partially or totally covers the sky (Figure 7.11). **Cirrocumulus** (Cc) clouds consist of small, white, rounded patches arranged in a wavelike or mackerel pattern (Figure 7.12). Rarely do these clouds cover the entire sky. No high cloud is thick enough to prevent objects on the ground from casting shadows during daylight hours.

## MIDDLE CLOUDS

The bases of middle clouds are at altitudes between 2000 and 7000 m (6600 and 23,000 ft). Their names include the prefix alto from the Latin meaning "high." These clouds, which feature temperatures generally between 0 and −25 °C (32 and −13 °F), are composed of supercooled water droplets or a mixture of supercooled water droplets and ice crystals. **Altostratus** (As) clouds occur as uniformly gray or white layers that totally or partially cover the sky (Figure 7.13). They are usually so thick that the sun is only dimly visible, as if it were viewed through frosted glass. **Altocumulus** (Ac) clouds consist of roll-like patches or puffs that often form waves or parallel bands (Figure 7.14). They are distingushed from cirrocumulus by the larger size of the cloud patches and by sharper edges. Sharp cloud boundaries indicate the presence of water droplets rather than ice crystals. Altocumulus may occur in sev-

**FIGURE 7.10**
Cirrus are high, thin wispy clouds that are composed almost exclusively of ice crystals. [Photograph by J. M. Moran]

**FIGURE 7.11**
Cirrostratus are high clouds that form a thin, transparent veil over the sky. [Photograph by J. M. Moran]

eral distinct layers simultaneously and rarely produce precipitation that reaches the ground. If they appear early on a warm, humid summer day, the odds favor an afternoon thunderstorm.

## LOW CLOUDS

The bases of low clouds range from the Earth's surface (fog) up to altitudes of perhaps 2000 m (6500 ft). Low clouds that form at temperatures above −5 °C (23

°F) are composed mostly of water droplets. **Stratocumulus** (Sc) clouds consist of large, irregular puffs or rolls separated by areas of clear sky (Figure 7.15). Rarely does precipitation fall from stratocumulus clouds. **Stratus** (St) appears as a uniform gray layer that stretches from horizon to horizon (Figure 7.16). Fog occurs where stratus meets the ground, and when fog "lifts," the stratus cloud deck may break up into stratocumulus.

Usually, only drizzle may fall from stratus clouds, but significant amounts of rain or snow may fall from

**FIGURE 7.12**
Cirrocumulus are high clouds that exhibit a wavelike or mackerel pattern of small, white puffs. [Photograph by Arjen and Jerrine Verkaik/SKYART]

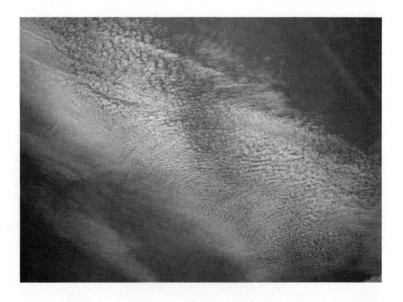

**FIGURE 7.13**
Altostratus are middle-level clouds composed of ice crystals or water droplets or a mixture of both. They form a uniform gray or white layer through which the sun is dimly visible. [Photograph by Arjen and Jerrine Verkaik/SKYART]

the much thicker **nimbostratus** (Ns) clouds. Nimbostratus resemble stratus except that they are darker gray and have a less uniform, more ragged base. Precipitation from nimbostratus tends to be light to moderate and continuous for upward of 12 hr or longer. By contrast, relatively brief but heavy precipitation is often associated with a cumulonimbus (thunderstorm) cloud. In fact, gentle rainfall from nimbostratus more readily infiltrates the soil, whereas the intense downpour of a thunderstorm quickly saturates the upper soil and then mostly runs off and perhaps causes flooding (Chapter 13).

**FIGURE 7.14**
Altocumulus are middle-level clouds consisting of patches arranged in a wavelike pattern. [Photograph by J. M. Moran]

**FIGURE 7.15**
Stratocumulus are low-level clouds consisting of relatively large, irregular rolls that form a layer. [Photograph by J. M. Moran]

## VERTICAL CLOUDS*

Air surging upward as convection currents can give rise to cumulus, cumulus congestus, and cumulonimbus clouds. The altitude at which condensation begins to occur through convection is known as the **convective condensation level (CCL)** and coincides with the

*Typically, this category is reserved for cumuliform clouds, that is, convective clouds exhibiting vertical development. However, we may also include altostratus and nimbostratus as vertical clouds. Although usually classified as middle and low clouds, respectively, because of cloud-base altitude, altostratus and nimbostratus often are sufficiently thick that they extend through more than one level.

altitude of the cloud base, typically between 1000 and 2000 m (3600 and 6600 ft). **Cumulus** (Cu) clouds resemble puffs of cotton dotting the sky on a fair weather day (Figure 7.17). In Figure 7.18, cumulus clouds are viewed from above, from an aircraft window. Because convection is driven by solar heating, not surprisingly cumulus cloud development often follows the daily variation of insolation. On a fair day, cumulus clouds begin forming by middle to late morning, after the sun has warmed the ground and initiated convection. Cumulus sky cover is most extensive by midafternoon, usually the warmest time of day. If cumulus clouds

**FIGURE 7.16**
Stratus is a low, gray, continuous cloud layer from which drizzle may fall. [Photograph by J. M. Moran]

**FIGURE 7.17**
Cumulus clouds resemble puffs of cotton floating in the sky. They develop as a consequence of convection currents. Airflow is upward where there are clouds and downward where the air is clear. [Photograph by J. M. Moran]

show some vertical growth, these normally *fair-weather* clouds may produce a brief, light shower of rain or snow. As sunset approaches, convection weakens, and cumulus clouds begin dissipating (that is, they vaporize).

Where convection is suppressed, so too is development of cumulus clouds. Relatively cold surfaces chill and stabilize the overlying air and inhibit convection, and hence, cumulus clouds do not readily form over snow-covered surfaces. Cold water also suppresses convection. This effect is sometimes observed along the

shores of the Great Lakes during late spring and early summer afternoons when, on average, lake surface temperatures are lower than those of the adjacent land surface. Fair-weather cumulus clouds develop over the land but not over the lake. In fact, rows of cumulus clouds may form a distinct boundary along the shoreline.

Once cumulus clouds form, the stability profile of the troposphere determines the extent of vertical cloud development and whether cumulus clouds build into more ominous thunderstorm clouds. If the ambient air

**FIGURE 7.18**
Cumulus clouds viewed from an aircraft window. [Photograph by J. M. Moran]

**FIGURE 7.19**
Cumulus clouds that exhibit significant vertical development are called cumulus congestus; they resemble a cauliflower. [Photograph by J. M. Moran]

aloft is stable, vertical motion is inhibited and cumulus clouds exhibit little vertical growth. Under these conditions, the weather is likely to remain fair. On the other hand, if the ambient air aloft is unstable for saturated (cloudy) air, then vertical motion is enhanced, and the tops of cumulus clouds surge upward. If the ambient air is unstable to great altitudes, the entire cloud mass takes on a cauliflower appearance as it builds into a **cumulus congestus** (Figure 7.19) and then a **cumulonimbus** (Cb) cloud (Figure 7.20).

## SKY WATCHING

We sometimes observe both cumuliform and stratiform clouds in the sky at the same time. For example, as shown in Figure 7.21, a layer of high cirrus overlies low-level cumulus clouds. Cirrus has developed in ad-

vance of an approaching surface warm front and is too thin to significantly weaken the incoming solar radiation. The sun heats the ground triggering convection and cumulus clouds develop.

When more than one cloud layer appears in the sky, it is instructive to visually track the direction of movement of each cloud layer. Often clouds at different altitudes move in different directions at different speeds. Because clouds are displaced with the wind, this observation indicates that, at least within the troposphere, the horizontal wind changes direction and magnitude with altitude. Any change in wind direction or wind speed with distance is known as **wind shear.** We have more to say about wind shear in later chapters.

Another feature to watch for are well-defined holes or canals in an otherwise uniform cloud layer. Holes in clouds are not unusual; numerous published reports

**FIGURE 7.20**
A cumulonimbus cloud billows upward and spreads laterally, forming an anvil that reaches to the tropopause. [Photograph by J. M. Moran]

**FIGURE 7.21**
High, thin cirrus clouds overlying cumulus clouds. [Photograph by J. M. Moran]

and photographs of them have appeared over the past 70 or so years. Professor Peter V. Hobbs of the University of Washington argues that at least some holes are caused by aircraft flying through thin clouds (probably altocumulus) composed of supercooled water droplets. Turbulence produced in the wake of an aircraft induces expansional cooling of air; water droplets freeze to ice crystals that fall out of the cloud, thereby creating a hole in the cloud. Aircraft ascending or descending through a cloud produce nearly circular holes, whereas aircraft flying horizontally through a cloud layer create canals.

## Unusual Clouds

Chances are that all of us have seen most of the clouds described above because the circumstances leading to their development are quite common. Other, more unusual clouds are formed by rarer atmospheric conditions.

The clouds in Figure 7.22 are striking in appearance. Not surprisingly, many people have reported them as UFOs. Observation of this type of cloud over a period of time reveals another strange characteristic: They tend to remain nearly stationary. They are **altocumulus lenticularis** clouds, that is, lens-shaped altocumulus clouds. These and other so-called **mountain-wave clouds** are generated by winds that are disturbed by a mountain range.

As strong horizontal winds encounter a mountain range, the wind is deflected up the windward slopes and then down the leeward slopes. This is a common occurrence as the prevailing westerlies, flowing from west to east, cross the Front Range of the Rocky Mountains. If the air ascending the windward slopes is stable, then the winds describe a wavelike pattern that extends many tens of kilometers downwind of the mountain crest (Figure 7.23). Within the wave, where air flows upward, expansional cooling leads to cloud formation. Where air flows downward within the wave, compressional warming causes clouds to vaporize. Lenticular clouds thus occur at the crests of the waves and are absent in the troughs. A lens-shaped cloud may also

**FIGURE 7.22**
Altocumulus lenticularis clouds are nearly stationary mountain-wave clouds that are generated by airflow disturbed by passage over a mountain range. [Photograph by Arjen and Jerrine Verkaik/SKYART]

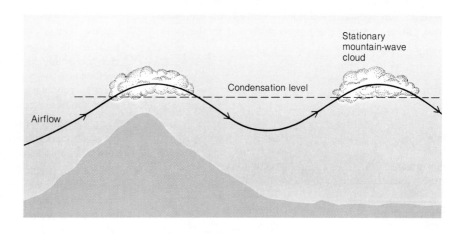

**FIGURE 7.23**
Mountain-wave clouds form when a mountain range deflects the horizontal wind into a wavelike pattern. Clouds develop on the wave crests where the airflow is upward and expansional cooling takes place. Clouds are absent in the wave troughs where airflow is downward and compressional warming occurs.

form over and extend just downwind of a mountain peak; such a cloud is known as a **banner cloud.**

Mountain-wave clouds do not move, even though the winds are strong, because the wave itself is stationary. A stationary wave within the atmosphere is known as a *standing wave,* and a standing wave produced by mountains is called a **mountain wave.** A waterfall on a river is a good analog. In the plunge pool at the base of the waterfall, the water is very turbulent, but as we follow the river downstream, we quickly leave the turbulent region behind. The river water flows ceaselessly, but the disturbed segment remains stationary at the foot of the waterfall, just as a mountain wave remains stationary to the lee of a mountain range.

Interestingly, patterns exhibited by stratocumulus and cirrocumulus (as well as by varieties of altocumulus in addition to lenticularis) are also caused by waves traveling within the atmosphere. These waves, however, are not linked to mountain ranges and propagate horizontally at different altitudes within the troposphere. Atmospheric waves produce bands of clouds that are aligned either parallel or perpendicular to the wind direction.

Because almost all water vapor is confined to the troposphere, so too, are most clouds. One exception is the occasional intense cumulonimbus cloud, the top of which may penetrate the tropopause and enter the lowest portion of the stratosphere. Another exception is the colorful **nacreous clouds,** which occur in the upper stratosphere. Temperatures at these altitudes (25 to 30 km or 16 to 19 mi) favor water in either the solid or supercooled state. Because of their soft, pearly luster, these rarely seen clouds are also called *mother-of-pearl*

*clouds.* They are best viewed at high latitudes in winter when illuminated by the setting sun.

Somewhat mysterious are the wavy, cirruslike **noctilucent clouds** that occur in the upper mesosphere. Some scientists suggest that noctilucent clouds are composed of ice deposited on meteoric dust particles. They are rarely seen, and then only at high latitudes during the twilight just after sunset or before sunrise. One of the most widely reported occurrences of noctilucent clouds was on 18–19 July 1990 over a large area of western Canada.* For unknow reasons, since 1981 noctilucent clouds have become brighter and more common.

## Conclusions

Nocturnal radiational cooling may lead to the development of dew, frost, or radiation fog. Other low-level processes that can bring air to saturation include advection of warm, humid air over a relatively cold surface or cold air advection over a relatively warm, wet surface. Within the atmosphere, as air approaches saturation, the abundance of nuclei (and especially hygroscopic nuclei) favors deposition or condensation, that is, the development of clouds. We classify clouds on the basis of appearance and altitude of occurrence.

*For more on this event, see the article by M. S. Zalcik and T. W. Lohvinenko, *Bulletin of the American Meteorological Society,* 72 (1991):1001–1003.

Clouds may or may not yield precipitation—most do not. The special conditions required for rain and snowfall are among the topics discussed in the next chapter.

## Key Terms

| | |
|---|---|
| dew | cirrus |
| frost | cirrostratus |
| dew point | cirrocumulus |
| frost point | altostratus |
| hoarfrost | altocumulus |
| fog | stratocumulus |
| mist | stratus |
| radiation fog | nimbostratus |
| air mass advection | convective condensation |
| advection fog | level (CCL) |
| arctic sea smoke | cumulus |
| steam fog | cumulus congestus |
| upslope fog | cumulonimbus |
| nuclei | wind shear |
| hygroscopic nuclei | altocumulus lenticularis |
| cloud condensation nuclei | mountain-wave clouds |
| (CCN) | banner cloud |
| ice-forming nuclei | mountain wave |
| stratiform clouds | nacreous clouds |
| cumuliform clouds | noctilucent clouds |

## Summary Statements

☐ Nocturnal radiational cooling chills the Earth's surface, which, in turn, cools the air in immediate contact with the surface. If the temperature of that surface drops below the dew point (or frost point), then water vapor condenses as dew (or deposits as frost).

☐ For unsaturated air, the dew point is lower than the wet-bulb temperature, which, in turn, is lower than the ambient air temperature. When the relative humidity is 100%, the dew point, wet-bulb temperature, and ambient air temperature are equivalent.

☐ Fog is a visibility-restricting suspension of tiny water droplets or ice crystals in an air layer in contact with the Earth's surface. Essentially, fog is a cloud at ground level. Based on mode of origin, fog is classified as radiation fog, advection fog, steam fog, or upslope fog.

☐ Clouds, the visible manifestations of condensation or de-

position within the atmosphere, are composed of minute water droplets, ice crystals, or a combination of the two.

☐ As the relative humidity within the atmosphere approaches 100%, condensation and deposition occur on nuclei, tiny solid and liquid particles. Cloud condensation nuclei are much more abundant than ice-forming nuclei. Most ice-forming nuclei are active at temperatures well below the freezing point.

☐ Many condensation nuclei are hygroscopic, that is, they have a chemical affinity for water molecules and promote condensation at relative humidities under 100%.

☐ Clouds are classified by appearance and altitude of occurrence. Based on appearance, clouds are cirrus, stratus, or cumulus; and, based on altitude, clouds are high, middle, low, or vertical.

☐ High, middle, and low clouds are caused by relatively gentle uplift of air over a broad area. Hence, these clouds are layered, that is, stratiform. Clouds with vertical development are the consequence of more vigorous uplift and are heaped or puffy in appearance, that is, cumuliform.

☐ High clouds are composed of mostly ice crystals, whereas middle and low clouds are mostly water droplets.

☐ Nimbostratus and cumulonimbus are the principal precipitation-producing clouds. Precipitation from nimbostratus is typically lighter and of longer duration than precipitation from cumulonimbus.

☐ Atmospheric stability determines whether cumulus clouds build vertically into cumulus congestus or cumulonimbus clouds. The convective condensation level corresponds to the base of cumuliform clouds.

☐ Mountain-wave clouds develop downwind of a prominent mountain range and are nearly stationary. Nacreous and noctilucent clouds are among the very few clouds that occur above the troposphere. Nacreous clouds form in the upper stratosphere and are composed of ice crystals and/or supercooled water droplets. Noctilucent clouds develop in the upper mesosphere and may be composed of ice deposited on meteoric dust particles.

## Review Questions

1. List the atmospheric conditions that are most favorable for extreme nocturnal radiational cooling.
2. Dew and frost are not forms of precipitation. Explain this statement.
3. How is the dew point a measure of the water vapor concentration of air?
4. Describe the circumstances that favor development of ra-

diation fog. Why does this type of fog often form in low-lying areas such as marshes or river valleys?

5. Why is radiation fog usually short-lived? Under what conditions might radiation fog persist for many days?

6. Describe two different situations in which fog forms as a consequence of warm air advection.

7. How does steam fog develop? Why does steam fog appear as rising streamers?

8. What is a cloud?

9. List several sources of cloud condensation nuclei.

10. What is the significance of *hygroscopic* nuclei?

11. Distinguish between cloud condensation nuclei and ice-forming nuclei.

12. Distinguish between freezing nuclei and deposition nuclei.

13. How are clouds classified?

14. Distinguish between stratiform and cumuliform clouds.

15. How and why does the composition of clouds vary with altitude?

16. What is the significance of the convective condensation level (CCL)?

17. Why do fair-weather cumulus clouds tend to vaporize toward sunset?

18. How does air stability influence the vertical development of cumuliform clouds?

19. What is the effect of a snow cover on cumulus cloud development?

20. What causes the ripple or banded appearance of cirrocumulus, altocumulus, and stratocumulus clouds?

## Quantitative Questions

1. If the air temperature at the convective condensation level is 0 °C (32 °F) and the air temperature at the ground is 20 °C (68 °F), determine the approximate altitude of the cumulus cloud base above the ground.

2. Determine the dew point if
   a. the air temperature is 10 °C (50 °F) and the wet-bulb depression is 2 Celsius degrees (3.6 Fahrenheit degrees).
   b. the air temperature is 15 °C (59 °F) and the relative humidity is 70%.
   c. the relative humidity is 100% and the wet-bulb temperature is 17 °C (63 °F).

3. How much latent heat is released (in calories) when 500 g of dew form on a cold surface.

4. How much latent heat is released (in calories) when 500 g of frost form on a cold surface.

## Questions for Critical Thinking

1. Is it possible for upslope fog, orographic rainfall, and wave clouds to form simultaneously on the same mountain range? Elaborate on your answer.

2. Speculate on what might cause a cumuliform cloud to tilt or lean with altitude.

3. Advection fog can have a catastrophic effect on a snow cover. Explain why.

4. Is the formation of dew an adiabatic process? Explain your response.

5. On a clear, calm night, would you expect a coating of frost to be thicker on the topside (facing the sky) or bottomside (facing the ground) of a plant leaf? Justify your choice.

## Selected Readings

Bohren, C. F. "All That Glistens Isn't Dew," *Weatherwise* 43, No. 5 (1990):284–287. Argues that water drops that appear at the tips of plant leaves may not be dew but rather water exuded by the plant (a process known as guttation).

Bohren, C. F. "An Essay on Dew," *Weatherwise* 41, No. 4 (1988):226–231. Provides a fascinating explanation of the formation of dew and frost.

Day J. A., and V. J. Schaefer. *Peterson's First Guide to Clouds and Weather.* Boston, MA: Houghton Mifflin Company, 1991, 128 pp. Presents an exceptionally well-illustrated survey of clouds, optical effects, and precipitation processes.

Gedzelman, S. D. "In Praise of Altocumulus," *Weatherwise* 41, No. 2 (1988):143–149. Discusses the origins of the various forms of altocumulus clouds and associated optical effects.

Hobbs, P. V. "Holes in Clouds?" *Weatherwise* 38, No. 5 (1985):254–258. Explores proposed explanations for holes and canals in clouds.

Richards, C. J. "Focus on the Sky—The Black and White Way," *Weather* 41 (1986):214–224. Advises on the advantages and techniques of photographing clouds with black and white film.

Scorer, R., and A. Verkaik. *Spacious Skies.* London: David and Charles, 1989, 192 pp. Includes outstanding cloud photography.

# 8 Precipitation, Weather Modification, and Atmospheric Optics

*Be thou the rainbow to the storms of life,*
*The evening beam that smiles the clouds away,*
*And tints to-morrow with prophetic ray!*

LORD BYRON
*BRIDE OF ABYDOS*

Snow and all other forms of precipitation originate in clouds where a special set of conditions triggers rapid growth of ice crystals or water droplets. [Photograph by Tranquality/Philip Chaudoir]

T HIS CHAPTER continues our study of water in the atmosphere by considering several closely related topics. In the first section, we focus on how precipitation develops in clouds and the various forms and measurement of precipitation. This discussion serves as the basis for our subsequent consideration of techniques to induce precipitation (cloud seeding) and to dissipate fog. The chapter closes with a description of various optical phenomena that are produced when sunlight or moonlight interacts with clouds or precipitation.

## Precipitation Processes

Appearance of clouds in the sky is no guarantee that it will rain or snow. Nimbostratus and cumulonimbus clouds produce the bulk of precipitation, but most clouds—even most of those associated with a large storm system—do not yield any rain or snow. This is because a special combination of circumstances, as yet not fully understood, is required for precipitation to develop.

### TERMINAL VELOCITY

The water droplets or ice crystals that compose clouds are so minute that they remain suspended indefinitely, unless they vaporize or undergo considerable growth. Updrafts within clouds are usually strong enough to prevent cloud particles from leaving the base of a cloud and falling to the Earth's surface. Even if droplets or ice crystals descend from a cloud, their downward drift is so slow that they travel only a short distance before vaporizing in the unsaturated air beneath the cloud.

The speed of a falling cloud droplet or ice crystal (or any other particle) in calm air is regulated by two forces: (1) the force of gravity, which accelerates the particle downward, and (2) an opposing force caused by the resistance of the air through which the particle descends. As the particle accelerates downward, it meets with increasing air resistance. (Meanwhile, the force of gravity remains essentially constant.) Eventually, the resisting force equals (balances) the force of gravity, and the particle drifts downward at a constant speed known as the **terminal velocity.***

For a particle to remain suspended in air, updrafts must be strong enough to counter the particle's terminal velocity. Generally, terminal velocity increases with the size of the particle (Figure 8.1). Hence, the larger the particle, the more vigorous the updraft must be in order to keep the particle in suspension. Cloud droplets and ice crystals are so small (most having diameters of 10 to 20 micrometers) that their very low terminal velocities (typically only 0.3 to 1.2 cm/sec) are readily countered by even weak updrafts. Even if there were no updrafts, such low terminal velocities would mean that it would take 24 hr or longer for cloud particles to reach the ground. Long before that happened, all the cloud particles would vaporize in the unsaturated air below cloud base.

Cloud particles somehow must grow large enough to counter updrafts so that they survive a descent to

---

*The concept of balanced forces is discussed in greater detail in the next chapter.

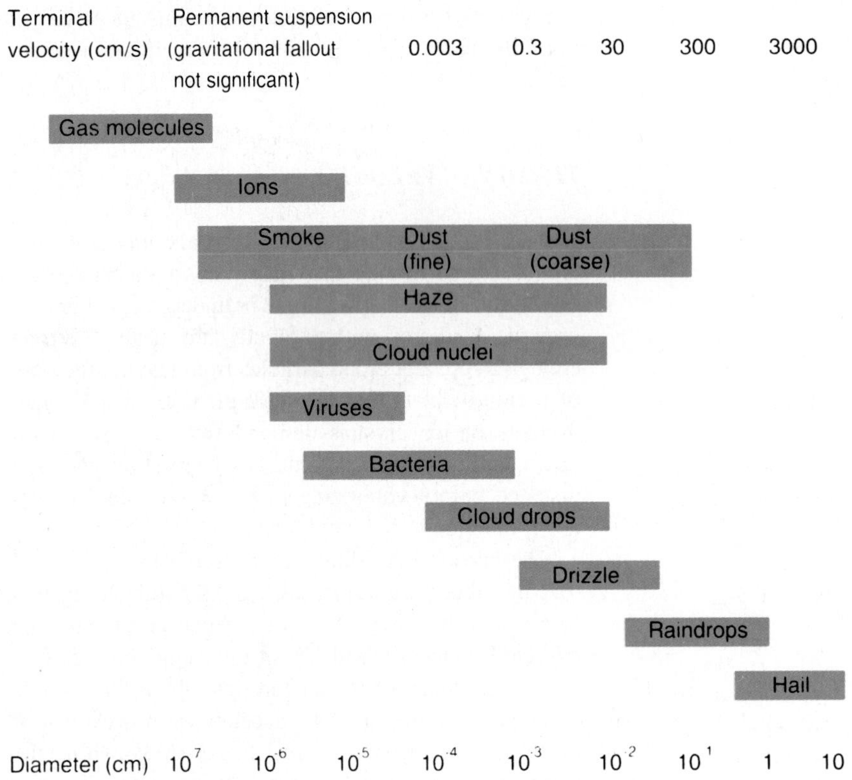

| Terminal velocity (cm/s) | Permanent suspension (gravitational fallout not significant) | | 0.003 | 0.3 | 30 | 300 | 3000 |

Gas molecules

Ions

Smoke    Dust (fine)    Dust (coarse)

Haze

Cloud nuclei

Viruses

Bacteria

Cloud drops

Drizzle

Raindrops

Hail

| Diameter (cm) | $10^7$ | $10^6$ | $10^5$ | $10^{-4}$ | $10^{-3}$ | $10^{-2}$ | $10^1$ | 1 | 10 |

**FIGURE 8.1**
The terminal velocity of a particle falling through air increases with the size of the particle. [Adapted from R. A. Anthes, *Meteorology*, 6th ed. New York: Macmillan Publishing Company, 1992, p. 6.]

the Earth's surface as raindrops or snowflakes without completely vaporizing. This is no minor task! It takes about 1 million droplets (having diameters of 10 to 20 micrometers) to form a single raindrop (about 2 mm in diameter). How does this growth take place?

Cloud physicists have determined that condensation alone cannot cause cloud droplets to grow into raindrops or snowflakes. They have identified two important processes by which cloud particles grow large enough to precipitate: the collision–coalescence process and the Bergeron process.

## COLLISION–COALESCENCE PROCESS

Within a **warm cloud,** that is, a cloud at temperatures above the freezing point of water, droplets may grow by colliding and coalescing (merging) with one another. The **collision–coalescence process** requires that droplets composing a cloud have a range of diameters. Droplets of uniform diameter have essentially the same terminal velocity, and so collisions between

such droplets are rare. In contrast, cloud droplets of unequal diameters have different terminal velocities, and collisions are much more frequent. Unequal droplet diameters within the same cloud usually stem from the presence of *giant* sea-salt nuclei that produce relatively large droplets (greater than 40 micrometers in diameter).

A larger (heavier) droplet descends more rapidly than a smaller (lighter) droplet. As it falls, a larger droplet intercepts and coalesces with smaller (slower) droplets in its path (Figure 8.2) By repeated collisions and coalescence, a droplet grows, its terminal velocity increases, and collisions become more frequent. A droplet may become so large that it splits into smaller droplets that go on to coalesce with other droplets in a kind of chain reaction.* Eventually, droplets are large enough to fall from the cloud and reach the Earth's surface as raindrops.

---

*Not all collisions between droplets result in coalescence. Sometimes small droplets bounce off larger droplets just as small insects sweep past the windshield of a moving auto.

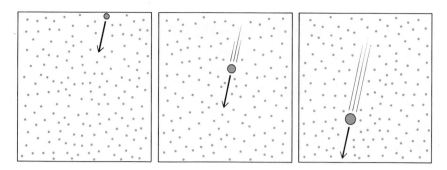

**FIGURE 8.2**
A relatively large water droplet falls within a cloud of much smaller droplets. The large droplet collides with smaller droplets in its path and grows by coalescence.

## *BERGERON PROCESS*

Although cloud droplet growth through collision–coalescence alone is important, especially in the tropics, most precipitation that falls in the middle and high latitudes originates through the Bergeron process. Named for the Scandinavian meteorologist Tor Bergeron, who first described it in 1933, the **Bergeron process** applies to **cold clouds,** which are at temperatures below 0 °C (32 °F). The process requires the coexistence of water vapor, ice crystals, and supercooled water droplets. Before getting into the specifics of the Bergeron process, we first need to examine the composition of cold clouds.

As we noted in Chapter 7, most ice-forming nuclei are not active at temperatures higher than −9 °C (16 °F). Consequently, clouds at temperatures between 0 and −9 °C (32 and 16 °F) are typically composed of supercooled water droplets exclusively. At temperatures between about −10 and −20 °C (14 and −4 °F), clouds are mixtures of mostly supercooled water droplets and some ice crystals. Below −20 °C (−4 °F), the

activation temperature for many deposition nuclei, clouds usually consist of ice crystals only.

The distribution of supercooled water droplets and ice crystals is somewhat more complicated in clouds that have significant vertical development. Cumulonimbus clouds, for example, have different components at different altitudes depending on the vertical temperature profile within the cloud. Typically, they are composed of ice crystals aloft where temperatures are very low and water droplets near the milder cloud base. In between is a mixture of supercooled water droplets and ice crystals. Furthermore, vigorous convection currents within a cumulonimbus cloud transport liquid water droplets aloft where they freeze. This is an important source of ice crystals within thunderstorm clouds.

The Bergeron process takes place in clouds (or portions of clouds) that consists of a mixture of ice crystals and supercooled water droplets (Figure 8.3). Initially, supercooled water droplets far outnumber ice crystals because cloud condensation nuclei are much more abundant than ice-forming nuclei. Very quickly, however, ice crystals grow at the expense of supercooled

**FIGURE 8.3**
Within a cold cloud, ice crystals grow at the expense of supercooled water droplets. As they grow larger, ice crystals fall faster and collide with droplets and other ice crystals in their paths. Eventually, they grow large enough to fall out of the cloud as snowflakes.

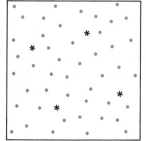
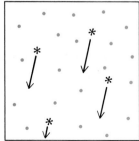
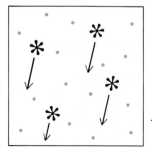

Supercooled water droplets

✱ Ice crystals

**FIGURE 8.4**
Some of the rain falling from the base of a distant thundershower never reaches the ground because the precipitation vaporizes in the relatively dry air beneath the thundercloud. [Photograph by J. M. Moran]

water droplets primarily because the saturation vapor pressure is greater over water than over ice.

At subfreezing temperatures, water molecules vaporize more readily from a liquid water surface than from solid ice because water molecules are bonded more strongly in the solid phase than in the liquid phase. Consequently, the saturation vapor pressure is greater over supercooled water than over ice (refer back to Table 6.3). It follows that in clouds composed of a mixture of ice crystals and supercooled water droplets, a vapor pressure that is *saturated* for water droplets is *supersaturated* for ice crystals. Suppose, for example, that the vapor pressure is 2.86 mb in a cloud at a temperature of $-10$ °C (14 °F). From Table 6.3, this vapor pressure translates into a relative humidity of 100% (saturation) for the air surrounding water droplets and a relative humidity of 110% (supersaturation) for the air surrounding ice crystals. In response to supersaturated conditions, water vapor deposits on the ice crystals causing them to grow. Deposition removes water vapor from the cloud so that the relative humidity of air surrounding the water droplets dips below 100%, and the droplets vaporize. Thus, in the Bergeron process, ice crystals grow at the expense of supercooled water droplets.

As ice crystals grow larger and heavier, their terminal velocities increase, and they collide and coalesce with smaller, slower moving supercooled water droplets and ice crystals that are in their path. Thereby ice crystals grow still larger. Eventually, some ice crystals

become so heavy that they fall out of the cloud base. If air temperatures are below freezing at least most of the way to the ground, ice crystals reach the Earth's surface as snowflakes. If the air below the cloud is above freezing, snowflakes melt and fall as raindrops.

Once a raindrop or a snowflake leaves a cloud, it enters unsaturated air—an environment that favors either evaporation or sublimation of precipitation. In general, the longer the journey to the ground and the lower the relative humidity of the air beneath the clouds, the greater the quantity of rain or snow that vaporizes. Figure 8.4 shows rain falling from a distant thunderstorm; some of the rain evaporates as it descends through drier air. The shaft of precipitation appears as a dark curtain against a bright background and is known as **virga.** Virga may consist of snow melting to rain, precipitation that has not yet reached the ground, or precipitation that completely vaporizes prior to reaching the ground.*

## Forms of Precipitation

**Precipitable water** is the amount of water produced when all the water vapor in a column of air condenses. The air column is assumed to extend from the Earth's

---

*For more information on virga, refer to the article by C. F. Bohren cited at the end of this chapter.

surface to the tropopause, and the condensed water is described in units of depth. Condensing all the water vapor within the troposphere would produce a layer of water covering the entire Earth's surface to an average depth of only about 25 mm (1.0 in.). With the poleward decline in air temperature, precipitable water amounts also decline. (Lower temperatures reduce the rate at which water vaporizes into the atmosphere.) Average precipitable water depths decrease with latitude from more than 40 mm (1.6 in.) in the humid tropics to less than 5 mm (0.2 in.) near the poles.

As we have seen, within the troposphere some water vapor changes phase and becomes the water droplets and ice crystals of clouds, and through the collision–coalescence and Bergeron processes, some clouds yield precipitation. **Precipitation** is water in solid or liquid form that falls from clouds to the Earth's surface. Besides the familiar rain and snow, precipitation also occurs as drizzle, freezing rain, ice pellets, and hail.

**Drizzle** consists of small water drops from 0.2 to 0.5 mm in diameter that drift very slowly toward the Earth's surface. Drizzle drops are relatively small because they originate in stratus clouds. Stratus clouds are so low and thin (compared to nimbostratus and cumulonimbus clouds) that droplets have only a limited opportunity to grow by collision–coalescence. Drizzle is associated with fog and poor visibility, but never with convective clouds.

**Rain** falls mostly from nimbostratus and cumulonimbus clouds, and the bulk of rain originates as snowflakes (or hailstones), which melt on the way down as they descend through air that is above 0 °C (32 °F). Because rain originates in thicker clouds that have higher bases, raindrops travel farther than drizzle and undergo more growth by collision–coalescence. Most commonly, raindrop diameters are in the range of 1 to 6 mm; at greater diameters, drops are unstable and break up into smaller drops.

Raindrops that are produced in warm clouds are usually smaller than those falling from cold clouds. Rarely do warm cloud raindrops reach 2 mm in diameter. Precipitation growth via a combination of the Bergeron and collision–coalescence processes (in cold clouds) produces larger drops than through collision–coalescence alone (in warm clouds). Recently, research meteorologists sampling warm cloud raindrops over Hilo, Hawaii discovered an exception to this rule. They re-

ported numerous raindrops of 4 to 5 mm in diameter and some as large as 8 mm in diameter (about the size of a pea). In this case, strong updrafts in convective clouds prolonged the period of collision–coalescence, thereby enabling warm drops to grow to extraordinary sizes.

**Freezing rain** (or freezing drizzle) forms a coating of ice (Figure 8.5) that sometimes grows thick and heavy enough to bring down tree limbs, snap power lines, and totally disrupt traffic. Freezing rain develops when rain falls from a relatively mild air layer aloft into a shallow layer of subfreezing air at ground level. The drops become supercooled and then freeze immediately on contact with cold surfaces.

Freezing rain can be localized and persistent, especially in hilly or mountainous terrain such as the Appalachians of Pennsylvania and West Virginia. Weather events may unfold as follows. During a clear winter night extreme radiational cooling chills a shallow layer of air in contact with the ground to temperatures well

**FIGURE 8.5**
Freezing rain coats subfreezing surfaces of branches, power lines, and pavement. [Photograph by Mike Brisson]

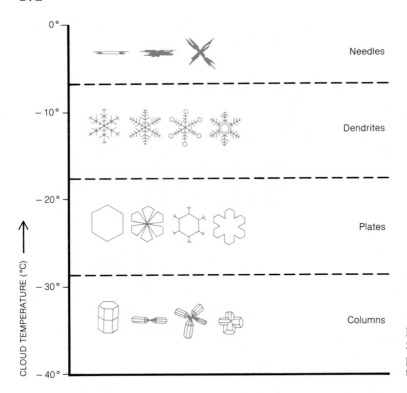

**FIGURE 8.6**
Snowflakes take on a variety of forms depending at least partially on cloud temperature.

below freezing. The cold, dense air drains downslope and settles into deep river valleys. (Under these circumstances, a temperature inversion develops, and the lowest air temperatures occur at the lowest elevations.) Then, after sunrise, an approaching storm system spreads clouds into the region. The wind strengthens but blows perpendicular to the mountain range so that shallow pools of subfreezing air are trapped in the valley bottoms. Meanwhile, at higher elevations along the mountain slopes, air temperatures rise above freezing. Rain then develops and is chilled as it falls through subfreezing air in the valley bottoms. Raindrops freeze on contact with cold surfaces such as roads, tree branches, and utility wires. Such a situation often persists until the wind shifts parallel to the valleys and flushes out the layer of subfreezing air.

**Snow** is an assemblage of ice crystals in the form of flakes. Although it is said that no two snowflakes are identical, all snowflakes are hexagonal (six-sided), just as the constituent ice crystals are hexagonal. Snowflake form varies with water vapor concentration and temperature and may consist of plates, stars, columns, or needles (Figure 8.6). Snowflake size depends in part on the availability of water vapor during the crystal growth process. At very low temperatures, water vapor concentrations are low, so snowflakes are relatively small. Snowflake size also depends on collision efficiency as the flakes drift toward the ground. At relatively high temperatures, snowflakes are wet and readily stick together after colliding, so that aggregate diameters may exceed 5 to 10 cm (2 to 4 in.). The appearance of such large flakes is usually a signal that the snow is about to turn to rain.

Snow pellets and snow grains are closely related to snowflakes. **Snow pellets** are soft conical or spherical white particles of ice with diameters of 1 to 5 mm. They form when supercooled cloud droplets collide and freeze together. **Snow grains** originate in much the same way as drizzle except that they are frozen; diameters usually are less than 1 mm.

Is it ever too cold to snow? When is it too warm to snow? For answers to these questions, see the Special Topic.

**Ice pellets,** commonly called *sleet,* are actually frozen raindrops 5 mm or less in diameter. They develop in much the same way as freezing rain with this difference: The surface layer of cold air is so deep that raindrops freeze *prior* to striking the ground. We can

SPECIAL TOPIC

# When Is It Too Cold or Too Warm to Snow?

During an episode of particularly frigid winter weather, some people argue that it's too cold to snow. Indeed, the coldest weather is often accompanied by fair skies, and the climate records of the northern United States and Canada indicate that snowfall totals decline with falling temperature. For some midwestern United States cities (such as Bismarck and Minneapolis), March on average is both the snowiest and the mildest of the winter months (December through March). In Canada, outside of the mountains, average annual snowfall declines from more than 400 cm (157 in.) in parts of Newfoundland to less than 100 cm (40 in.) on the frigid shores of the Arctic Ocean. Even though total snowfall does indeed decrease with falling temperature, snow is possible even at extremely low air temperatures. In bitter cold air, however, snowflakes are small and accumulations are usually meager.

The relatively small amount of water vapor in very cold air means that comparatively little water is available for precipitation. Recall from Chapter 6 that the saturation vapor pressure drops rapidly as air temperature falls. For example, the water vapor concentration in saturated air at −30 °C (−22 °F) is only about 12% of the water vapor concentration in saturated air at −5 °C (23 °F). Hence, the amount of water available for precipitation decreases with falling air temperature.

The heaviest snowfalls typically occur when the temperature of the lower atmosphere is within a few degrees of the freezing point because at that temperature the potential amount of water that can precipitate in the solid form (as snow) is at a maximum. On the other hand, moderate or heavy snowfall is very unlikely when the temperature of the lower atmosphere falls below −20 °C (−4 °F). But even in the coldest regions of the globe where precipitable water amounts are lowest, some snow falls. For example, an estimated average annual 5 cm (2 in.) of snow falls on the high interior plateau of Greenland where average annual temperatures are below −30 °C (−22 °F).

When is it too warm to snow? Surprisingly, snow can fall even when the near-surface air temperature is as high as 10 °C (50 °F)! The only requirement is that the wet-bulb temperature remains below 0 °C (32 °F), which also means that the relative humidity is very low. For example, if the air temperature is 5 °C (41 °F), the relative humidity must be under 32% for the wet-bulb temperature to be subfreezing (see the psychrometric tables inside the back cover). The first snowflakes vaporize or melt as they fall through relatively dry, above-freezing air and reach the ground as raindrops. Vaporization and melting of snowflakes tap heat from the ambient air; that is, sensible heat is converted to latent heat, and the air temperature drops. With sufficient vaporization and melting, the air eventually cools to the wet-bulb temperature, that is, below the freezing point. Hence, at the Earth's surface, what started out as rain (or a mixture of rain and snow) turns to snow. This is most likely to happen if the precipitation is moderate to heavy—more snowflakes that are melting and vaporizing means more cooling.

readily distinguish ice pellets from freezing rain, because ice pellets bounce when striking the ground and freezing rain does not. Accumulations of ice pellets or freezing rain can cause very hazardous walking and driving conditions.

**Hail** consists of rounded or jagged lumps of ice, often characterized by concentric layering resembling the internal structure of an onion (Figure 8.7). Hail develops within intense thunderstorms as strong convection currents transport ice pellets upward, into the middle and upper reaches of a cumulonimbus cloud. Along the way, ice pellets grow larger by collecting supercooled water droplets, and eventually become too heavy to be supported by convective updrafts. Ice pellets then descend through the cloud, exit the cloud base, and enter air that is typically above the freezing point. The pellets begin to melt, but if large enough initially, some ice will survive the journey to the ground as hailstones. Most hail consists of harmless granules of ice less than 1 cm in diameter, but violent thunderstorms may spawn destructive hailstones the size of golf balls or larger. Hail is usually a spring and summer phenomenon that

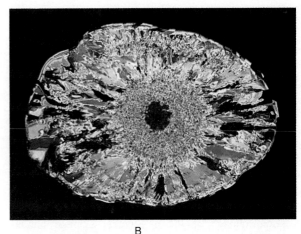

A                                                          B

**FIGURE 8.7**
(A) Hailstones sometimes grow to the size of golf balls or larger. (B) A hailstone consists of
concentric layers of clear and opaque ice as shown in this cross section photographed in polar-
ized light. [Photographs from NOAA (A) and National Center for Atmospheric Research/
University Corporation for Atmospheric Research/National Science Foundation (B)]

is particularly devastating to crops. We have more to
say about hail in Chapter 13.

## Precipitation Measurement

Today we collect and measure precipitation using essentially the same device that was used as long ago as
the fifteenth century: a container open to the sky. The
standard National Weather Service **rain gauge** is
equipped with a cone-shaped funnel at the top that di-
rects rainwater into a long, narrow cylinder, which is
seated inside a larger, outer cylinder (Figure 8.8). The
funnel and narrow cylinder magnify the scale so that
the instrument resolves rainfall to increments of 0.01
in. Total rainfall of less than 0.005 in. is recorded as a
*trace*. Rainwater that accumulates in the inner cylinder
is measured by a stick, which, in the United States, is
graduated in inches. Canada and most other nations use
metric units. Rainfall is measured at some fixed time
once every 24 hours, and the gauge is then emptied.

With regard to snow, we are interested in measuring (1) the depth of snow that falls during each 24-hour
period between observations, (2) the meltwater equivalent of that snowfall, and (3) the depth of snow on the
ground at each observation time. New snowfall is usu-

ally collected on a simple board that is left on top of
the old snow cover. When new snow falls, the depth is
measured to the board; the board is then swept clean
and moved to a new location. The meltwater equivalent of new snowfall can be determined by melting the
snow collected in a rain gauge (with the funnel and inner cylinder removed). Snow depth is usually measured
with a yardstick graduated in tenths of an inch or a
meterstick. Snow depth is determined at several representative locations and then averaged.

As a general rule, 10 cm of fresh snow melts down
to 1 cm of water, although this ratio varies considerably depending on the temperature at which the snow
falls. Wet snow falling at surface air temperatures at or
above freezing has a much greater water content (per
centimeter) than dry snow falling at very low surface
air temperatures. The ratio of snowfall to meltwater
may vary from 3 to 1 for very wet snow to 30 to 1 for
dry fluffy snow.

Monitoring the rate of rainfall is often desirable, especially in flood-prone areas. Accordingly, some rain
gauges are designed to provide a cumulative record of
rainfall with time. A **weighing-bucket rain gauge** consists of a continuously recording scale that calibrates
the weight of accumulating rainwater as water depth.
Cumulative rainfall is recorded by either a device that
punches a paper tape or a pen that marks a chart on a
clock-driven drum. During subfreezing weather, anti-

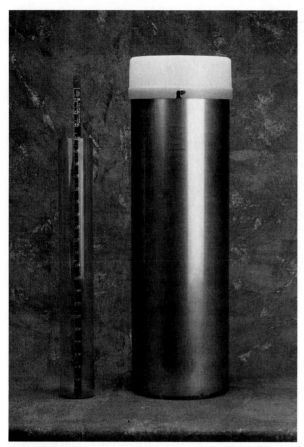

**FIGURE 8.8**
A standard National Weather Service rain gauge. [Courtesy of Qualimetrics]

**FIGURE 8.9**
A tipping-bucket rain gauge is designed to provide a cumulative record of rainfall in increments of 0.01 in. [Courtesy of Qualimetrics]

freeze in the collection bucket melts snow as it falls into the gauge so that a cumulative meltwater record is produced.

A **tipping-bucket rain gauge** (Figure 8.9) is somewhat more accurate than a weighing-bucket rain gauge but does not perform well in subfreezing weather. This instrument features two small, free-swinging containers, each of which can collect the equivalent of 0.01 in. of rainfall. Alternating with one another, each container fills with water, tips and spills its contents, and thereby trips an electric switch that either marks a clock-driven chart or sends an electric pulse for recording by a computer or magnetic tape.

Both rainfall and snowfall are notoriously variable from one place to another, especially in convective showers. Therefore, the siting of precipitation gauges is particularly important in order to ensure accurate and representative readings. The site must be sheltered from strong winds and well away from buildings and vegetation that might shield the instrument. Generally, obstacles should be no closer than four times their height.

## Weather Modification

**Weather modification** is any change in weather that is induced by human activity. The activity may be either intentional or inadvertent. This section describes two principal types of intentional weather modification: cloud seeding and fog dispersal. Later, we discuss hail suppression (Chapter 13), hurricane modification (Chapter 15), and the impact of air pollution on weather and climate (Chapters 17 and 20).

## *CLOUD SEEDING*

Since World War II, considerable research has gone into methods of enhancing precipitation by cloud seeding. **Cloud seeding** is an attempt to stimulate natural precipitation processes by injecting nucleating agents into clouds. Most cloud-seeding experiments are directed at cold clouds.

The objective of seeding cold clouds is to stimulate the Bergeron process in clouds that are deficient in ice crystals. The seeding (nucleating) agent is either silver iodide (AgI), a substance with crystal properties similar to those of ice, or dry ice, solid carbon dioxide ($CO_2$) at a temperature of about $-80$ °C ($-112$ °F). Silver iodide crystals are freezing nuclei that are active at $-4$ °C (25 °F) and below. Dry ice pellets are so cold that within a cloud they cause surrounding supercooled water droplets to freeze. Frozen droplets then function as nuclei that grow into snowflakes. Furthermore, latent heat released when supercooled water droplets freeze increases buoyancy and thereby stimulates additional cloud growth.

Cloud seeding is done from aircraft by either firing flares containing silver iodide (Figure 8.10) or dispensing dry ice pellets from a hopper. Cloud seeding from ground-based generators can be less satisfactory because the seeding agent (silver iodide) may not diffuse adequately or reach sufficient altitudes to be effective. Nonetheless, both methods continue to be used to seed cold clouds.

In warm clouds of relatively uniform droplet size, sea-salt crystals and other hygroscopic substances can be injected to trigger the development of relatively large cloud droplets. Such seeding stimulates the collision–coalescence process.

The *Sierra Cooperative Pilot Project (SCPP)* is an example of a long-term, ongoing study of cloud seeding. So far, investigators have compiled more than a decade of data on efforts to seed winter orographic clouds on the windward western slopes of California's Sierra Nevada. The goal of this seeding is to enhance snowfall and thereby thicken the mountain snowpack. The consequent increase in spring runoff is intended to help California meet its growing domestic and agricultural water demands.

The American River Basin, just west of Lake Tahoe, is the principal SCPP study site, and January through March is the primary seeding season. Clouds targeted for seeding are those rich in supercooled water droplets and deficient in ice crystals. In order to identify clouds best suited for precipitation enhancement through seeding, SCPP scientists employ an aircraft that is outfitted with sophisticated instruments that measure the size and concentration of cloud and precipitation particles. Another aircraft seeds clouds with silver iodide crystals or dry ice pellets, and, on the ground, an array of precipitation gauges, radar, and other weather instruments monitor the effectiveness of seeding.

Does cloud seeding work? In the 1970s and early 1980s, NOAA scientists conducted a statistically rigorous experiment designed to test the effectiveness of weather modification. The experiment was carried out over southern Florida and involved the seeding of cumulus clouds.* Test days were about evenly divided between days when clouds were seeded with silver iodide crystals and days when clouds were seeded with inert (chemically inactive) sand grains. Only after the experiment was completed and the results analyzed were participating scientists told of the specific days

---

*Florida Area Cumulus Experiment (FACE).

**FIGURE 8.10**
This aircraft, operated by the University of Wyoming, is injecting clouds with nucleating agents (silver iodide or dry ice) in an effort to stimulate precipitation formation via the Bergeron process. [Photograph by National Center for Atmospheric Research/University Corporation for Atmospheric Research/ National Science Foundation]

when silver iodide was used. This procedure was designed to ensure both an unbiased selection of clouds to be seeded and an unbiased interpretation of results.

The experiment was divided into two phases. Results from an initial *exploratory* phase were to be either verified or rejected by a later *confirmatory* phase when seeding was repeated. Results of the first phase were very encouraging, showing a 25% increase in rainfall on days when silver iodide was the seeding agent, compared to days when sand was the bogus seeding agent. This finding was statistically significant at the 90% level, which means that there is only a 10% probability that the rainfall increase was a chance occurrence. The success of the initial phase was not repeated during the second phase of the experiment, however, when seeding showed no statistically significant increase in rainfall.

The same type of experimental design, involving separate exploratory and confirmatory phases, was carried out in Israel from 1961 to 1967 and from 1969 to 1975. In these studies, however, statistically significant rainfall enhancement during the first phase was verified during the second phase of seeding. The contrasting findings of the Florida and Israeli experiments underscore the uncertainties of cloud seeding and the need for a better understanding of precipitation processes.

Although cloud seeding is probably successful in some instances, the actual volume of additional precipitation produced by cloud seeding and the advisability of large-scale seeding efforts are matters of considerable controversy. Some cloud seeders claim to increase precipitation by 15% to 20% or more, but the question remains: Would the rain or snow that follows cloud seeding have fallen anyway? Even if apparently successful, cloud seeding on a large geographical scale may merely redistribute a fixed supply of precipitation, so that an increase in precipitation in one area might mean a compensating reduction in another. For example, rainmaking might benefit agriculture on the high plains of eastern Colorado but might also deprive wheat farmers of rain in the adjacent downwind states of Kansas and Nebraska. Such conflicts can cause legal wrangles between adjoining counties, states, and provinces.

In some experiments, seeding may actually have reduced precipitation. An example is a well-documented cloud-seeding experiment carried out over south central Missouri during five consecutive summers in the 1950s (called *Project Whitetop*). In this case, clouds

may have been overseeded. Prior to seeding, cumulus clouds apparently contained just enough ice crystals for precipitation. The addition of more nuclei by seeding probably produced too many ice crystals competing for too few supercooled water droplets. Consequently, Project Whitetop seeding generated a large number of ice pellets that were so small that they either remained suspended in the clouds or vaporized before reaching the ground.

## *FOG DISPERSAL*

Fog can be hazardous for both surface and air travel. Many chain-reaction motor vehicle accidents and some ship collisions and aircraft crashes have been attributed, at least in part, to visibility restrictions caused by dense fog. Fog frequently forces flight delays, reroutings, and cancellations that cost airlines millions of dollars each year and inconvenience hundreds of thousands of passengers. Although the need for an effective method of **fog dispersal** is great, especially at airports, little progress has been made in this area for both technical and economic reasons.

During World War II, the British had considerable success in clearing warm radiation fogs* from runways at 15 military airfields. Heat from fuel burners deployed alongside runways raised the air temperature and thereby lowered the relative humidity to below saturation so that fog droplets vaporized. First operational in late 1943, this fog dispersal system assisted in the safe landing of an estimated 2500 Royal Air Force planes, which returned a total of 15,000 airmen from combat missions. Although this thermal approach to fog dispersal was revived from time to time after the war, the fuel costs incurred were excessive.

Today, the only warm fog dispersal systems operated routinely are at Orly and De Gaulle airports outside Paris. In this so-called *Turboclair* system, underground jet engines direct streams of warm exhaust over runways. Warming lowers the relative humidity causing fog droplets to vaporize.

With shallow radiation fogs at airports, some clearing may be achieved by using helicopters to induce ver-

*Warm fog is composed of water droplets at temperatures above 0 °C (32 °F) and is the most common type of fog. The British fog dispersal system was known by the acronym FIDO, for Fog Investigation Dispersal Operation.

tical circulation of air. The rotor blades force the warmer unsaturated air that is just above the fog layer to mix with the cooler saturated air below. Consequently, the relative humidity of the mixture falls below 100% and fog droplets vaporize.

Another approach to fog dispersal is cloud seeding. Warm fogs are seeded with hygroscopic substances that absorb water vapor and thereby reduce the relative humidity by reducing the vapor pressure. Fog droplets vaporize, and hygroscopic droplets grow into raindrops that fall to the surface. Cold fogs (0 to −20 °C or 32 to −4 °F) are seeded with dry ice pellets that stimulate the Bergeron process. For the most part, however, fog dispersal by seeding is an experimental technology that requires further research and development, very little of which is now taking place.

# Atmospheric Optics

As the sun's rays travel through the atmosphere, they may be reflected or refracted by cloud droplets or ice crystals, or by raindrops. The consequence is a variety of optical phenomena, including halos, rainbows, coronae, and glories. The characteristics and origins of these optical effects are the subjects of this section. As a Special Topic, we also examine mirages, an optical effect arising from density gradients within the atmosphere.

## HALOS

A **halo** is a whitish ring of light surrounding the sun or the moon. It forms when the sun's rays are refracted by the tiny ice crystals that compose high, thin clouds such as cirrostratus. **Refraction** is the bending of light as it passes from one transparent medium (such as air) into another transparent medium (such as ice or water). The light rays bend because the speed of light is greater in air than in ice or water. Refraction occurs whenever light rays strike the interface between two different transparent media at any angle other than 90 degrees (Figure 8.11). (Beams of light that are perpendicular to a water or ice surface are not refracted.)

An analogy may help to explain refraction. Suppose you are driving your auto down a highway and suddenly encounter a patch of ice along the right side of the road. The right wheels travel over the ice, while

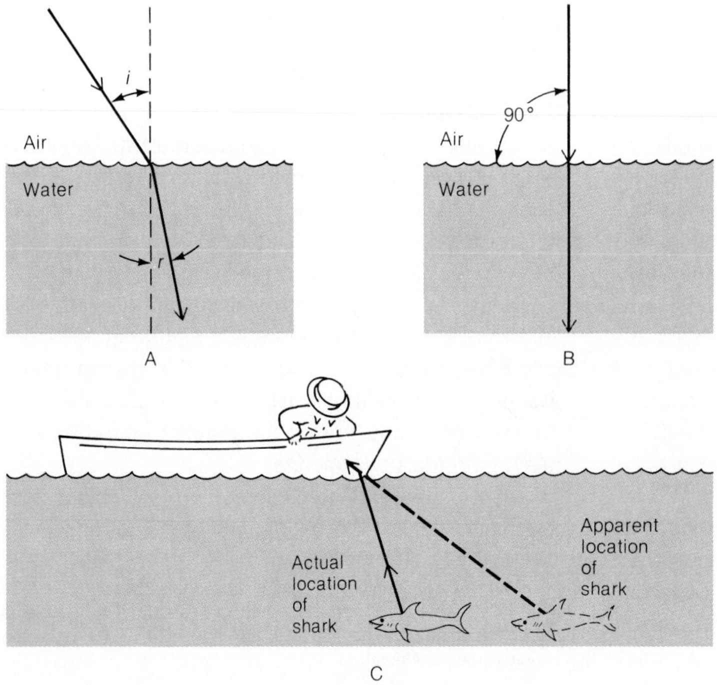

**FIGURE 8.11**
Light rays may be refracted (bent) as they travel from one transparent medium into another. (A) A light ray is refracted as it travels from air into water. The speed of light is less in water than in air, so the light ray is bent toward a line drawn perpendicular to the water surface, and angle $r$ is less than angle $i$. (B) Light rays that are perpendicular to the water surface are not refracted. (C) Refracted light can be deceptive. The shark appears to be farther away from the boat than it really is. This is because of refraction and the fact that humans assume that light always travels in straight lines.

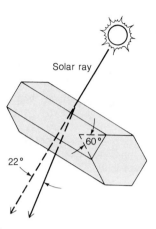

**FIGURE 8.12**
A ray of sunlight is refracted as it passes through an ice crystal from one side (the top side in the drawing) to another side (the bottom side in the drawing). The angle between the two sides of a hexagonal ice crystal is 60 degrees. This type of refraction produces a 22-degree halo about the sun or moon.

**FIGURE 8.13**
A 22-degree halo about the sun is caused by refraction of sunlight by ice crystals. [Photograph by Arjen and Jerrine Verkaik/SKYART]

the left wheels remain on dry pavement. You slam on the brakes. The left side of the car slows, while the right side slips on the ice. Your auto consequently swerves to the left. The swerving of the auto is analogous to a light ray bending toward the medium in which light travels more slowly.

Light is refracted twice as it passes through an ice crystal, that is, upon entering and upon exiting. In clouds, ice crystals occur as hexagonal (six-sided) plates or columns. If crystals are very small and randomly oriented, the light ray is refracted from side to side as it passes through an ice crystal (Figure 8.12). This side-to-side refraction focuses the light in a circle with a radius* of about 22 degrees (Figure 8.13). For reference, the halo's radius appears to be the same as the width of this page when held at arm's length.

A much rarer halo has a radius of about 46 degrees about the sun (or moon). In this case, refraction is caused by columnar ice crystals with diameters in the 15 to 25 micrometer range. Light rays travel through an ice crystal from side to top or from side to base, rather than from side to side (Figure 8.14).

*The radius of a halo is expressed as an angle. Suppose, for example, that you are viewing a halo about the moon. Now visualize two lines: one line joins you with the moon, and the other line joins you with any point on the halo. Depending on the type of halo, the angle formed by the two imaginary lines will be either 22 or 46 degrees.

**FIGURE 8.14**
In a rarer situation than shown in Figure 8.12, a solar ray is refracted from side to top or from side to base as it passes through a columnar ice crystal. The angle between side and top (or bottom) is 90 degrees. This type of refraction produces a 46-degree halo about the sun or moon.

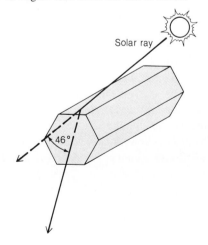

**SPECIAL TOPIC**

# Mirages

Appearances can be deceiving—especially in the case of mirages. Distant buildings or hills may appear higher or lower than they really are. A nonexistent pool of water may suddenly appear on the highway ahead, or a sailboat viewed from shore may appear upside down. These mirages are optical illusions that are caused by the refraction (bending) of light rays within the lower atmosphere.

Light travels at different speeds through different transparent substances. Light changes speed, for example, as it travels from air into an ice crystal. For reasons discussed elsewhere in this chapter, a light ray is refracted at the interface between two different transparent media. The speed of light also varies within a single medium if the density of that medium is not uniform. Hence, light rays bend as they pass through a substance of varying density, such as the atmosphere.

If the density of the atmosphere were the same throughout, then light rays would always travel along straight paths at constant speed, and no refraction would occur. As described in Chapter 5, however, air density varies with changes in temperature and pressure. Because our concern here is with optical phenomena that occur within the lowermost troposphere and involve relatively short viewing distances, we need not be concerned with the influence of horizontal air pressure gradients on air density. For the same reasons, we can also ignore the effect of horizontal temperature gradients on air density, but we cannot ignore the effect of vertical temperature profiles on the change of air density with altitude.

As a rule, light rays traveling through the atmosphere are refracted such that the denser air is on the inside (concave side) of the bend, and the less dense air is on the outside (convex side) of the bend. Air density decreases with altitude so that light reflected from a distant object follows a curved path to the observer (Figure 1). Because human perception is based on the

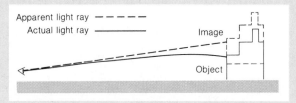

**FIGURE 1**
Because air density declines with altitude, objects appear to be slightly higher than they really are.

assumption that light always travels in straight paths, to the observer the object *appears* to be higher than it really is. Hence, for example, the setting or rising sun appears slightly higher in the sky than it actually is.

Now, if the air temperature decreases with altitude at less than the usual rate or if there is a temperature inversion (an increase of temperature with altitude), air density decreases with altitude faster than normal. Hence, light rays reflected from a distant object bend more sharply than usual before reaching the viewer, and objects appear higher than normal (Figure 2). This is known as a *superior mirage*.

If, on the other hand, the lowest air layer features a greater than usual temperature lapse rate, light rays are refracted less than normal and objects appear lower

**FIGURE 2**
Within a temperature inversion, air density declines with altitude more rapidly than usual. Hence, objects appear significantly higher than normal. This is known as a superior mirage.

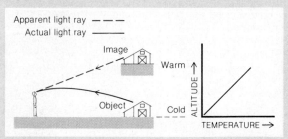

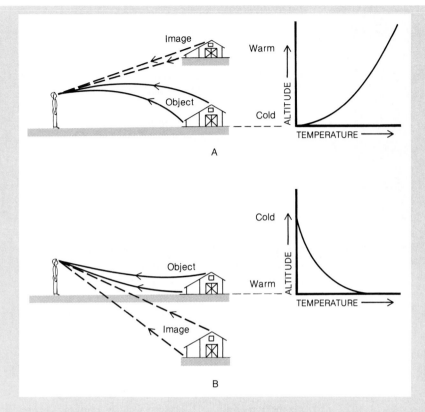

**FIGURE 3**

(A) When the vertical temperature gradient decreases with altitude while the air temperature increases with altitude (a temperature inversion), an object appears elevated but shorter than it really is. (B) When the temperature lapse rate is steepest near the Earth's surface, an object appears taller than it really is.

than we usually see them. This is known as an *inferior mirage*.

What happens when the vertical temperature gradient changes with altitude? As warm air is advected over a cold, snow-covered surface, the warm air mass is chilled from below, and a temperature inversion develops. The vertical temperature gradient is greatest in the lowest layer of air (Figure 3A). Consequently, a distant object appears to be displaced upward (superior mirage), but because the bottom of the object is within a greater air temperature gradient than is the top of the object, the bottom appears to be lifted more than the top. The object, a building, for example, not only appears to be uplifted, but it also appears shorter than it really is.

A different optical effect is observed when both the temperature and the vertical temperature gradient decrease with altitude (Figure 3B). By midday a desert surface is heated intensely by the sun, and the greatest lapse rate develops just above the ground. With such a temperature profile, a distant object appears to be displaced downward (inferior mirage), but because the bottom of the object is within a greater air temperature gradient than is the top of the object, the bottom is displaced downward more than the top. The object not only appears to sink, but it also appears to be stretched vertically.

These are only a few examples of the many possible types of mirages. As the vertical air temperature profile becomes more complex, so too do the types of mirage that can appear. For example, the familiar oasis mirage in a desert is an inverted image of the sky seen below the horizon. All mirages, then, are displacements or distortions of something real.

**FIGURE 8.15**
Bright spots appearing on either side of the sun, sometimes called sundogs, are caused by refraction of sunlight by ice crystals in the atmosphere. [Photograph by Arjen and Jerrine Verkaik/SKYART]

Sometimes light is concentrated in two brilliant spots situated on either side of the sun (Figure 8.15). These spots are called **sundogs** (because they appear to follow the sun around the sky) or **parhelia** (from the Greek meaning *beside the sun*). Cirrus clouds composed of relatively large platelike ice crystals are responsible for sundogs. Because of air resistance, these crystals are oriented with top and bottom surfaces nearly horizontal. Sunlight is refracted twice as it enters one side of a hexagonal crystal and exits another side. The net effect is for the sun's rays to deviate by 22 degrees from its original path. A cloud of such crystals focuses the sun's rays in two spots about 22 degrees to the right and left of the sun.

Ice crystals also act like prisms and disperse sunlight into its component colors. The more energetic violet end of the visible spectrum is refracted the most, so that the violet portion of parhelia is farthest from the sun. The less energetic red end of the visible spectrum is refracted the least, so that the red portion of parhelia is closest to the sun.

## RAINBOWS

A **rainbow** is caused by a combination of refraction and reflection of sunlight (or, rarely, of moonlight) by raindrops. Sunlight striking a shaft of falling raindrops is refracted and internally reflected by each drop of rain. As shown in Figure 8.16, a solar ray is refracted as it enters a raindrop; then the ray is reflected by the inside back of the drop before being refracted again as it exits the drop.

A rainbow appears to an observer who has his or her back to the sun and is facing a distant rain shower (Figure 8.17). A rainbow never forms when the sky is completely cloud covered—the sun must be shining. Because weather usually progresses from west to east, the appearance of a rainbow in the evening signals improving weather. Rain showers to the east are moving away, and clearing skies are approaching from the west, where the sun is setting.*

Like a prism, raindrop refraction disperses sunlight into its component colors forming the concentric bands of color of a *primary* rainbow. From outer to innermost band, the colors are red, orange, yellow, green, blue, and violet (Figure 8.18). In most cases, a much dimmer, *secondary* rainbow appears about 8 degrees above the primary rainbow. This secondary rainbow is produced by double reflection within raindrops (Figure 8.19). The order of colors in the secondary rainbow is the reverse of that in the primary rainbow.

Because ice crystals refract light just as raindrops do, you may wonder why halos are not colored. In fact, ice crystals do disperse light into its component colors, but because the size and shape of ice crystals vary more than the size and shape of raindrops, the colors produced by an assemblage of ice crystals tend to overlap one another rather than form discrete bands. In halos, colors therefore wash out, although occasionally a reddish tinge is visible on the inside of a halo.

## CORONAE

A **corona** consists of a series of alternating light and dark rings that surround the moon or, less often, the sun. Typically, a corona is only a few degrees in radius

---

*You can create your own rainbow with the spray from a garden hose. Simply direct the spray so that you observe the spray with the sun at your back.

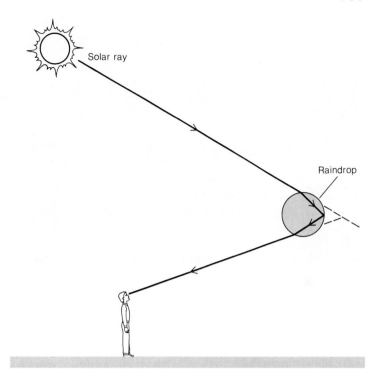

**FIGURE 8.16**
A solar ray is refracted and internally reflected by a raindrop. This is the basic optical process that causes a rainbow.

and hence, is far smaller than a halo. It is caused by diffraction of light around water droplets that compose a thin, translucent veil of altocumulus, altocumulus lenticularis, or cirrocumulus clouds.

**Diffraction** is the slight bending of a light wave as it moves along the boundary of an object such as a water droplet. As light waves bend, they interfere with one another. When the crests of one wave coincide with the crests of another wave, interference is constructive and a larger wave results. On the other hand, when the crests of one wave coincide with the troughs of another wave, the interference is destructive and the waves can-

cel each other. Where light waves interfere constructively, we see a ring of bright light, and where they interfere destructively, we perceive darkness.

If cloud droplets are of uniform size, then a corona is colored with blue-violet on the inside and red on the outside of each ring. Dependency of diffraction on wavelength causes this separation of color: the longer wavelength red light is diffracted more than the shorter wavelength blue-violet light.

**FIGURE 8.17**
A rainbow appears to an observer who has his/her back to the sun and faces a distant rain shower.

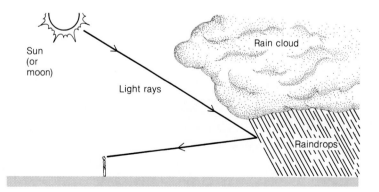

**FIGURE 8.18**
A rainbow is caused by refraction and internal reflection of sunlight by falling raindrops. Refraction disperses visible light into its component colors. [Photograph by J. M. Moran]

## GLORIES

Before the age of aircraft travel, about the only way to view a glory was from the vantage point of a lofty mountain peak. To see a glory an observer must be in bright sunshine above a cloud or fog layer, and the sun must be situated so as to cast the observer's shadow on the clouds below. The observer then sees a **glory** as concentric rings of color around the shadow of his or her head. Although less distinct, the colors of a glory are the same as those of the primary rainbow, with violet being the innermost band and red

the outermost band. Today, aircraft pilots and observant passengers often see a glory around the shadow of their aircraft on a cloud deck below (Figure 8.20).

The glory results from much the same optics as the primary rainbow. One difference, however, is the size of the reflecting and refracting particles. Whereas rainbows occur when sunlight strikes a mass of raindrops, glories are the consequence of sunlight interacting with a mass of smaller water droplets of uniform size that form a cloud. In both cases, the sun's rays undergo refraction upon entering the droplet, followed

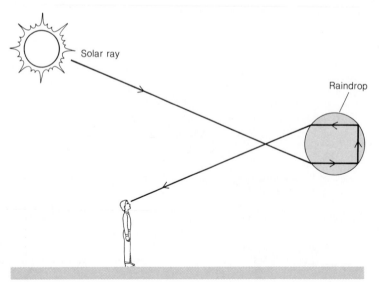

**FIGURE 8.19**
Refraction of sunlight by raindrops plus double reflection within raindrops produces a dimmer secondary rainbow just above the primary rainbow.

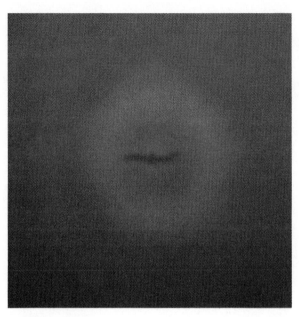

**FIGURE 8.20**
The glory, photographed from an aircraft, is a consequence of refraction, internal reflection of sunlight, and diffraction by cloud droplets of uniform size. [Photograph by Steve Dutch]

by a single internal reflection, and another refraction upon exiting.

The optics of rainbows and glories have one other important difference besides droplet size. In the glory, sunlight is refracted and reflected *directly* back toward the sun, but this is not the case with rainbows. With the glory, a spherical cloud droplet diffracts light rays ever so slightly toward the droplet, so that light rays incident on a cloud droplet parallel those returning from the cloud droplet (Figure 8.21).

The special optics of a glory explain why it appears about the shadow of the observer who is situated in the

**FIGURE 8.21**
In the special optics that produce a glory, both the incident and returning solar rays are diffracted slightly toward the cloud droplet. Consequently, the incident and returning rays follow parallel paths.

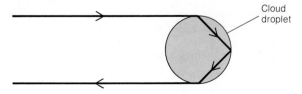

direct path of both the incident solar rays and the returning (refracted and reflected) solar rays. This also explains an observation that must have fascinated ancient mountain mystics. Suppose that you and a friend are side by side high on a mountain slope at a site favorable for viewing a glory. On the cloud deck below, you see your own shadow beside that of your friend, but a glory appears about your head and not about your friend's head. Lest you presume that you have been singled out, note that your friend has the opposite observation, for your friend sees a glory only about his or her head. The fact is that each observer is in a position to view only one glory.

## Conclusions

Cloud droplets and ice crystals are much too small to reach the Earth's surface as precipitation. Through collision–coalescence and the Bergeron process, cloud droplets and ice crystals grow large enough to survive a fall to the Earth's surface. An understanding of the circumstances necessary for clouds to yield precipitation has enabled scientists to develop techniques to stimulate natural precipitation processes through cloud seeding and to disperse radiation fog. However, these efforts are not always successful.

We have seen that interactions of sunlight (or moonlight) with clouds or rainfall produce a variety of optical effects including halos, rainbows, coronae, and glories. These interactions involve reflection, refraction, and/or diffraction.

Atmospheric circulation plays a key role in bringing air to saturation and triggering cloud development. The next seven chapters describe the many atmospheric circulation systems and their associated weather conditions. We begin in Chapter 9 with a discussion of the forces that drive and shape atmospheric circulation.

## *Key Terms*

| | |
|---|---|
| terminal velocity | collision–coalescence |
| warm cloud | process |

Bergeron process
cold cloud
virga
precipitable water
precipitation
drizzle
rain
freezing rain
snow
snow pellets
snow grains
ice pellets
hail
rain gauge

weighing-bucket rain
    gauge
tipping-bucket rain gauge
weather modification
cloud seeding
fog dispersal
halo
refraction
sundogs
parhelia
rainbow
corona
diffraction
glory

## Summary Statements

☐ The relatively low terminal velocities of cloud droplets and ice crystals mean that they will remain suspended in the atmosphere indefinitely unless they vaporize or undergo significant growth.

☐ The Bergeron and collision–coalescence processes are mechanisms whereby cloud particles grow large enough to counter updrafts and fall to the Earth's surface as precipitation. The bulk of precipitation on Earth originates in cold clouds composed of a mixture of ice crystals and supercooled water droplets.

☐ The collision–coalescence process occurs in warm clouds and requires the presence of relatively large cloud droplets that grow through collisions and coalescence with smaller cloud droplets.

☐ The Bergeron process requires the coexistence of ice crystals, water droplets, and water vapor. In the same cloud, the saturation vapor pressure surrounding a supercooled water droplet is higher than the saturation vapor pressure surrounding an ice crystal. Hence, air that is saturated for droplets is supersaturated for ice crystals. Consequently, ice crystals grow at the expense of water droplets.

☐ The principal forms of precipitation are rain, drizzle, freezing rain, snow, ice pellets (sleet), and hail. The form of precipitation depends on the source cloud and the temperature profile of the air beneath the cloud.

☐ Rain and snow fall from thicker clouds (nimbostratus or cumulonimbus) than drizzle (stratus). Ice pellets are raindrops that freeze prior to reaching the ground while freezing rain consists of supercooled drops that freeze on contact with cold surfaces. Hail is produced by thunderstorms.

☐ Cloud seeding is a type of weather modification intended to increase rainfall or snowfall by stimulating natural

precipitation processes. Cloud seeding primarily targets cold clouds by injecting nucleating agents (silver iodide or dry ice pellets) in an attempt to spur the Bergeron process.

☐ Fogs can be dispersed by raising the air temperature (thus lowering the relative humidity) or by seeding.

☐ A halo forms about the sun or moon when light is refracted by ice crystals composing high, thin clouds. A rainbow is the consequence of the refraction and internal reflection of sunlight by raindrops. Coronae are produced when water droplets composing a thin cloud layer diffract either moonlight or sunlight. Glories are caused by the same optical effect as a primary rainbow, except that the refracting and reflecting particles are cloud droplets instead of raindrops.

## Review Questions

1. What is meant by *terminal velocity?*
2. Distinguish between warm clouds and cold clouds. Compare their composition.
3. Describe the collision–coalescence process of precipitation formation.
4. Describe the Bergeron process of precipitation formation.
5. Once a raindrop or snowflake falls from a cloud, it travels through a *hostile* environment. What is meant by this statement?
6. Why are drizzle drops smaller than raindrops? Why do cold clouds generally produce larger raindrops than do warm clouds?
7. Distinguish among these frozen forms of precipitation: ice pellets, freezing rain, and hail.
8. Under what conditions does freezing rain develop? Explain why freezing rain can be destructive.
9. Explain how hail reaches the ground even when air temperatures are well above the freezing point.
10. Define *weather modification.*
11. What is the basic objective of cloud seeding? How is cloud seeding carried out?
12. Comment on the effectiveness of cloud seeding. Does it always work?
13. Identify some of the benefits and costs of successful cloud seeding.
14. How might cloud seeding actually reduce the amount of precipitation?
15. Describe the various methods for dispersing fog.
16. Why do raindrops and ice crystals refract light rays?
17. What type of cloud produces a halo about the sun or moon?

18. What causes the colors of a rainbow? Also, distinguish between a *primary* and *secondary* rainbow.
19. What does the appearance of a morning rainbow suggest about the weather later in the day?
20. Compare and contrast the optics of a rainbow with those of a glory. Explain why you can view only your own glory.

## Quantitative Questions

1. What is the usual (average) meltwater equivalent of 25 cm (10 in.) of fresh snowfall?
2. Determine the average density of fresh-fallen snow.
3. How many calories of heat are required per square centimeter to melt 10 cm of snow? Assume the snow density to be 0.1 $g/cm^3$.
4. A cold cloud consists of mostly supercooled water droplets and some ice crystals. The temperature is $-20\ °C$ ($-4\ °F$). If the relative humidity surrounding water droplets is 100%, what is the relative humidity surrounding the ice crystals?

## Questions for Critical Thinking

1. Speculate on why in some localities the exterior of a rain gauge is painted white. Why is it desirable to shield a rain or snow gauge from strong winds?
2. Why is the saturation vapor pressure surrounding supercooled water droplets greater than that surrounding ice crystals at the same temperature?
3. Give two reasons why rainfall is likely to be heavier on top of a mountain than in a nearby valley.
4. The changeover from rain to snow is usually faster if precipitation is moderate to heavy rather than light. Explain why.
5. Explain why dispersal of advection fog is likely to be a much greater (if not impossible) challenge than dispersal of radiation fog.

## Selected Readings

BOHREN, C. F. "Virga: A Heretical View," *Weatherwise* 44, No. 5 (1991):31–35. Argues that virga may be snow melting to rain or precipitation that has not yet reached the ground.

BOHREN, C. F., and M. L. Sowers. "Crepuscular Rays," *Weatherwise* 45, No. 2 (1992):34–38. Discusses the origin of alternating light and dark streaks that appear to diverge from the sun across the sky.

BLANCHARD, D. C. "Science, Success and Serendipity," *Weatherwise* 32, No. 6 (1979):236–241. Recounts the pioneering efforts of V. Schaefer, I. Langmuir, and T. Bergeron in cloud physics research.

COLBECK, S. C. "What Becomes of a Winter Snowflake?" *Weatherwise* 38, No. 6 (1985):312–315. Describes processes taking place within a snowbank.

GEDZELMAN, S. D., AND E. LEWIS. "Warm Snowstorms, A Forecaster's Dilemma," *Weatherwise* 43, No. 5 (1990):265–270. Describes atmospheric conditions favorable for late spring and early fall snowfall when surface temperatures are above the freezing point.

GREENLER, R. *Rainbows, Halos,* and *Glories.* New York: Cambridge University Press, 1980, 195 pp. Discusses the causes of a wide variety of atmospheric optical phenomena; includes spectacular photography.

OGDEN, R. J. "Fog Dispersal at Airfields—Part 1," *Weather* 43 (1988):20–25; "Fog Dispersal at Airfields—Part 2," *Weather* 43 (1988):34–38. Recounts the highly successful British effort to disperse radiation fogs at military airfields during World War II.

REYNOLDS, D. W., AND A. S. DENNIS. "A Review of the Sierra Cooperative Pilot Project," *Bulletin of the American Meteorological Society* 67, No. 5 (1986):513–523. Provides a detailed account of a long-term study of seeding of orographic clouds.

SNOW, J. T., AND S. B. HARLEY. "Basic Meteorological Observations for Schools: Rainfall," *Bulletin of the American Meteorological Society* 69, No. 5 (1988):497–507. Discusses methods of rainfall measurement and evaluates inexpensive rain gauges suitable for classroom use.

WALKER, J. *Light from the Sky.* New York: W. H. Freeman, 1980, 78 pp. Reprints of *Scientific American* articles on various atmospheric optical phenomena including mirages, halos, rainbows, and glories.

# 9 The Wind

*No one can tell me,*
*Nobody knows,*
*Where the wind comes from,*
*Where the wind goes.*

A. A. MILNE
*"Wind on the Hill" Now We*
*Are Six*

Many forces interact to initiate and shape the wind. The sun is the ultimate source of energy that sustains atmospheric circulation. [Photograph by J. M. Moran

SOME WEATHER systems favor clear skies, light winds, and frosty mornings, whereas others bring ominous clouds, precipitation, and biting winds. Some weather systems trigger brief showers, and others are accompanied by persistent fog and drizzle. Certain weather systems are highly localized and short-lived; others dominate the weather over thousands of square kilometers for prolonged periods. Different weather systems bring different types of weather depending on the air circulation pattern that characterizes each system.

The atmosphere is coupled to the planet and rotates with it. Once every 24 hr, every point on the Earth's surface and in the atmosphere describes a circular path. The circumference of that path decreases with increasing latitude. Hence, at the equator, the atmosphere moves eastward at 1670 km (1035 mi) per hour; at 60 degrees N or S, the speed is 835 km (517 mi) per hour. Why is it that we are not aware of this rapid motion of the air? The reason is that we also move with the rotating planet and its atmosphere at the same speed. In meteorology, we are interested in air motion measured *relative* to the Earth's surface. This is what is meant by **wind.**

In this chapter, we describe the various forces that initiate and control the wind. We begin by examining each force separately as if each force acted independently of all the other forces. We then show how these forces combine to initiate and modify atmospheric circulation. In the Special Topic "Wind Power," we also consider how modern technology is tapping the energy of air in motion.

# The Forces

It is useful at the outset of our discussion of the forces that initiate and control the wind to distinguish between the horizontal (east–west and north–south) and the vertical (up–down) components of the wind. Except in small, intense weather systems such as thunderstorms, the magnitude of vertical air motion is typically only 1% to 10% of the horizontal wind speed. Nonetheless, as we saw in Chapter 6, the vertical component plays the key role in the formation of clouds. Furthermore, as we will see later in this chapter, vertical and horizontal wind components are linked so that a change in one may be accompanied by a change in the other.

For convenience of study, imagine the wind as a continuous stream of air composed of discrete *air parcels*. Assume that any force acting on an air parcel represents the influence of that same force on a stream of air parcels, in other words, on the wind. Now assume also that each parcel consists of a unit mass of air—a single gram, for example. Hence, in examining each force that influences air motion, we examine the force per unit mass of air.

A force per unit mass is numerically equivalent to an acceleration. This equivalency follows from **Newton's second law of motion,**\* where

$$\text{force} = \text{mass} \times \text{acceleration}$$

For this reason, we sometimes use the terms *force* and *acceleration* interchangeably when we consider the

---

\*This is the second of three fundamental laws of motion first formulated in 1687 by Sir Isaac Newton.

SPECIAL TOPIC

## Wind Power

Harnessing the energy of the wind is a technology that was well established as early as the twelfth century in portions of the Middle East where water power was not available. In North America, the energy crisis of the 1970s spurred renewed interest in this ancient technology. Today scientists are employing modern aerodynamic principles and space-age materials in designing and constructing modern wind-driven turbines that convert some of the wind's kinetic energy into electricity (Figure 1).

In Chapter 4, we saw how the sun drives the atmosphere. Although only about 2% of the solar energy that reaches the Earth is ultimately converted to the kinetic energy of wind, that is still a tremendous quantity of energy. Theoretically, windmill blades can convert a maximum of 60% of the wind's energy into mechanical energy. In practice, however, wind generators extract only about 25% of the wind's energy. Furthermore, average wind speeds must be at least 19 km (12 mi) per hour before most wind-powered electricity-generating systems can operate economically.

The power that a windmill can extract from the wind is directly proportional to (1) air density, (2) the area swept out by the windmill blade, and (3) the cube of the instantaneous wind speed ($v^3$). Wind speed is by far the most important consideration in evaluating a region's wind energy potential. Even small changes in wind speed translate into large changes in energy output. For example, doubling the wind speed (a common occurrence) multiplies the available wind power by a factor of eight ($2 \times 2 \times 2$).

Tapping the wind's energy faces several difficulties. Both wind speed and direction vary continuously with time (see Figure 9.27), and wind speed also varies with the exposure of the site, roughness of the terrain, elevation above the surface, and season of the year. As a general rule, a minimum of several years of detailed wind monitoring is needed for a preliminary evaluation

**FIGURE 1**
This 200-kW wind turbine generator supplies part of the electrical power used by residents of Clayton, New Mexico. [U.S. Department of Energy photograph]

of wind power potential at any site. The long-term climate record should also be consulted to check for the frequency of potentially destructive winds. Wind data from nearby weather stations can be very useful, as long as care is taken in extrapolating wind data between localities and from one elevation to another.

At ordinary speeds, wind is a relatively diffuse energy source, comparable in magnitude to insolation.

Hence, a wind turbine's power generation potential also depends on the area swept out by the windmill blades. Larger windmill blades harvest more energy. However, as windmill blades get larger, design problems develop, and costs soar and soon become prohibitive. So far, the largest experimental wind turbine systems have blade diameters approaching 100 m (325 ft).

The most formidable obstacle to the development of wind power potential stems from the inherent variability of the wind. The electrical output of wind turbines varies as a consequence, and a wind power system must include a means of storing the energy generated during gusty periods for use when the wind is light or calm. A 3- to 5-kW wind turbine is needed to meet the total electrical requirements, including heating, of a typical household. Today, such systems are commercially available, but the cost of materials and construction—including a tower, storage batteries, and generator—ranges from about $5,000 to $20,000.

Economy of scale suggests that centralized arrays of wind turbines, called *windmill farms,* are preferable to individual household wind turbines. Windmill farms consist of 50 or more super wind-power generators, each capable of producing one or more megawatts of electricity. Windmill arrays now operating in California, for example, supplement conventional energy sources by feeding electricity to existing power grids (Figure 2). Since 1981, more than 16,000 turbines have been installed in windy mountainous sites in California. By 1990, California windmill farms were supplying about 1% of the state's total electrical demand; this figure is expected to climb to 10% by 2005.

Given current technical and economic limitations, wind power has its greatest immediate potential in those regions where average winds are relatively strong and consistent in direction. In North America, such regions include the western High Plains, the Pacific Northwest coast, portions of coastal California, the eastern Great Lakes, the south coast of Texas, and exposed summits and passes in the Rockies and Appalachians. Low-power wind systems have potential in small, isolated communities and on individual farms

**FIGURE 2**
This windmill farm in southern California feeds electricity into existing power grids. [Courtesy of Southern California Edison Company]

and ranches. The environmental impact of wind systems is usually minimal: They may be somewhat noisy, detract some from the beauty of the landscape, and kill birds that fly into the blades.

Recent technological advances coupled with new tax incentives have given a boost to the wind power industry. U.S. Windpower, the nation's major wind turbine manufacturer, has developed a turbine that operates over a wind-speed range of 10 to 97 km/hr (6 to 60 mi/hr). Previous turbines were limited to a range of about 19 to 73 km/hr (12 to 45 mi/hr). The 1992 Federal Energy Bill granted wind turbine operators a special tax break of $0.015 per kilowatt-hour. These events have prompted many major electrical utilities to announce plans to install wind turbines during the latter part of this decade.

motion of air parcels. Although the terms are numerically equivalent, a change in velocity is actually a response to a force. A force acts on an air parcel to bring about an acceleration or deceleration of that parcel.

Forces acting on air parcels, which either initiate or modify motion, are the consequence of (1) air pressure gradients, (2) the centripetal force, (3) the Coriolis effect, (4) friction, and (5) gravity. Actually, the centripetal force is not an independent force but a consequence of other forces. Nonetheless, it is instructive to examine the centripetal force along with the other forces.

### PRESSURE GRADIENT FORCE

A *gradient* is simply a change in some property with distance. An **air pressure gradient** exists whenever air pressure varies from one place to another. As noted in Chapter 5, spatial variations in air pressure can arise from contrasts in air temperature (principally), from differing water vapor concentrations, or from both. An air pressure gradient thus develops between a mass of cold, dry air and a mass of warm, humid air. In addition, diverging and converging winds can bring about air pressure changes and thereby induce an air pressure

gradient. This important process is covered in Chapter 11, where we examine the origin of highs and lows.

Air pressure gradients develop both horizontally and vertically within the atmosphere. A horizontal pressure gradient refers to air pressure changes along a surface of constant altitude (mean sea level, for example). A vertical pressure gradient is a permanent feature of the atmosphere because air pressure is greatest at the Earth's surface and decreases with altitude.

In order to represent horizontal air pressure gradients on a weather map, the air pressure measured at each weather station is first reduced to sea level, as discussed in Chapter 5. Lines are then drawn joining localities that have the same air pressure reading; some interpolation between weather stations is always necessary. These lines are called **isobars,** and the interval between successive isobars is usually 4 mb.

An isobaric analysis is used to locate centers of high and low pressure and to determine the magnitude of the horizontal air pressure gradient between weather systems. Closely spaced isobars (Figure 9.1A) mean that air pressure changes rapidly with distance, and the pressure gradient is described as steep or strong. More widely spaced isobars (Figure 9.1B) indicate that air pressure changes less with distance, and the pressure gradient is weaker. Note that air pressure gradients are

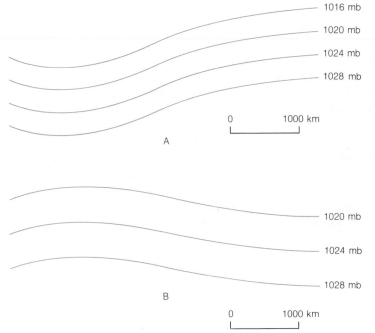

1016 mb
1020 mb
1024 mb
1028 mb

A

0    1000 km

1020 mb

1024 mb

1028 mb

B

0    1000 km

**FIGURE 9.1**
The horizontal air pressure gradient is relatively steep where isobars are close together (A) and weaker where isobars are farther apart (B). Isobars are lines of equal air pressure. Here the contour interval (the difference between successive isobars) is 4 mb.

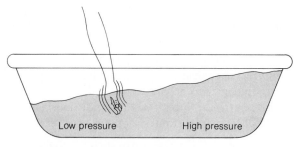

**FIGURE 9.2**
Sloshing water in a bathtub back and forth creates a horizontal pressure gradient on the bottom of the tub. The water pressure gradient in the tube is analogous to horizontal air pressure gradients in the atmosphere. That is, in response to a pressure gradient, the water (or air) flows from an area of higher pressure toward an area of lower pressure.

always measured in a direction perpendicular to the isobars.

How do air pressure gradients influence the movement of air? Let us examine an analogous situation. Suppose a bathtub is partially filled with water, as shown in Figure 9.2. As we slosh the water back and forth from one end of the tub to the other, a water pressure gradient develops along the bottom of the tub. At any instant, the water pressure is high where the water level is high, and low where the water level is low. If we stop agitating the water, the water level gradually returns to a horizontal surface and thus creates a uniform water pressure everywhere along the tub

bottom. Hence, in response to a water pressure gradient, water flows from one end of the tub (where the water pressure is greater) to the other end (where the water pressure is less), thus eliminating the pressure gradient.

Similarly, when an air pressure gradient develops, air flows in such a way as to eliminate the pressure gradient. Thus, the wind blows away from regions where air pressure is relatively high and toward locales where air pressure is relatively low. The wind is strong where the pressure gradient is steep (closely spaced isobars), and light or calm where the pressure gradient is weak (widely spaced isobars). The force that causes air parcels to move as the consequence of an air pressure gradient is known as the **pressure gradient force.**

### *CENTRIPETAL FORCE*

We may illustrate centripetal force by a simple demonstration. We tie a rock to a string and whirl it about so that the rock describes a circular orbit of constant radius (Figure 9.3). We then cut the string and the rock flies off in a straight line. The behavior of the rock illustrates **Newton's first law of motion;** that is, an object in straight-line, unaccelerated motion remains that way unless acted upon by an unbalanced force. Prior to being cut, the string exerts a net force on the rock

**FIGURE 9.3**
A rock attached to a string describes a circular path. Centripetal (for "center seeking") is the name we give to the force that confines an object to a curved path. If the string is cut, the centripetal force is eliminated and the rock flies off in a straight line (tangent to the circular path).

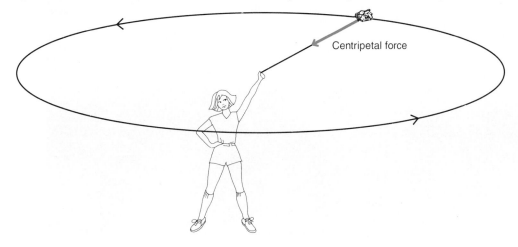

by confining it to a curved (circular) path. The net force is directed inward, perpendicular to the direction of motion, and toward the center of the circular orbit. For this reason, the net force is known as the **centripetal** *(center-seeking)* **force.** When we cut the string, the centripetal force no longer operates and the rock follows a straight path.

A net force causes an acceleration. We usually think of an acceleration as a change in speed, as when an automobile speeds up or slows down. But acceleration is a vector quantity; that is, it has both magnitude and direction. Hence, an acceleration may consist of a change in either speed or direction, or both. In our rock-on-a-string example, centripetal force is responsible only for a continuous change in the direction of the rock (a curved rather than straight path); the rock neither speeds up nor slows down.*

The centripetal force is not itself an independent force; rather, it arises from the action of other forces

---

*It can be demonstrated that the acceleration imparted to a unit mass by the centripetal force is directly proportional to the square of the velocity and inversely proportional to the radius of curvature.

**FIGURE 9.4**
In this visible satellite image, an intense midlatitude cyclone is readily identified by its swirling mass of clouds. The storm center is located just east of New Jersey. [NOAA National Environmental Satellite, Data, and Information Service]

**FIGURE 9.5**
When viewed from space, our north–south, east–west frame of reference changes as the Earth rotates on its axis. Here a south wind in the Northern Hemisphere is deflected (relative to the frame of reference) to the right and becomes a southwest wind. [Modified from F. K. Lutgens and E. J. Tarbuck, *The Atmosphere: An Introduction to Meteorology,* 4th ed., © 1989, p. 169. Adapted by permission of Prentice-Hall, Inc., Englewood Cliffs, NJ]

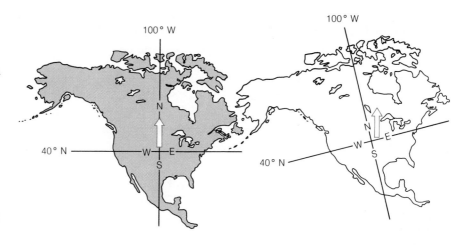

and may be the consequence of imbalances in other forces. In our rock-on-a-string example, the string is responsible for the centripetal force. Consider another example. Suppose that you are a passenger in an auto that rounds a curve at a high rate of speed. You feel a force that pushes you outward from the curve. Actually, what you experience is the tendency of your body to continue in a straight path while the auto is following a curved path. In this case the frictional resistance of the tires against the pavement provides centripetal force.

It follows from this discussion that a centripetal force operates whenever the wind follows a curved path. As we will see later in this chapter, however, the centripetal force results from an imbalance of other forces operating in the atmosphere.

## CORIOLIS EFFECT

Imagine that you are located far away at some fixed point in space and are looking back at planet Earth. Over many hours, you follow the track of a storm, which is clearly identifiable by its slowly swirling mass of clouds (Figure 9.4). From your perspective, the storm center appears to move in a straight line at constant speed. At the same time, an observer on Earth is tracking the storm. From that observer's perspective, the storm center appears to follow a curved path. Surprisingly, both descriptions of the storm's track are correct!

The two descriptions are correct because the two observers used different frames of reference in following the storm's movements. The Earthbound observer's frame of reference is the familiar north–south, east–west, and up–down coordinate system that rotates with the Earth. To the Earthbound observer, it is not obvious that this coordinate system is rotating because it and the observer rotate together. Viewed from space, however, the Earthbound coordinate system actually shifts as the Earth rotates (Figure 9.5). It is as if the Earth and the coordinate system rotate under the storm (or any other object moving over the Earth's surface). Meanwhile, from your distant vantage point in space, you followed the storm's movement with respect to a nonrotating coordinate system, fixed in space. In summary, then, the difference in storm track (straight versus curved) is due to the difference in coordinate system (nonrotating versus rotating).

Recall our earlier discussion of Newton's first law of motion and the centripetal force. We saw that curved motion implies that a net force is operating, whereas unaccelerated, straight-line motion implies a balance of forces. If we apply this law to our storm track example, we conclude that a net force operates when we use the Earthbound rotating coordinate system, whereas forces are balanced when we use the nonrotating coordinate system fixed in space. Hence, changing our frame of reference (coordinate system) from nonrotating to rotating gives rise to a net force responsible for curved motion. This deflective force is known as the **Coriolis effect,** named for Gustav Gaspard de Coriolis, who first described the phenomenon mathematically in 1835.

Wind direction and wind speed are measured with respect to the north–south, east–west, and up–down frame of reference that rotates with the planet. Therefore, we must take the Coriolis effect into account in

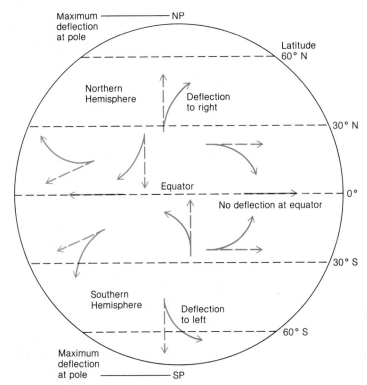

**FIGURE 9.6**
Large-scale winds are deflected to the right in the Northern Hemisphere and to the left in the Southern Hemisphere. This deflection is due to the Coriolis effect.

any explanation of air circulation. Because of the Coriolis effect, the wind is deflected to the right of its initial direction in the Northern Hemisphere and to the left of its initial direction in the Southern Hemisphere (Figure 9.6).

Although the Coriolis effect influences the wind regardless of its direction, the magnitude of the effect varies significantly with latitude.* This is because the Coriolis effect stems from the rotation of the Earth on its axis, which imparts a rotation to our Earthbound frame of reference. The rotation of our frame of reference is maximum at the poles and declines with decreasing latitude to zero at the equator. This variation can be understood by visualizing the daily rotation of towers situated at different latitudes. In a 24-hr day, the Earth makes one complete rotation, as would a tower situated at the North or South Pole. In the same period, a tower at the equator would not rotate at all, but instead would describe an end-over-end motion. For a tower located at any latitude in between, some rotation of the tower occurs as the Earth rotates but not as much as at the poles. The Coriolis effect is thus latitude-dependent:

*The magnitude of the Coriolis effect varies with the sine of the latitude.

the Coriolis effect is zero at the equator and increases with latitude to a maximum at the poles.

The Coriolis effect also varies with wind speed; that is, the deflection increases as the wind strengthens. This is because, in the same period of time, fast air parcels cover greater distances than slow air parcels. The longer the trajectory, the greater is the shift of the coordinate system with respect to a moving air parcel. For practical purposes, the Coriolis effect has an important influence only on the air circulation within large-scale weather systems, that is, systems larger than thunderstorms.

A rotational motion usually accompanies the draining of water from a sink or a bathtub. A popular misconception is that the direction of this rotation (clockwise or counterclockwise) is consistently in one direction in the Northern Hemisphere and in the opposite direction in the Southern Hemisphere, presumably because of the Coriolis effect. At the very small scale represented by the sink or bathtub, however, the magnitude of the Coriolis effect is simply too small to have a significant influence on the direction of rotation. The drainage direction is more likely a consequence of some residual motion of the water when the sink or

**FIGURE 9.7**
Rocks in a riverbed break the current into turbulent eddies that appear as white water to the lee of rocks. [Photograph by J. M. Moran]

bathtub was first filled with water and can be either clockwise or counterclockwise.

It is useful to know why the Coriolis effect reverses direction between the hemispheres and causes large-scale winds in the Southern Hemisphere to swerve to the left rather than to the right. This reversal is related to the difference in an observer's sense of the Earth's rotation in the two hemispheres. To an observer at the North Pole, the planet rotates counterclockwise, whereas to an observer at the South Pole, the planet rotates clockwise. This rotation reversal translates into a reversal in Coriolis deflection between the two hemispheres.

## FRICTION

We normally think of **friction** as the resistance that an object encounters as it moves in contact with other objects. Furthermore, we commonly associate friction with solids, as when we attempt to slide a heavy appliance across the floor. But friction also affects fluids, both liquids and gases. The friction of fluid flow is known as **viscosity** and is of two types, *molecular* and *eddy*.

One source of fluid friction is the random motion of molecules composing a liquid or gas; this type of fluid friction is called **molecular viscosity.** Considerably more important, however, is fluid friction that arises from much larger irregular motions, called eddies, which develop within fluids; this type of fluid friction is known as **eddy viscosity.**

The stream in Figure 9.7 illustrates the effects of eddy viscosity. Rocks in the streambed obstruct the flow of water and cause the stream to break into eddies to the lee of the rocks. Eddies, visible as white water, tap some of the stream's kinetic energy so that the stream slows. In an analogous manner, obstacles on the Earth's surface such as trees, houses, and telephone poles break the wind into eddies of various sizes to the lee of each obstacle (Figure 9.8). Consequently, the wind slows.

A snow fence (Figure 9.9) provides a practical illustration of the frictional slowing of the wind. Snow fences are designed to trap wind-blown snow, in some instances to prevent snow from drifting onto a nearby highway and in others to keep the soil snow covered.* A snow fence breaks the wind into small eddies, thereby tapping some of the wind's kinetic energy. The wind diminishes and loses some of its snow-transporting ability. Consequently, snow accumulates on the downwind side of the fence.

The rougher the surface of the Earth, the greater is the eddy viscosity of the wind. A forest thus offers more frictional resistance to the wind than does the

*Snow accumalated downwind of a snow fence protects the underlying soil from deep penetration of subfreezing temperatures. The snow accumulation also ensures a supply of soil moisture.

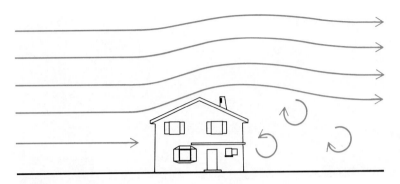

**FIGURE 9.8**
Turbulent eddies develop in the wind on the leeward side of a house.

smoother surface of a freshly mowed lawn. The eddy viscosity diminishes rapidly with altitude above the Earth's surface, away from the obstacles mainly responsible for frictional resistance. Hence, wind speed increases with altitude. This explains the advantage of siting a windmill at as high an elevation as possible (Figure 9.10). Above an average altitude of about 1 km (0.62 mi), friction is a minor force that merely acts to smooth the flow of air. The atmospheric zone to which frictional resistance (eddy viscosity) is essentially confined is called the **friction layer.**

**Turbulence** is fluid flow that is characterized by eddy motion. So far we have been considering obstacles on the Earth's surface as sources of eddies that exert a drag on the wind. Irregular fluid flow that originates in this way is known as *mechanical turbulence.* In addition, eddies develop in air as a consequence of solar heating of the ground. Irregular fluid flow that originates in this way is known as *thermal turbulence.* Convection currents are examples of thermal turbulence within the atmosphere. In actual practice, it is virtually impossible to distinguish between the two sources of turbulent eddies.

Regardless of source, we experience turbulent eddies as gusts of wind. You may have noticed that the gustiness of the wind often varies with the time of day; that is, gusts tend to be strongest during the warmest hours of the day. The reasons for this relationship are discussed in the Special Topic "Wind Gusts, Wind Shear, and Atmospheric Stability."

## GRAVITY

The atmosphere is subject to the same force that holds all objects on the Earth's surface, **gravity.** The

**FIGURE 9.9**
A snow fence slows the wind, thereby decreasing the wind's ability to transport snow. Hence, snow accumulates to the lee of the snow fence. [Photograph by J. M. Moran]

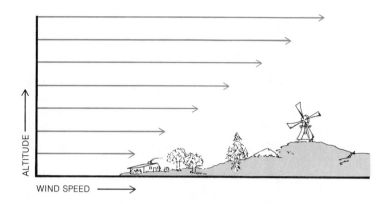

ALTITUDE →

WIND SPEED ⟶

**FIGURE 9.10**
The horizontal wind strengthens with altitude, away from the frictional resistance offered by objects on the Earth's surface.

force of gravity is actually the net effect of two other forces working together: (1) the force of attraction between the Earth and all other objects, called **gravitation,** and (2) a much weaker centripetal force imparted to all objects because of their rotation with the Earth on its axis. The two forces combine to produce the force of gravity, which accelerates a unit mass of any object downward at the rate of 9.8 m/sec each second.

The force of gravity always acts downward and perpendicular to the Earth's horizontal surface. For this reason, gravity, unlike the Coriolis effect and frictional forces, does not modify the horizontal wind. Gravity does, however, affect air that is ascending or descending, such as in convection currents, and gravity is responsible for the downhill drainage of cold, dense air.

### SUMMARY

We have now examined the various forces that affect horizontal and vertical air motion, and we can draw the following conclusions.

1. The *pressure gradient force* accelerates air away from regions of high air pressure and toward areas of low air pressure. Acceleration is directly proportional to the pressure gradient; that is, the closer the spacing of isobars, the greater is the acceleration.

2. A *centripetal force* is an imbalance of actual forces and operates whenever the wind describes a curved path.

3. The *Coriolis effect* deflects large-scale winds to the right of the initial direction in the Northern Hemisphere and to the left of the initial direction in the Southern Hemisphere. Its magnitude varies from zero at the equator to a maximum at the poles and is directly proportional to wind speed.

4. *Friction* slows winds within about 1 km (0.62 mi) of the Earth's surface.

5. *Gravity* accelerates air downward but does not modify horizontal winds.

## Joining Forces

To this point, we have examined forces operating in the atmosphere as if each force acted independently of all the others. In reality, these forces interact with one another in governing both the direction and speed of the wind. In some cases, two or more forces achieve a balance or equilibrium. From Newton's first law of motion, when the forces acting on an air parcel are in balance, there is no net force, and the parcel either remains stationary or continues to move along a straight path at constant speed. Therefore, when forces balance, the net acceleration is zero.

Let us now examine how forces interact in the atmosphere to control the vertical and horizontal flow of air, that is, the wind. These interactions result in (1) hydrostatic equilibrium, (2) the geostrophic wind, (3) the gradient wind, and (4) surface winds, the winds within the friction layer (Table 9.1).

## Wind Gusts, Wind Shear, and Atmospheric Stability

You may have noticed that the day's strongest and gustiest winds often occur during the afternoon hours. This phenomenon is perhaps most apparent following a night of extreme radiational cooling. By dawn, a temperature inversion develops in the lowest air layer, the surface air temperature has reached its diurnal minimum, and surface winds are very light or calm. In the hours following sunrise, the surface temperature rises steadily in response to increasing insolation, and through sensible heating eventually the lowest air layer is destabilized. By late morning or early afternoon, surface winds strengthen and become gusty, only to die down by sunset. These observations suggests a relationship between the gustiness of the wind and atmospheric stability.

Wind gusts are caused by irregular whirls of air called eddies. We are all familiar with the analogous whirls or eddies that form in rapidly flowing streams of water (see Figure 9.7). Eddies characterize turbulent motion of fluids and, as noted elsewhere in this chapter, are either mechanical or thermal in origin.

As a rule, as the wind strengthens, it becomes more turbulent and eddies become more energetic. Within the friction layer, horizontal wind speeds increase with altitude. Hence, strong eddies develop aloft, where winds are strong, and weaker eddies form near the Earth's surface, where winds are lighter. Eddies are transported upward and downward within the friction layer depending on atmospheric stability.

As shown in Chapter 6, atmospheric stability influences the vertical motion of air; that is, stable air suppresses vertical motion, and unstable air enhances vertical motion. When the friction layer is unstable, strong eddies are transported from higher altitudes downward toward the surface. Consequently, winds aloft weaken because they lose some kinetic energy, and winds at the surface strengthen and become gusty because they gain some kinetic energy from eddies descending from aloft. On the other hand, when the friction layer is stable, strong eddies generated aloft remain there, and weak eddies that form near the surface remain near the surface.

It follows that vertical wind shear and stability are related. A *wind shear* occurs whenever the wind changes speed (or direction) with distance or, as in this case, with altitude. Hence, a vertical wind shear characterizes the friction layer because horizontal winds strengthen with altitude. The magnitude of this wind shear, however, varies and is influenced by air stability. As noted, when the air is unstable, eddy transport reduces the vertical wind shear (the difference in horizontal wind speed between the surface and aloft). When the air is stable, the lack of eddy transport increases the vertical wind shear. In fact, although winds are typically light or calm within a low-level temperature inversion, winds are often quite strong in the air layer just above the inversion.

In summary, the greater the stability of the atmosphere's friction layer, the stronger is the vertical wind shear and the less gusty are surface winds. An unstable friction layer features a weaker vertical wind shear and relatively energetic and gusty surface winds. Thus, in our example, as the lower troposphere is destabilized in the hours following sunrise, surface winds strengthen and become gusty.

### HYDROSTATIC EQUILIBRIUM

We noted earlier that the atmosphere features a vertical pressure gradient. As shown schematically in Figure 9.11, the force due to this pressure gradient is directed upward from high pressure at the Earth's surface toward lower air pressure aloft. If this force acted alone, the vertical pressure gradient force would accelerate air away from the Earth, and we would be left gasping for breath. However, except in some small-scale violent weather systems (tornadoes, for example), the atmosphere's vertical pressure gradient force is almost balanced by the equal but oppositely directed force of gravity. An actual balance of these two forces is known as **hydrostatic equilibrium.**

Whenever forces are in balance, no net acceleration occurs; that is, there is no change in velocity. Hydrostatic equilibrium, then, does not preclude vertical (up

**TABLE 9.1**
**Forces Involved in Large-Scale Atmospheric Circulation—A Summary**

| Forces | Hydrostatic equilibrium | Geostrophic wind | Gradient wind | Surface winds |
|---|---|---|---|---|
| Pressure gradient | x | x | x | x |
| Centripetal | | | x | x |
| Coriolis | | x | x | x |
| Friction | | | | x |
| Gravity | x | | | |

or down) motion of air. Because of balanced forces, upward-moving air parcels continue upward at *constant* velocity, and downward-moving air parcels continue downward at *constant* velocity. Slight deviations from hydrostatic equilibrium, however, cause air to change speed (accelerate) vertically.

## GEOSTROPHIC WIND

The **geostrophic wind** is an unaccelerated horizontal wind that flows along a straight path at altitudes above the friction layer. It results from a balance be-

tween the horizontal pressure gradient force and the force due to the Coriolis effect.

When a horizontal pressure gradient $(P_H)$ develops, air parcels at first accelerate directly across isobars, away from high pressure and toward low pressure (Figure 9.12). As air parcels speed up, however, the Coriolis effect *(C)* strengthens and causes air parcels to swerve gradually to the right of their initial flow direction (in the Northern Hemisphere). The two forces eventually attain a balance, so that the wind blows at a constant speed in a straight path parallel to the isobars with the lowest air pressure to the left of air motion. Because the Coriolis effect is a large-scale phenomenon, the geostrophic wind develops only in large-scale weather systems.

## GRADIENT WIND

The **gradient wind** has many characteristics in common with the geostrophic wind. It also is large-scale, horizontal, and frictionless and blows parallel to isobars. The important distinction between the two is that the geostrophic wind blows in a straight path, whereas the path of the gradient wind is curved. Because a net centripetal force constrains air parcels to a curved trajectory, the gradient wind is not the consequence of balanced forces. Recall that the centripetal force changes only the air parcel's direction of motion, not its speed. The horizontal pressure gradient force, the Coriolis effect, and the centripetal force thus interact in the gradient wind.

A gradient wind develops at altitudes above the friction layer around a dome of high air pressure, called an **anticyclone** (or *high*), or around a center of low air pressure, called a **cyclone** (or *low*). In an ideal anticyclone, isobars form a series of concentric circles about the location of highest air pressure, as shown in Figure 9.13. The horizontal pressure gradient force $(P_H)$ is directed radially outward, away from the center of the high. The Coriolis effect *(C)* is directed inward. The Coriolis effect is slightly greater than the pressure gradient force, with the difference giving rise to the inward-directed centripetal force *(Ce)*. (This is what was meant earlier when we indicated that a centripetal force results from an imbalance of other forces.) In a Northern Hemisphere anticyclone above the friction layer, the gradient wind consequently blows clockwise and parallel to isobars.

**FIGURE 9.11**
With hydrostatic equilibrium, the upward-directed vertical pressure gradient force acting on an air parcel is balanced by the downward-directed force of gravity.

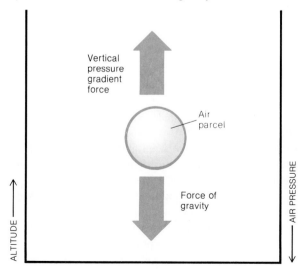

Vertical pressure gradient force

Air parcel

Force of gravity

ALTITUDE

AIR PRESSURE

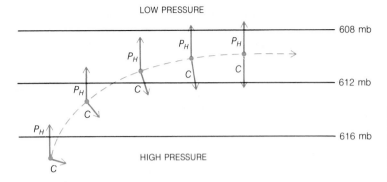

C = Coriolis effect

$P_H$ = Horizontal pressure gradient force

---→ Geostrophic wind

**FIGURE 9.12**
The horizontal air pressure gradient causes air parcels to accelerate across isobars from areas of high pressure toward areas of low pressure. The Coriolis effect then deflects air parcels to the right in the Northern Hemisphere. The Coriolis effect increases in magnitude until it balances the pressure gradient force. The result is an unaccelerated horizontal wind blowing parallel to isobars, that is, the geostrophic wind.

In an ideal cyclone, isobars form a series of concentric circles about the location of lowest air pressure. As indicated in Figure 9.14, the horizontal pressure gradient force $(P_H)$ is directed inward toward the cyclone center, and the Coriolis effect $(C)$ is directed radially outward from the center of the low. The pressure gradient force is slightly greater than the Coriolis effect, with the difference being equal to the net inward-directed centripetal force $(C_e)$. In a Northern Hemisphere cyclone above the friction layer, the gradient wind consequently blows counterclockwise and parallel to isobars.

The geostrophic and gradient winds are models (Chapter 1) that only approximate the actual behavior of horizontal winds above the friction layer. These approximations are nonetheless quite useful, and meteorologists rely on such approximations in their analysis of isobaric patterns on weather maps. A quantitative description of geostrophic and gradient winds is presented in the Mathematical Note at the end of this chapter.

**FIGURE 9.13**
In a Northern Hemisphere anticyclone above the friction layer, the gradient wind blows clockwise and parallel to isobars. In this idealized situation, the isobars form a series of concentric circles.

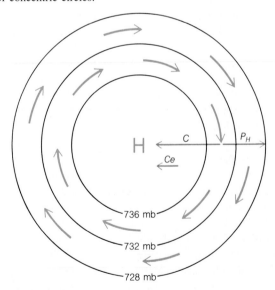

$P_H$ = Horizontal pressure gradient force

C = Coriolis effect

$C_e$ = Centripetal force

↰ Gradient wind

*SURFACE WINDS*

Geostrophic winds and gradient winds are frictionless; that is, they occur at altitudes above the friction layer. What is the effect of friction on the horizontal winds within the friction layer, that is, how does friction influence surface winds? Intuitively, we know that friction should slow the wind, but in addition, friction interacts with the other forces to change the wind direction.

For large-scale air motion in a straight path, the frictional force $(F)$ combines with the Coriolis effect $(C)$ to balance the horizontal pressure gradient force $(P_H)$, as shown in Figure 9.15. Note that the frictional force and Coriolis effect are at right angles to one another.

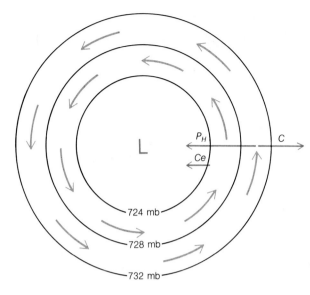

$P_H$ = Horizontal pressure gradient force

$C$ = Coriolis effect

$Ce$ = Centripetal force

Gradient wind

**FIGURE 9.14**

In a Northern Hemisphere cyclone above the friction layer, the gradient wind blows counterclockwise and parallel to isobars. In this idealized situation, the isobars form a series of concentric circles.

**FIGURE 9.15**

At the Earth's surface and elsewhere within the friction layer, the Coriolis effect combines with friction to balance the horizontal pressure gradient force. As a consequence, the horizontal wind blows across isobars and toward lower air pressure. Recall that the Coriolis effect acts at right angles to the initial wind direction and friction acts in a direction opposite that of the wind.

Friction acts opposite (180 degrees) to the direction of motion, and the Coriolis effect is at a right angle (90 degrees) to the direction of motion. Consequently, the horizontal wind slows down and shifts direction across isobars and toward low pressure. The deflection angle of surface winds crossing isobars varies from about 10 degrees over relatively smooth surfaces, where friction is low, to almost 45 degrees over rough terrain, where friction is greater.

As we noted earlier, friction's influence on the horizontal wind diminishes with altitude and is negligibly small at the top of the friction layer. Hence, horizontal winds strengthen with altitude. Furthermore, the angle between the wind direction and the isobars is maximum near the Earth's surface, decreases with altitude, and is essentially zero at the top of the friction layer (Figure 9.16).

What is the effect of friction on the horizontal surface winds blowing in an anticyclone and cyclone? As with straight-line surface winds, friction slows cyclonic and anticyclonic winds and combines with the Coriolis effect to shift winds so that they blow across isobars and toward low pressure. At the Earth's surface, therefore, anticyclonic winds blow clockwise and outward, as shown in Figure 9.17, and surface cyclonic winds blow counterclockwise and inward, as shown in Figure 9.18.

The characteristic circulation of a cyclone enables us to formulate a simple rule of thumb for locating cyclone centers. If you stand with your back to the wind and then turn approximately 45 degrees to your right,

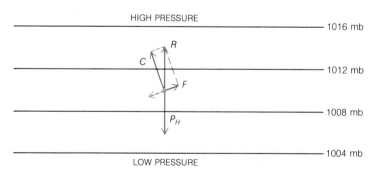

$C$ = Coriolis effect

$F$ = Force of friction

$R$ = Coriolis effect + force of friction

$P_H$ = Horizontal pressure gradient force

Direction of surface wind

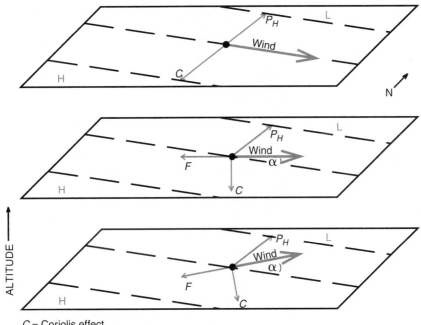

C = Coriolis effect
F = force of friction
$P_H$ = horizontal pressure gradient force
α = angle between wind direction and isobars

**FIGURE 9.16**
For the same horizontal pressure gradient, the angle between the wind direction and the isobars (dashed lines) decreases with altitude. [Adapted from R. A. Anthes, *Meteorology*, 6th ed. New York: Macmillan Publishing Company, 1992, p. 79]

**FIGURE 9.17**
Surface winds blow clockwise and outward in a Northern Hemisphere anticyclone.

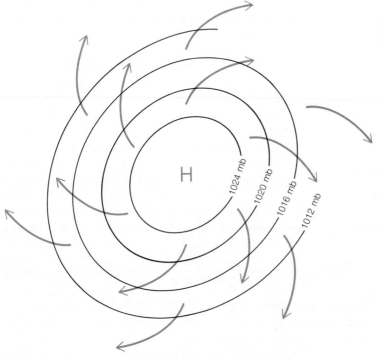

Surface winds within friction layer

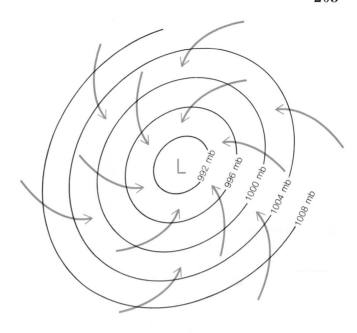

**FIGURE 9.18**
Surface winds blow counterclockwise and inward in
a Northern Hemisphere cyclone.

→ Surface winds within friction layer

the cyclone center will be located to your left. This rule
is a modification of an observation first stated in 1857
by the Dutch meteorologist Christopher H. D. Buys-
Ballot. It must be applied with caution, however, because large-scale surface winds may be modified by local air circulation such as a sea breeze.

In the Southern Hemisphere, cyclonic and anticyclonic circulations are opposite their Northern Hemisphere counterparts. This contrast is due to the change
in direction of the Coriolis deflection between the two
hemispheres. In the Southern Hemisphere, surface
winds in a cyclone blow in a clockwise and inward direction, and surface winds in an anticyclone blow in a
counterclockwise and outward direction. Above the
friction layer, Southern Hemisphere cyclonic winds are
clockwise and parallel to isobars, and Southern Hemisphere anticyclonic winds are counterclockwise and
parallel to isobars.

A glance at almost any national weather map (Figure 9.19) reveals that isobars seldom describe lengthy
straight segments or circular patterns. Instead, isobars
often form patterns of ridges and troughs. Nonetheless,
in ridges and troughs, winds tend to parallel isobars
above the friction layer and cross isobars toward low
pressure within the friction layer.

An additional consideration in analyzing isobaric
patterns for wind is the spacing of isobars. As noted

earlier, the steeper the air pressure gradient, the faster
is the wind. Where isobars are closely spaced, the
geostrophic and gradient winds are relatively strong.
Where isobars are widely spaced, these winds are weak.
The same rule applies to surface winds.

## Continuity of Wind

Air is a continuous fluid, and this continuity implies a
link between the horizontal and vertical components of
the wind. For example, surface winds are forced to follow the undulating topography of the Earth's surface
and ascend hills and descend into valleys. In addition,
uplift occurs along frontal surfaces as one air mass
moves horizontally and either overrides or pushes under another air mass (Chapter 6). Having examined the
horizontal circulation of anticyclones and cyclones, we
can identify other important connections between the
horizontal and vertical components of the wind.

As noted earlier, surface winds in a Northern Hemisphere anticyclone spiral clockwise and outward. Consequently, the horizontal surface winds diverge away
from the center of the high. A vacuum does not develop
at the center, however, because air slowly descends to-

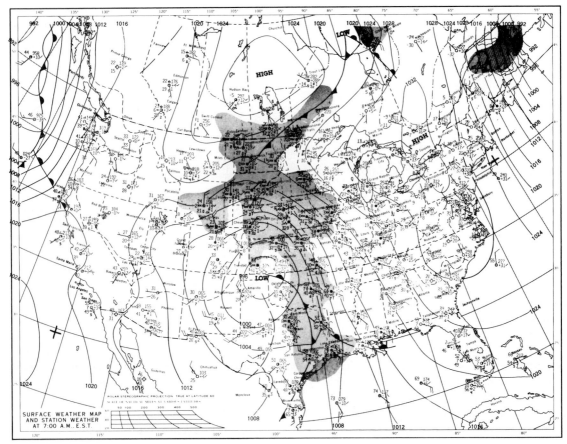

**FIGURE 9.19**
Note how isobars on a typical surface weather map describe trough and ridge patterns. Shaded
areas indicate precipitation. [NOAA weather map]

ward the Earth's surface and replaces the air that is diverging. Air is driven downward by air that converges
toward the high center aloft (Figure 9.20). Recall that
adiabatic compression raises the temperature and lowers the relative humidity of descending air. Skies therefore tend to be fair within anticyclones, and anticyclones are appropriately described as *fair weather*
systems.

Furthermore, within an anticyclone, the horizontal
air pressure gradient is typically very weak over a broad
area around the center of the system. The resulting light
or calm winds coupled with clear night skies favor intense radiational cooling. Hence, the ground and the air
adjacent to the ground may be chilled to the point that
dew, frost, or radiation fog develops.

Surface winds in a Northern Hemisphere cyclone
spiral counterclockwise and inward. Surface winds
therefore converge toward the center of a low. Air does
not simply pile up at the center; rather, the air ascends
in response to diverging air aloft (Figure 9.21). Recall
that adiabatic expansion lowers the temperature and increases the relative humidity of ascending air. Clouds
and precipitation may eventually develop, so that cyclones are typically *stormy weather* systems.

Continuity of the wind also means that vertical motion can be induced by downwind changes in frictional
resistance. The rougher the Earth's surface, the more
resistance it offers to horizontal winds. When the horizontal wind blows from a rough surface to a relatively
smooth surface—as when it blows from land to sea—
the wind accelerates. As Figure 9.22 shows, this acceleration causes the wind to diverge (stretch), thereby in-

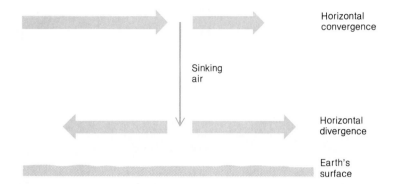

**FIGURE 9.20**
In this idealized vertical cross section of an anticyclone, air converges aloft, sinks, and diverges at the Earth's surface.

ducing downward motion of air. In contrast, when the horizontal wind blows from a smooth to a rough surface, the wind slows and converges (piles up), thereby inducing upward air motion. This is one reason why, along a coastline, cumuliform clouds (cumulus) tend to develop with an onshore wind (from sea to land) and tend to dissipate with an offshore wind (from land to sea).

## Scales of Weather Systems

Although the atmosphere is a continuous fluid, for convenience of study we subdivide atmospheric circulation into discrete weather systems that operate at various spatial and temporal scales (Table 9.2). The large-scale wind belts encircling the planet (polar easterlies, westerlies, and trade winds) are global or **planetary-scale systems. Synoptic-scale systems** are continental or oceanic in scale; migrating cyclones, hurricanes, and air masses are examples. **Mesoscale systems** include thun-

derstorms and sea and lake breezes—phenomena that are so small that they may influence the weather in only a portion of a large city. A circulation system covering a very small area—a tornado, for example—represents the smallest spatial subdivision of atmospheric motion, **microscale systems.**

Circulation systems not only differ in spatial scale, they also contrast in life expectancy. Thus, patterns in the planetary-scale circulation may persist for weeks or even months. Synoptic-scale systems typically last for several days to a week or so. Mesoscale systems usually complete their life cycles in a matter of hours to perhaps a day, whereas microscale systems persist for minutes or less.

There are other differences among the various scales of atmospheric circulation. At the micro- and mesoscale, vertical velocities may be comparable in magnitude to horizontal velocities. At the synoptic and planetary scales, however, horizontal winds are considerably stronger than vertical flow. Furthermore, at the micro- and mesoscale, the Coriolis effect is usually negligibly small. In contrast, the Coriolis effect is very important in synoptic- and planetary-scale circulation systems.

**FIGURE 9.21**
In this idealized vertical cross section of a cyclone, air converges at the Earth's surface, rises, and diverges aloft.

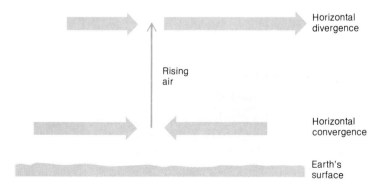

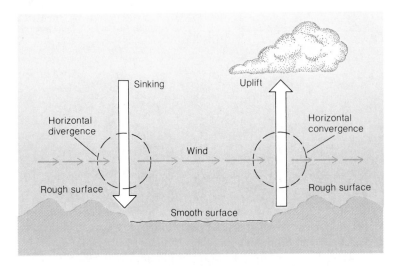

**FIGURE 9.22**
Surface winds undergo horizontal divergence when blowing from a rough to a smooth surface, and horizontal convergence when blowing from a smooth to a rough surface. Horizontal divergence causes air to sink and horizontal convergence causes air to rise.

Each smaller scale weather system is part of and dependent on larger scale atmospheric circulation. That is, the various scales of atmospheric motion form a kind of hierarchy. For example, extreme nocturnal radiational cooling requires a synoptic weather pattern that favors clear skies and light or calm winds. At the microscale, such weather conditions may be accompanied by frost formation or radiation fog.

In Chapters 10 through 15, we examine weather systems of all scales starting with the planetary scale.

## Wind Pressure

As wind speed increases, the potential for wind damage to trees, buildings, and other structures also increases. In this regard, the concept of wind pressure is useful. **Wind pressure** is defined as the force per unit area produced by the wind on an object in its path. Wind pressure is directly proportional to the square of the wind speed $(v^2)$. Hence, each doubling of wind speed increases the wind pressure by a factor of four $(2^2 = 4)$.

In designing buildings to withstand high winds, engineers and architects consult the climatic record to determine the probability of potentially destructive winds. Because wind speed increases with altitude, a building's upper stories are subject to greater wind pressure than its lower stories. A building's exposure is also an important consideration in planning for wind pressure. For example, a building situated on a coastal plain and facing the open ocean is exposed to stronger winds than a building situated among many other buildings within a congested city center.

In the reverse application of this approach, wind pressure, and hence wind speed, can be reconstructed from storm damage (Figure 9.23). In this way, for example, meteorologists have been able to estimate wind speeds in tornadoes based on the damage they produce. Such reconstructions are necessary because tornadic winds are so strong that they destroy conventional weather instruments.

**TABLE 9.2**
**Scales of Atmospheric Circulation**

| Circulation | Space scale | Time scale |
|---|---|---|
| Planetary-scale | 10,000–40,000 km | weeks–months |
| Synoptic-scale | 100–10,000 km | days–weeks |
| Mesoscale | 1–100 km | hours–days |
| Microscale | 1 m–1 km | seconds–hours |

**FIGURE 9.23**
In the absence of direct measurements of wind, it is possible to reconstruct the wind speed from storm damage based on wind pressure estimates. This tree was twisted by winds estimated at 80 km per hour. [Photograph by J. M. Moran]

## Wind Measurement

Meteorologists are interested in monitoring both the speed and direction of wind. Most wind-monitoring instruments are designed to measure only the horizontal component of the wind because it is usually considerably stronger than the vertical component. For some specialized research purposes, very sensitive instruments are available that measure vertical wind speeds or a combination of vertical and horizontal wind components.

A traditional **wind vane,** like the one shown in Figure 9.24, consists of a free-swinging horizontal shaft with a flat vertical plate at one end and a counterweight (arrow) at the other end. The counterweight always points directly into the wind. Another design is the airport **wind sock,** which consists of a cone-shaped cloth bag that is open at both ends. The larger end of the sock is held open by a metal ring that is attached to a pole and is free to rotate. Air enters the larger opening and stretches the sock downwind (Figure 9.25).

Wind direction is always designated as the *direction from which* the wind blows. For example, a wind blowing from the east toward the west is described as an

**FIGURE 9.24**
Horizontal wind direction is monitored by a wind vane. The instrument's arm points in the direction from which the wind blows. [Photograph by J. M. Moran]

*east* wind and a wind blowing from the northwest toward the southeast is a *northwest* wind. A wind vane may be linked electronically or mechanically to a dial that is calibrated to read in points of the compass or in degrees. Measured clockwise from true north, an east wind is specified as 90 degrees, a south wind as 180 degrees, a west wind as 270 degrees, and a north wind

**FIGURE 9.25**
An airport wind sock gives wind direction and a general indication of wind speed. The sock points downwind. [Photograph by J. M. Moran]

as 360 degrees. The wind is recorded as 0 degrees only under calm conditions.

Wind speed can be estimated by observing the wind's effect on lake or ocean surfaces or on flexible objects such as trees. Such observations are the basis of the **Beaufort scale,** which is a graduated sequence of wind strength ranging from 0 for calm conditions to 12 for hurricane-strength winds (Table 9.3). The scale bears the name of Sir Francis Beaufort, who developed it in the early 1800s while a ship's commander in the British Navy. Beaufort's goal was to standardize terms used by sailors in describing the state of the sea under various wind conditions. In 1838, after some revision, the British Navy adopted the Beaufort scale, and in 1853 it was sanctioned for international use by seafarers. Later, when the scale was extended from sea to land, it was necessary to develop wind speed equivalents for each Beaufort number; this was done in 1926. The scale is still used today. In fact, some modern-day mariners prefer Beaufort numbers to onboard instrument measurements of wind speed.

As a rule, a **cup anemometer,** such as the one shown in Figure 9.26, accurately monitors horizontal wind speed. This device works on the same principle as a

**FIGURE 9.26**
Wind speed is measured by a cup anemometer. The faster the wind speed, the faster the cups spin. [Courtesy of Belfort Instrument Company]

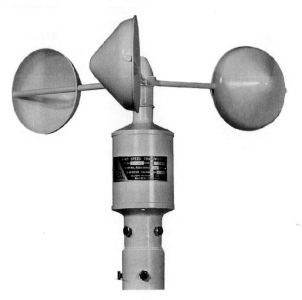

**TABLE 9.3**
**Beaufort Scale of Wind Force**

| Beaufort number | General description | Land and sea observations for estimating wind speeds | Wind speed 10 m above ground (km/hr) |
|---|---|---|---|
| 0 | Calm | Smoke rises vertically. Sea like mirror. | Less than 1 |
| 1 | Light air | Smoke, but not wind vane, shows direction of wind. Slight ripples at sea. | 1–5 |
| 2 | Light breeze | Wind felt on face, leaves rustle, wind vanes move. Small, short wavelets. | 6–11 |
| 3 | Gentle breeze | Leaves and small twigs moving constantly, small flags extended. Large wavelets, scattered whitecaps. | 12–19 |
| 4 | Moderate breeze | Dust and loose paper raised, small branches moved. Small waves, frequent whitecaps. | 20–28 |
| 5 | Fresh breeze | Small leafy trees swayed. Moderate waves. | 29–38 |
| 6 | Strong breeze | Large branches in motion, whistling heard in utility wires. Large waves, some spray. | 39–49 |
| 7 | Near gale | Whole trees in motion. White foam from breaking waves. | 50–61 |
| 8 | Gale | Twigs break off trees. Moderately high waves of great length. | 62–74 |
| 9 | Strong gale | Slight structural damage occurs. Crests of waves begin to roll over. Spray may impede visibility. | 75–88 |
| 10 | Storm | Trees uprooted, considerable structural damage. Sea white with foam, heavy tumbling of sea. | 89–102 |
| 11 | Violent storm | Very rare; widespread damage. Unusually high waves. | 103–118 |
| 12 | Hurricane | Very rare; much foam and spray greatly reduce visibility. | 119 and over |

bicycle or an automobile speedometer. The wind spins the cups (usually 3 or 4) and thus generates a weak electric current, which is calibrated on a dial in meters per second, kilometers per hour, miles per hour, or knots.* Several other types of anemometer are available, including the very sensitive **hotwire anemometer.** In this instrument, the wind blows past a heated wire, or wires, and the heat lost to moving air is calibrated in terms of wind speed.

Recording a continuous trace of wind speed and direction is sometimes informative. Wind vanes and anemometers can be linked to pens that record on a clock-driven drum. As shown in Figure 9.27, the trace indicates a considerable variation in both wind direction and wind speed with time. The spectrum of wind gusts and lulls indicate turbulence. More commonly today, the output from wind vanes and anemometers is recorded by computer or on magnetic tape.

*One knot = 1 nautical mile (1.85 km) per hour; 1 knot = 0.51 meter per second; 1 mile per hour = 0.44 meter per second.

Ideally, a wind vane or anemometer system should be mounted on a tower so that the instruments monitor horizontal winds 10 m (33 ft) above the ground. It is best to avoid rooftop locations because winds tend to accelerate over buildings. In addition, the system should be sited well away from (1) structures that might shelter the instruments and (2) any obstacles that might channel (and thus accelerate) the wind.

# Conclusions

In this book so far, we have seen that unequal rates of radiational heating and cooling in the Earth–atmosphere system give rise to gradients in temperature. In response to those gradients, the atmosphere circulates and thereby, heat energy is converted to kinetic energy.

WEATHER FACT

## The Windiest Place on Earth

On 12 April 1934, the weather station at the 1910-m (6262-ft) summit of Mount Washington, New Hampshire, recorded a peak wind gust of 373 km/hr (231 mi/hr). This is the highest wind speed ever recorded.* That same day, the wind averaged 303 km/hr (188 mi/hr) over a 5-min span. At 57 km/hr (35 mi/hr) the average annual wind speed on Mount Washington is the greatest of any location in the United States. Average monthly wind speed ranges from about 40 km/hr (25

*There is no doubt that stronger winds occur in violent tornadoes, but so far they have not been recorded by anemometers.

mi/hr) in July to about 73 km/hr (45 mi/hr) in January. Winds in excess of hurricane force (119 km/hr) are common in winter.

Elevation, exposure, and reduced friction factor into the windiness of mountaintops. In addition, as air flows up and over a mountain, it is squeezed between the summit and overlying air layers. This constriction accelerates the wind. The prevailing westerlies are normally quite strong over New England and the constricted flow over Mount Washington accelerates them to extreme magnitudes.

In this chapter, we have examined the various forces that initiate and shape atmospheric circulation (the wind). Note that the pressure gradient force and gravity would exist even if air were not in motion, and the other forces (centripetal, Coriolis, and friction) come into play only after air is in motion.

From everyday experience, we are aware of the end product of the forces described in this chapter, namely, the wind, but we are not readily aware of the individual forces. In learning about atmospheric forces and how they interact, we see that each force is bound by certain constraints. For example, friction is

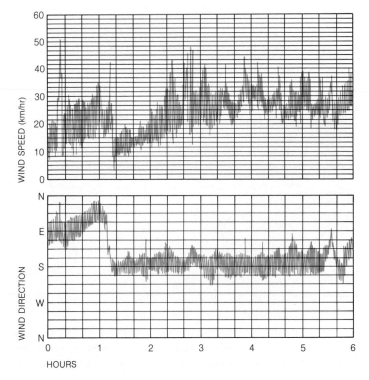

**FIGURE 9.27**
Continuous trace of time variations in wind speed and wind direction over a six-hour period.

important only in the lower troposphere, and the Coriolis effect always shifts the wind to the right in the Northern Hemisphere. In the following chapters, our awareness of these and other constraints will aid in understanding the characteristics of the various weather systems—for example, why hurricanes do not form at the equator, and why winds in a tornado may blow in either a clockwise or counterclockwise direction. We are now ready to begin our discussion of the characteristics of the various circulation systems, beginning with those operating at the planetary scale.

## MATHEMATICAL NOTE

## Geostrophic and Gradient Winds

The geostrophic wind is the result of a balance between the horizontal pressure gradient force and the Coriolis effect. Here we examine the geostrophic motion of a unit mass of air (1 g, for example). The acceleration imparted to this air parcel by a horizontal pressure gradient is given by

$$\frac{1}{\rho}\left(\frac{\Delta P}{\Delta N}\right)$$

where $\rho$ is the density of the air and $\Delta P$ is the change in air pressure over a horizontal distance, $\Delta N$, measured perpendicular to isobars. (Pressure gradients are always measured perpendicular to isobars.)

The acceleration imparted to the air parcel by the Coriolis effect is given by

$$(2\Omega \sin \phi) v$$

where $v$ is the speed of the air parcel, $\Omega$ is the angular velocity of the Earth as it rotates on its axis ($\Omega = 7.29 \times 10^{-5}$ radian per second), and $\phi$ is the latitude.

For the geostrophic wind,

$$\frac{1}{\rho}\left(\frac{\Delta P}{\Delta N}\right) = (2\Omega \sin \phi)v$$

Solving for $v$, the speed of the geostrophic wind, we have

$$v = \frac{1}{(2\Omega \sin \phi)\rho}\left(\frac{\Delta P}{\Delta N}\right)$$

The geostrophic wind thus strengthens with an increasing pressure gradient (that is, closer spacing of isobars) and decreasing latitude. Recall that the geostrophic wind blows parallel to isobars and at right angles to the pressure gradient force and Coriolis effect (which are directed one opposite the other).

The gradient wind results from a slight imbalance between the horizontal pressure gradient force and the Coriolis effect. This imbalance is realized as a centripetal force. Again, assume that we are examining the motion of a unit mass of air. The acceleration imparted to the air parcel by the centripetal force is

$$v^2/r$$

where $v$ is the speed of the air parcel, and $r$ is the radius of curvature of the path described by the air parcel. The centripetal force is the force that constrains the air parcel to a curved trajectory. This force is strong where the curvature is sharp (small $r$) and weak where the curvature is gradual (large $r$). The three forces (pressure gradient, Coriolis, and centripetal) interact such that

$$\frac{v^2}{r} + (2\Omega \sin \phi) v - \frac{1}{\rho}\left(\frac{\Delta P}{\Delta N}\right) = 0$$

The net force, which is the centripetal force, operates on the gradient wind only to change the direction of the wind as it follows a curved path, and not to change the wind's speed. We could solve the above equation for $v$ to determine the speed of the gradient wind for some radius of curvature, latitude, and horizontal pressure gradient (measured perpendicular to isobars). Note that for wind blowing in a straight line,

$$v^2/r = 0$$

and the equation reduces to that presented earlier for geostrophic flow.

## *Key Terms*

| | |
|---|---|
| wind | gravitation |
| Newton's second law of motion | hydrostatic equilibrium |
| | geostrophic wind |
| air pressure gradient | gradient wind |
| isobars | anticyclone |
| pressure gradient force | cyclone |
| Newton's first law of motion | planetary-scale systems |
| | synoptic-scale systems |
| centripetal force | mesoscale systems |
| Coriolis effect | microscale systems |
| friction | wind pressure |
| viscosity | wind vane |
| molecular viscosity | wind sock |
| eddy viscosity | Beaufort scale |
| friction layer | cup anemometer |
| turbulence | hotwire anemometer |
| gravity | |

## *Summary Statements*

☐ Wind is the movement of air measured relative to the Earth's surface. Wind is the consequence of interactions of the pressure gradient force, the Coriolis effect, friction, and gravity. The centripetal force is the resultant of other forces.

☐ The pressure gradient force initiates air motion and arises in part from spatial variations in air temperature and, to a lesser extent, water vapor concentration. In response to gradients in pressure, air accelerates from areas of relatively high pressure toward areas of relatively low pressure.

☐ Wind is a vector quantity, that is, it has both direction and magnitude. Hence, an acceleration of the wind may consist of a change in speed or direction or both.

☐ The centripetal force is the imbalance in actual forces that operates whenever the wind follows a curved path. The centripetal force is responsible for a change in direction of the wind and not a change in speed.

☐ The Coriolis effect arises from the Earth's rotation on its axis. It deflects the wind to the right of its initial direction in the Northern Hemisphere and to the left in the Southern Hemisphere. The deflective force is zero at the equator and increases with latitude to a maximum at the poles. The Coriolis effect is important only in large-scale (planetary- and synoptic-scale) circulation systems.

☐ Friction affects horizontal winds blowing within about 1 km (0.62 mi) of the Earth's surface. Obstacles on the Earth's surface slow the wind by breaking it into turbulent eddies.

☐ Gravity always accelerates objects downward and perpendicular to the Earth's surface. Gravity is important for vertical motion of air.

☐ Hydrostatic equilibrium is the balance between the upward-directed pressure gradient force and the downward-directed force of gravity. Slight deviations in hydrostatic equilibrium cause air to accelerate upward or downward.

☐ The geostrophic wind is an unaccelerated, horizontal wind that blows in a straight path parallel to isobars at altitudes above the friction layer. The geostrophic wind results from a balance between the horizontal pressure gradient force and the Coriolis effect.

☐ The gradient wind is a horizontal wind that parallels curved isobars at altitudes above the friction layer. The centripetal force operates in the gradient wind and is the result of an imbalance between the horizontal pressure gradient force and the Coriolis effect. In the Northern Hemisphere, the gradient wind blows clockwise in anticyclones and counterclockwise in cyclones.

☐ In large-scale (synoptic- and planetary-scale) circulation systems, friction slows the wind and combines with the Coriolis effect to shift the wind direction across isobars and toward low pressure.

☐ Within the friction layer, horizontal winds blow clockwise and outward in Northern Hemisphere anticyclones and counterclockwise and inward in Northern Hemisphere cyclones.

☐ In an anticyclone, horizontal divergence of surface winds causes descending air. Hence, an anticyclone is a fair-weather system. In a cyclone, horizontal convergence of surface winds causes ascending air. Hence, a cyclone is a stormy weather system.

☐ Along a coastline, offshore winds undergo horizontal divergence (inducing descending air), whereas onshore winds undergo horizontal convergence (inducing ascending air).

☐ Atmospheric circulation is divided into four spatial/temporal scales: planetary, synoptic, mesoscale, and microscale.

## Review Questions

1. Provide a definition for *wind*. Compare the magnitude of the vertical wind versus the horizontal wind in large-scale weather systems. Do the same for meso- and microscale systems.
2. What causes horizontal gradients in air pressure?
3. What is the relationship between horizontal air pressure gradients and wind speed and direction?
4. Why must a centripetal force operate whenever the wind follows a curved trajectory?
5. What causes the Coriolis effect and why does its magnitude increase with increasing latitude? Why is the Coriolis deflection important only in large-scale (planetary- and synoptic-scale) weather systems?
6. Why does the Coriolis deflection reverse between the Northern and Southern hemispheres?
7. Distinguish between *molecular viscosity* and *eddy viscosity.*
8. How does the horizontal wind speed change with altitude within the friction layer?
9. State Newton's first and second laws of motion.
10. Provide an example of how gravity influences the motion of air.
11. Define *hydrostatic equilibrium*. Does hydrostatic equilibrium imply that there can be no ascent or descent of air within the atmosphere? Explain your response.
12. Distinguish between the geostrophic wind and the gradient wind.
13. What is responsible for centripetal forces in cyclones and anticyclones?
14. How does the roughness of the Earth's surface affect the horizontal wind direction?
15. Describe the horizontal circulation in a cyclone (a) within the friction layer and (b) above the friction layer.
16. Describe the horizontal circulation in an anticyclone (a) within the friction layer and (b) above the friction layer.
17. Why is fair weather usually associated with anticyclones and stormy weather with cyclones?
18. Horizontal air pressure gradients are usually weak over a wide area around the center of an anticyclone. What does this imply about the weather at the center of an anticyclone?
19. How do downwind changes in surface roughness (frictional resistance) induce divergence and convergence of the horizontal wind? Provide specific examples.
20. Present examples of planetary-scale, synoptic-scale, mesoscale, and microscale circulation systems.

## Questions for Critical Thinking

1. Suppose that the magnitude of the horizontal pressure gradient changes with altitude. How would this affect the horizontal wind?
2. In view of Newton's first law of motion, is gradient wind the consequence of *balanced* forces? Explain your response.
3. How does the horizontal pressure gradient compare in magnitude to the vertical pressure gradient?
4. Predict how wind direction and speed at your locality change as a cyclone approaches. Do the same for an approaching anticyclone.
5. Suppose that a synoptic-scale cyclone is centered over central Kansas. Describe the air mass advection to the southeast, west, and northeast of the storm center.

## Selected Readings

BLACKADAR, A. "Simple Motions on the Rotating Earth," *Weatherwise* 39, No. 2 (1986):99–103. Provides a clear explanation of the Coriolis effect.

FORRESTER, F. H. "How Strong is the Wind?" *Weatherwise* 39, No. 3 (1986):147–151. Describes the origin of the Beaufort scale.

HIGBIE, J. "Simplified Approach to Coriolis Effects," *The Physics Teacher* 18 (1980):459–460. Describes ways to demonstrate the Coriolis effect.

KARAPIPERIS, P. P. "The Tower of the Winds," *Weatherwise* 39, No. 3 (1986):152–154. Concerns a building in Athens, probably dating to the first century B.C., that demonstrates that the people of the time understood that wind direction and weather were linked.

SNOW, J. T., ET AL. "Basic Meteorological Observations for Schools: Surface Winds," *Bulletin of the American Meteorological Society* 70 (1989):493–508. Examines the principles of wind-measuring instruments and includes suggestions for construction of a wind vane and cup anemometer for school use.

# 10 Planetary-Scale Circulation

*The fair breezes blew, the white
  foam flew,
The furrow follow'd free;
We were the first that ever burst
Into that silent sea.
Down dropt the breeze, the sails
  dropt down,
'Twas sad as sad could be
And we did speak only to break
The silence of the sea!
All in a hot and copper sky,
The bloody Sun, at noon,
Right up above the mast did stand,
No bigger than the Moon.
Day after day, day after day,
We stuck, nor breath nor motion;
As idle as a painted ship
Upon a painted ocean.*

SAMUEL TAYLOR COLERIDGE
*The Rime of the Ancient Mariner*

Planetary-scale circulation encompasses wind belts that encircle the globe and pressure systems that cover large areas of the Earth's oceans. [NOAA satellite photograph]

P LANETARY-SCALE air circulation ultimately is responsible for the development and displacement of most smaller scale weather systems. It is therefore, appropriate for us to begin our description of weather systems with those that are global in scale. In this chapter, we describe the wind belts and pressure systems that operate at the global scale, with special emphasis on circulation patterns of the midlatitude westerlies.

## Idealized Circulation Pattern

Planetary-scale air circulation is a complex pattern of winds and pressure systems. To better understand how this circulation is shaped, we start with an idealized model of planet Earth. Picture the Earth as a nonrotating sphere with a uniform solid surface. As on the real Earth, assume that the sun heats the equatorial regions more intensely than the polar regions. On the sunlit side

of the globe, a temperature gradient develops between the equator and poles. In response, two huge convection cells form—one in each hemisphere (Figure 10.1A). Cold, dense air sinks at the poles and flows at the surface toward the equator, where warm, light air rises and aloft flows toward the poles.

If our idealized planet begins to rotate, the Coriolis effect comes into play (Figure 10.1B). In the Northern Hemisphere, surface winds shift to the northeast (that is, northeast to southwest), and in the Southern Hemisphere, surface winds become southeasterly (that is, southeast to northwest). Hence, on our hypothetical planet, surface winds blow counter to the planet's rotation direction, which is from west to east. This is an impossible situation because the surface winds would have a braking effect on the rotating Earth.

Circulation is maintained in the atmosphere of our idealized Earth because the global winds divide into three belts in each hemisphere, so that some winds blow with and some winds blow against the planet's rotational direction (Figure 10.1C). In the Northern Hemisphere, average surface winds are northeasterly from the equator to 30 degrees latitude, southwesterly from 30 to 60 degrees, and northeasterly from 60 degrees to the North Pole. In the Southern Hemisphere, surface winds are southeasterly from the equator to 30 degrees, northwesterly from 30 to 60 degrees, and southeasterly from 60 degrees to the South Pole.

Surface winds converge along the equator and along the 60-degree latitude circles. By causing air to rise, convergence leads to expansional cooling, cloud development, and precipitation. These convergence zones are therefore belts of low pressure (Figure 10.1D). On the other hand, surface winds diverge at the poles and along the 30-degree latitude circles. In these regions, air descends and leads to compressional warming and fair weather. These are zones of high pressure.

If we add the continents and oceans to our idealized Earth, the temperature characteristics of the Earth's surface become more complex and more realistic, and so do the air pressure patterns and the planetary-scale circulation. Some of the pressure belts break into separate cells, and important contrasts in air pressure develop over land versus sea. Now our idealized model of planetary-scale circulation more closely approximates the actual circulation. In the next section, we describe the principal circulation features of the planetary scale.

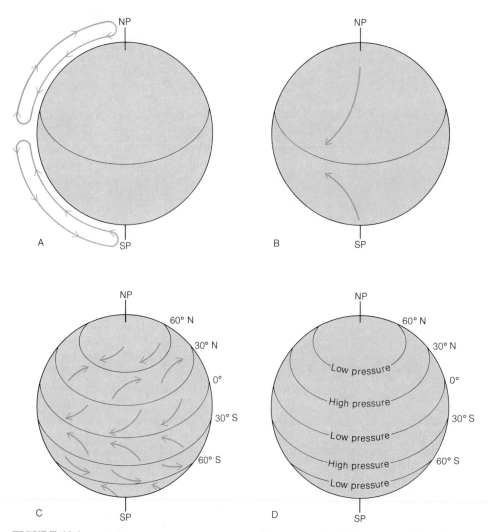

**FIGURE 10.1**
Planetary-scale air circulation on an idealized model of the Earth featuring a uniform solid surface. (A) If the sphere is nonrotating, huge convection currents develop in the atmosphere on the sunlit portion of the planet so that air circulates between the hot equator and the cold poles. (B) With a rotating Earth, surface winds become northeasterly in the Northern hemisphere and southeasterly in the Southern Hemisphere owing to the Coriolis effect. (C) In reality, surface winds divide into three zones in each hemisphere. (D) Zones of converging and diverging surface winds give rise to east–west belts of low and high pressure.

## Pressure Systems and Wind Belts

Maps of average air pressure at sea level for January and July (Figure 10.2) reveal several areas of relatively high and low atmospheric pressure. These are the **semi-permanent pressure systems.** Although these systems are persistent features of the planetary-scale circulation, they undergo important seasonal changes in both location and strength—hence the modifier *semipermanent.* Pressure systems include subtropical anticyclones, the intertropical convergence zone (ITCZ), and subpolar lows.

**Subtropical anticyclones** are imposing features of

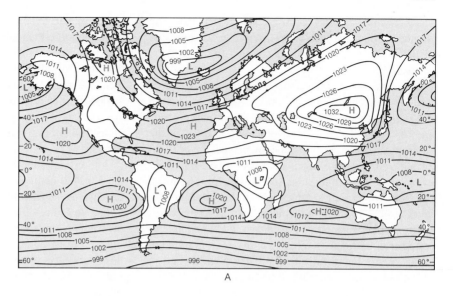

A

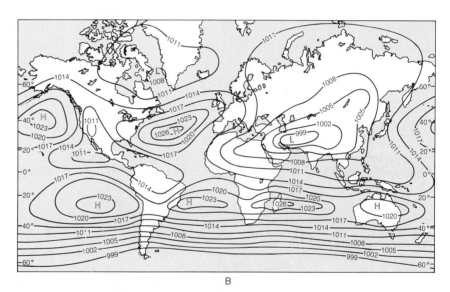

B

**FIGURE 10.2**
Mean sea-level air pressure for (A) January and (B) July. Contours are isobars in millibars.

the planetary-scale circulation that are centered over subtropical latitudes (on average, near 30 degrees N and S) of the North and South Atlantic, the North and South Pacific, and the Indian Ocean. These highs extend vertically from the ocean surface to the tropopause and exert a strong influence on weather and climate.

Stretching from the center of each subtropical high, outward over its eastern flank, are extensive areas of subsiding stable air. Subsiding air undergoes compressional warming, which produces low relative humidities and sunny skies. Thus the world's major deserts,

including the Sahara of North Africa and the Sonora of Mexico and southwest United States, are located under the eastern flanks of subtropical anticyclones. On the far-western portions of the subtropical highs, however, subsidence is less and the air is not as stable. Consequently, episodes of cloudy, stormy weather are more frequent in these regions. This contrast in climate between the eastern and western flanks of a subtropical high is particularly apparent across southern North America. The weather of the American Southwest (on the eastern side of the North Pacific high, also called the *Hawaiian high*) is considerably drier than the weather of the American Southeast (on the western side of the North Atlantic high, also called the *Bermuda–Azores* high).

As is typical of anticyclones, a subtropical high features a weak horizontal air pressure gradient over a broad area at the system's center. Hence, surface winds are very light or even calm over extensive areas of the subtropical oceans. This situation played havoc with ancient sailing ships, which were becalmed for days or even weeks at a time. Ships setting sail from Spain to the New World were often caught in this predicament, and crews were forced to jettison their cargo of horses when supplies of food and water ran low. For this reason, early mariners referred to this region of calm winds as the **horse latitudes,** a name now applied to all latitudes under subtropical highs. The stanzas from Coleridge's *Rime of the Ancient Mariner* at the beginning of this chapter aptly describe the weather conditions of horse latitudes.

In the Northern Hemisphere, surface winds blow clockwise and outward, away from the centers of the subtropical highs, forming the westerlies and trade winds (Table 10.1). Surface winds north of the horse latitudes constitute the highly variable **midlatitude westerlies** (which on average actually blow from the southwest). Surface winds blowing out of the southern flanks of the anticyclones are known as the northeast **trade winds.** The trades are the most persistent winds

on the planet, in some regions blowing from the same direction more than 80% of the time. Analogous winds develop in the Southern Hemisphere. Recall, however, that the Coriolis deflection reverses in the Southern Hemisphere. A counterclockwise and outward surface airflow thus causes southeast trade winds on the northern flanks of the Southern Hemisphere subtropical highs and a belt of northwest winds on the southern flanks.

Interestingly, the belts of westerlies and trades over the North Atlantic were known to mariners at least as far back as the fifteenth century. On his venture to find a route to India, Christopher Columbus took advantage of what we now know is the circulation about the Bermuda–Azores high. In his westward voyage, Columbus first sailed southward and picked up the northeast trade winds that eventually took his ships to the Caribbean. On his return trip to Spain, he sailed north and picked up the westerlies.

The trade winds of the two hemispheres converge into a broad east–west belt of light and variable winds, called the **doldrums.** In that belt, ascending air induces cloudiness and rainfall. The most active weather develops along the **intertropical convergence zone (ITCZ),** a discontinuous belt of thunderstorms paralleling the equator (Figure 10.3). The mean position of the ITCZ is at the latitude of the Earth's highest mean surface temperature, the so-called **heat equator.** For reasons discussed in Chapter 18, the heat equator is in the Northern Hemisphere, just north of the geographical equator.

On the poleward side of the subtropical highs, surface westerlies flow into regions of low pressure. In the Northern Hemisphere, there are typically two separate **subpolar lows**—the *Aleutian low* over the North Pacific Ocean and the *Icelandic low* over the North Atlantic. These pressure cells mark the convergence of the midlatitude southwesterlies with the polar northeasterlies. In contrast, in the Southern Hemisphere the midlatitude northwesterlies and the polar southeasterlies converge along a nearly continuous belt of low pressure surrounding the Antarctic continent.

The surface westerlies meet and override the polar easterlies along the **polar front.** Recall that a *front* is a narrow zone of transition between air masses of contrasting density, that is, air masses of different temperature, different water vapor concentration, or both. In this case, dense, cold air masses flowing toward the

**TABLE 10.1**
**Features of Planetary-Scale Circulation**

| Latitude | | Planetary-scale systems |
|----------|------|-------------------------|
| 90° N | High | Polar anticyclones |
| | ↙↙ | Polar northeasterlies |
| 60° N | Low | Subpolar cyclones |
| | ↗↗ | Westerlies |
| 30° N | High | Subtropical anticyclones |
| | ↙↙ | Northeast trades |
| 0° | Low | ITCZ |
| | ↖↖ | Southeast trades |
| 30° S | High | Subtropical anticyclones |
| | ↘↘ | Westerlies |
| 60° S | Low | Subpolar cyclones |
| | ↖↖ | Polar southeasterlies |
| 90° S | High | Polar anticyclones |

**FIGURE 10.3**
This satellite photograph shows a discontinuous east–west band of cumulonimbus clouds (white blotches) near the equator. The cloud band marks the convergence of the surface trade winds of both hemispheres and is known as the intertropical convergence zone (ITCZ). [NOAA satellite photograph]

equator meet milder, lighter midlatitude air masses moving poleward. The polar front is not continuous around the globe; rather, it is well-defined in some areas and not in others, depending on the temperature contrast across the front. Where that temperature gradient is steep, the front is well-defined and is a poten-

tial site for development of synoptic-scale storms (Chapter 11). On the other hand, where the air temperature contrast is minimal, the polar front is poorly defined and inactive.

This brief description is a generalized view of the major components (winds and pressure systems) of the

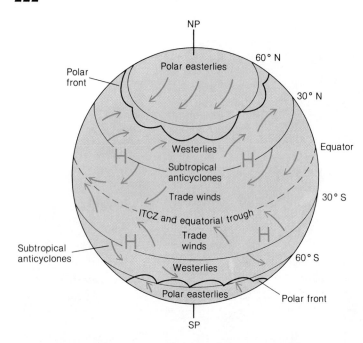

**FIGURE 10.4**
A schematic representation of the planetary-scale surface circulation of the atmosphere.

planetary-scale circulation at the Earth's surface. The distribution of these circulation systems is summarized schematically in Figure 10.4.

## WINDS ALOFT

What is the pattern of the planetary-scale winds aloft, that is, in the middle and upper troposphere? As noted earlier, air subsides in subtropical anticyclones, sweeps toward the equator as the surface trade winds, and then ascends in the ITCZ. Aloft, in the middle and upper troposphere, air flows poleward, away from the

doldrums and into the subtropical highs. The Coriolis effect shifts these upper-level winds toward the right in the Northern Hemisphere (southwest winds), and toward the left in the Southern Hemisphere (northwest winds). In the tropics, therefore, the winds aloft actually blow in a direction opposite that of the surface trade winds. The vertical profile of this circulation, shown schematically in Figure 10.5, resembles a huge convective cell and is known as the **Hadley cell,** named for the scientist who first proposed its existence in 1735. Hadley cells are thus situated on either side of the ITCZ and extend poleward to the subtropical highs.

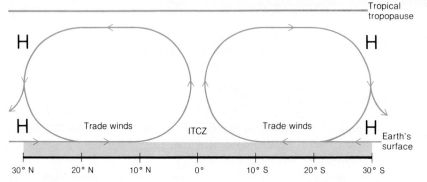

H = Subtropical anticyclones

**FIGURE 10.5**
An idealized representation of the Hadley cell circulation in tropical latitudes of the Southern and Northern hemispheres. Air rises over the intertropical convergence zone (ITCZ) and sinks in the subtropical anticyclones.

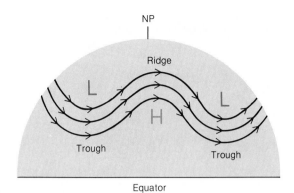

**FIGURE 10.6**
Aloft, in the middle and upper troposphere, the wester-lies flow in a wavelike pattern of ridges and troughs.

It was once proposed that separate cells, similar to Hadley cells, occurred in both midlatitudes and polar latitudes. Detailed upper-air monitoring, however, has failed to provide definitive evidence of such cells. Aloft, midlatitude winds blow from west to east in a wavelike pattern of ridges and troughs, as shown in Figure 10.6. These winds are responsible for the de-velopment and displacement of the synoptic-scale weather systems (highs, lows, and air masses) dis-cussed in Chapter 11 and for the poleward heat trans-port described in Chapter 4. Because these winds are so important for midlatitude weather, we examine the upper-air westerlies more extensively in a separate sec-tion of this chapter.

In polar regions, air subsides and flows away at the surface from shallow, cold anticyclones. In the North-ern Hemisphere, these highs are well-developed only in winter over the continental interiors. In the South-ern Hemisphere, cold highs persist over the glacier-bound Antarctic continent year-round. Aloft, polar winds are westerly.

Figure 10.7 shows the vertical profile of Northern Hemisphere winds in the troposphere from the equa-tor to the poles. In this perspective, we are viewing only the north–south and up–down components of the wind and are neglecting the west–east component. Note that the tropopause is not continuous from pole to equator but occurs in discrete segments. Thus, the polar tropopause is at a lower altitude than the midlatitude tropopause, which, in turn, is at a lower altitude than the tropical tropopause. The altitude of the tropopause is directly proportional to the mean temperature of the troposphere; that is, the lower the mean temperature, the lower is the altitude of the tropopause.

## SEASONAL SHIFTS

Between winter and summer important changes take place in the planetary-scale circulation. Pressure sys-tems, the polar front, the global wind belts, and the ITCZ follow the sun, shifting toward the poles in spring and toward the equator in autumn. Because the seasons are reversed in the two hemispheres, the planetary-scale systems of both hemispheres move north and south in tandem. In addition, the strength of pressure cells var-ies between seasons. Subtropical anticyclones exert

**FIGURE 10.7**
Vertical cross section showing the north–south (me-ridional) winds in the Northern Hemisphere tropo-sphere. Note that the tropopause occurs in three seg-ments and that the vertical scale is greatly exaggerated.

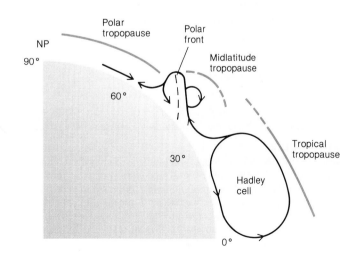

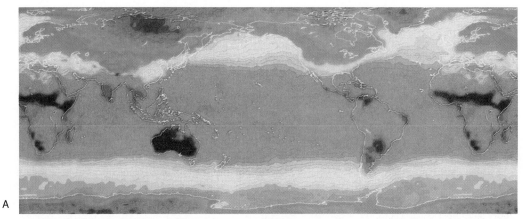

A

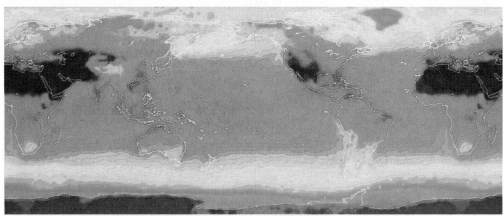

B

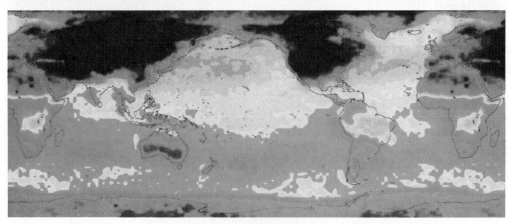

C

**FIGURE 10.8**
Infrared sensors aboard weather satellites measure surface temperature patterns over the globe.
(A) January 1979; (B) July 1979. Temperature increases from violet to blue to green to yellow
to red to black-brown. Subfreezing temperatures are indicated by green, blue, and violet. In (C)
the color scale represents the magnitude of mean temperature change from January to July. Red
and black-brown signify areas of greatest warming, whereas blue and violet indicate areas of
greatest cooling. Yellow and green represent relatively little temperature contrast between January and July.

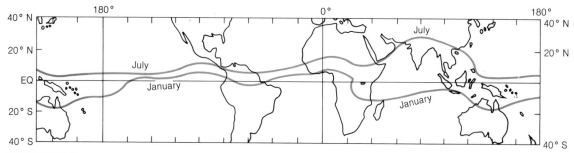

**FIGURE 10.9**
Seasonal shifts of the intertropical convergence zone (ITCZ). The ITCZ follows the sun, reaching its most northerly location in July and its most southerly location in January. As we will see in Chapter 12, these shifts contribute to the monsoon circulations of tropical latitudes.

higher surface pressures in summer than in winter. The Icelandic low deepens in winter and greatly weakens in summer, and, though well-developed in winter, the Aleutian low disappears in summer.

Seasonal reversals in surface air pressure also occur over the continents. These pressure shifts stem from the contrast in solar heating between land and sea. For reasons presented in Chapter 3, the sea surface exhibits smaller temperature variations over the course of a year than does the Earth's land surface. Maps of global surface temperature in Figure 10.8 vividly display this land–sea contrast in thermal stability. Temperatures were derived from infrared radiation sensors aboard weather satellites.* Figures 10.8 A and B are color-enhanced images of January and July temperature patterns in 1979. In the Northern Hemisphere, these two months are usually the coldest and warmest months of the year respectively. The third image (Figure 10.8C) is the temperature difference between the two months. Note that the seasonal temperature contrast is considerably greater over the continents (up to 30 Celsius degrees or 54 Fahrenheit degrees) than over the oceans (typically 8 to 10 Celsius degrees or 14 to 18 Fahrenheit degrees).

As a consequence of this seasonal temperature contrast, the continents are dominated by relatively high pressure in winter and relatively low pressure in summer. In winter, cold anticyclones appear over northwestern North America and over the interior of Asia, the most prominent of which is the massive *Siberian*

*Recall from Chapter 2 that the band of radiation emitted by any object is temperature-dependent, so temperature can be measured remotely by infrared sensors aboard Earth-orbiting satellites.

*high.* In summer, belts of low pressure form across North Africa and from the Arabian peninsula eastward into Southeast Asia. Warm low-pressure cells also develop in summer over arid and semi-arid regions of Mexico and southwestern United States.

Seasonal shifts in the location of the planetary-scale windbelts, pressure systems, and the ITCZ leave their mark on the world's climates. For example, northward migration of the ITCZ triggers summer monsoon rains in Central America, North Africa, India, and Southeast Asia (Chapter 12). As shown in Figure 10.9, north–south movements of the ITCZ are greater over continents than over the oceans. Over the oceans the mean latitude of the ITCZ varies by only about 4 degrees through the year (from 4 degrees N in April to 8 degrees N in September). The anchoring of the ITCZ over the ocean is a consequence of the ocean's greater thermal stability.

As a subtropical anticyclone shifts north and south with the sun, its dry eastern flank influences some localities in winter and other localities in summer, thereby imposing a pronounced seasonality on rainfall. This is the reason, for example, that southern California has wet winters and dry summers. We have more to say about the influence of the planetary circulation on world patterns of climate in Chapter 18.

### SINGULARITIES

A **singularity** is a weather event that occurs on or near a certain date with unusual regularity. The frequency of these events in the climatic record is greater than would be expected on the basis of chance alone. Most singularities are linked to regular changes in features of the planetary-scale atmospheric circulation. For example, in regions of seasonal precipitation, the onset and ending of the rainy season may constitute singularities. Consider some other examples.

The **January thaw** is the most widely recognized and perhaps the only real singularity in a statistically rigorous sense. It is a period of relatively mild weather around January 20 to 23 and occurs primarily in the New England states. The thaw is caused by a flow of warm air on the back side (western flank) of the Bermuda–Azores anticyclone, which, for some unexplained reason, temporarily shifts north of its usual midwinter location.

Weather episodes that do not fit precisely the definition of a singularity, but nonetheless are fairly regular, are the July rainfall maximum in Arizona and the so-called Indian summer weather in the northeastern United States. In late June, the North Pacific subtropical anticyclone shifts abruptly northward and the Bermuda–Azores high extends its influence westward, across subtropical North America. The anticyclone's clockwise circulation pumps warm, humid air into the southwestern United States and brings a rainy end to Arizona's dry spring. For example, in Phoenix, Arizona, the mean monthly rainfall for April, May, and June is only 8 mm (0.30 in.), 3 mm (0.12 in.), and 2 mm (0.08 in.), respectively. For July, however, the mean monthly rainfall jumps abruptly to 20 mm (0.80 in.).

There is no precise date for **Indian summer,** but it usually develops in October or November after the autumn's first freeze. Large, warm anticyclones stagnate over the eastern United States and displace the principal storm track northward along the St. Lawrence River Valley. Typical Indian summer weather consists of an episode of mild, sunny days, hazy skies, cool nights, and frosty mornings.

## Upper-Air Westerlies

The midlatitude westerlies of the Northern Hemisphere merit special attention here because they govern the weather in the United States and Canada. As we noted earlier, in the middle and upper troposphere, the westerlies flow about the hemisphere in wavelike patterns of ridges and troughs (refer back to Figure 10.6). Winds exhibit a clockwise (anticyclonic) curvature in the ridges, and counterclockwise (cyclonic) curvature in the troughs. Between two and five waves typically encircle the hemisphere at any one time. These long waves are called **Rossby waves,** after Carl G. Rossby, the Swedish-American meteorologist who discovered them in the late 1930s.

Rossby waves characterize the tropospheric westerlies above the 500-mb level, that is, above the altitude where the pressure drops to 500 mb. Below this, the waves are distorted somewhat by friction and topographic irregularities of the Earth's surface.

The wavelike configuration of the circulation allows us to describe the westerlies by wavelength (distance between successive troughs or, equivalently, successive ridges), amplitude (north–south extent), and the number of waves encircling the hemisphere. The westerlies exhibit changes in all three characteristics, and, as a direct consequence, the weather changes.

The westerlies are more vigorous in winter than in summer. In winter, they strengthen and exhibit fewer waves of longer length and greater amplitude. This seasonal difference stems from the north–south pressure gradient, which is steeper in winter because of the greater temperature contrast between north and south at that time of year. In summer, north–south temperature differences are less, pressure gradients are weaker, and, as a consequence, so are the westerlies.

## LONG-WAVE PATTERNS

The *weaving westerlies* consist of two components of motion: a north–south wind superimposed on a west-to-east wind. We refer to the north–south airflow as the westerlies' *meridional component* and the west-to-east airflow as its *zonal component*. The meridional component of Rossby waves brings about a north–south exchange of air masses and poleward transport of heat. In the Northern Hemisphere, winds from the south carry warm air masses northward, and winds from the north transport cold air masses southward. Cold air is thus exchanged for warm air, and heat is transported poleward. As Rossby waves change in length, amplitude, and number, however, concurrent changes take place in the advection of air masses. Consider some examples.

Occasionally, the westerlies flow almost directly from west to east, nearly parallel to latitude circles, with a weak meridional component (Figure 10.10). This is a **zonal flow pattern** in which the north–south exchange of air masses is minimal. Cold air masses stay to the north, and warm air masses remain in the south. At the same time, the United States and southern Canada are flooded by air that originated over the Pacific Ocean. Pacific air dries out to some extent as it passes over the western mountain ranges, and then the air is compressed and warmed adiabatically as it descends onto the Great Plains—spreading relatively mild and generally fair weather east of the Rocky Mountains.

At other times, the westerlies exhibit considerable amplitude and flow in a pattern of deep troughs and sharp ridges (Figure 10.11). In this **meridional flow pattern,** masses of cold air surge southward, and warm air streams northward. Greater temperature contrasts develop over the United States and southern Canada. Where contrasting air masses collide, warm air overrides cold air and the stage is set for the development of cyclones that are then swept along by the westerlies.

These two illustrations of Rossby wave configurations are the opposite extremes of a wide range of possible westerly wind patterns, each featuring different components of meridional and zonal flow. The situation may be further complicated by a **split flow pattern,** in which westerlies to the north have a wave configuration that differs from that of westerlies to the south. For example, winds may be zonal over central Canada and meridional over the United States.

The westerly wind pattern typically shifts back and forth between dominantly zonal and dominantly meridional flow. For example, zonal flow might persist for a week and then give way to a more meridional flow that lasts for a few weeks, and then it's back to zonal flow again. The transition from one wave pattern to another is usually abrupt and sometimes takes place within a single day. This abruptness poses a challenge to weather forecasters because a sudden shift in the upper-air winds may divert a storm toward or away from a locality, or cause an unanticipated influx of colder air that could, for example, change rain to snow.

Unfortunately for the long-range weather forecaster, the shifts between westerly wave patterns have no regularity. That is, there is no predictable zonal/meridional cycle. The only observation useful to forecasters is that meridional patterns tend to persist for longer periods than zonal patterns. During the winter of 1976–77, for example, the strong meridional wave pattern shown in Figure 10.12 persisted from late October through mid-February. Northwesterly flow aloft brought surge after surge of bitterly cold arctic air into the midsection of the United States and resulted in one of the coldest winters of this century for that area. Meanwhile, southwesterly winds aloft brought unseasonably mild air to the far-western United States, including Alaska.

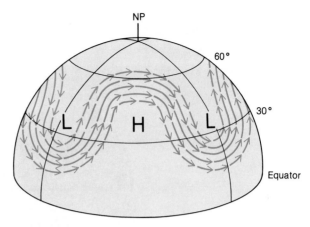

**FIGURE 10.11**
Midlatitude westerlies exhibit a meridional flow pattern aloft when west–east winds have a strong meridional (north–south) component.

**FIGURE 10.10**
Midlatitude westerlies exhibit a zonal flow pattern aloft when winds blow almost directly from west to east, with only a small meridional (north–south) component.

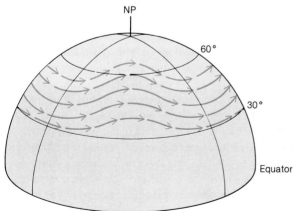

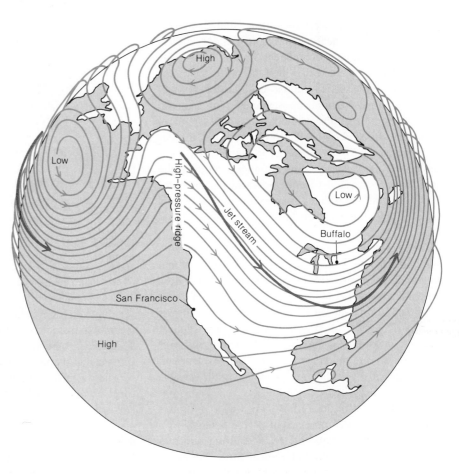

**FIGURE 10.12**
The strong meridional westerly wind pattern that persisted through much of the winter of 1976-77 brought record cold to the Midwest and East and record drought and warmth to the West. [National Weather Service data]

## BLOCKING SYSTEMS

For the continent as a whole, North American weather is more dramatic when the westerly wave pattern is strongly meridional. Sometimes undulations of the westerlies become so great that huge, whirling masses of air actually separate from the main westerly air flow. This situation, shown schematically in Figure 10.13, is analogous to the whirlpools that form in rapidly flowing rivers. In the atmosphere, cutoff masses of air whirl in either a cyclonic or an anticyclonic direction. A *cutoff low* or a *cutoff high* that prevents the usual west-to-east progression of weather systems is referred to as a **blocking system.** Because blocking systems tend to persist for extended periods of time (usually several weeks or longer), extremes of weather such as drought or anomalous heat or cold often result.

In July and August 1976, a large blocking high, stationed to the west of the British Isles, deflected the flow of cool, humid maritime air far to the north of its usual path. As a consequence, residents of Great Britain and most of Europe experienced one of the hottest and driest summers on record. Fires raged through the parched forests of Germany and Italy, and drought severely cut the yields of many food crops.

Blocking circulation patterns were also responsible for the serious 1972 shortfall in the wheat harvest of the former Soviet Union. In winter, stationary anticyclones over the Soviet wheatlands favored prolonged periods of dry, cold weather. Because of unusually light snowfall and thin snow cover, subfreezing temperatures penetrated deeply into the soil and killed much of the winter wheat. Wheat replanted in the spring did not fare much better because of drought. Through much of the summer of 1972, a blocking high diverted storm systems away from the wheat-growing region, so rainfall totals for May, July, and August were less than half of the long-term average. This drought, following a rela-

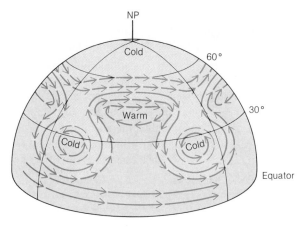

**FIGURE 10.13**
Aloft, the midlatitude westerlies sometimes exhibit an extreme meridional flow pattern in which huge pools of rotating air are cut off from the main west–east circulation. The pool of cold air rotating counterclockwise is a blocking cyclone, and the pool of warm air rotating clockwise is a blocking anticyclone. The latter is sometimes referred to as an *omega block* because of its resemblance to the Greek letter.

tively dry winter, meant that much of the wheat withered in desiccated soil.

The 1972 failure of the Soviet wheat crop had worldwide economic ramifications because it forced the Soviets to enter the international grain market. As a result, the price of wheat rose sharply, and the reserves of the grain-exporting nations were depleted. In the United States, this increased demand caused the price of wheat to more than triple between early 1972 and early 1974.

Another example of the association of weather extremes with a blocking circulation pattern comes from the southeastern United States. In the spring and early summer of 1986, an unusually severe and prolonged drought affected a ten-state region. In the hardest hit areas precipitation totals for the eight months preceding August 1986 were less than 40% of the long-term average (Figure 10.14). The primary reason for the May-to-early-July episode of moisture deficit was a persistent westerly long-wave pattern that featured a trough anchored off the East Coast near longitude 65 degrees W and, upstream (to the west), a ridge over

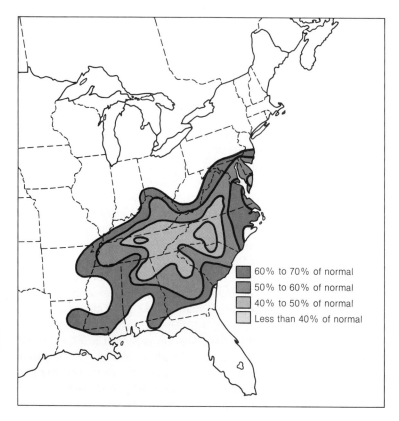

60% to 70% of normal
50% to 60% of normal
40% to 50% of normal
Less than 40% of normal

**FIGURE 10.14**
Percentage of the long-term (1951–1980) average December–July precipitation over portions of the southeastern United States for the period December 1985 through July 1986. [From K.H. Bergman et al., "The Record Southeast Drought of 1986," *Weatherwise* 39 (1986):262]

the Southeast. In effect, the center of the Bermuda–Azores high extended its influence well west of its usual location. The persistence of this circulation pattern meant day after day of subsiding stable air over the Southeast and suppression of the convective activity that normally brings the region abundant spring and summer rainfall.

The drought that affected the Midwest and Great Plains during the spring and summer of 1988 is a more recent example of the linkage between a weather extreme and a blocking weather pattern. In Figure 10.15, the major upper-air circulation features that dominated the summer of 1988 are contrasted with long-term average conditions. Between early May and mid-August, prevailing westerlies were more meridional than usual and featured a huge stationary warm high-pressure system over the nation's midsection and troughs over both the west and east coasts. The belt of strongest westerlies was displaced north of its usual position so that moisture-bearing weather systems were diverted into central Canada and well north of their usual paths. In the Corn Belt, the May through June period was the driest since 1895. By late July, drought was categorized as either severe or extreme over 43% of the land area of the contiguous United States. By the end of the growing season, the impact on the nation's grain harvest was severe: Corn production was down by 33%, soybeans by 20%, and spring wheat by more than 50%.

A blocking circulation pattern was responsible for weather extremes and record flooding in portions of the United States during the summer of 1993. A cold upper-air trough stalled over the Pacific Northwest and northern Rockies brought unusually cold weather to that region. Meanwhile, the Bermuda–Azores high was located west of its usual location, causing the worst drought since 1986 over the southeastern United States. Between the northwestern trough and the southeastern high, the jet stream and principal storm track weaved over the central United States. As this circulation pattern persisted through June and July, a nearly continuous procession of disturbances brought record rainfall to the Missouri and Upper Mississippi River valleys. The consequence was saturated soils, excessive runoff, record river crests, and the worst flooding in United States history. As of this writing, total property damage from flooding and its aftermath is expected to top $12 billion, making this second to Hurricane Andrew (August, 1992) as the most costly natural disaster in the nation's history.

In addition to drought, stagnant anticyclones in summer bring extended periods of searing heat that can take many human lives. In the summer of 1980, more than 1200 people in the southern Mississippi River Valley died from heat-related stress. A stalled, warm anticyclone caused maximum daily air temperatures to hover near 38 °C (100 °F) for over a month. A similar episode in the Midwest two years later contributed to an estimated 200 deaths.

## JET STREAM

Embedded in the upper-level westerlies is a relatively narrow corridor of very strong winds called the **jet stream.** The jet stream is not always a single stream of air but may splinter into separate ribbons. In midlatitudes, the jet stream is situated in the upper troposphere between the midlatitude tropopause and the polar tropopause, and directly over the polar front (Figure 10.16). For this reason, it is known as the **polar front jet stream.** This jet stream, which follows the meandering path of the planetary westerly waves, attains wind speeds that frequently exceed 160 km (100 mi) per hour. Westbound aircraft understandably avoid the jet stream because it is a head wind, whereas eastbound flights seek it because it is a tail wind.

Why is a jet stream associated with the polar front? Recall that air temperature influences air density and hence air pressure. Where horizontal temperature contrasts are great, horizontal pressure gradients are steep and winds are strong. The polar front, situated between the cold polar easterlies and the mild westerlies, is a zone of significant temperature contrast, which causes a relatively steep horizontal air pressure gradient aloft. The strongest winds (that is, the jet stream) would be anticipated over the zone of greatest surface temperature contrast, the polar front. The relationship between the jet stream and the polar front is examined in more detail in the Mathematical Note at the end of this chapter.

Like the polar front, the jet stream is not uniformly well-defined around the globe. Where the polar front is well-defined, that is, where surface temperature gradients are particularly steep, jet stream winds accelerate. Such a segment, in which the wind may accelerate

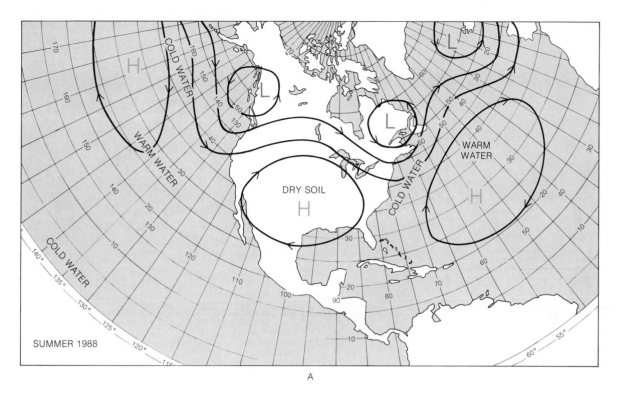

A

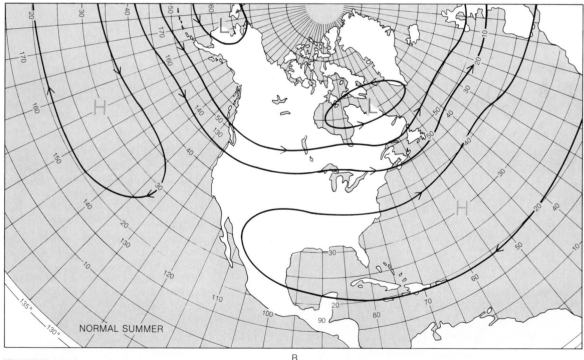

B

**FIGURE 10.15**
Prevailing upper-air winds (A) during the summer of 1988 as contrasted with (B) the long-term average pattern. A blocking warm anticyclone over the nation's midsection contributed to a severe drought during much of the summer of 1988. [From J. Namias, "Written in the Winds, The Great Drought of '88," *Weatherwise* 42 (1989):86–87]

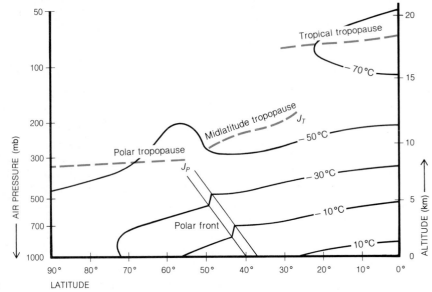

$J_P$ = Polar front jet stream

$J_T$ = Subtropical jet stream

**FIGURE 10.16**
A vertical cross section (meridional profile) through the Northern Hemisphere troposphere showing the location of the polar front jet stream $(J_P)$ above the polar front and near the break in the tropopause between midlatitudes and polar latitudes. Also, the subtropical jet $(J_T)$ is located near 25 degrees N and near the break in the tropopause between midlatitudes and tropical latitudes. In this perspective, both jet streams blow perpendicular to and into the page. Solid lines are isotherms in Celsius degrees. Note the considerable vertical exaggeration.

by as much as 100 km (62 mi) per hour, is known as a **jet maximum.** The strongest jet maxima develop in winter along the east coasts of North America and Asia where the land-to-sea temperature contrast is particularly great. In those areas, wind speeds in the jet maximum on rare occasions have topped 350 km (217 mi) per hour. A typical jet maximum is 160 km (100 mi) wide, 2 to 3 km (1 to 2 mi) thick, and several hundred kilometers in length.

What, then, is the role of the polar front jet stream in the generation and maintenance of synoptic-scale storms? Air flowing through a jet maximum changes both speed and direction, and these changes induce a complex pattern of horizontal divergence and horizon-

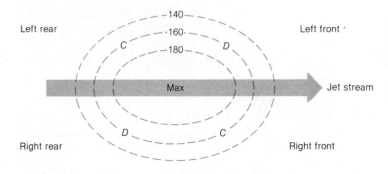

$C$ = Convergence

$D$ = Divergence

**FIGURE 10.17**
A jet stream maximum is divided into quadrants. The jet maximum is outlined by isotachs, lines (dashed) of equal wind speed (in km per hr). Air movement into and out of the jet maximum induces areas of horizontal divergence $(D)$ and horizontal convergence $(C)$.

tal convergence aloft. In Figure 10.17, a jet maximum (viewed from above) is divided into four quadrants: left rear, right rear, left front, and right front. Horizontal divergence occurs in the left front and right rear quadrants, and horizontal convergence occurs in the right front and the left rear quadrants. Now recall from Chapter 9 that a synoptic-scale cyclone is characterized by horizontal convergence of air at the surface and by horizontal divergence aloft. Where the jet stream triggers upper-air horizontal divergence, the jet stream contributes to the development and maintenance of cyclones that form and travel along the polar front. The strongest horizontal divergence develops in the left front quadrant, so it is under this sector of the jet stream maximum that a cyclone typically develops.

Although the polar front jet stream weaves with the westerlies, a jet maximum typically progresses from west to east at a faster pace than does the west to east displacement of the troughs and ridges in the westerlies (Figure 10.18). The strongest divergence aloft occurs when a jet maximum is on the east side of a trough so that cyclone development is much more likely when the jet maximum is in that position.

Like the polar front, the jet stream undergoes important seasonal shifts. The jet stream strengthens in winter, when north–south temperature contrasts are great, and weakens in summer, when temperature contrasts are less. As shown in Figure 10.19, the average summer location of the jet stream is across southern Canada, and the average winter position is across the southern United States. These locations represent long-term averages; the jet stream actually weaves over a considerable range of latitude from week to week, and even from one day to the next. As a general rule, when the polar front jet stream is south of us, the weather tends to be relatively cold, and when the polar front jet stream is north of us, the weather tends to be relatively warm.

The polar front jet stream is not the only jet stream. The **subtropical jet stream** occurs near the break in the tropopause between tropical and middle latitudes and on the poleward side of the Hadley cell. It is stronger and less variable with latitude than its northerly counterpart. Jet streams also occur in the Southern Hemisphere, but so far have not received as much study as those to the north.

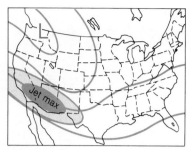

Monday

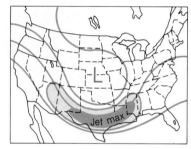

Tuesday

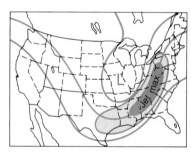

Wednesday

**FIGURE 10.18**

An upper-level trough is shown progressing from west to east across the United States. Meanwhile, a jet maximum (outlined in dark blue) travels eastward at a more rapid pace than does the trough. [From F.S. Sechrist and E.J. Hopkins, *Meteorology: Weather and Climate*. Madison, WI: University of Wisconsin Extension, 1974, p. 146.]

### SHORT WAVES

**Short waves** are another feature of the upper-level westerlies. These waves, which are ripples superimposed on Rossby long waves, are important to surface weather systems. Although Rossby waves drift very slowly eastward, short waves travel rapidly through the Rossby waves. Whereas five or fewer long waves en-

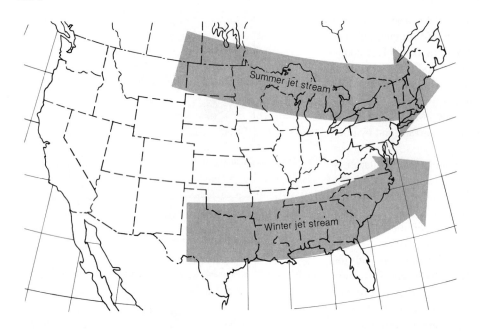

**FIGURE 10.19**
Approximate average locations of the polar front jet stream in winter (December through March) and in summer (June through October).

circle the hemisphere, there may be a dozen or more short waves.

Both short waves and long waves in the westerlies can contribute to cyclone development. For the same isobar spacing (air pressure gradient), gradient and geostrophic wind speeds are not equal. Anticyclonic gradient winds are stronger than geostrophic winds, and cyclonic gradient winds are weaker than geostrophic winds. Hence, as shown in Figure 10.20, westerly winds tend to strengthen in a ridge and weaken in a trough. The result is horizontal divergence of air to the east of a trough (and west of a ridge). Short and long waves thus favor storm development in the same way

as the polar front jet stream—by inducing horizontal divergence aloft. Consequently, conditions aloft are most favorable for storm development when the jet stream maximum appears on the east side of a trough (and west of a ridge).

## El Niño/Southern Oscillation (ENSO)

Earlier we saw how blocking patterns in the westerlies can trigger extremes in weather. In addition, some mid-

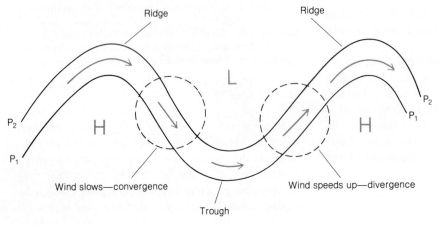

**FIGURE 10.20**
Westerly gradient winds speed up in ridges and slow down in troughs. This induces horizontal convergence aloft ahead of ridges and horizontal divergence aloft ahead of troughs. Solid lines are isobars; $P_1$ is greater than $P_2$.

latitude weather extremes are linked to circulation anomalies that develop in tropical latitudes. One of the most extensively studied of these circulation anomalies involves both the tropical atmosphere and ocean.

For most of the year, coupling of the southeast trades with the ocean drives surface waters westward off the northwest coast of South America. As warm, surface water moves away from the continent, it is replaced by cold, nutrient-rich water that wells up from below, from depths of 200 to 1000 m (660 to 3280 ft). This process is known as **upwelling** (Figure 10.21). Nutrients spur plankton growth, which, in turn, supports the Peruvian anchovy harvest, the world's largest fishery. In good years, the anchovy catch can exceed 11 million metric tons.

Just about every year in December, warm surface waters spread eastward into the eastern equatorial Pacific and then southward along the coast of southern Ecuador and northern Peru. The warm water suppresses upwelling and, deprived of nutrients, phytoplankton, and anchovy production decline, and fishing is disrupted. Peruvian fishermen refer to this period of unusually high sea-surface temperatures and reduced fish catch as **El Niño.\***

A typical El Niño is short-lived, lasting only a few months, and affects primarily the waters off Ecuador and northernmost Peru. Occasionally, however, an intense El Niño develops that persists for a year or longer and spreads anomalously warm surface waters along

*El Niño is the Spanish equivalent of Christ child.

### FIGURE 10.21
Schematic cross section of upwelling in a coastal area whereby cold, nutrient-rich bottom waters circulate up toward the surface and replace surface waters that are blown offshore.

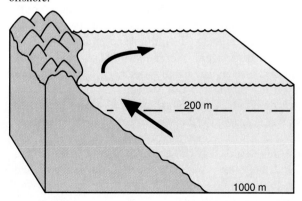

the entire coast of Peru and over large areas of the eastern and central equatorial Pacific. During an intense El Niño, the anchovy fishery collapses and weather extremes occur in widely separated regions of the world.

An important step in understanding the link between an intense El Niño and weather extremes came in 1966 when Jacob Bjerknes, a researcher at the University of California at Los Angeles, demonstrated a relationship between El Niño and the southern oscillation. The **southern oscillation** is a seesaw variation in air pressure between the eastern and western tropical Pacific.* Specifically, air pressure over Indonesia and northern Australia varies inversely with air pressure over the tropical east Pacific. The horizontal pressure gradient thus changes as air pressure to the west rises and air pressure to the east falls (and vice versa). Bjerknes found that El Niño begins when the air pressure gradient between the eastern and western Pacific starts to weaken.

Under normal conditions, a steep air pressure gradient between the eastern and western Pacific maintains strong trade winds and, hence, vigorous upwelling along South America's northwest coast. However, when the air pressure gradient declines as part of the southern oscillation, the southeast trades weaken and so does upwelling. Slackening trades allow the warm surface water that had been driven westward to drift slowly eastward along the equator and then southward off the Peruvian coast. Upwelling is suppressed, and El Niño is in progress.

During an intense El Niño, the southeast trades eventually shift direction and become equatorial westerlies. The westerlies drive warm surface waters eastward over a much larger area of the Pacific than is typical during an ordinary El Niño. The scientific community often refers to such an event as **ENSO,** the acronym for El Niño/Southern Oscillation.

Persistence of anomalously high sea-surface temperatures in the eastern Pacific plus reversal of the trades alter the planetary-scale circulation. Consequently, in some parts of the world, weather extremes accompany an ENSO (Figure 10.22). Consider some examples. Normally, prevailing winds over Indonesia are onshore (southeast trades) so that warm, humid air

*Sir Gilbert Walker discovered the southern oscillation in 1924. Walker noted that when air pressure is low at Darwin (on the north coast of Australia), it is high at Tahiti (a south Pacific island at about 18 degrees S and 149 degrees W), and vice versa.

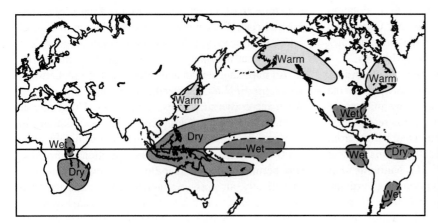

**FIGURE 10.22**
Locations of anomalous weather conditions that usually accompany an ENSO. [Climate Analysis Center, National Meteorological Center, NOAA]

SPECIAL TOPIC

## The ENSO Event of 1982–83

In 1982, an ENSO event began in May, and by December, warming of surface waters of the Pacific was unprecedented in both magnitude and extent (Figure 1).In the equatorial Pacific, unusually high sea-surfacetemperatures spread east of the international dateline (180 degrees W), and, in places, sea-surface temperature anomalies (departures from normal) reached +6 Celsius degrees (+10.8 Fahrenheit degrees). This anomalously warm water over vast areas triggered major changes in atmospheric circulation patterns over tropical and middle latitudes. Circulation changes, in turn, led to weather extremes in many areas.

The winter storm track was displaced hundreds of kilometers southeast of its usual position, which resulted in destructive high winds and heavy rains in California. Excessive rains between mid-November 1982 and late January 1983 caused the worst flooding of the century in Ecuador. No less than six tropical storms struck French Polynesia in a span of only three months. At the other extreme, drought parched Australia, Indonesia, and southern Africa. Australia's drought was its worst in 200 years, resulting in a $2 billion loss in crops and the deaths of millions of sheep and cattle. Meanwhile, drought in the African Sahel grew worse.

The 1982–83 ENSO also had a devastating impact on marine life in the eastern tropical Pacific, particu-larly off the coast of South America. With suppression of upwelling, the growth of microscopic plankton that normally flourish in upwelling areas diminished sharply. The 1982–83 ENSO was so intense that it triggered a 20-fold reduction in the quantity of plankton off the west coast of South America. This decline reduced the numbers of anchovy, which feed heavily on plankton, to a record low. Other fish dependent on the plankton, such as jack mackerel, also decreased in number. With the decline of fish populations, marine birds (such as frigate birds and terns) and marine animals (such as fur seals and sea lions) also suffered major population declines because of breeding failures and loss of food sources.

Even wildlife on remote Kiritimati (Christmas Island), at 2 degrees N, 157 degrees W in the Central Pacific, suffered from the effects of the 1982–83 ENSO. An estimated 17 million sea birds, which feed on fish and squid, normally nest on this island. However, scientists arriving on the island in November 1982 discovered that virtually all the adult birds had deserted the site and left their young behind to starve. Apparently, the sharp decline in food sources forced fish to search for better feeding grounds, and the adult birds left in search of fish. By July 1983, with the return of more normal atmospheric and oceanic conditions near Kiritimati, the few surviving birds began to nest again.

surging inland brings abundant rainfall. During an ENSO, however, winds become offshore (westerly) and the weather is generally dry. Drought also afflicts northern Australia and India. Meanwhile, on the other side of the tropical Pacific, anomalously high sea-surface temperatures off the northwest coast of South America spur convection and heavy rainfall in normally arid coastal localities. Consequently, the desert blooms.

An ENSO is also accompanied by weather extremes in midlatitudes. Circulation changes associated with sea-surface temperature anomalies shift the polar front jet stream north of its usual latitude. This shift alters-storm tracks and brings relatively mild winters to the northern United States and southern Canada. Also, a northward shift of the subtropical jet stream brings

abundant rainfall to the Gulf Coast. On the other hand, in southern California, heavy winter rains may or may not occur during an ENSO.

We can expect an ENSO about every 3 to 7 years. Eight moderate to intense ENSO events have occurred since the early 1950s, the most recent of which was a moderate episode that began in mid-1991 and continued into mid-1993. The next previous ENSO was in 1986–87. Probably the most intense ENSO of this century took place in 1982–83. For more on the impact of that event, refer to the Special Topic.

Recently, scientists have coined the term **La Niña** (the girl) for oceanic–atmospheric conditions opposite those of ENSO. La Niña features exceptionally strong trade winds and unusually low sea-surface tempera-

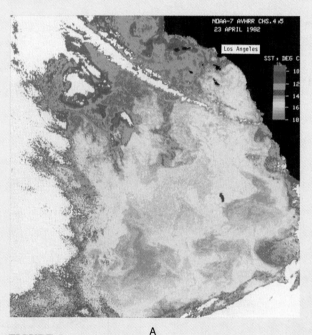
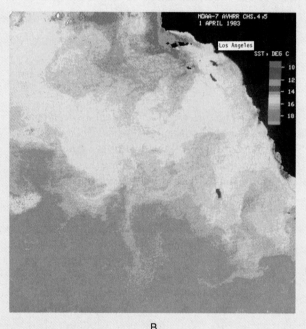

A                                      B

**FIGURE 1**
Sea-surface temperature patterns as determined by infrared sensors aboard the NOAA-7 satellite. (A) 23 April 1982. The black area in the upper right is the coast of southern California and the white patches are clouds. Note the red area where sea-surface temperatures are near or above 18 °C (65 °F). (B) The same area nearly one year later (1 April 1983) shows a greatly expanded area where sea-surface temperatures are near or above 18 °C (65 °F). [Courtesy of NOAA]

The 1982–83 ENSO event is a remarkable example of the vast complex of interconnections between atmosphere and ocean, and the many forms of life, not only in the equatorial Pacific, but around the

entire globe as well. Scientists will be studying these interconnections for many years in an attempt to better understand and predict the origin and impact of future ENSO events.

tures in the central and eastern tropical Pacific. An intense La Niña is accompanied by weather extremes that are usually opposite those of an ENSO. For example, the persistent atmospheric circulation pattern that led to the United States drought of 1988 (discussed earlier in this chapter) coincided with an intense La Niña. ENSO is not always followed by La Niña. We have more to say about efforts to predict ENSO and La Niña in Chapter 16.

## Conclusions

We have now examined the time-averaged characteristics of the planetary-scale circulation. The main features of that circulation are the ITCZ, trade winds, sub-tropical highs, the westerlies, polar front, subpolar lows, and polar easterlies. We saw that aloft (in the middle and upper troposphere), winds blow counter to the trades in tropical latitudes and in a west-to-east wave pattern at mid and high latitudes. Through the course of a year, these components of the planetary-scale circulation shift in tandem north and south with the sun.

In this chapter, we also focused on the linkages between the westerly winds aloft and the weather of midlatitudes. Specifically, the westerlies (1) are the source of horizontal divergence aloft for the development of cyclones, (2) steer storms, and (3) control air mass advection and poleward heat transport. Our examination of atmospheric circulation continues in the next chapter with a focus on synoptic-scale weather systems.

---

**MATHEMATICAL NOTE**

## The Polar Front and the Midlatitude Jet Stream

Why is a jet stream found over the polar front? To understand the reasons for this important association, we must return to the concept of *hydrostatic equilibrium*. We learned in Chapter 9 that hydrostatic equilibrium is the balance between the upward-directed pressure gradient force and the downward-directed force of gravity. We now express this relationship mathematically.

The air parcel in Figure 1 has density $\rho$, a cross-sectional area $A$, and a volume $V$. The air pressure acting on the top of the parcel, $P_2$ (at altitude $Z_2$), is due to the weight of the total atmospheric column above $Z_2$. The air pressure acting on the bottom of the parcel, $P_1$ (at altitude $Z_1$), is due to the weight of the total atmospheric column above $Z_1$. The pressure difference between the parcel top and the parcel bottom is the vertical pressure gradient and depends on the weight of the parcel itself. The parcel's weight, $W$, is equal to its mass, $M$, times the acceleration of gravity, $g$, and is distributed over the area $A$. Hence,

$$P_2 - P_1 = W/A$$

but since

$$W = Mg = \rho V g$$

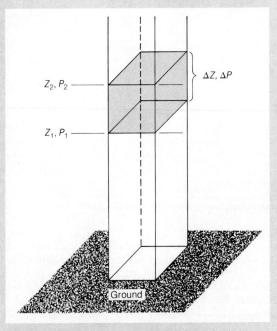

**FIGURE 1**
In a column of air, a change in altitude $(Z_2 - Z_1)$ is accompanied by a change in air pressure $(P_1 - P_2)$.

and

$$V = A(Z_2 - Z_1)$$

then

$$P_2 - P_1 = \rho g(Z_2 - Z_1)$$

That is,

$$\Delta P = -\rho g \Delta Z$$

This is known as the *hydrostatic equation*. The minus sign on the right side of the equation accounts for the inverse relationship between air pressure and altitude. That is, pressure drops as altitude increases.

According to the hydrostatic equation, the pressure change ($\Delta P$) that accompanies an altitude change ($\Delta Z$) depends on the density of air and on the acceleration of gravity. For practical purposes, we may assume that gravity is constant with altitude, at least for several kilometers. Air density is thus the utlimate determinant of the rate at which air pressure drops with increasing altitude. Because cold air is denser that warm air, air pressure drops more rapidly with altitude in cold air than in warm air.

The hydrostatic equation allows us to compute the thickness ($\Delta Z$) of an air layer bounded above and below by constant pressure (isobaric) surfaces; say, for example, the thickness of the 700- to 500-mb layer. Air layer thickness is inversely proportional to the average air density within the layer. Air layer density, in turn, depends on temperature and humidity (Chapter 5). Vertical variations in humidity are not nearly as important as temperature in determining air density so that we can formulate this rule of thumb: the thickness of an air layer between successive isobaric surfaces is directly proportional to the average temperature within the air layer. Hence, for example, the 700- to 500-mb thickness is greater in warm air than in cold air.

Now suppose that a sharp horizontal air temperature gradient develops at the Earth's surface, as across a well-defined segment of the polar front. The hydrostatic equation tells us that on the warm (less dense) side of the front the air pressure drops less rapidly with altitude than it does on the cold (more dense) side of the front. Equivalently, an air layer bounded by isobaric surfaces is thicker on the warm side than on the cold side of the front. Consequently, isobaric surfaces slope

as shown in Figure 2, and the slope steepens with altitude. The steepening slope means that the horizontal air pressure gradient strengthens with altitude, so that the westerly wind speed must increase with altitude. The wind reaches its maximum speed just below the tropopause and over the area of maximum horizontal temperature gradient (the polar front); this is the polar front jet stream.

Westerly wind speeds peak near the tropopause because, within the stratosphere, the latitudinal (north–south) temperature gradient reverses. (That is, within the stratosphere, air temperatures rise from equator to pole.) This weakens the latitudinal pressure gradient so that the westerlies slow with altitude within the stratosphere.

In summary, a jet stream is associated with steep horizontal air temperature gradients at the Earth's surface. Hence, where the temperature contrast across a front is particularly great, we would expect to encounter a jet maximum aloft. As noted elsewhere in this chapter, a jet stream maximum provides upper-air support (divergence) for cyclone development.

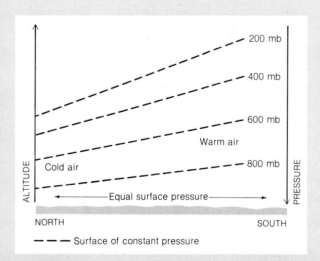

**FIGURE 2**
Air pressure drops more rapidly with altitude within a column of cold air than within a column of warm air. Hence, the horizontal air pressure gradient steepens with altitude when cold air is situated next to warm air. A strengthening of the horizontal pressure gradient, in turn, means that the horizontal wind speed increases with altitude.

## Key Terms

semipermanent pressure
  systems
subtropical anticyclones
horse latitudes
midlatitude westerlies
trade winds
doldrums
intertropical convergence
  zone (ITCZ)
heat equator
subpolar lows
polar front
Hadley cell
singularity
January thaw
Indian summer

Rossby waves
zonal flow pattern
meridional flow pattern
split flow pattern
blocking system
jet stream
polar front jet stream
jet maximum
subtropical jet stream
short waves
upwelling
El Niño
southern oscillation
ENSO
La Niña

## Summary Statements

☐ The principal features of the planetary-scale circulation
are the intertropical convergence zone (ITCZ), trade
winds, subtropical anticyclones, westerlies, subpolar
lows, polar front, and polar easterlies. These features follow the sun and shift poleward during spring and toward
the equator during autumn.

☐ Trade winds blow out of the equatorward flank of the
subtropical anticyclones, and the westerlies blow out of the
poleward flank. The east side of a subtropical anticyclone
features subsiding air and dry conditions, whereas the
west side is more humid and receives more rainfall.

☐ Aloft, in the middle and high troposphere, the trade winds
reverse direction thus completing the Hadley cell circulation. At higher latitudes, upper-air winds weave from west
to east as Rossby long waves.

☐ Contrasts in the Earth's surface temperatures in winter favor relatively high pressure over continents and low pressure over the ocean. In summer, this pattern reverses with
low pressure prevailing over continents and high pressure
over the ocean.

☐ Over time, the wave pattern of the upper-air westerlies
changes in length, amplitude, and number. At one
extreme, westerlies are mostly zonal, that is, west-to-east
with little amplitude. At the other extreme, westerlies are
strongly meridional, that is, west-to-east with considerable amplitude. Shifts between zonal and meridional
flow patterns affect north-south air mass exchange,

poleward heat transport, and storm development and
movement.

☐ Cutoff cyclones and cutoff anticyclones block the usual
west to east progression of weather systems and may
lead to extreme weather conditions such as drought,
excessive rainfall, or periods of unusually high or low
temperature.

☐ The midlatitude jet stream is a narrow corridor of strong
winds within the westerlies; it is located near the tropopause and above the polar front. A jet maximum is a region of particularly strong jet stream winds situated over
a well-defined segment of the polar front.

☐ Cyclone development is most likely under the left front
quadrant of a jet stream maximum and to the east of an
upper-level trough. Horizontal divergence is strongest in
that quadrant.

☐ ENSO refers to a lengthy period of unusually high sea-
surface temperatures over a vast area of the equatorial and
eastern tropical Pacific. An ENSO is accompanied by
weather extremes in tropical latitudes and in other areas
of the globe. La Niña is the term coined for conditions
opposite an ENSO.

## Review Questions

1. Why are the subtropical anticyclones described as *semipermanent?*
2. Why do the world's major deserts occur on the eastern
   side of subtropical anticyclones?
3. Distinguish between *horse latitudes* and *doldrums.* What
   is the origin of these names?
4. What is the relationship between the trade winds and subtropical anticyclones?
5. What is the relationship between the westerlies and subtropical anticyclones?
6. Describe the linkage among the intertropical convergence
   zone (ITCZ), trade winds, and Hadley cells.
7. What is the weather along the ITCZ?
8. In what sense is a Hadley cell like a huge convection
   current?
9. How does the altitude of the tropopause change with latitude?
10. Through the course of the year, the principal features of
    the planetary-scale circulation follow the sun. Explain
    this phenomenon.
11. What is a *singularity?* Provide an example of a well-
    established singularity.
12. Distinguish between zonal flow and meridional flow in
    the westerlies.

13. Explain why the weather is more extreme when the westerly wind pattern is strongly meridional.
14. What type of weather is associated with a zonal flow pattern?
15. What is a jet stream?
16. Describe the seasonal changes in the location of the polar front jet stream.
17. How does a jet stream contribute to the development of a synoptic-scale cyclone?
18. How does an upper-air trough contribute to the development of a synoptic-scale cyclone?
19. Describe the relationship between El Niño and the southern oscillation.
20. Why are farmers and fishermen concerned when an intense ENSO develops?

## Questions for Critical Thinking

1. Speculate on how a split flow pattern in the westerlies might influence the weather over North America.
2. Explain how you can determine the location of the mid-latitude jet stream by examining surface temperatures across North America.
3. The Southern Hemisphere westerlies are stronger than their Northern Hemisphere counterparts. Explain why.
4. How does an intense ENSO influence atmospheric stability over the eastern tropical Pacific? How might this affect thunderstorm development?
5. How does a blocking anticyclone affect the Bowen ratio of the area beneath the system? What is the implication of this for air temperatures of the lower atmosphere?

## Selected Readings

BERGMAN, K. H., ET AL. "The Record Southeast Drought of 1986," *Weatherwise* 39, No. 5 (1986):262–266. Provides some background on a severe drought caused by a blocking anticyclone.

CERVENY, R. S., AND J. S. HOBGOOD. "Meteorological Implications of the First Voyage of Christopher Columbus," *Bulletin of the American Meteorological Society* 73 (1992):173–178. Concludes that during the period of Columbus's first voyage, the Bermuda–Azores high was in about the same location as today but with somewhat stronger winds over the eastern North Atlantic.

HARMAN, J. R. *Synoptic Climatology of the Westerlies: Process and Patterns.* Washington, DC: Association of American Geographers, 1991, 80 pp. Presents an in-depth analysis of characteristics of the planetary-scale westerlies.

KNOX, P. N. "A Current Catastrophe: El Niño," *Earth* 1, No. 5 (1992):30–37. Describes in detail the ocean/atmosphere changes during an ENSO.

LORENZ, E. N. "A History of Prevailing Ideas About the General Circulation of the Atmosphere," *Bulletin of the American Meteorological Society* 64 (1983):730–733. Surveys the principal scientists and concepts involved in developing our understanding of planetary-scale circulation.

RAMAGE, C. S. "El Niño," *Scientific American* 254, No. 6 (1986):77–83. Reviews the ENSO mechanism and some of the problems associated with modeling the phenomenon. Also see followup letters to the editor in the November 1986 issue of *Scientific American.*

# 11

# Air Masses, Fronts, Cyclones, and Anticyclones

*The Westerly Wind asserting his sway from the southwest quarter is often like a monarch gone mad, driving forth with wild imprecations the most faithful of his courtiers to shipwreck, disaster, and death.*

JOSEPH CONRAD
*The Mirror of the Sea*

*Chapter written by*

## Professor Patricia M. Pauley

*Department of Meteorology*
*Naval Postgraduate School, Monterey, California*

Color-enhanced infrared satellite image of an intense cyclone over the North Atlantic. [NOAA satellite photograph]

T HE CEASELESS succession of synoptic-scale weather systems is directly responsible for day-to-day variations in weather. This chapter examines the weather conditions that accompany the major weather systems of midlatitudes: air masses, fronts, cyclones, and anticyclones. We learn how these systems are interrelated and how their development and movements are governed by the planetary-scale circulation.

# Air Masses

An **air mass,** a huge volume of air covering thousands of square kilometers, is relatively uniform horizontally in temperature and water vapor concentration, although both quantities typically decrease with altitude. The properties of an air mass are determined by the type of surface over which it develops, that is, its source region. A source region features nearly homogeneous surface characteristics over a wide area, such as a great expanse of snow-covered ground or a vast stretch of ocean. In order for an air mass to become uniform in temperature and water vapor concentration, it must reside in its source region for several days to weeks.

Simply put, air masses are either cold (polar, abbreviated as *P*) or warm (tropical or *T*), and either dry (continental or *c*) or humid (maritime or *m*). Air masses that form over the cold, often snow-covered surfaces of high latitudes are relatively cold, and those that develop over the warm surfaces of low latitudes are relatively warm. Air masses that form over land tend to be relatively dry, and those that develop over the oceans are relatively humid. Thus, we have four basic types of air masses: cold and dry continental polar *(cP)*, cold and humid maritime polar *(mP)*, warm and dry continental tropical *(cT)*, and warm and humid maritime tropical *(mT)*. A fifth type, arctic *(A)* air, is dry like continental polar air but distinguished by its bitter cold.

## NORTH AMERICAN TYPES

All of these air mass types occur over North America and form over the continent and surrounding waters in certain characteristic regions. Figure 11.1 shows the locations of the principal source regions for each type.

**Continental tropical air** develops over the subtropical deserts of Mexico and southwestern United States primarily in summer and is hot and dry. **Maritime tropical air** is very warm and humid, and its source regions are over tropical and subtropical oceans. It retains these properties year-round and is responsible for oppressive summer heat and humidity east of the Rocky Mountains. The source regions for **maritime polar air** are over the cold ocean waters of the North Pacific and North Atlantic. Along the West Coast, *mP* air brings heavy winter rains (snows in the mountains) and persistent coastal fogs in summer. Dry **continental polar air** develops over the northern interior of North America. In winter, *cP* air is typically very cold because the ground in its source region is often snow covered, days are short, insolation is weak, and radiational cooling is extreme. In summer, when the snow-free source region warms in response to long days of bright sunshine, *cP* air is quite mild and pleasant.

**Arctic air** forms over the snow- or ice-covered re-

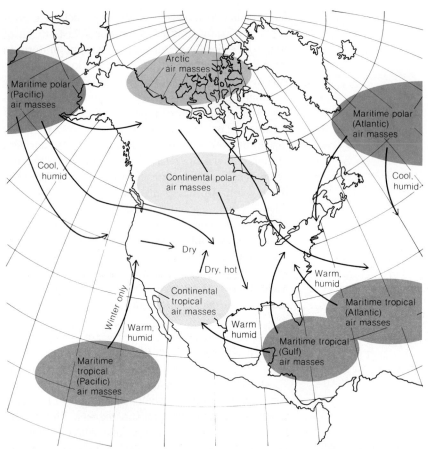

**FIGURE 11.1**
Air mass source regions for North America.

gions of Siberia, the Arctic Basin, Greenland, and North America north of about 60 degrees N in much the same way as continental polar air but in a region that receives almost no insolation in winter, although it still radiates strongly in the infrared. These air masses are exceptionally cold and dry and are responsible for the bone-numbing winter cold waves that sweep across the Great Plains and at times penetrate as far south as the Gulf of Mexico and Florida. For example, the so-called Siberian Express is an extremely cold Arctic air mass that forms over Siberia, crosses over the North Pole into Canada, and plunges into the Great Plains and on southward.

Air masses differ not only in temperature and humidity but also in stability. As we noted in Chapter 6, stability is an important property of air because it influences vertical motion and the consequent development of clouds and precipitation. Table 11.1 lists the usual stability, temperature, and humidity charac-

teristics of North American air masses within their source regions.

## MODIFICATION

Air masses do not remain in their source regions indefinitely but move from place to place. As they travel, their properties modify, those of some air masses changing more than those of others. Changes may occur in temperature, humidity, and/or stability. **Air mass modification** occurs primarily by (1) exchanging heat or moisture, or both, with the surface over which the air mass travels; (2) radiational cooling; and (3) processes associated with large-scale vertical motion. Let us now examine examples of these mechanisms.

In winter, as a mass of *cP* air travels southeastward from Canada into the United States, its temperature usually modifies quite rapidly. Although temperatures

**TABLE 11.1**
**Stability, Temperature, and Humidity Characteristics of North American Air Masses**

| Air mass type | Source region stability | | Characteristics | |
|---|---|---|---|---|
| | *Winter* | *Summer* | *Winter* | *Summer* |
| A | Stable | | Bitter cold, dry | |
| cP | Stable | Stable | Very cold, dry | Cool, dry |
| cT | Unstable | Unstable | Warm, dry | Hot, dry |
| mP (Pacific) | Unstable | Unstable | Mild, humid | Mild, humid |
| mP (Atlantic) | Unstable | Stable | Cold, humid | Cool, humid |
| mT (Pacific) | Stable | Stable | Warm, humid | Warm, humid |
| mT (Atlantic) | Unstable | Unstable | Warm, humid | Warm, humid |

in the Northern Plains might dip well below $-18$ °C (0 °F), temperatures may not drop much below the freezing point by the time the polar air reaches the southern United States. This rapid air mass modification occurs because, outside of its source region, polar air is usually colder than the ground over which it travels. The sun heats snow-free ground, and the warmer ground heats the bottom of the air mass, destabilizing it and triggering convective currents that distribute heat vertically throughout the air mass.

A similar process of heating from below and destabilization also occurs when a cP air mass crosses the Atlantic Coast and travels over the warm waters of the western Atlantic. In addition, evaporation from the sea surface increases the water vapor concentration of the air mass. Saturation is easily achieved in the cold and relatively unstable air and leads to extensive low-level cloudiness and fog.

When cP air travels over snow-covered ground, however, modification is less rapid because much of the incoming solar radiation is reflected, rather than being absorbed and heating the ground. The relatively cold surface increases the stability of the air and thereby reduces convection. Radiational cooling can play a key role in this situation and lead to a drop in temperature when days are short and solar heating and convection are minimal.

A tropical air mass does not modify as readily as a polar air mass because, outside of its source region, tropical air is often warmer than the ground over which it travels. The bottom of the air mass cools by contact with the ground; this cooling stabilizes the air and suppresses convective currents. As a result, cooling is restricted to the lowest portion of the air mass. In contrast, if the ground is warmer than the tropical air mass moving over it, the air mass can become even warmer. Thus, a cold wave loses much of its punch as it pushes southward, but a summer heat wave can retain its warmth as it journeys from the Gulf of Mexico well into Canada.

As a consequence of orographic lifting, air masses undergo significant changes in temperature and humidity. When cool and humid mP air sweeps inland off the Pacific Ocean, the air is forced up the windward slopes of coastal mountain ranges and so cools adiabatically. Cooling leads to condensation (or deposition) and development of clouds and precipitation. Latent heat released during condensation (and deposition) partially offsets adiabatic cooling. Then, as the air descends the leeward slopes into the Great Basin, it warms adiabatically and clouds dissipate. Some decrease in temperature through evaporative cooling is associated with the dissipation of clouds, but net heating (and drying) occurs because the water vapor that was condensed and precipitated out is no longer available for evaporation. The same processes are repeated as the air mass is forced to flow up and over the Rockies. Eventually, the air mass emerges on the Great Plains considerably milder and drier than the original mP air. East of the Rockies, such an air mass is described simply as **Pacific air.**

When the westerly wave pattern aloft is predominantly zonal (Chapter 10), Pacific air floods the eastern two-thirds of the United States and southern Canada. Polar air masses stay far to the north, and tropical air masses keep to the south. Consequently, much of that region experiences an episode of mild and generally dry weather.

## Frontal Weather

A **front** is a narrow zone of transition between air masses that differ in density. Density differences are usually due to temperature contrasts; for this reason we use the nomenclature *cold* and *warm* fronts. However, density differences may also arise from contrasts in humidity or some combination of temperature and humidity. Although the transition zone associated with an actual front often occurs over a distance of a few hundred kilometers, traditionally we draw the front on a map along the warm (less dense) side of the transition zone. Temperatures are therefore nearly constant on the warm side of the front and decrease behind the front to a region of nearly constant temperatures in the cold air mass.

A front is also associated with a trough in the sea-level pressure pattern, a corresponding wind shift, and convergence. As we saw in Chapter 6, the warmer air ascends where contrasting air masses meet, and uplift may cool the air sufficiently for clouds and precipitation to develop along the front. Depending on the slope of the front in the vertical plane and the air motions relative to the front, frontal weather may be confined to a very narrow band, or it may extend over a broad

region. In this section, we discuss the four basic fronts: stationary, warm, cold, and occluded.

### STATIONARY FRONT

A **stationary front** is just that—a front that exhibits essentially no movement. For example, a stationary front frequently develops along the Front Range of the Rocky Mountains when a shallow pool of polar air surges south and southwestward out of Canada. The cold dry air mass abuts the mountain range and can push no further westward; therefore, its leading edge is marked by a stationary front paralleling the mountain range. Similarly, a stationary front can form when any type of preexisting front loses its motion as it becomes parallel to the upper-level flow pattern. Under the proper conditions, a stationary front can also form along a boundary in surface thermal characteristics, such as a coastline or a boundary between a regional snow cover and bare ground.

Many features of a stationary front are common to all fronts. As seen in a vertical cross section such as that shown in Figure 11.2, a front slopes back from the Earth's surface toward colder air, or more precisely toward denser air. A front lies in a trough in the pressure

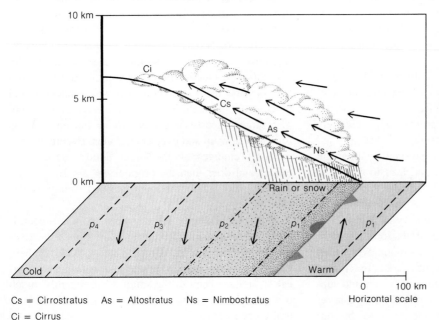

**FIGURE 11.2**
A stationary front has surface winds essentially parallel to the front and often a wide region of clouds and rain or snow to the cold side of the front. The clouds and precipitation result from overrunning. Note the considerable vertical exaggeration in the cross section.

Cs = Cirrostratus    As = Altostratus    Ns = Nimbostratus
Ci = Cirrus

pattern on any horizontal surface intersecting the front; this is especially evident in the sea-level isobars. Recall that the wind is approximately parallel to the isobars, with friction turning the wind somewhat toward lower pressure. The wind will typically change direction rather abruptly across the front; this condition is known as a *wind shift*. The change in wind direction and speed across a front is usually associated with convergence, which leads to upward motion, clouds, and perhaps precipitation, as explained in Chapter 9.

A stationary front does not always have a broad region of associated clouds and precipitation as depicted in Figure 11.2. The associated weather can vary considerably from case to case, depending on the supply of moisture and the specifics of the motion relative to the front. However, in the cases that do produce precipitation, the rain or snow falls mostly on the cold side of the stationary front. Warm humid air flows upward over the cooler air mass, more or less along the frontal surface. The air cools through adiabatic expansion, which triggers condensation and precipitation. This situation is often referred to as **overrunning** and can result in an extended period of relatively widespread cloudiness and drizzle, light rain, or light snow.

## WARM FRONT

If a stationary front begins to move in such a way that the warm (less dense) air advances while the cold (more dense) air retreats, the front changes in character and becomes a **warm front.** The overall characteristics of a warm front are very similar to those of the stationary front, as shown in Figure 11.3.

The typical differences between warm fronts and stationary fronts can be identified by comparing Figures 11.3 and 11.2. The slope (the ratio of vertical rise to horizontal distance) of the warm front is shallower near the Earth's surface because friction has retarded the front. The airflow on the warm side of the front is quite similar in both instances, but the air on the cold side of the front is retreating in the instance of the warm front. Thus, the warm air can advance relative to the Earth's surface, rather than just gliding up and over the cold air.

As a warm front approaches some locality, clouds develop and gradually lower and thicken in the following sequence: cirrus, cirrostratus, altostratus, nimbostratus, and stratus. The initial wispy cirrus clouds (Figure 11.4) may appear more than 1000 km (620 mi) in advance of the surface warm front. Slowly, clouds

**FIGURE 11.3**
As with a stationary front, overrunning along a warm front triggers cloud development. However, the warm front has a shallower slope at low levels, and the surface winds on the cold side of the front are retreating.

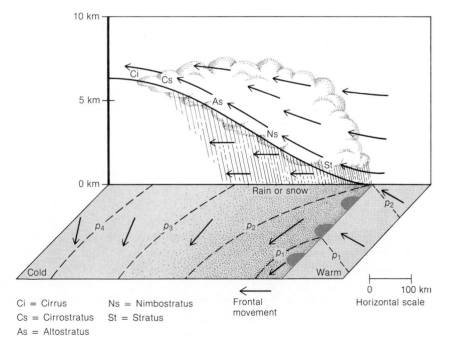

Ci = Cirrus    Ns = Nimbostratus
Cs = Cirrostratus    St = Stratus
As = Altostratus

**FIGURE 11.4**
Thin, wispy cirrus clouds may appear more than 1000 km (620 mi) in advance of a surface warm front. [Photograph by J. M. Moran]

spread laterally and form thin sheets of cirrostratus that turn the sky a bright milky white. The tiny ice crystals comprising these high clouds (bases above 7 km) may reflect and refract sunlight to produce halos or sundogs (Chapter 8). Appearance of these optical phenomena may herald the approach of stormy weather several days in advance.

In time, cirrostratus clouds give way to altostratus, thin clouds with bases at altitudes of 2 to 7 km. Light rain or snow begins soon after altostratus clouds thicken enough to block out the sun. Steady precipitation falls from low, gray nimbostratus clouds and persists until the warm front finally passes, a period that may exceed 24 hr. Thus, copious amounts of rain may fall ahead of the surface warm front, and because the precipitation intensity is usually only light to moderate, much of the water infiltrates the soil (as long as the ground is unfrozen and not already saturated with water). This is the type of rain that farmers appreciate. If it is cold enough for the precipitation to fall in the form of snow, then accumulations may be substantial.

Just ahead of the surface warm front, steady precipitation usually gives way to drizzle falling from low stratus clouds (bases below 2 km) and sometimes to fog. So-called **frontal fog** develops when rain falling into the shallow layer of cool air at the ground evapo-

rates and increases the water vapor concentration to saturation. After the warm front finally passes, frontal fog dissipates and skies at least partially clear because the zone of overrunning has also passed. The weather typically turns warmer and more humid.

The cloud and precipitation sequence just described for a warm front applies when the advancing warm air is stable. The sequence changes somewhat when the warm air is unstable. In that situation, uplift is more vigorous and often gives rise to cumulonimbus clouds (thunderstorms) embedded within the zone of overrunning ahead of the surface warm front. Hence, brief periods of heavy rainfall, or perhaps snowfall, may punctuate the otherwise steady fall of light-to-moderate precipitation.

## COLD FRONT

A stationary front becomes a **cold front** if it begins to move in such a way that colder (more dense) air displaces warmer (less dense) air. Over North America, the temperature contrast across a cold front is typically greater than that across stationary or warm fronts. However, in the summer, air temperatures on either side of the cold front can sometimes be nearly identical. When this occurs, the density contrast results from differences in water vapor concentration rather than differences in temperature. Following the passage of a cold front, we usually notice a drop in temperature in the winter and a drop in humidity in the summer.

Although friction causes the slope of a warm front to become more shallow, it steepens a cold front close to the Earth's surface into a characteristic nose shape, shown in Figure 11.5. The slope of a cold front is steeper (1:50 to 1:100) than the slope of a warm front (1:150 on average). Because of the steep frontal slope and the typical flow aloft across the front from the cold to the warm side, uplift is restricted to a narrow zone at or near the front's leading edge. Low-level air motion is also quite different from that in warm and stationary fronts; the low-level motion in the cold air is at least in part toward the front and forces the warm air aloft.

If the cold front advances slowly but steadily, say 30 km (19 mi) per hour, the type of frontal weather depends on the stability of the warmer air. Any precipitation is likely to be showery and brief and to oc-

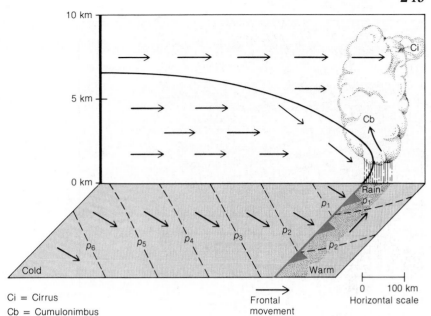

**FIGURE 11.5**
A cold front has a characteristic "nose" shape in cross section. The surface winds on the cold side of the front blow toward the front, and clouds and precipitation are limited to a narrow band at or just ahead of the front.

Ci = Cirrus
Cb = Cumulonimbus

cur in a narrow band at or just ahead of the front. If the warm air is relatively stable, then nimbostratus and altostratus clouds may form. If the warm air is unstable, the uplift is more vigorous, giving rise to cumulonimbus clouds towering above nimbostratus with cirrus clouds blown downstream from the cumulonimbus by winds at that altitude (Figure 11.5). These thunderstorms can be accompanied by strong and gusty surface winds, hail, or other violent weather.

If the cold front moves along at a rapid pace, say 45 km (28 mi) per hour, then a **squall line,** a band of intense thunderstorms, may develop either right at the front, or up to 300 km (180 mi) ahead of the front. Squall lines are discussed in more detail in Chapter 13, where we consider severe weather systems.

A typical cold front trails southward from an extratropical cyclone, as is discussed later in this chapter, and sweeps along from west to east. In the winter, many cold fronts also drop southward out of Canada into the Great Plains, even traveling as far south as the Gulf of Mexico. Southward or southwestward propagating cold fronts occur east of the Appalachians in New England, but in that region they are seen most frequently in the summer and fall and are referred to as **back-door cold fronts.** These fronts often usher in welcome relief from hot weather, in the form of *cP* air from Canada or *mP* air from the Atlantic. The Appalachians tend to impede the westward progress of the cooler air, which is then free to move farther south on the eastern side of the Appalachians along the piedmont and coastal plains (Figure 11.6).

## OCCLUDED FRONTS

A cold front typically travels about twice as fast as a warm front and eventually may catch up with the warm front to form an **occluded front,** sometimes called simply an **occlusion.** If the air behind the advancing cold front is colder than the cool air ahead of the warm front, the cold air slides under and lifts the warm air, the cool air, and the warm front (Figure 11.7). The resulting **cold-type occlusion** has the characteristics of a cold front at the surface, but the temperature contrast between the cold air mass and the cool air mass is typically less than that for a cold front. The weather ahead of the occlusion is similar to that which occurs in advance of a warm front; the actual frontal passage may be marked by more showery conditions, such as those associated with a cold front.

A second and less common type of occlusion occurs when the air behind the advancing cold front is not as cold as the air ahead of the warm front (Figure 11.8). Such a **warm-type occlusion** often occurs in the northerly portions of western coasts, such as in Europe or in the Pacific Northwest. In this case, the air behind the

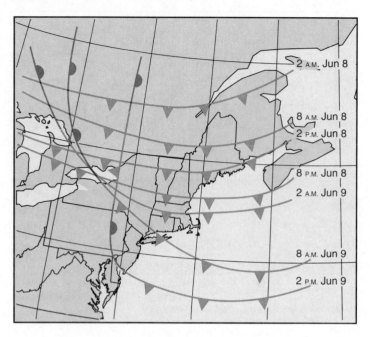

**FIGURE 11.6**
A back-door cold front advances toward the south and southwest.

cold front is relatively mild *(mP)*, having traversed ocean waters, whereas the air ahead of the warm front is relatively cold *(cP)*, having traveled over land. With this type of occlusion, the cool air behind the cold front slides under the warm air but rides over the cold air.

The weather ahead of a warm-type occlusion is similar to that ahead of a warm front, with the surface front behaving as a warm front. Both types of occlusions can be difficult to locate from surface weather observations, because the temperature contrast across the front is often small, precipitation occurs over a broad region

**FIGURE 11.7**
Schematic vertical cross section of a cold-type occlusion. This is the more common type of occlusion over North America. The vertical scale is greatly exaggerated.

**FIGURE 11.8**
Schematic vertical cross section of a warm-type occlusion. The vertical scale is greatly exaggerated.

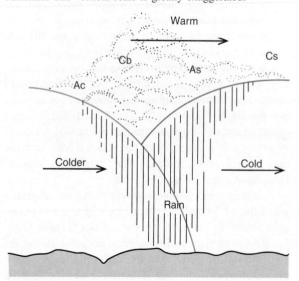

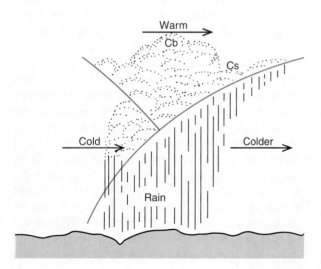

masking the front, and the trough in the pressure pattern associated with the front is at times not as pronounced as that for cold and warm fronts. However, satellite imagery shows that an occluded front can be as sharply defined as a cold front in oceanic weather systems.

### CHANGES IN FRONTS

The characteristics that define a front—differences in temperature and moisture content, wind shift, convergence, and the like—change with time in much the same way that air masses modify with time. We define a front primarily by the density contrast across the front. If processes in the atmosphere act to increase the density contrast, for example, through convergence, then a front grows stronger. This process is called **frontogenesis.** On the other hand, if the density contrast weakens, the front also weakens, a process called **frontolysis.** Precipitation associated with the front also tends to increase or diminish in intensity as the front strengthens or weakens.

### SUMMARY

In midlatitudes, the troposphere tends to be comprised of air masses that differ in temperature and water vapor concentration. These air masses are separated by transition zones called fronts, which are classified primarily by the temperature behind the front as it passes by, relative to the temperature ahead of the front. The motion of a front is related to the low-level motion on the cold side of the front. Cold air retreats ahead of a warm front, advances behind a cold front, and moves parallel to a stationary front. Clouds and precipitation develop along fronts only when (1) a significant density contrast exists between air masses, and (2) adequate water vapor is present. If there is little difference in temperature and humidity across the front, then the front may pass virtually unnoticed except for a shift in wind direction. The strength of the density contrast marking a front varies with time, leading to the formation and dissipation of fronts.

Frontal weather occurs in combination with cyclones. We explore this relationship next.

# Midlatitude Cyclones

The extratropical **cyclone,** or low-pressure system, is the principal weathermaker of midlatitudes. The counterclockwise (in the Northern Hemisphere) and inward low-level flow of air is associated with convergence and therefore rising motion, cloudiness, and precipitation. We describe the life cycle and characteristics of these synoptic-scale storms in this section.

### LIFE CYCLE

If conditions are favorable in the middle and upper troposphere, that is, if there is upper-air support, an extratropical cyclone can form and grow. **Cyclogenesis,** the birth of a cyclone, usually takes place along the polar front directly under an area of strong horizontal divergence aloft. As noted in Chapter 10, strong divergence aloft occurs to the east of an upper-level trough and under the left-front quadrant of an upper-level jet stream maximum. If divergence aloft removes more mass from a column of air than is brought in by convergence near the surface, the surface air pressure at the bottom of the column drops, a process also referred to as **deepening.** Consequently, a horizontal pressure gradient develops, and a cyclonic circulation begins. In other words, a storm is born. The westerly flow aloft then steers and supports the cyclone as it progresses through its life cycle (Figure 11.9).

Just prior to the formation of our prototypical cyclone, the polar front is stationary, and surface winds are directed parallel to the front. As the surface air pressure drops, surface winds converge and the front begins to move (Figure 11.9A). West of the low center,* the front advances toward the southeast as a cold front. East of the low center, the front moves northward as a warm front. The minimum pressure for the incipient low might be 1000 mb, with a single closed isobar on a standard surface weather map, which uses a 4-mb increment between isobars. Satellite imagery shows that the narrow cloud band associated with the stationary front develops a bulge at the low center and extends

---

*The low center is at the point of lowest sea-level pressure. The minimum sea-level pressure value is known as the *central pressure* of the cyclone.

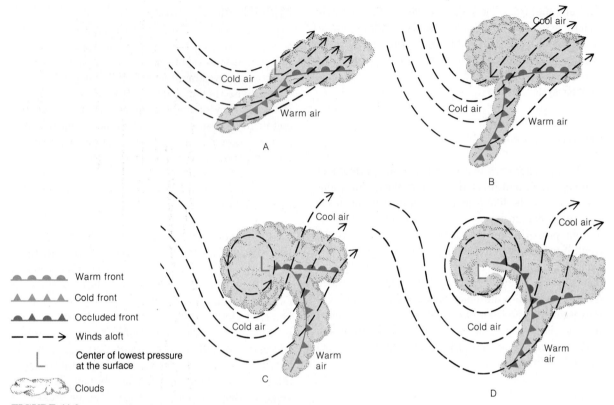

~~~~ Warm front

▲▲▲▲ Cold front

~▲~▲ Occluded front

— — → Winds aloft

L    Center of lowest pressure
     at the surface

Clouds

**FIGURE 11.9**
A midlatitude cyclone passes through its life cycle: (A) incipient cyclone, (B) wave cyclone, (C) beginning of occlusion, and (D) bent-back occlusion. As a wave cyclone, the storm is east of the upper-level trough; at occlusion, the storm is under the upper-level trough. [Modified from G. T. Trewartha and L. H. Horn, *An Introduction to Climate,* 5th ed. New York: McGraw-Hill, 1980, p. 165.]

along the warm front. The upper-level pattern depicted in Figure 11.9A has a trough to the west of the surface low, a position that favors further development of the system.

An example of a cyclone at this stage can be seen over Nevada in Figure 11.10A. A standard surface weather map, such as the one in this figure, includes plotted observations, isobars, high- and low-pressure centers, fronts, and shading to represent areas of precipitation. Before the isobars are constructed, the surface air pressure is reduced to sea level to remove the influence of topography (Chapter 5). These isobars portray high and low pressure centers, troughs, and ridges and also can be used to infer geostrophic and gradient winds (Chapter 9). The behavior of the cyclone system in Figure 11.10A is somewhat complicated by the Rocky Mountains with two separate low centers, and

so neither a warm front nor a simple cloud pattern is evident in the satellite image at this time.

Figure 11.11A shows the corresponding upper-air pattern. Such upper-air charts depict conditions on constant pressure surfaces, in this case 500 mb. Pressure surfaces are nearly flat, but the topography of the surface that does exist is important because of its relationship to the gradient wind. On a pressure surface, the gradient wind is parallel to the contours (lines of equal height above sea level) and faster where contours are closer together. Figure 11.11A shows that the observed wind, as described by the plotted wind observations, is close to gradient at this level, as shown by the wind blowing essentially parallel to the contours (solid lines). The contours also portray the ridges and troughs that characterize the upper-level flow. The 500-mb surface, located at an average altitude of roughly 5.5 km

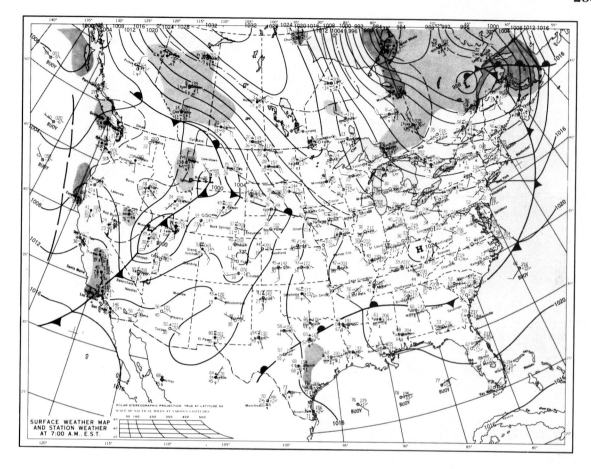

A

**FIGURE 11.10**
Surface weather maps and GOES visible
satellite imagery for (A) 1 April 1982, (B)
2 April 1982, and (C) 3 April 1982. [From
NOAA]

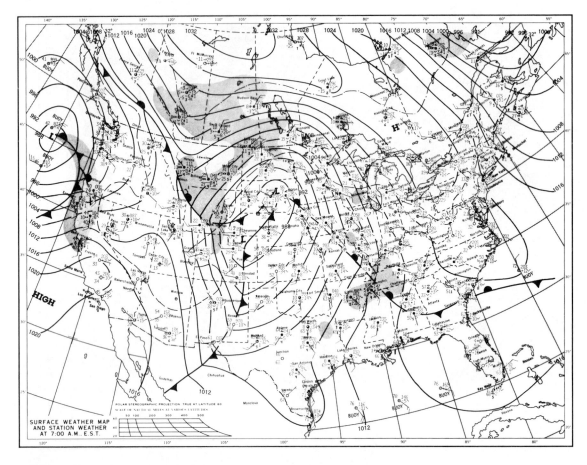

B

**FIGURE 11.10 (continued)**

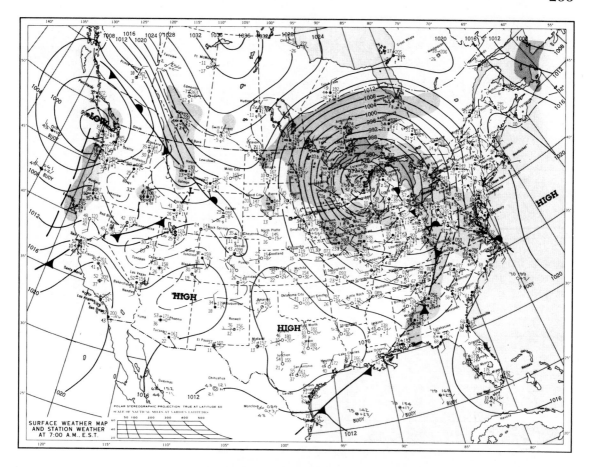

C

**FIGURE 11.10 (continued)**

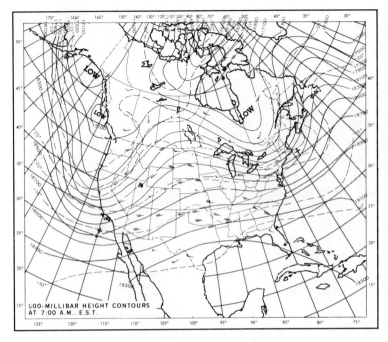

A

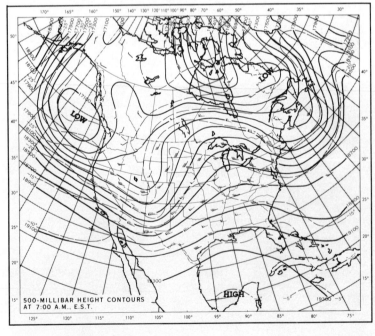

B

**FIGURE 11.11**
The 500-mb analyses for (A) 1 April 1982,
(B) 2 April 1982, and (C) 3 April 1982.
Contour lines correspond to the height of the
500-mb surface in feet; dashed lines are iso-
therms in C°.

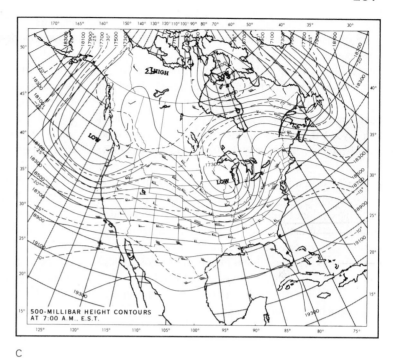

500-MILLIBAR HEIGHT CONTOURS
AT 7:00 A.M., E.S.T.

C

**FIGURE 11.11 (continued)**

(3.4 mi), is commonly used to examine the degree to which the upper-level flow favors development and displacement of surface high and low pressure systems. The pattern in Figure 11.11A has a deep trough along the Pacific Coast in a position to yield divergence aloft over the incipient low center and future deepening of the surface cyclone.

If the upper-air circulation is not favorable for further development of the cyclone, the low center typically ripples along the stationary front without deepening further, producing light precipitation along the way. Such cyclones affect only a small area and typically travel along the front at 50 to 70 km (31 to 43 mi) per hour.

If conditions are favorable for the incipient cyclone to continue to mature, its central pressure drops further and the associated horizontal pressure gradient and counterclockwise winds strengthen. The upper-level trough also frequently deepens while remaining to the west of the low center. Because the cold front generally moves faster than the warm front, the angle between the two fronts gradually closes. That is, the so-called **warm sector** of the storm, situated between the warm and cold fronts (and occupied by warm air at the surface), becomes better defined. At this stage in the storm's life cycle, fronts form a pronounced wave pattern (Figure 11.9B)—hence the descriptive name **wave cyclone.** The surface cyclone may now have a central pressure of 992 mb, with perhaps three closed isobars, and is moving eastward or northeastward at 40 to 55 km (25 to 34 mi) per hour. A **comma cloud** is apparent in satellite images at this stage, reflecting the strengthening of the storm's circulation. The head of the comma extends from the low center to the northwest, with its tail trailing along the cold front. Clouds associated with overrunning are also present north of the warm front. Figure 11.10B shows the storm from Figure 11.10A one day later; a well-defined warm sector is evident in the southern Plains. The secondary front coming in behind the cold front marks the boundary between *mP* air to the south and an arctic air mass to the north. The upper-air trough (Figure 11.11B) is very well defined and positioned to favor further development of the system.

As the cold front continues to advance toward the warm front, an occluded front begins to form near the low center, displacing the warm air aloft and causing the warm sector at the surface to occupy a smaller area.

Figure 11.9C represents the beginning of the occlusion stage of the cyclone. Note that the upper-level pattern now features a closed circulation almost above the surface cyclone. When the upper-level low center is directly over the surface low center, the system is said to be *vertically stacked.* Dry air descending behind the cold front is drawn nearly into the center of the cyclone as a *dry slot* that separates the cloud band along the cold front from the comma head, now more west than northwest of the low center. The central pressure of the storm has dropped significantly to perhaps 985 mb in the time that has elapsed between Figures 11.9B and 11.9C.

Some cyclones continue to deepen after an occluded front forms at the surface. Typically, the upper-air low center is not directly over the surface low center while deepening is occurring, but it may be fairly close. These storms develop a closed, vertically stacked (or nearly so) circulation that is troposphere-deep and typically move much more slowly than previously, say 30 km (19 mi) per hour, and may even stall completely. The circulation tends to draw the occluded front around the low center into a configuration that is sometimes called a **bent-back occlusion** (Figure 11.9D). The central pressure of the low may now be 980 mb or less, with a half dozen or more closed isobars quite close together, associated with winds that can exceed 75 km (47 mi) per hour. At the Earth's surface, the warm sector is still present but detached from the cyclone itself.

The cold, warm, and occluded fronts all come together at the point of occlusion, or **triple point,** where conditions are favorable for formation of a new cyclone, sometimes called a *secondary cyclone.* (Note the similarity in appearance between the triple point with its fronts in Figure 11.9D and the incipient low with its fronts in Figure 11.9A.) At this stage of development, the cloud pattern typically becomes a spiraling swirl with enhanced bands associated with the fronts. A very intense system with a central pressure of 960 mb or lower can cause the spiraling cloud band to circle the low center several times. Figure 11.10C shows our sample cyclone one day after Figure 11.10B. Note the large number of closed isobars, the tight pressure gradient indicating strong winds, and the large swirling cloud mass in the satellite image. A low center is also evident aloft (Figure 11.11C), but not quite directly over the surface low.

Eventually, the cyclone weakens, its winds lessen, and its central pressure rises; this process is called **cy-** **clolysis** or **filling.** The low can weaken during any of the stages described above if its upper-air support diminishes. As the central pressure rises, the storm may lose its identity in the sea-level pressure field and be indicated only by cloudy skies and drizzle. Such weakening is inevitable once the system is vertically stacked, but the filling process may not begin for many hours and may proceed slowly, allowing an intense circulation to persist for days.

These basic stages in the life cycle of cyclones were first formulated during World War I by Norwegian researchers in Bergen. This conceptual model is therefore referred to as the **Norwegian cyclone model.** Major advances in weather monitoring, especially those involving remote sensing by satellite, have verified the Norwegian model with only minor alterations. Amazingly, it remains a close approximation of our current understanding of midlatitude cyclones, even though individual cyclones may not follow the model exactly as we saw in our sample cyclone.

The life cycle described above may occur over several days as in the example in Figures 11.10 and 11.11, or it may develop over a much shorter period. If upper-air support is less favorable, the storm may spend a longer time in any one of the early stages, even weaken temporarily, and still become fully occluded. Sometimes storms develop with meager upper-air support (weak divergence aloft) and are weak and poorly defined. At other times, widespread cloudiness and precipitation are linked to an upper-air or surface trough, and not to a closed cyclonic circulation at the surface. When upper-level conditions are ideal, the entire life cycle from incipient cyclone to bent-back occlusion can occur in less than 24 hr.

A rapidly intensifying extratropical cyclone, or *bomb,* is defined as a cyclone whose central pressure drops by at least 24 mb in 24 hr adjusted to 60 degrees N, the latitude of Bergen, Norway, where the criterion was developed. Few cyclones meet this criterion, and most of those that do satisfy it occur over warm ocean currents such as the Kuroshio Current off the coast of Japan and the Gulf Stream off the East Coast of the United States. Because these oceanic storms are not closely observed on a day-to-day basis, recent field experiments have used research aircraft and other special instrumentation to investigate them. The *Genesis of Atlantic Lows Experiment (GALE),* sponsored by a number of agencies including the National Science Foundation, the U.S. Office of Naval Research (ONR), and the National Oceanic and Atmospheric Administration, took place from January to March 1986; its primary goal was to measure the initial formation of these oceanic cyclones in a region off the coast of North Carolina. The *Experiment on Rapidly Intensifying Cyclones over the Atlantic (ERICA)* was sponsored primarily by ONR and was conducted from December 1988 to February 1989 over the North Atlantic. The goal of ERICA was to monitor rapidly intensifying cyclones throughout their life cycle.

An extreme example of a developing cyclone was observed during ERICA on 4–5 January 1989. An incipient cyclone with a central pressure of 996 mb was first identified off Cape Hatteras, North Carolina, at approximately 7:00 P.M. EST on 4 January. Twenty-four hours later, the storm was located 700 km (430 mi) south of Newfoundland with a central pressure of 936 mb. The storm deepened by 60 mb in 24 hr—2.5 times the criterion for a *bomb*. Satellite images for this storm as it neared peak intensity are shown at the beginning of this chapter and in Figure 11.12. The color-enhanced image at the beginning of the chapter is an infrared satellite image, which depicts temperatures of the relatively warm surface of the Earth in cloud-free areas or the relatively cold tops of clouds, with higher clouds

generally colder. The visible image in Figure 11.12 depicts the Earth's surface and clouds as they would appear to a black and white camera in space. Note that the cold (blue and magenta regions in the infrared image) mid- and upper-level clouds in the comma cloud have a relatively smooth appearance in the visible image and a shape similar to the shaded area in Figure 11.9D. The dark magenta shading in the infrared image indicates thunderstorms embedded in the comma tail along the cold front. The streaky or mottled appearance of the cumulus clouds in the visible image, seen streaming off the coast, passing south of the comma head, and entering the dry slot, are considerably warmer and therefore at a lower altitude than the comma cloud itself, as indicated by the orange to green colors in the infrared satellite image. These clouds form as very cold air sweeps off the coast and is warmed, moistened, and destabilized from below by the much warmer ocean surface; they are a hallmark of oceanic cyclones.

### CYCLONE WEATHER

As an illustration of typical cyclone weather, consider a winter wave cyclone that developed a well-defined warm sector as it moved into the upper Mid-

**FIGURE 11.12**
Visible satellite image of the cloud pattern associated with an intense occluded midlatitude cyclone centered over the North Atlantic. [NOAA satellite photograph]

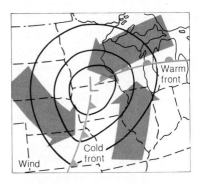 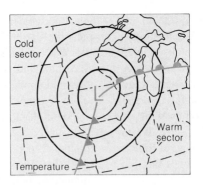 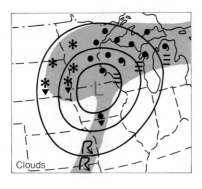

**FIGURE 11.13**
A wave cyclone showing typical patterns of (A) surface winds, (B) surface air temperatures, and (C) clouds and precipitation.

| | | | |
|---|---|---|---|
| ☰ | Fog | ✳ | Snow shower |
| ● | Rain | ▼ | Rain shower |
| ❥ | Drizzle | R | Thunderstorm |
| ✳ | Snow | | |

west. Although the storm is still intensifying, its circulation, clouds, and precipitation already affect a wide region. Figure 11.13 is a schematic representation of the winds, temperature distribution, cloud shield, and precipitation pattern associated with our typical midlatitude low.

Ideally, based on surface weather, we can divide a wave cyclone into four sectors about the storm center. The lowest air temperatures are to the northwest of the storm center, where strong and gusty northwest winds advect cP air or A air southward and eastward. Strong winds make the air feel colder than the temperature might suggest (windchill effect). West of the low center, precipitation tapers off to showers, and there is a tendency toward clearing skies as the storm center moves toward the northeast. The leading edge of the cold air mass (cold front) is south of the storm center and is accompanied by a narrow band of showers and embedded thunderstorms. The southwest sector has generally clear skies and sinking motion. The mildest air is in the southeast (warm) sector of the storm, where south and southeast winds advect mT air northward from over the Gulf of Mexico. Skies are generally partly cloudy, dew points are high, and scattered convective showers are possible, especially during the afternoon. Cloudiness increases, and showers become more prevalent, as the cold front progresses eastward. To the north and northeast of the

storm center is a zone of extensive overrunning as mT air surges over a wedge of cool air maintained by east and northeast winds at the surface. Skies in the northeast sector are cloudy, and precipitation is steady and substantial.

Note the weather conditions in the various sectors of a wave cyclone are consistent with our earlier description of cold and warm frontal weather. Although the flow in and around cyclones is mostly horizontal, the uplift of air along frontal surfaces leads to the characteristic cloud and precipitation patterns shown in Figure 11.9.

On taking a second look at the cyclone case in Figures 11.10 and 11.11, the NOAA *Weekly Weather and Crop Bulletin* reported the following weather summary for each of the three days. Notice how the major weather events associated with this cyclone and its fronts correspond to the previous description of cyclone weather.

Thursday, 1 April 1982—A warm front reached from South Carolina to Arkansas and through central Texas. Warm, moist air from the Gulf of Mexico flowed into Texas and then veered northeastward into Arkansas. Light showers fell in Texas, but moderate to heavy showers and thunderstorms battered Arkansas and northern Mississippi. A complex frontal system through the West produced light to moderate rain and heavy snow in the mountains. The precipitation covered the area west of the Rockies and spread into the northern Plains.

Friday, 2 April 1982—An intense low pressure system deepened in the central Plains. Severe weather moved ahead of the storm through the entire Mississippi Valley and spread slowly eastward to a line from the upper Ohio Valley through Alabama. Light showers and thunderstorms extended to the mid-Atlantic states by the end of the day. The low pressure center deepened to record intensity as it reached the Great Lakes area. Tornadoes, hail, high winds, and heavy rain lashed the area from northeastern Texas into Minnesota and Wisconsin. Rain, wind, and snow from another storm hit the Pacific Northwest.

Saturday, 3 April 1982—The intense low pressure system moved slowly into the central lakes area. The cold front, marking the line of severe weather, moved from the central Great Lakes, through the Appalachians to central New York State, and along the east coast to northern Florida. Thunderstorms were not as intense along the front as on Friday, but golf-ball-size hail and high winds were reported from New York to Georgia. Rain showers, with snow in the mountains, spread into central California over the Plateau to the northern Plains.

## *CONVEYOR BELT MODEL*

A conceptual model that combines horizontal and vertical air motions into a three-dimensional depiction describes fronts and cyclones in terms of three interacting airstreams, often referred to as *conveyor belts.* Just as mechanical conveyor belts transport goods (or even people) from one location to another, these atmospheric conveyor belts transport air with certain properties from one location to another and are named for the type of air they transport. A schematic of this conceptual model is shown in Figure 11.14.

According to the **conveyor-belt model,** the warm conveyor belt (broad red arrow in Figure 11.14) originates in the cyclone's warm sector and follows the warm side of the cold front near the Earth's surface, south and east of the storm center. Typically, this warm and humid airstream ascends slightly as it progresses northward in the cyclone's warm sector at low levels, then ascends rapidly as it glides northward over the

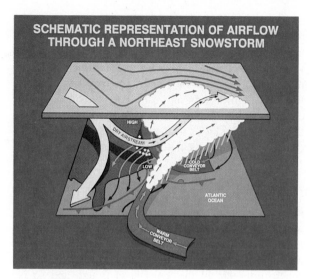

**FIGURE 11.14**
Schematic representation of the circulation within an intense cyclone showing the cold and warm conveyor belts and the dry airstream. [From P. J. Kocin and L. W. Uccellini, *Snowstorms Along the Northeastern Coast of the United States: 1955–1985.* Boston: American Meteorological Society, 1990, p. 71.]

sloping warm frontal surface north of the surface warm front (compare Figures 11.14 and 11.3). As the air ascends, it cools adiabatically, its water vapor condenses forming clouds, and rain or snow falls. Latent heat released during this process causes the air in the warm conveyor belt to ascend even more rapidly. However, recall that the flow aloft at this stage of the cyclone's development is more or less from the southwest, with the surface cyclone located between the upper-level trough and ridge. The warm conveyor belt thus turns from a southerly to a southwesterly or even westerly direction as it ascends over the warm front and follows the upper-level flow. This airstream therefore helps explain the broad region of cloudiness and precipitation north of the warm front.

While the warm conveyor belt is gliding up and over the warm frontal surface, the flow underneath it is colder and directed from the east or northeast. This flow forms the cold conveyor belt (broad blue arrow in Figure 11.14) and at low levels is located just to the cold (north) side of the warm front. Like the warm conveyor belt, this airstream also ascends as it progresses toward the west. The air is moistened by the evaporation of part of the warm frontal precipitation falling through it and so becomes saturated and also contributes to the cyclone's precipitation. As the cold conveyor belt ascends, it comes under the influence of the mid- and upper-level flow and turns clockwise to follow the southwesterly or westerly flow aloft. The ascending saturated cold conveyor belt produces the distinctive *comma cloud* to the northwest of the cyclone center as depicted by the thin blue arrows in Figure 11.14.

The third airstream or conveyor belt is the dry airstream (broad yellow arrow in Figure 11.14). Whereas the cyclone center and fronts are typically regions of upward motion, the region west of the cold front experiences downward motion. This descending flow originates aloft in the upper troposphere and lower stratosphere upstream of the upper-level trough and so is quite dry, especially in comparison to the humid air in the warm conveyor belt. As the dry conveyor belt descends, part of it turns southward and descends behind the cold front to give the typically clear skies behind the surface cold front. The other branch of the dry airstream first descends as it moves northward toward the surface low center, then ascends as it comes under the influence of the upper-level flow east of the upper-level trough. As the dry airstream moves toward the surface low center, it forms the dry slot that separates the head and tail of the comma cloud and becomes especially

prominent in the latter phases of the cyclone's development (compare Figures 11.14 and 11.9).

## CYCLONE TRACKS

A midlatitude wave cyclone has a warm side and a cold side. Note that in Figure 11.13, the coldest air is located to the northwest of the low center (where winds are northwesterly), and the warmest air is to the southeast of the low center (where winds are southerly). Hence, as a cyclone traverses the continent, the weather to the left (cold side) of the storm track is quite different from the weather to the right (warm side) of the track.

Consider, for example, Figure 11.15. A winter cyclone develops in eastern Colorado. As the storm matures, it moves northeastward toward the Great Lakes region and may take track A or track B. In either case, the storm center passes within 150 km (93 mi) of Chicago, but track A takes the storm west of Chicago and track B takes it east of Chicago. The storm's effect on Chicago's weather depends on its track, as summarized briefly in Table 11.2.

If the storm follows track A, Chicago residents experience the warm side of the storm. Steady rain (or perhaps snow, briefly, at the onset) gives way to drizzle and fog after 12 to 24 hr. As the surface warm front

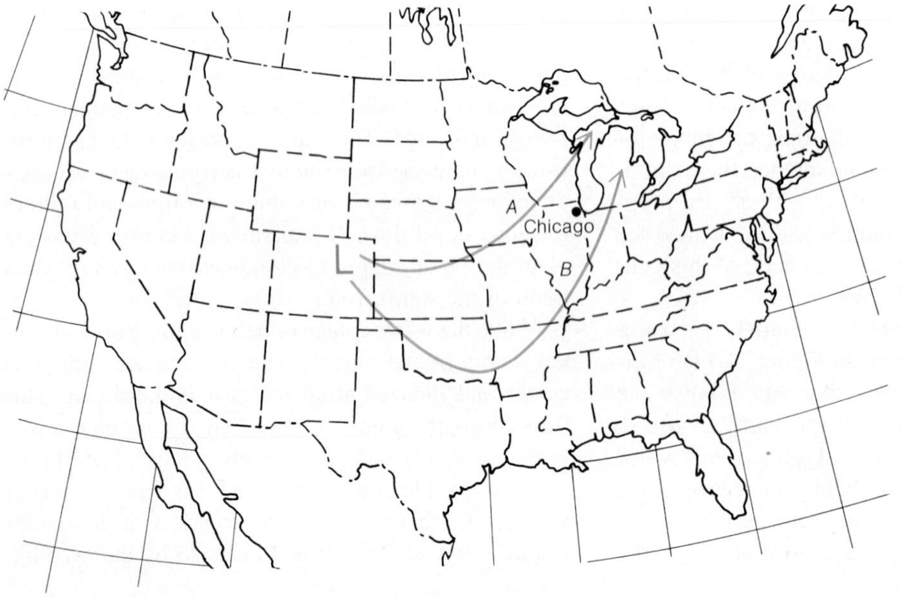

**FIGURE 11.15**
A cyclone that developed over eastern Colorado follows a path that takes the storm center either to the west (track A) or to the east (track B) of Chicago. For track A, Chicago is on the relatively warm side of the storm; for track B, Chicago is on the relatively cold side of the storm.

**TABLE 11.2**
**Sequence of Surface Weather Conditions in Chicago as Mature Winter Storm Tracks West (Track A) and East (Track B) of City**
WF = Warm front
CF = Cold front
F = Pressure falling
R = Pressure rising

| | Track A | | | | | | Track B | | | |
|---|---|---|---|---|---|---|---|---|---|---|
| Wind direction | E | SE | S | SW | W | NW | E | NE | N | NW |
| Frontal passage | – | WF | – | CF | – | – | – | – | – | – |
| Advection | – | Warm | Warm | Cold | Cold | Cold | – | – | Cold | Cold |
| Air pressure tendency | F | F | F | R | R | R | F | F | R | R |
| | Time ———————→ | | | | | | Time ———————→ | | | |

passes over the city, skies partially clear and winds shift abruptly from east to southeast, advecting warm and humid (*mT*) air at the surface. Clearing is short-lived, however, as scattered showers and thunderstorms herald the arrival of colder air. As the surface cold front passes through the city, winds *veer* (turn clockwise) so that they blow first from the southwest, then west, and finally northwest. Skies clear again, and the air temperature and dew point fall.

In contrast, if the storm takes track *B*, Chicago residents experience the cold side of the storm and no frontal passages. Gusty east and northeast winds drive steady snow or rain (depending on how cold the air is) for 12 hr or longer. Then winds gradually *back* (turn counterclockwise) to a northerly direction, precipitation tapers off to snow flurries or showers, and air temperatures begin to drop. Finally, winds shift slowly to the northwest, skies begin to clear, and temperatures continue to drop.

In summary, then, if you are on the warm side of a cyclone, the wind direction veers with time and a warm front is followed by a cold front. If you are on the cold side of a cyclone, the wind direction backs with time without the passage of fronts.

The specific track taken by any synoptic-scale cyclone depends on the pattern of the upper-level westerlies in which the storm is embedded. The storm center tends to move in the direction of the wind blowing directly above the storm at the 500-mb level. As a rule, the forward motion of the storm is about one-half of the 500-mb wind speed. However, keep in mind that the upper-level flow pattern also changes with time, as is evident in Figure 11.9.

Figure 11.16 shows the principal storm tracks across the lower 48 states of the United States. Cyclones following some of these tracks have nicknames. For example, the Panhandle Hook starts to the lee of the Rockies and *hooks* almost due north, whereas the Alberta Clipper starts to the lee of the Canadian Rockies and moves very rapidly across the northern tier of states. All storms tend to converge toward the northeast; their ultimate destination is usually the Icelandic low* of the North Atlantic or Western Europe. Although many storm tracks appear to begin just east of the Rocky Mountains, in reality they have their origins over the Pacific Ocean. As cyclones travel through the mountains, they often lose their identity temporarily but redevelop on the Great Plains just east of the Front Range of the Rockies. What actually happens to these cyclones is explained in more detail in the Special Topic "The Case of the Missing Storm."

---

*Actually, the Icelandic low is a statistical feature of the large-scale circulation. On any given day, it may or may not be evident on a weather map, and it may or may not be over Iceland.

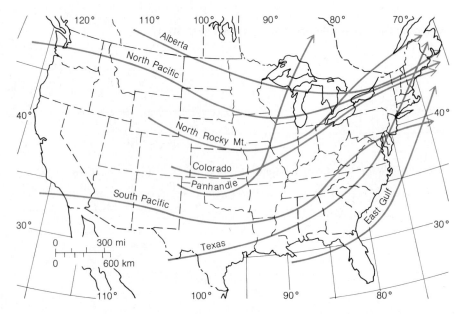

**FIGURE 11.16**
Principal storm tracks across
the lower 48 states of the
United States.

The notion that storms in middle latitudes generally move from west to east was first suggested in 1703 by Daniel Defoe, the English journalist and novelist. Defoe drew this conclusion from his study of a great storm that lashed the British Isles on 7–8 December of that year. He received reports that, several days earlier, a similar storm had ravaged the East Coast of North America.

In the United States, Benjamin Franklin is generally credited with discovering that storms usually move in an easterly or northeasterly direction. On 21 October 1743, storm clouds over Philadelphia prevented Franklin from viewing a lunar eclipse. Later, through correspondence, Franklin learned that his brother had observed the eclipse in Boston (500 km northeast of Philadelphia) and the next day was stormy. Franklin concluded that a storm and its associated cloud cover had tracked northeastward up the coast.

It may be confusing to think of a storm as moving *toward* the northeast, while the winds in the northeast sector of the storm blow *from* the northeast. In fact, New Englanders long assumed that the powerful *nor'easters* that pounded their shores moved down the coast from the northeast because the storm winds were northeasterly. Actually, nor'easters usually are storms similar to the ERICA storm described earlier; they intensify off the North Carolina coast and then track in a

northeasterly direction up the coast. In effect, there are two motions: (1) the movement of the storm center up the coast and (2) the counterclockwise circulation of winds about the storm center. The circulation of a storm is for the most part independent of the storm's path, much as the spin of a frisbee is independent of its trajectory.

Generally, storms that form in the south yield more precipitation than those that develop in the north because southern storms are closer to the primary source of moisture, maritime tropical air. For example, Alberta cyclones typically yield only light amounts of rain or snow, whereas Colorado and Gulf-track storms (Figure 11.16) often produce heavy accumulations of rain or snow. The speed of the storm also affects the total precipitation it produces at a particular location. A storm that is fast-moving may yield rain or snow for only a few hours; a slower moving storm may for 12 hr or more.

From our earlier discussion of the linkage between winds aloft and migratory synoptic-scale cyclones, we can deduce that storms should exhibit seasonal variability. In summer, when the mean position of the polar front and jet stream is across southern Canada, very few well-organized cyclones occur in the United States, and the Alberta storm track shifts northward across central Canada. In winter, however, when the mean positions

# The Case of the Missing Storm

After cyclones sweep ashore along the west coast of North America, they disappear in the sea-level pressure pattern as they move inland over the mountainous West. A few days later, cyclogenesis occurs to the east of the Front Range, typically on the plains of Alberta or eastern Colorado. What actually happens to these storms as they cross the mountains?

Visualize a storm moving ashore as a huge cylinder of air spinning about a vertical axis in a cyclonic (counterclockwise) direction (Figure 1). As the cylinder moves up the windward slopes of a mountain range, the column is forced to shrink in height. As it shrinks it also widens, and the spin of the cylinder slows so that, in effect, the storm's circulation weakens. As the cylinder of air then descends the leeward slopes, the air column stretches vertically and, contracts horizontally, and the cyclonic spin strengthens. The weakening of cyclonic circulation upslope and the strengthening of the circulation downslope account for the

seeming disappearance and reappearance of a storm as it crosses mountainous terrain.

The changes in the storm's circulation are analogous to an ice skater performing a spin. The skater changes the rate of spin by extending or drawing in her arms. When the arms are extended (analogous to horizontal divergence and the widening of our cylinder), the spin rate slows. When the arms are brought as close as possible to the skater's body (analogous to horizontal convergence and the contracting of our cylinder), the spin rate increases. In effect, both the spinning skater and the storm passing through mountainous terrain conserve angular momentum. Simply put, *conservation of angular momentum* means that a change in the radius of a rotating mass is balanced by a change in its rotational speed. An increase in radius (horizontal divergence) is accompanied by a reduction in rotational rate, and a decrease in radius (horizontal convergence) is accompanied by an increase in rotational rate.

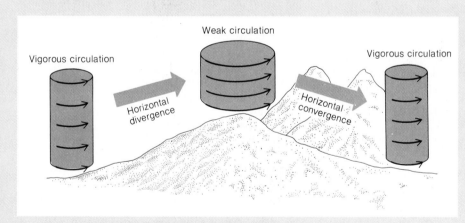

**FIGURE 1**
The cyclonic spin (circulation) of a cylinder of air weakens with horizontal divergence on the windward slopes of the mountain range and strengthens with horizontal convergence on the leeward slopes of the mountain range.

of the polar front and jet stream shift southward, cyclogenesis is more frequent in the United States. Alberta-track storms are the most common because they occur year-round, whereas storms with more southerly tracks develop primarily in winter. In fact, one sure sign of the beginning of winter circulation patterns is the appearance of Colorado lows.

## COLD- AND WARM-CORE SYSTEMS

An occluded cyclone such as shown in Figure 11.9D is a cold-core system; that is, the lowest temperatures on a horizontal surface occur at the center of the low. In vertical cross section, constant pressure (isobaric) surfaces within a **cold-core cyclone** are concave up

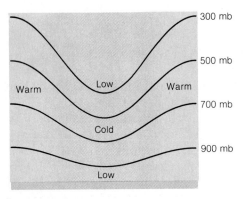

**FIGURE 11.17**
Vertical cross section shows sloping isobaric (constant-pressure) surfaces within a cold-core low-pressure system. Vertical scale is greatly exaggerated.

ward (Figure 11.17). Furthermore, the depth of the low increases with altitude, implying that the circulation is cyclonic throughout the atmosphere and most intense at higher levels. Recall from the Mathematical Note in Chapter 10 that the thickness between pressure surfaces is directly proportional to the layer-mean temperature. The requirement that the thickness (mean temperature) be lowest at the center of the low produces the characteristic isobaric pattern.

A wave cyclone, on the other hand, has the lowest temperatures to the northwest of the storm and the highest temperatures to the southeast. Figure 11.18 portrays a cross section through the low from northwest to

southeast. Thickness (mean temperature) arguments lead to the conclusion that the low center aloft is not located above the low center near the surface, but rather is displaced to the cold side of the storm, implying a tilt with altitude for the system. This condition is consistent with Figure 11.9B, which shows the upper-level trough lagging behind the surface cyclone.

A different type of low that sometimes appears on surface weather maps has characteristics markedly different from those of cold-core or tilted lows. Cyclones of this type are stationary, have no fronts, and are associated with fair weather. They develop over arid or semiarid deserts, including the interior of Mexico and the southwestern United States, when the summer sun heats the ground, which in turn heats the overlying air. Intense heating lowers the air density over an area wide enough for a synoptic-scale low to appear. This **warm-core cyclone** (or *thermal low*) is very shallow, and its circulation weakens rapidly with altitude, that is, away from the heat source (the ground). The surface counterclockwise circulation frequently reverses at some altitude, and the thermal low is overlain by an anticyclone, as indicated by the upper-level pressure surface bulging upward in Figure 11.19.

In this section, we have focused on the general weather conditions associated with cyclones. In the daily march of weather in midlatitudes, cyclones are followed by anticyclones. We now examine these fair-weather systems.

**FIGURE 11.18**
Vertical cross section through a wave cyclone. Vertical scale is greatly exaggerated.

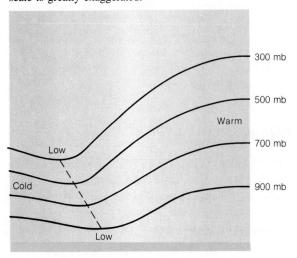

**FIGURE 11.19**
Vertical cross section shows sloping isobaric (constant-pressure) surfaces within a warm-core low-pressure system. The shallow near-surface cyclonic circulation weakens and then reverses with altitude. Vertical scale is greatly exaggerated.

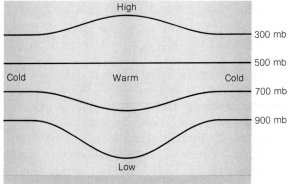

# Anticyclones

An **anticyclone** (or *high*) is, in many ways, the opposite of a cyclone (or *low*). Cyclonic circulation favors the convergence of contrasting air masses and the development and maintenance of fronts. In anticyclones, by contrast, subsiding air and diverging surface winds favor the formation of a uniform air mass and fair skies. Like cyclones, however, anticyclones, can have either cold or warm cores.

## COLD- AND WARM-CORE SYSTEMS

A **cold-core anticyclone** is actually a dome of continental polar *(cP)* or arctic *(A)* air and, depending on which air mass type, is labeled either a **polar high** or an **arctic high.** They are products of extreme radiational cooling over the often snow-covered continental interior of North America well north of the polar front. Cold anticyclones are shallow systems in which the clockwise circulation weakens with altitude and often reverses. A cold trough is therefore typically situated over a cold anticyclone.

Cold anticyclones are most intense (that is, they exhibit highest surface pressures) in winter when the associated air mass is coldest. The air is extremely stable in these systems, and soundings indicate a temperature inversion in the lowest kilometer or two, associated with strong subsidence and adiabatic heating above the inversion. Massive arctic or polar anticyclones with very high central pressures tend to remain stationary over their source regions. However, lobes of cold air (smaller cold highs) often break out of the source region and slide southeasterly across Canada and into the United States. These cold air surges interact with the circulation of migrating cyclones and help to maintain and strengthen the temperature contrast across the cyclone's cold front. This explains why winter storms are usually followed by clearing skies and sharply lower temperatures.

On rare occasions in winter, a particularly strong arctic high brings a surge of bitterly cold air that sweeps as far south as southern Florida and can even traverse the Gulf of Mexico into Central America. The resulting subfreezing temperatures can spell disaster for citrus growers. As an illustration, the synoptic weather pattern responsible for the costly Florida freeze of De-

cember 1983 is shown in Figure 11.20. Note that the cold front has already pushed all the way south to Cuba. Cloudiness over the Gulf of Mexico results from the warming and moistening of the cold air by the warm Gulf waters, which generate fog and low clouds. Much of what appears to be white over the continent east of the Rockies is not cloudiness but, rather, snow cover.

A **warm-core anticyclone** forms south of the polar front and consists of extensive areas of subsiding warm, dry air. Thickness arguments show that, like cold-core cyclones, warm-core anticyclones strengthen with altitude. They are massive systems with a circulation extending from the Earth's surface up to the tropopause. The semipermanent subtropical anticyclones, such as the Bermuda–Azores high, are examples of warm-core highs, but other warm-core anticyclones may develop over the interior of North America, especially in summer.

A cold anticyclone produces high surface air pressures because cold air is relatively dense, but how does a warm anticyclone produce high surface pressures? After all, in equal volumes, warm air is lighter than cold air. The answer is that the volumes are not equal. The high surface pressure of warm-core anticyclones results from essentially a larger volume of air over the high center, related to a higher tropopause. A layer 200 mb deep contains the same amount of mass regardless of its thickness. In a warm-core anticyclone individual layers have the greatest thickness at the center of the high, and the intensity of the high increases with altitude.

Cold-core anticyclones modify as they travel and may eventually become warm-core systems. As noted earlier, a cold-core anticyclone is actually a dome of relatively cold air, and as that air mass traverses land that is bare of snow, it is heated from below and moderates considerably, its pressure decreasing significantly. As a cold-core high drifts southeastward over the United States, air mass modification may be sufficient so that the pressure system eventually merges with the warm-core subtropical high over the Atlantic.

## ANTICYCLONE WEATHER

As noted in Chapter 9, an anticyclone is a fair-weather system. This is because surface winds blowing in a clockwise and outward pattern (in the Northern Hemisphere) induce subsidence of air over a broad area. Because subsiding air is compressionally warmed,

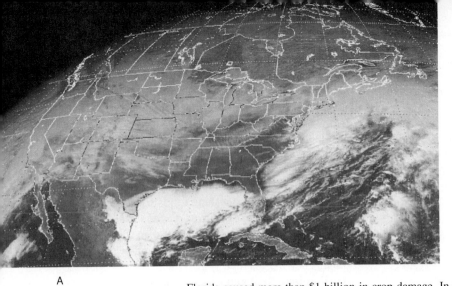

A

**FIGURE 11.20**

Visible satellite image of much of North America on 24 December 1983 (A) and the corresponding pattern of surface air pressure (B). Solid lines are isobars labeled in millibars. An unusually extensive arctic high-pressure system is centered over eastern Montana where surface pressure readings exceed 1060 mb. The frigid air mass associated with the high covers almost the entire United States except the Southwest. Early that morning, temperatures plunged to record low levels over the Plains, Midwest, and Deep South. Subfreezing temperatures in the citrus- and vegetable-growing areas of South Texas and South Florida caused more than $1 billion in crop damage. In the satellite image, much of the white in the northern United States is snow cover rather than clouds. The bright white over the Gulf is low cloudiness caused by cold air streaming over the relatively warm ocean water. (Dashed line off the East Coast is a trough.) [Photograph from NOAA/Satellite Data Services; map from NOAA]

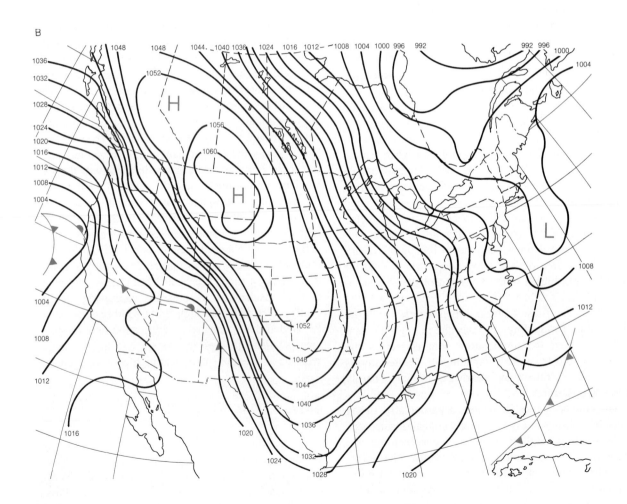

B

the relative humidity drops and clouds usually dissipate or fail to develop. In addition, as noted earlier, the horizontal pressure gradient is weak over a wide region about the center of an anticyclone; prevailing winds are therefore very light or calm. At night, clear skies and light winds favor intense radiational cooling and, in some instances, development of dew, frost, or fog (Chapter 7).

The horizontal pressure gradient strengthens away from the central region of an anticyclone, and so does the wind. With stronger winds, significant advection occurs. Typically, well to the east of the high center, northerly winds advect cold air southward, whereas to the west of the high center, southerly winds advect warm air northward. Air mas advection helps to increase the temperature contrast across the trough that separates highs and thereby influence frontogenesis.

An understanding of the basic circulation characteristics of an anticyclone helps us to anticipate the sequence of weather events as an anticyclone travels into and out of a midlatitude location. Consider what happens in winter as a cold anticyclone slides southeastward out of southern Canada and into the northeastern United States. The following sequence may take several days to a week, depending on the anticyclone's forward speed.

Ahead of the anticyclone, strong northwest winds bring a surge of cold continental polar or arctic air. Strong winds and falling temperatures produce low windchill temperatures and make more work for home furnaces. To the lee of the Great Lakes, heavy lake-effect snow showers break out (discussed in Chapter 12). Even hundreds of kilometers downwind of the lakes, instability showers* bring light accumulations of snow. However, as the center of the anticyclone drifts closer, winds slacken, skies clear, and nocturnal radiational cooling produces very low surface temperatures. Under these conditions, air temperatures dip to their lowest readings, especially if the ground is snow-covered. Then, as the anticyclone center moves away toward the southeast, winds again strengthen, but this time from the south, and warm air advection begins. The first sign of warm air advection is the appearance of high, thin cirrus clouds in the western sky.

*As cold air flows over the relatively warm surface of the lakes, it is warmed and moistened and its stability decreases. The rain or snow showers that develop can be advected for a long distance before their moisture supply is depleted and they dissipate.

In summer, a Canadian high-pressure system causes the same advection patterns as in winter except that the temperature contrast between air masses is considerably less. Air advected ahead of the high on northwesterly winds may not be much cooler than the air advected behind the high on southwesterly winds. Often, the most noticeable difference between the northwesterlies and southwesterlies in summer is a contrast in water vapor concentration. Air advected ahead of the high is often less humid, and therefore more comfortable, than the air advected behind the high.

The pattern of air mass advection associated with an anticyclone also applies to a ridge. Cold air advection usually occurs ahead (to the east) of a ridge, and warm air advection occurs behind (to the west of) a ridge.

The circulation of an anticyclone (or ridge) does not occur in isolation from that of a cyclone (or trough). The atmosphere is, after all, a continuous fluid, with anticyclones following cyclones and cyclones following anticyclones. In our illustration, therefore, northwest winds develop ahead of the high and on the back (west) side of a retreating low. The southerly airflow behind the high develops to the east of a low, which is approaching from the west. In both cases, winds are caused by horizontal pressure gradients that develop between migrating anticyclones and cyclones. We consider the circulation patterns of cyclones and anticyclones in more detail in the Mathematical Note, "Vorticity."

## Conclusions

We have now examined the features of the principal weathermakers of midlatitudes: air masses, fronts, cyclones, and anticyclones. We have seen how these synoptic-scale systems interact with one another and how they are linked to the planetary-scale circulation described in Chapter 10. Also recall from our discussion in Chapter 4 that synoptic-scale systems play an important role in poleward heat transport. In Chapter 12, we continue our analysis of atmospheric circulation by focusing on some special regional and local systems.

MATHEMATICAL NOTE

## Vorticity

We have been examining the circulation of air in planetary-scale and synoptic-scale weather systems. A powerful but challenging concept used in discussing such circulation patterns is vorticity. *Vorticity* is a measure of the rotation of an air parcel about an axis passing through an air parcel and the center of the Earth. Mathematically, this is expressed as

$$\zeta = \frac{\Delta v}{\Delta x} - \frac{\Delta u}{\Delta y}$$

where, $\zeta$ is vorticity, $u$ is the west-to-east component of the wind, $v$ is the south-to-north component, $x$ is positive toward the east, and $y$ is positive toward the north.

Examine a simple case where cyclonic flow follows a circle with a radius of curvature of 1000 km at a constant speed of 10 m/s (36 km/hr) (Figure 1). By the above formula,

$$\frac{\Delta v}{\Delta x} = \frac{20 \text{ m/s}}{2000 \text{ km}} \quad \text{and} \quad \frac{\Delta u}{\Delta y} = \frac{-20 \text{ m/s}}{2000 \text{ km}}$$

yielding a vorticity of $2 \times 10^{-5}$ s$^{-1}$. This example shows that counterclockwise rotation (in the Northern Hemisphere) has a positive value for vorticity. If the rotation were clockwise (in the Northern Hemisphere), the vorticity would be negative.

If a group of neighboring air parcels all have positive vorticity, that does not mean that each is spinning-about its own axis like a group of spinning figure skaters. More likely, the group of air parcels is spinning around a common axis, such as a low-pressure center. To extend the analogy, this would be like a group of figure skaters holding hands and skating in a large circle. If our circle of skaters were skating at constant speed, each would have the same vorticity. If the skaters then decreased the radius of their circle by half but kept skating at the same speed, they would be completing one trip around the circle in half the time and so would double their vorticity.

In the atmosphere, vorticity arises in two ways. As shown in the previous calculation, if wind follows a

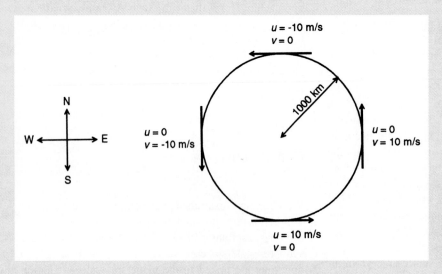

**FIGURE 1**
Counterclockwise circulation of air at constant speed in the horizontal plane.

curved path it will have either positive or negative vorticity. Flow around an anticyclone or ridge has anticyclonic or negative vorticity; flow around a cyclone or trough has cyclonic or positive vorticity. All other things being equal, vorticity values are greatest in the centers of cyclones or anticyclones, at the crest of a ridge, and in the base of a trough. These are all locations where the radius of curvature of the flow is relatively small, which indicates that air parcels are turning more sharply than in surrounding regions. Figure 2A depicts vorticity due to curvature. If pin-wheels were placed horizontally in the flow as shown, the one in the ridge would begin to rotate anticyclonically (indicating negative vorticity), whereas the one in the trough would begin to rotate cyclonically (indicating positive vorticity).

The second way to produce vorticity is to have a horizontal change in the wind, also referred to as *wind shear,* in the direction perpendicular to the direction of the wind. Figure 3 shows an idealized jet stream maximum, a feature typical of midlatitudes at altitudes near 10 km, with solid lines portraying isotachs (lines of constant wind speed). In this example, the maximum wind speed in the jet stream is 70 m/s (250 km/hr), with stronger wind shear across the flow and weaker wind shear in the direction of the flow. Westerly flow is present everywhere, so according to the above formula the vorticity is given by the change in the $u$-component of the wind in the $y$-direction. North of the windspeed maximum, $u$ decreases in the positive $y$-direction from 70 m/s (250 km/hr) to 40 m/s (145 km/hr) in 500 km, yielding a vorticity of $6 \times 10^{-5}$ s$^{-1}$. South of the wind speed maximum, $u$ increases in the positive $y$-direction from 40 m/s (145 km/hr) to 70 m/s in 500 km, yielding a vorticity of $-6 \times 10^{-5}$ s$^{-1}$. Because the wind shear north of the jet stream maximum yields a positive value for vorticity as does the flow around a cyclone, this type of wind shear is referred to as cyclonic shear. Similarly, the wind shear south of the jet stream maximum is said to be anticyclonic because it has negative vorticity.

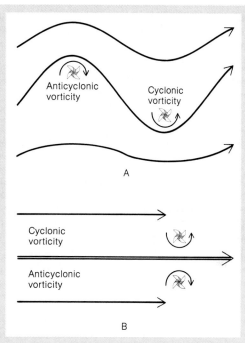

**FIGURE 2**
Vorticity is produced either by curvature of the wind (A) or by wind shear (B).

The relationship between shear and vorticity as a measure of local rotation can be seen from Figure 2B. If north is to the top of the page, the long arrow in the center depicts the greatest wind speed, whereas the arrows to the north or south depict slower wind speed. The pinwheel to the north would gain a counterclockwise rotation, which indicates positive or cyclonic vorticity, whereas the pinwheel to the south would gain a clockwise rotation, which indicates negative or anticyclonic vorticity. Note that there is no curvature in the flow in this example; the vorticity results entirely from the wind shear.

The previous discussion examines vorticity due to both curvature and wind shear, but leads to the conclusion that flow without either curvature or shear would have zero vorticity. This is because we are examining the concept relative to the Earth's surface. This type of vorticity is therefore more specifically called *relative vorticity*. However, the planet is also rotating about its own axis and so also exhibits vorticity. Thinking back to our discussion of the Coriolis effect (Chapter 9), we can understand that the Earth's vorticity, or *planetary vorticity* (about a vertical axis perpendicular to the Earth's surface), must be dependent on latitude and increase from zero at the equator to a maximum at the poles. As the globe turns, a telephone pole at the equator appears to an observer in space to sweep out a circle but does not spin about its own axis (zero planetary vorticity), whereas a telephone pole at the north pole spins with the Earth in a counterclockwise direction (maximum positive planetary vorticity). At the south pole, a telephone pole spins with the Earth in a clockwise direction (maximum negative planetary vorticity).

*Absolute vorticity* is defined as the sum of the relative vorticity plus the planetary vorticity. The concept of absolute vorticity helps explain why a trough fre-

quently develops in the westerlies just east of the Colorado Front Range (Figure 4). Assume that west of the mountains, winds aloft are zonal; that is, the wind blows directly from west to east, and the relative vorticity is zero. As the air is forced up the windward slopes, however, the tropopause acts as a lid on vertical motion, and the air is squeezed into a shallow layer, as discussed in the Special Topic "The Case of the Missing Storm." Then, as the wind descends the leeward slopes, the air layer increases in depth. From a theoretical perspective, the ratio of the absolute vorticity to the depth of this layer of air should remain constant for air parcels as they traverse the mountains. That is,

$$\frac{(\zeta + f)}{d} = \text{constant}$$

where $\zeta$ is the relative vorticity (defined earlier), $f$ is the planetary vorticity, and $d$ is the depth of the layer of air between the Earth's surface and the tropopause. Hence, as air parcels travel over the mountain barrier, the depth of the layer and absolute vorticity must alternately de-

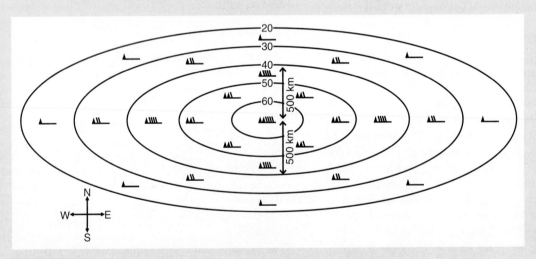

**FIGURE 3**
A jet stream maximum. Solid lines are isotachs, lines of equal wind speed.

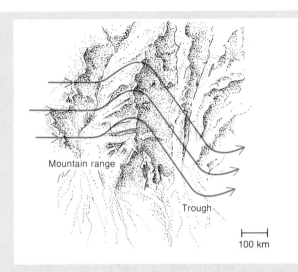

**FIGURE 4**

Development of a lee-of-the-mountain trough. Winds aloft are deflected first northeastward as the flow traverses the windward slopes of the mountain range, then southeastward as the winds blow over the leeward slopes, and then northeastward again over the western plains. Consequently, a ridge forms over the mountain range and a trough forms to the east of the mountain range.

crease and increase together to maintain a constant ratio.

On the windward slopes, a decrease in absolute vorticity (and air layer depth) means a decrease in relative vorticity, so that the initially zonal winds begin to exhibit negative or anticyclonic relative vorticity. That is, the wind turns clockwise, a ridge forms over the mountain range, and northwest winds descend the leeward slopes. On the leeward slopes, air layer depth increases so that the absolute vorticity also increases. However, the airflow is into lower latitudes so that the planetary vorticity decreases. The increase in relative vorticity prevails over the decrease in planetary vorticity, so that the northwest wind begins to exhibit positive or cyclonic vorticity. That is, the wind gradually shifts from northwest to southwest and a trough forms east of the mountain range. As winds continue toward the northeast over the western plains, there is little change in air layer depth, the planetary vorticity increases so that the relative vorticity must decrease (anticyclonic vorticity) thus completing the trough. This *lee-of-the-mountain trough* supplies upper-air support (horizontal divergence) for Colorado-track cyclones.

## Key Terms

air mass
continental tropical air
  (cT)
maritime tropical air (mT)
maritime polar air (mP)
continental polar air (cP)
arctic air (A)
air mass modification
Pacific air
front
stationary front
overrunning
warm front
frontal fog
cold front
squall line
back-door cold fronts
occluded front
occlusion
cold-type occlusion
warm-type occlusion

frontogenesis
frontolysis
cyclone
cyclogenesis
deepening
warm sector
wave cyclone
comma cloud
bent-back occlusion
triple point
cyclolysis
filling
Norwegian cyclone model
conveyor-belt model
cold-core cyclone
warm-core cyclone
anticyclone
cold-core anticyclone
polar high
arctic high
warm-core anticyclone

## Summary Statements

☐ Air masses are classified on the basis of their source region characteristics. Hence, there are four basic types of air masses: cold and dry, cold and humid, warm and dry, and warm and humid.

☐ Air masses form over land (continental) and the ocean (maritime), and at high latitudes (polar or arctic) and low latitudes (tropical). Arctic air is distinguished from polar air by its bitter cold.

☐ As an air mass travels from one place to another, it modifies. The degree of modification depends on air mass stability and the nature of the surface over which the air travels, namely, whether the surface is warmer or colder than the air mass.

☐ A front is a narrow zone of transition between air masses that contrast in temperature or water vapor concentration, or both. There are four types of fronts: stationary, cold, warm, and occluded. Occluded fronts may be of either the cold-type or the less common warm-type.

☐ Weather along or ahead of a cold front usually consists of a narrow band of clouds and brief rain or snow showers, or thunderstorms. Weather associated with warm or stationary fronts typically consists of a broad cloud and precipitation shield that may extend hundreds of kilometers ahead of the surface front.

☐ As a midlatitude cyclone progresses through its life cycle, it is supported and steered by the upper-level circulation toward the east and northeast. The storm typically begins as a wave along the polar front and deepens as surface air pressures continue to drop. Consequently, winds strengthen and frontal weather develops. The storm finally occludes as the faster moving cold front "catches up" with the slower-moving warm front and the upper-level low center becomes vertically stacked over the surface low center.

☐ A midlatitude cyclone's life cycle can vary significantly. A typical cyclone takes approximately one day in each of the stages leading up to maximum intensity. However, when conditions are favorable, the cyclone can progress from its birth to maximum intensity in 24 hr or less. Under unfavorable conditions, the cyclone might never progress to a fully occluded system, instead spending its lifespan as a weak cyclone rippling along a stationary front.

☐ The conveyor-belt conceptual model describes the three-dimensional structure of a cyclone in terms of three airstreams. The warm conveyor belt originates in the cyclone's warm sector near the Earth's surface and rises rapidly as it passes over the warm front and eventually joins the westerly flow aloft. The cold conveyor belt originates close to the surface east of the cyclone and north of the warm front in colder air. It passes under the warm conveyor belt and rises near the surface center, turning clockwise to joins the westerly flow aloft and forming the head of the comma cloud. The dry airstream originates aloft and descends both behind the cold front and near the cyclone center. The latter branch forms the dry slot in the comma cloud before rising to also join the westerly flow aloft.

☐ The track followed by a cyclone is critical to the type of weather experienced at a given locality. On the cold side of a storm track, winds back with time and there are no frontal passages. On the warm side of the storm track, winds veer with time and a warm front is followed by a cold front.

☐ Alberta cyclones occur most frequently, but Colorado and coastal storm tracks bring the heaviest precipitation.

☐ Warm-core cyclones (thermal lows) are stationary, have no fronts, and are associated with very hot, dry weather.

☐ Cold-core anticyclones are shallow systems that coincide with domes of continental polar or arctic air.

☐ Warm-core anticyclones, such as the semipermanent sub-

tropical highs, extend high into the troposphere and are accompanied by broad areas of subsiding, warm, dry air.

## Review Questions

1. What factors determine the humidity and temperature characteristics of an air mass?
2. Distinguish among the various air masses that regularly invade North America.
3. How do the properties of continental polar air change between summer and winter?
4. What causes air mass modification and in what ways do air masses modify?
5. Explain why continental polar air modifies more rapidly than does maritime tropical air. What is the significance of this difference for cold waves and heat waves?
6. Distinguish among a stationary front, a warm front, and a cold front.
7. What types of clouds are associated with a well-defined cold front?
8. Describe the sequence of clouds as a warm front approaches your locality.
9. Explain why the surface circulation about a cyclone favors development of fronts.
10. Why might the appearance of a halo about the sun or moon signal the approach of stormy weather?
11. Describe the stages in the life cycle of a midlatitude cyclone.
12. A wave cyclone has a cold side and a warm side. Please explain.
13. A winter wave cyclone tracks northeastward along the East Coast so that the storm center passes out to sea about 80 km to the east of Boston. Describe the wind shifts and air mass advection at Boston. Do the same for the case when the storm passes well to the west of Boston.
14. Describe the airstreams associated with a wave cyclone, according to the conveyor-belt model. What features of the cyclone's cloud pattern does each airstream explain?
15. An Alberta-track cyclone usually produces less precipitation than a Colorado-track cyclone. Why?
16. What happens when a cyclone occludes?
17. Explain why the site of principal cyclone activity is in the United States in winter and in Canada in summer.
18. What is the relationship between a cold-core anticyclone and a polar or arctic air mass?
19. Why is a warm-core cyclone stationary? Why are fronts not associated with a warm-core cyclone?

**20.** Explain why the circulation in a cold-core anticyclone favors development of frost or radiation fog.

## Questions for Critical Thinking

1. Describe how Pacific air modifies as it travels from west to east across the United States.
2. Why is the slope of a cold front steeper than the slope of a warm front?
3. In summer, we may experience passage of a cold front, and yet the air ahead of the front has about the same temperature as the air behind the front. In what sense is this a "cold" front?
4. Why is air mass stability an important factor in the type of weather that occurs along a front?
5. What is the significance of a lee-of-mountain trough for Ohio Valley weather?

## Selected Readings

BUSINGER, S. "Arctic Hurricanes," *American Scientist* 79 (Jan.-Feb. 1991):18–33. Describes development of intense small-scale polar cyclones that are structurally similar to hurricanes.

CARLSON, T. N. "Airflow Through Midlatitude Cyclones and the Comma Cloud Pattern," *Monthly Weather Review* 108 (1980):1498–1509. Describes the concept of conveyor belts as related to frontal cyclones.

DORR, B. "Bombs and Ultrabombs, Exploring Explosive Ocean Storms," *Weatherwise* 43, No. 2 (1990):76–83. Describes an ERICA mission into a rapidly developing extratropical cyclone.

HUGHES, P. "The Blizzard of '88," *Weatherwise* 40, No. 6 (1987):312–320. Describes one of history's most famous and disruptive East Coast snowstorms.

KOCIN, P. J., AND L. W. UCCELLINI. *Snowstorms Along the Northeastern Coast of the United States: 1955 to 1985.* Boston, MA: American Meteorological Society, 1990, 280 pp. Presents case studies of 20 of the most intense snowstorms to affect the urban Northeast.

MARTNER, B. E., ET AL. "Impacts of a Destructive and Well-Observed Cross-Country Winter Storm," *Bulletin of the American Meteorological Society* 73 (1992):169–172. Describes an intense cyclone that brought a variety of severe weather to 35 states and southeastern Canada in February 1990.

NAMIAS, J. "The History of Polar Front and Air Mass Concepts in the United States—An Eyewitness Account," *Bulletin of the American Meteorological Society* 64 (1983):734–755. Provides a historical perspective on a key concept of midlatitude weather systems.

# 12 Local and Regional Circulation Systems

*Let me snuff thee up, seabreeze!
and whinny in thy spray.*

HERMAN MELVILLE
*White Jacket*

Along a coastline such as this we can often anticipate that a cool sea breeze will bring relief on a hot summer afternoon. A sea breeze is one of a variety of local and regional circulation systems. [Photograph by J. M. Moran]

276

I N THIS chapter, we describe several specialized atmospheric circulation systems. Although these systems cover different spatial scales and persist for varying periods, each is strongly influenced by land-sea temperature contrasts, topography, or other features of the Earth's surface.

## Monsoons

Monsoon is derived from the Arabic word *(mausim)* for season. A **monsoon circulation** characterizes regions where seasonal reversals in prevailing winds cause wet summers and relatively dry winters. Although a weak monsoon develops over central and eastern North America, we are concerned here with the much more vigorous monsoon circulations over Africa and Asia, where 2 billion people depend on monsoon rains for their drinking water and food. Over much of India,

for example, monsoon rains (between June and September) account for 80% or more of total annual precipitation.

What causes monsoons? A complete explanation is not yet available, but monsoons are linked to seasonal shifts in the planetary-scale circulation, specifically north–south shifts of the intertropical convergence zone (ITCZ). As first proposed in 1686 by Edmund Halley, monsoons also depend on seasonal contrasts in the heating of land and sea. As we have seen, the ocean has a greater thermal stability than does the land (Chapter 3). Beginning in spring, relatively cool air over the ocean and relatively warm air over the land give rise to a horizontal air pressure gradient directed from sea to land,* that produces a flow of humid air inland. Over the land, intense solar heating triggers convection. Hot, humid air rises, and consequent expansional cooling leads to condensation, clouds, and rain. Release of latent heat intensifies the buoyant uplift, triggering even more rainfall. Aloft, the air spreads seaward and subsides over the relatively cool ocean surface, thus completing the monsoon circulation.

By early autumn, radiational cooling chills the land more than the adjacent sea, setting up a horizontal air pressure gradient directed from land to sea. Air subsides over the land, and dry surface winds sweep seaward. Air rises over the relatively warm sea surface; aloft, it drifts landward, completing the winter monsoon circulation. Over land, therefore, the summer monsoon is wet, and the winter monsoon is dry.

Monsoon winds have sufficiently long trajectories and persist long enough to be influenced by the Coriolis effect. As shown in Figure 12.1, January and July surface monsoon winds are deflected to the right in the Northern Hemisphere and to the left in the Southern Hemisphere.

Topography complicates the monsoon circulation and the distribution of rainfall. For example, the massive Tibetan plateau has a major influence on the Asian monsoon. The elevation of the plateau tops 4000 m (13,100 ft) over a large area. In winter, the westerly jet stream splits into two branches, one to the south and the other to the north of the plateau. The southerly branch steers cyclones that originate in the Mediterranean across northern India and brings significant pre-

---

*Recall that air pressure gradients are always directed from areas of high pressure toward areas of low pressure.

January

July

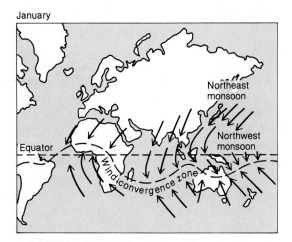
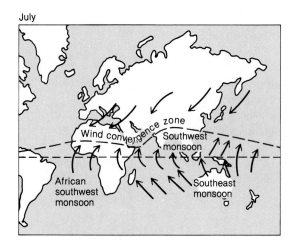

**FIGURE 12.1**
Surface air streams during the monsoon circulations of January and July are deflected by the Coriolis effect to the right in the Northern Hemisphere and to the left in the Southern Hemisphere. [After P. J. Webster, "Monsoons," *Scientific American* 245, No. 2 (1981):112. Copyright © 1981 by Scientific American, Inc. All rights reserved.]

cipitation to that region. Meanwhile, the rest of India experiences the dry monsoon. In spring, the southern branch weakens and by late May shifts northward over the plateau. It is not until this happens that the moist monsoon flow begins.

Monsoon rainfall is neither uniform nor continuous. On the contrary, the rainy season typically consists of a sequence of active and dormant phases. During a **monsoon active phase,** the weather is cloudy with frequent deluges of rain, but during a **monsoon dormant phase,** the weather is sunny and hot. The monsoon shifts from active to dormant phases as bands of heavy rainfall surge inland. Heavy rains first strike coastal areas and soak the ground. As the soil becomes saturated with water, more of the available solar radiation is used for evaporation and less for sensible heating. Coastal areas cool as a consequence, the uplift weakens, and skies partially clear. Meanwhile, the area of maximum heating, vigorous uplift, and heavy rains shifts inland. Back at the coastal areas, however, the hot sun eventually dries the soil, sensible heating intensifies, uplift strengthens, and the rains resume (monsoon active phase). This sequence of active and dormant phases is repeated about every 15 to 20 days during the wet monsoon.

Insolation, land and water distribution, and topography thus impose some regularity on the monsoon cir-

culation; that is, summers are wet and winters are dry. The planetary-scale circulation (especially shifts of the ITCZ) and the strength and distribution of convective activity, however, vary from one year to the next. These variations mean that the intensity and duration of monsoon rains change from year to year. As a consequence, drought is always possible in monsoon climates. The Special Topic, "Monsoon Failure and Drought in Sub-Saharan Africa," considers a particularly tragic example.

## Land and Sea (or Lake) Breezes

For those of us lucky enough to live near the ocean or a large lake, sea or lake breezes bring welcome respite from the oppressive heat of a summer afternoon.* On warm days, a weak synoptic-scale air pressure gradient allows a cool wind to sweep inland from the sea or from a large lake. Depending on the source, this refreshing wind is called either a **sea breeze** or a **lake breeze.** Both breezes are caused by differential heating of land and water.

*In tropical and subtropical localities, such as peninsular Florida, sea and lake breezes occur at any time of year.

SPECIAL TOPIC

# Monsoon Failure and Drought in Sub-Saharan Africa

Monsoon climates feature pronounced wet and dry seasons. The length of the rainy season and the quantity of rainfall vary both temporally, from year to year, and geographically. Perhaps nowhere in the world has this inherent variability of monsoon rains caused more human misery than in sub-Saharan Africa. A combination of low mean annual rainfall and high year-to-year variability in rainfall has meant frequent prolonged droughts and famine.

Sub-Saharan Africa (usually called the Sahel) is a transition zone between the Sahara Desert to the north and the humid savanna to the south. This zone stretches from the Atlantic ocean to the Red Sea and encompasses all or part of Mauritania, Senegal, Mali, Burkina Faso, Niger, Chad, and Sudan (Figure 1). Isolines of mean annual rainfall run essentially east–west and increase in magnitude southward, from about 200 mm (7.9 in.) at the Sahara's southern edge to more than 600 mm (23.6 in.) in the southern Sahel.

North–south shifts of the intertropical convergence-zone (ITCZ) govern the seasonality and amount of rainfall in the sub-Sahara. As the ITCZ follows the sun, its northward surge in spring triggers convective rains, and its southward shift in fall brings the rainy season to an end. However, as the northward decline in mean annual rainfall suggests, convection weakens and the rainy season shortens from south to north.

In a summer when the ITCZ does not penetrate as far north as usual, or arrives late, or moves southward

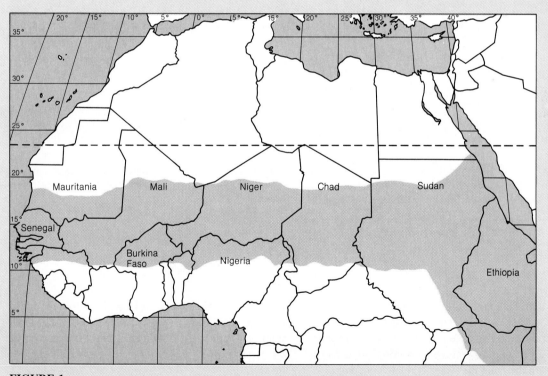

**FIGURE 1**
Sub-Saharan Africa is the transition zone between the Sahara desert to the north and humid savanna to the south.

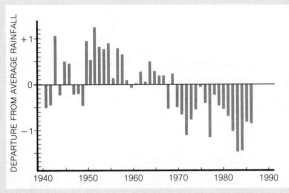

**FIGURE 2**

An index of the annual average April-to-October rainfall for 20 sub-Saharan localities situated between 11 and 18 degrees N and west of longitude 10 degrees E. Above-average rainfall plots above the horizontal line and below-average rainfall plots below the horizontal line. [From P. J. Lamb, "On the Development of Regional Climatic Scenarios for Policy-Oriented Climatic-Impact Assessment, *Bulletin of the American Meteorological Society* 68 (1987):1122]

early, rainfall is below average. A succession of such summers means drought. In this century, the Sahel has endured three major droughts, the most recent of which began in 1969 and persisted through 1990 albeit with a few brief interruptions (Figure 2). This latest dry episode followed a period of generally wet years from 1950 to 1968.

The cause of monsoon failure is much debated in the scientific community. The most recent episode of Sahelian drought is variously attributed to ENSO events (Chapter 10), the global increase in atmospheric $CO_2$ and other greenhouse gases (Chapter 3), or anoma-

lies in sea-surface temperatures of the tropical Atlantic. There is even some suggestion that modification of the land by human activities (overgrazing and deforestation) may contribute to aridity by altering the regional radiation balance.

*Desertification* is the name applied to the conversion of arable land to desert. Land degradation is often the consequence of some combination of climate change and overgrazing or other mismanagement of the land. For decades, scientists have been concerned that the northern Sahel is succumbing to desertification. Indeed, a recent study found that the average southern boundary of the Sahara desert in 1990 was about 130 km (81 mi) south of its 1980 location. This study relied upon satellite measurements of vegetation patterns to infer rainfall distribution and the position of the Sahara's southern edge.*

That same study, however, shows considerable year-to-year variability in the desert boundary, reflecting interannual fluctuations in rainfall. The desert boundary shifted northward 33 km (20 mi) in 1985–86, southward 55 km (34 mi) in 1987, northward 100 km (62 mi) in 1988, southward 44 km (27 mi) in 1989, and southward 33 km (20 mi) in 1990. Hence, we would be remiss in concluding that the Sahara is relentlessly expanding southward.

Although the cause of monsoon failure is unknown, in 1984 (the worst year of the drought) its devastating effects were strikingly brought home worldwide ontelevision news (Figure 3). Pictures of starving children, emaciated livestock, and withered crops spurred a massive international relief effort. The people of the

---

*See the article by Tucker et al. cited at the end of this chapter.

When both land and water are exposed to the same intensity of solar radiation, the land surface warms up more than the water surface. The relatively warm land heats the overlying air, thereby lowering air density. Compared to the land, the water is relatively cool, as is the air overlying the water. Consequently, as shown in Figure 12.2A, a local horizontal air pressure gradient develops between land and water, with the highest

pressure over the water surface. In response to this gradient, cool air sweeps inland. Aloft, continuity requires a return airflow directed from the land to the water, with air rising over the land and sinking over the water.

Sea (or lake) breezes are shallow systems, generally confined to the lowest kilometer of the troposphere. Typically, the breeze begins near the shoreline several

Sahel are particularly vulnerable to drought because 80% of them depend on agriculture for their livelihood. Prolonged drought forced them off lands that even in the best of times are marginal for crops and livestock. People migrated to urban areas in search of food and work, and many ended up in refugee camps.

In the Sahel and elsewhere, drought is a complex problem that involves more than just an anomalous atmospheric circulation pattern. Social, economic, and political factors also affect the impact of drought. As a further complication, in the sub-Sahara, the mix of these nonatmospheric factors varies between nations.

With famine a perennial risk in the Sahel, the U.S. Agency for International Development (USAID) is attempting to shorten the response time of international relief organizations. To this end, USAID operates the *Famine Early Warning System (FEWS)*. FEWS utilizes imagery from a NOAA polar-orbiting weather satellite (Chapter 16) to track crop and vegetation patterns over the sub-Sahara. The goal is to detect early signs of drought. This information is combined with various social data acquired from the ground (such as population movements and health problems) and is issued as monthly reports to African government officials and international planning agencies.

**FIGURE 3**
The human tragedy of the prolonged sub-Saharan drought. [Photograph courtesy of FAO]

hours after sunrise and gradually expands both inland and out over the body of water, reaching maximum strength by midafternoon. The inland extent of the breeze varies from only a few hundred meters to many tens of kilometers.

After sunset, the sea (or lake) breeze dies down. By late evening, however, surface winds begin to blow offshore as a **land breeze.** The change in wind direction is caused by a reversal in heat differential between land and water. At night, radiational cooling chills the land surface (and the air over the land) more than the water surface (and the air over the water). The land thus becomes cool relative to the water surface. A horizontal gradient in air density gives rise to a horizontal air pressure gradient that is directed from land to sea (or lake). A cool offshore breeze develops, along with a return

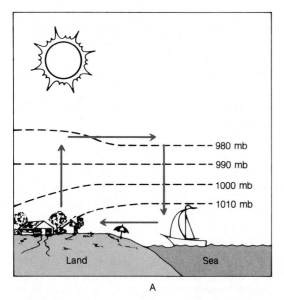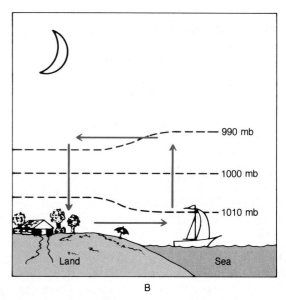

**FIGURE 12.2**

Vertical cross sections of (A) a sea (or lake) breeze and (B) a land breeze. Arrows indicate wind direction, and dashed lines are isobaric (constant-pressure) surfaces.

airflow aloft, and air subsides over the land and rises over the water (Figure 12.2B). A land breeze reaches maximum strength just before sunrise but tends to be weaker than a sea (or lake) breeze.

Land and sea (or lake) breezes typically develop and diminish so rapidly that they are not significantly influenced by the Earth's rotation. In some localities, however, the Coriolis effect is responsible for a gradual shift in the direction of a sea breeze through the course of a day, but the magnitude of the Coriolis effect is always weaker than the pressure gradient force.

The shoreline of Lake Michigan is often the site of lake breezes in spring and summer. During daylight hours from May through August, the surface waters of Lake Michigan, on average, are cooler than the surface of the adjacent land. If synoptic-scale winds are relatively weak, then a shallow mesoscale high-pressure system develops over the lake. As shown in Figure 12.3, surface winds diverge from the high center toward the shoreline. The leading edge of this surge of cool air, the lake breeze, forms a miniature cold front, which, by midafternoon, may push many kilometers inland. In so doing, the lake breeze front forces the warm air over the land to rise so that convective clouds may mark the leading edge of the lake breeze front.

**FIGURE 12.3**

On days when synoptic-scale winds are relatively weak and Lake Michigan surface waters are cooler than the adjacent land surface, a shallow mesoscale high develops over the lake. Surface winds diverge outward from the high and give rise to lake breezes. [From W. A. Lyons, "Some Effects of Lake Michigan upon Squall Lines and Summertime Convection." *Proceedings, Ninth Conference on Great Lakes Research.* Ann Arbor, MI: International Association for Great Lakes Research, 1966, p. 262.]

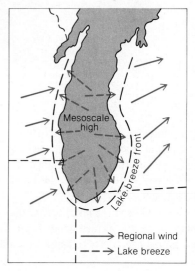

# Lake-Effect Snows

A highly localized fall of snow immediately downwind from an open lake is known as **lake-effect snow** (Figure 12.4). Typically, such snows extend inland only a few tens of kilometers and fall over such a small area that the event often is not detected by the regular network of weather stations. Residents of the affected area, however, may be swamped by snow. For example, in February 1976, Barnes Corner, New York, at the eastern end of Lake Ontario, received 137 cm (54 in.) of road-clogging snow in less than 24 hours. Such an extreme lake-effect snowfall is sometimes called a *snowburst*.

Lake-effect snow is most common in autumn and early winter when lake surface temperatures are still relatively mild. As early season outbreaks of cold (arctic) air stream over the lake, water readily evaporates and raises the vapor pressure of the lowest portion of the advecting air mass. (Recall from Chapter 7 that this same process may produce steam fog.) In addition, the warmer lake water heats the advecting cold air from below, reducing its stability and enhancing convection and cloud development. Often this is enough to trigger snowfall over the lake.

As the modified (milder, more humid, and less stable) air flows toward the lake's lee shore, the contrast in surface roughness between the lake and land becomes important. The rougher land surface slows onshore winds, and the consequent horizontal convergence (Chapter 9) induces ascent of air, further development of clouds, and lake-effect snow. The topography of the shore also affects the amount of precipitation: Hilly terrain forces greater uplift and heavier snowfalls than does flat terrain (Figure 12.5).

Ultimately, the frequency and intensity of lake-effect snows hinge on the degree of air mass modification, which, in turn, depends on (1) the temperature contrast between the mild lake surface and overlying cold air and (2) the over-water trajectory *(fetch)* of the advecting cold air. As the temperature contrast between lake surface and air increases, so too does the potential for lake-effect snow. Hence, to the lee of the Great Lakes, the bulk of lake-effect snow falls between mid-November and mid-January, the usual period of maxi-

**FIGURE 12.4**
Landsat imagery of a sharply defined snowfall pattern on Michigan's Upper Peninsula on 20 October 1972. Cold northwest winds blowing across Lake Superior induced lake-effect snows. [From U.S. Geological Survey, EROS Data Center, Sioux Falls, SD]

**FIGURE 12.5**
Some of the heaviest lake-effect snows fall in the rugged terrain of upper Michigan's Keweenaw Peninsula. This snow pole documents past seasonal snowfalls. [Photograph by Steven Dutch]

mum temperature difference between the lake surface and overlying air.

Cold air usually sweeps into the Great Lakes region on northwest winds. Hence, considering the maximum possible fetch, the greatest potential for substantial lake-effect snows is along the downwind southern and eastern shores of the Great Lakes (Figure 12.6). In these so-called **snowbelts,** lake-effect snow accounts for a substantial percentage of the total seasonal snowfall. Occasionally, however, lake-effect snow develops on the normally upwind western shores of the lakes. For example, an early winter cyclone tracking through the lower Great Lakes region may produce strong northeast onshore winds. Hence, lake-effect snows may supplement the snow produced by frontal overrunning, and the Milwaukee-Chicago metropolitan area, for example, ends up with paralyzing accumulations of snow.

The same snow-making mechanism also affects the valleys southeast of Utah's Great Salt Lake (Figure 12.7). Each year perhaps a half dozen significant lake-effect snowfalls occur in the Tooele and Salt Lake valleys, bowl-shaped depressions that slope up and away from Great Salt Lake. Both valleys parallel the northwest–southeast-trending long axis of the lake. Which of the two valleys receives the heavier snowfall depends on the wind direction. Cold air streaming over the lake from the northwest is moderated by contact with the relatively warm lake surface, and, downwind, the mountain valley topography forces horizontal convergence and uplift. Clouds billow upward and locally release bursts of heavy snow.

On 17–18 October 1984, for example, weather observers measured a record 46.7 cm (18.4 in.) of lake-effect snow at Salt Lake City International Airport, about 5 km (3.1 mi) southeast of the lake's southern shore. In the southeastern suburbs of Salt Lake City, snowfall topped 60 cm (24 in.) during the same period. Wet snow clung to trees (still in full leaf) and many tree limbs snapped under the weight of thick accumulations of snow. Falling limbs brought down power lines and caused an estimated $1 million in damage.

Great Salt Lake snowbursts are generally less intense than their Great Lakes counterparts. A major reason is the smaller area and shorter fetch of Great Salt Lake, so that air mass modification is less. The waters of Great Salt Lake cover an area that is only about 13% of that of Lake Ontario, the smallest of the five Great Lakes. The maximum fetch on Great Salt Lake is only 120 km (74 mi), whereas maximum fetches on the Great Lakes are many hundreds of kilometers.

**FIGURE 12.6**
Average yearly snowfall in the Great Lakes region (in inches). Note that the greatest snowfall totals occur downwind of the Great Lakes and are caused by frequent lake-effect snows.

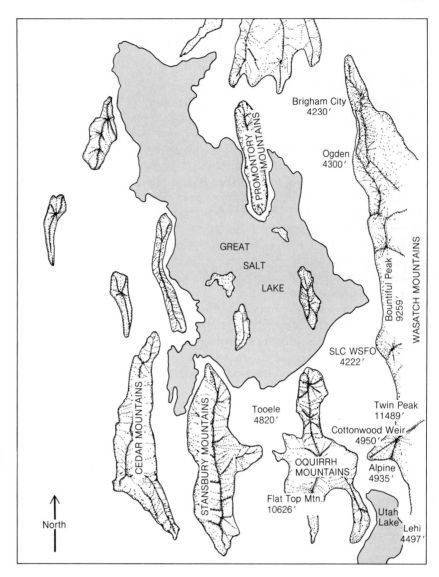

**FIGURE 12.7**
Great topographic relief contributes to the development of lake-effect snows in the Salt Lake Valley. (SLC WSFO = Salt Lake City Weather Service Forecast Office.) [From D. M. Carpenter, "Utah's Great Salt Lake—A Classic Lake-Effect Snowstorm," *Weatherwise* 38, No. 6 (1985):309]

Lake-effect snows also develop downwind of Canadian lakes (including Hudson Bay) north of the Great Lakes. Typically, snowfall peaks in autumn (October and November), prior to lake freeze-over. For example, Lake Winnipeg, in southern Manitoba, usually is ice covered by late November. Two weather stations downwind of the lake (Norway House and Berens River) usually receive more snowfall in November than in either December or January.

## Heat Island Circulation

An unsightly veil of dust, smoke, and haze may greet travelers approaching large cities. This dome of polluted air is the product of a convective circulation of air that concentrates pollutants over a city. The convective circulation, in turn, is related to the temperature contrast between a city and surrounding rural areas.

The average annual air temperature of a city typically is only slightly higher than that of the surrounding countryside, but on some days the temperature contrast may be 10 Celsius degrees (18 Fahrenheit degrees) or more. Consequently, snow melts faster and flowers bloom earlier in a city. This climatic effect is known as the **urban heat island.** Figure 12.8 illustrates the urban heat island of Washington, D.C., and Figure 12.9 is a satellite view of heat islands in southeastern New England.

Historical studies demonstrate that the intensity of an urban heat island increases with urbanization. For example, H. E. Landsberg found that as Boston grew, its heat island warmed significantly compared to its surroundings. Summers warmed at the rate of 2.6 Celsius degrees (4.7 Fahrenheit degrees) per 100 years, and winter warmed by 1.6 Celsius degrees (2.9 Fahrenheit degrees) per 100 years. As we will see in Chapter 19, warming due to urbanization complicates the search for large-scale trends in climate.

Several factors contribute to the development of an urban heat island. One is the relatively high concentration of heat sources in cities (for example, people, cars, industry, air conditioners, and furnaces). Because all the heat from every source eventually escapes to the

atmosphere, the air of a large city receives a considerable input of waste heat. On a very cold winter day in New York City, heat from urban sources may approach 100 W per m$^2$, about 8% of the solar constant.

The thermal properties of urban building materials also facilitate conduction of heat into city air. Concrete, asphalt, and brick conduct heat more readily than do the soil and vegetative cover of rural areas. The heat loss at night by infrared radiation to the atmosphere and to space is thus partially compensated for in cities by a release of heat from buildings and streets. In addition, during the day, a city's canyonlike terrain of tall buildings and narrow streets traps solar energy because of multiple reflections of sunlight by the sides of buildings.

A city's typically lower rate of evapotranspiration further accentuates the temperature contrast between city and countryside. Urban drainage (sewer) systems rapidly and efficiently remove most runoff from rain and snowmelt, so that less of the absorbed radiation is used for evaporation (latent heating), and more of the absorbed radiation is used to heat the ground and air directly (sensible heating). On the other hand, the mois-

**FIGURE 12.8**
Average winter low temperatures (in °F) in Washington, DC, and vicinity, illustrating the urban heat island effect. [From C. A. Woolum, "Notes from a Study of the Microclimatology of the Washington, DC, Area for the Winter and Spring Seasons," *Weatherwise* 17, No. 6 (1964):6]

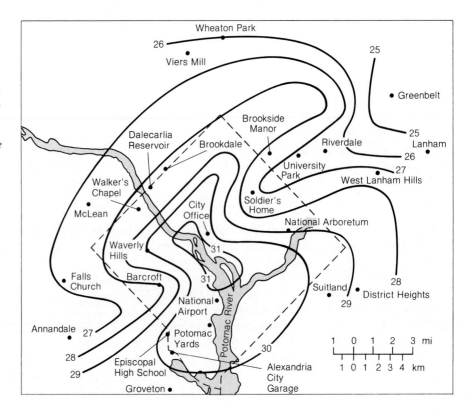

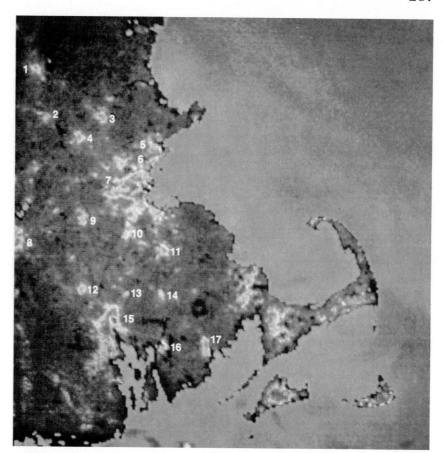

**FIGURE 12.9**
Color-enhanced IR satellite image shows 17 urban heat islands in southeastern New England. Number 7 is Metropolitan Boston. Temperature-color calibration: orange/yellow, 25.5 to 31.0 °C; light to dark green, 16.5 to 25.0 °C; light to dark blue, 6.0 to 16.0 °C. This image was taken by the NOAA-5 satellite on 23 May 1978 at 1400 GMT. Relatively warm areas indicated for Cape Cod and adjacent islands are likely due to land-use patterns or soil types. [Courtesy of Michael Matson, National Environmental Satellite Service, NOAA]

ter surfaces of the countryside (lakes, streams, soil, vegetation) increase the fraction of absorbed radiation that is used for latent heating. That is, the Bowen ratio is greater in the city than in the country.

An urban heat island is most likely to develop when synoptic-scale winds are weak. (Strong winds would mix city air and country air and diminish the tem-perature contrast.) Under such conditions, in some large metropolitan areas, the relative warmth of the city compared to its surroundings can promote a convective circulation of air, as shown in Figure 12.10. Relatively warm air rises over the city's center and is replaced by cooler, denser air converging from the countryside. The rising columns of air gather aerosols into a **dust dome**

**FIGURE 12.10**
The heat island of a large metropolitan area may give rise to a convective circulation that transports air pollutants into the city. [After W. P. Lowry, "The Climate of Cities," *Scientific American* 217, No. 2 (1967):20. Copyright © 1967 by Scientific American, Inc. All rights reserved.]

over the city. In this way, aerosols may become as much as a thousand times more concentrated over an urban-industrial area than in the air over the rural countryside. If regional winds strengthen to more than about 15 km (9 mi) per hour, the dust dome elongates downwind in the form of a **dust plume** and spreads the city's pollutants over the countryside. The Chicago dust plume, for example, is sometimes visible 240 km (150 mi) from its source.

## Katabatic Winds

Under the influence of gravity, a shallow mass of cold, dense air slides downhill. This **katabatic wind** usually originates in winter over extensive snow-covered plateaus or other highlands. Although adiabatic compression warms the air to some extent, the air is so cold to start with that these winds are still quite cold when they reach the lowlands.

Among the best-known katabatic winds are the **mistral,** which descends from the snow-capped Alps down the Rhone River Valley of France and into the Gulf of Lyons along the Mediterranean coast, and the **bora,** which originates in the high plateau region of Yugosla-via and cascades onto the narrow Dalmatian coastal plain along the Adriatic Sea. Both the mistral and the bora are winter phenomena.

Most katabatic winds are weak, usually under 10 km (6.2 mi) per hour. But, in some places, such as inlets of the coastal mountain ranges of British Columbia and Alaska, katabatic winds are channeled by narrow valleys, and this constricted flow sometimes accelerates the wind to potentially destructive speeds. Steep slopes also accelerate katabatic flow. Along the edge of the massive Greenland and Antarctic ice sheets, for example, katabatic winds frequently top 100 km (62 mi) per hour. The bora also may produce gusts of 50 to 100 km (31 to 62 mi) per hour.

## Chinook Winds

Like the katabatic wind, the chinook is a downslope breeze. But whereas the katabatic wind is cold and dry, the chinook wind is warm and dry.

A **chinook wind** develops when relatively mild air aloft is adiabatically compressed as it descends the leeward slopes of mountain ranges. For every 1000 m of descent, the air temperature rises about 10 Celsius de-

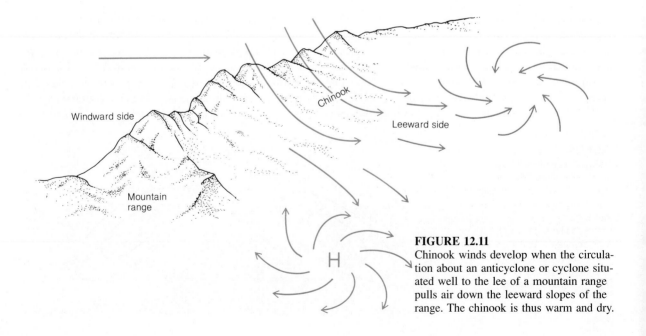

**FIGURE 12.11**
Chinook winds develop when the circulation about an anticyclone or cyclone situated well to the lee of a mountain range pulls air down the leeward slopes of the range. The chinook is thus warm and dry.

grees (the dry adiabatic lapse rate). Air that flows down the slopes of high mountain ranges such as the Rockies thus undergoes considerable warming.

Because the air is relatively warm, the chinook does not flow downslope under the influence of gravity, as does a katabatic wind; rather, the chinook wind is pulled downslope. Typically, a chinook develops when strong winds force a layer of stable air in the lower troposphere to ascend the windward slopes of a mountain range. When the air reaches the leeward slopes, its stability causes the air to return (descend) to its original altitude. Further descent of the air is caused by the larger scale circulation. For example, chinook winds descending the leeward slopes of the Rocky Mountain Front Range are drawn downslope by strong west winds associated with cyclones and anticyclones located well east of the mountains (Figure 12.11).

At the onset of a chinook, surface air temperatures often climb abruptly tens of degrees in response to compressional warming. For example, on 6 January 1966, at Pincher Creek, Alberta, a chinook sent the temperature soaring 21 Celsius degrees (38 Fahrenheit degrees) in only 4 minutes. An even more dramatic temperature surge was recorded at Spearfish, South Dakota, on 22 January 1943: The air temperature rose from $-20\ °C$ ($-4\ °F$) at 7:30 A.M. to $7\ °C$ ($45\ °F$) at 7:32 A.M., that is, 27 Celsius degrees (49 Fahrenheit de-

grees) in only 2 minutes! The sudden springlike warmth may just as quickly give way to severely cold conditions. For example, at the foot of the Rockies, a shift of synoptic-scale winds from west to north brings an abrupt end to the chinook and the return of polar or arctic air.

Chinook is a Native American word that, according to tradition, means *snow eater*. The term is appropriate because of the wind's catastrophic effect on a snow cover. As noted in Chapter 6, air ascending on the windward slopes loses much of its water vapor to condensation and deposition, that is, cloud formation. Then, on the leeward slopes, as the air descends and is compressionally warmed, the relative humidity drops dramatically. Because the chinook is both warm and very dry, a snow cover melts and vaporizes rapidly. It is not unusual, for example, for half a meter of snow to disappear in this way in only a few hours.

Chinook winds are gusty and, locally, may reach destructive velocities, especially along the foothills of the Front Range of the Rocky Mountains. At Boulder, Colorado, in the foothills just northwest of Denver (Figure 12.12), violent downslope winds, sometimes gusting to 160 km (100 mi) per hour or higher, unroof buildings and topple power poles. On average, the community experiences $1 million in property damage each year because of these destructive winds. Currently, re-

**FIGURE 12.12**
Boulder, Colorado, in the foothills of the Rockies, experiences some particularly violent downslope winds. [Photograph by J. M. Moran]

search is under way at the National Center for Atmospheric Research (in Boulder) to improve prediction of violent chinook winds.

Researchers do not agree on the precise mechanism that triggers destructive chinook winds, but according to one view, downslope winds are linked to the interaction of the Rockies with the planetary-scale westerlies. Recall from Chapter 7 that a north–south mountain range such as the Rockies deflects a strong westerly airflow into a **standing wave,** a stationary pattern of vertically oriented crests and troughs that stretches downwind of the mountain range and gives rise to mountain-wave clouds. The chinook is actually a segment of the wave that dips down the leeward slopes of the mountain range. In a violent chinook, very energetic turbulent eddies that are generated aloft are transported downward into the foothills so that surface winds become very strong and gusty.

Chinook-type winds are not restricted to the leeward slopes of the Rockies. A similar wind, called the **foehn,** flows into the Alpine valleys of Austria and Germany. The same type of warm, dry wind is drawn down the eastern slopes of the Andes in Argentina, where it is known as the **zonda.** In southern California, the notorious **Santa Ana wind** is another chinook wind that typically develops in autumn and winter. A strong high-pressure system centered over the Great Basin sends northeast winds over the southwestern United States, driving air downslope from the desert plateaus of Utah and Nevada, around the Sierra Nevada, and as far west as coastal southern California (Figure 12.13). The consequent adiabatic compression produces hot and dry, gusty winds that desiccate vegetation and contribute to outbreaks of forest and brush fires. Santa Ana winds sometimes gust to 130 to 145 km (80 to 90 mi) per hour and almost always cause some property damage.

## Desert Winds

Deserts typically are windy places primarily because of intense solar heating of the ground. In deserts, most of the absorbed solar radiation goes into sensible heating because little is used to evaporate the scarce water. In some spots, the midday temperature of the surface exceeds 55 °C (131 °F). Such a hot surface produces a

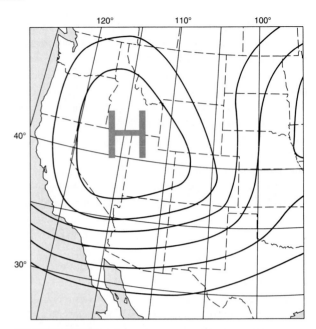

**FIGURE 12.13**
Schematic representation of the synoptic-scale weather pattern that favors development of Santa Ana winds over Southern California. Solid lines are isobars.

*superadiabatic*\* lapse rate in the lowest air layer. A superadiabatic lapse rate, in turn, means great instability (Chapter 6), vigorous convection, and gusty surface winds. The strength and gustiness of the wind vary with the intensity of solar radiation so that wind speeds and gustiness usually peak in the early afternoon and during the warmest months.

Dust devils are a common sight over flat, desert terrain (Figure 12.14). They develop during sunny days as a consequence of local variations in surface characteristics (albedo, moisture, topography) that give rise to localized hot spots. Air over hot spots is heated and rises and is replaced by converging surface winds. Shear in the horizontal wind causes the column of rising hot air to spin about a nearly vertical axis. The source of wind shear may be nearby obstacles that disrupt the horizontal wind or the overturning of air induced by extreme instability. In the process, dust is lifted off the ground, and the circulation is visible as a whirling mass of dust-laden air known as a **dust devil.** Note that unlike a tornado, a dust devil is not linked to a cloud.

\*A superadiabatic lapse rate is greater than 10 Celsius degrees per 1000 m and indicates absolute instability (Chapter 6).

**FIGURE 12.14**
A large dust devil at Silopio, Turkey. [Photograph by Steven Dutch]

Dust devils are microscale systems. The most common ones are small whirls less than 1.0 m in diameter that typically last less than 1 minute. On rare occasions, a dust devil may exceed 100 m (328 ft) in diameter and whirl about for 20 minutes or longer. Such dust devils may be visible to altitudes topping 900 m (3000 ft), but the rising air column may reach 4500 m (14,750 ft).

Most dust devils are too weak to cause serious property damage. The larger ones, however, are known to produce winds in excess of 73 km (45 mi) per hour and may cause damage. According to the National Weather Service (NWS), every year in New Mexico, several large dust devils cause substantial damage to mobile homes, travel trailers, and buildings under construction. In the spring of 1991, a powerful dust devil fortuitously passed over anemometers at the NWS office at Albuquerque and produced a wind gust of 113 km (70 mi) per hour.

Thunderstorms or migrating cyclones produce larger scale winds in deserts. Surface winds associated with these weather systems can give rise to either dust storms or sand storms, the difference between the two hinging on the size range of the loose surface sediment that is lifted by the wind. Dust consists of very small particles (less than 0.06 mm in diameter) that can be carried by the wind to great altitudes. Sand, which typically covers only a small fraction of desert terrain, consists of larger particles (0.06 to 2.0 mm in diameter) that are transported by winds within about a meter of the ground.

The strong, gusty downdraft of a thunderstorm generates one of the most spectacular dust storms, known as a **haboob.** In a desert, rains falling from thunderstorm clouds often evaporate completely in the dry air beneath cloud base and do not reach the ground. The thunderstorm downdraft, however, exits the thunderstorm as a surge of cool, gusty air. Dust picked up from the ground fills the air, severely restricting visibility, and the mass rolls along the ground as a huge ominous black cloud. A haboob may be more than 100 km (62 mi) wide and may reach altitudes of several kilometers. These dust storms are most common in the Sudan of North Africa but also occur in the American Southwest deserts.

Under some conditions, winds blow out of deserts and into quite different climatic regions. For example, in spring, when the subtropical anticyclones shift poleward, hot and dry winds stream northward from the Sahara Desert of North Africa, out over the Mediterranean Sea, and then into southern Europe. These desert winds are little modified after crossing the Mediterranean, except in the west, where winds approaching Spain have a longer fetch over water and consequently become more humid.

## Mountain and Valley Breezes

In summer, a localized circulation system may develop in wide, deep mountain valleys that face the sun (Figure 12.15). After winter snows have melted, bare valley walls absorb solar radiation and sensible heating

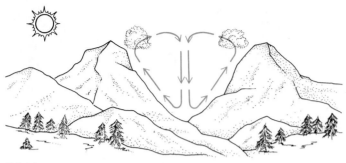

Valley breeze

Mountain breeze

**FIGURE 12.15**
A schematic representation of valley and mountain breeze circulation.

raises the temperature of the air in contact with the walls. Consequently, the air adjacent to the valley wall becomes warmer and less dense than air at the same altitude over the valley floor. Air over the valley sinks as air adjacent to the valley walls flows upslope as a **valley breeze.** The ascending valley breeze expands and cools and may trigger development of cumulus clouds near the summit (Figure 12.16).

**FIGURE 12.16**
Cumulus clouds developing over a mountain ridge as a consequence of the upslope flow of air associated with a valley breeze. [Photograph by J. M. Moran]

A valley breeze develops best between late morning and sunset. By midnight, the circulation reverses direction and continues until about sunrise. With clear skies, nocturnal radiational cooling chills the valley walls and the air in contact with the walls. Now the air adjacent to the valley walls is colder and denser than the air at the same altitude over the valley. Air over the valley ascends as the cold, gusty **mountain breeze** flows downslope. Cold air accumulates in the valley bottom where further radiational cooling may lead to formation of fog or low stratus clouds.

Mountain and valley breezes are most likely to develop during fair weather and when synoptic-scale winds are light or calm. Hence, these localized winds typically occur when mountainous regions are under the influence of a slow-moving anticyclone.

# Conclusions

In this chapter we described several local and regional circulation systems, and in subsequent chapters we will consider others. The domination of these systems by larger scale circulation is apparent; that is, the planetary- and synoptic-scale patterns set boundary conditions for any smaller scale circulation. Hence, these circulation systems are vulnerable to changes in large-scale patterns. In some cases, synoptic-scale winds reinforce mesoscale winds, as when regional winds blow in the same direction as a sea or lake breeze. In other cases, synoptic-scale winds overwhelm mesoscale winds, as when northerly winds sweep along the edge of the Rocky Mountains and eliminate the possibility of chinook winds. Even the large-scale monsoon winds are influenced by shifts of the ITCZ and the subtropical anticyclones.

## *Key Terms*

| | |
|---|---|
| monsoon circulation | lake breeze |
| monsoon active phase | land breeze |
| monsoon dormant phase | lake-effect snow |
| sea breeze | snowbelts |

| | |
|---|---|
| urban heat island | foehn |
| dust dome | zonda |
| dust plume | Santa Ana wind |
| katabatic wind | dust devil |
| mistral | haboob |
| bora | valley breeze |
| chinook wind | mountain breeze |
| standing wave | |

## *Summary Statements*

☐ Monsoons in tropical latitudes are the consequence of the interplay among several climatic features. Differences in solar heating of land and sea, and seasonal shifts in the planetary-scale circulation (especially the ITCZ) play important roles.

☐ Although summers are wet and winters are dry, the wet monsoon is punctuated by active and dormant phases. The intensity and duration of monsoon rains can vary significantly from one year to the next.

☐ When synoptic-scale winds are weak, a localized horizontal pressure gradient develops between land and sea (or lake). In response, winds blow onshore during the day (a sea or lake breeze) and offshore at night (land breeze).

☐ Lake-effect snows develop on the downwind (leeward) shores of the Great Lakes and Great Salt Lake when cold air streams across a relatively mild lake surface. Fetch and lake shore topography play important roles in governing the intensity of lake-effect snows.

☐ Human activities influence the climate of large cities by altering the local radiation balance. This gives rise to an urban heat island and, in large metropolitan areas, perhaps to a convective circulation that concentrates air pollutants in a dust dome over the city.

☐ A katabatic wind is a gravity-driven downslope flow of dense, cold air. A chinook wind, on the other hand, consists of compressionally warmed air that is pulled down the leeward slopes of mountains by the circulation about cyclones or anticyclones.

☐ Intense solar heating in desert terrain results in a steep lapse rate in the lowest air layer. Dust devils may develop in such unstable air. In addition, strong winds associated with thunderstorms or migrating cyclones may produce dust storms or sand storms.

☐ Mountain and valley breezes develop in summer in deep, wide mountain valleys that face the sun. A valley breeze is an upslope wind that forms during the day and a mountain breeze is a downslope wind that forms at night.

## Review Questions

1. Distinguish between the wet and dry monsoon circulations.
2. In what areas of the globe are monsoon circulations dominant features of the climate?
3. Why does the land surface warm up more than the sea surface in response to the same intensity of solar radiation?
4. Explain the shifts between the *active* and *dormant* phases of a wet monsoon.
5. How is the monsoon circulation linked to seasonal shifts of the intertropical convergence zone (ITCZ) and the subtropical anticyclones?
6. The Coriolis effect influences the monsoon circulation but not most land and sea breezes. Why the difference?
7. Why does a sea breeze develop on some summer days but not on others?
8. Speculate on how topography might affect the inland extent of a sea breeze or lake breeze.
9. Sea breezes, lake breezes, and land breezes are driven by a horizontal air pressure gradient. What causes this pressure gradient to develop?
10. Describe the surface air circulation about a mesoscale high. Why does it form only during the day.
11. Explain why a lake breeze may develop along the Lake Ontario shoreline in summer but not in winter.
12. Explain why the appearance of steam fog just off the western shore of Lake Michigan suggests that it may be snowing on the lake's eastern shores.
13. Identify and describe the various factors that contribute to the development of an urban heat island.
14. What steps might be taken by urban planners to reduce the intensity of urban heat islands?
15. What force drives a katabatic wind?
16. Why is a chinook wind both dry and warm?
17. Compare a katabatic wind with a chinook wind.
18. Describe the development of a dust devil.
19. Explain how a dust storm forms in a desert.
20. Distinguish between a mountain breeze and a valley breeze. What special conditions are required for development of a mountain breeze and a valley breeze?

## Questions for Critical Thinking

1. Speculate on the type of weather that might develop along a sea breeze front.
2. How does the Bowen ratio of an urban area compare to that of the surrounding countryside?
3. A mountain valley is oriented so that at midday, one slope is in bright sunshine and the other is in shade. If synoptic-scale winds are light, describe the circulation that develops in the valley.
4. Would you expect a monsoon circulation over North America? Support your answer.
5. In view of your understanding of atmospheric stability, explain the following observation: In spring, in the Great Lakes region, cumulus clouds develop over the land but not over the adjacent lake waters.

## Selected Readings

CARPENTER, D. M. "Utah's Great Salt Lake—A Classic Lake Effect Snowstorm," *Weatherwise* 38, No. 6 (1985): 309–311. Summarizes conditions required for lake-effect snows to the lee of Great Salt Lake.

EICHENLAUB, V. *Weather and Climate of the Great Lakes Region.* Notre Dame, IN: University of Notre Dame Press, 1979, 335 pp. Presents a lucid discussion of local circulation systems of the Great Lakes region.

ELLIS, W. S. "Africa's Sahel, The Stricken Land," *National Geographic* 172, No.2 (1987):140–179. Vividly portrays the plight of people trying to survive the erratic monsoon.

GLANTZ, M. H. "Drought in Africa." *Scientific American* 256, No. 6 (1987):34–40. Argues that drought in sub-Saharan Africa entails social, economic, and political as well as meteorological aspects.

KERR, R. A. "Chinook Winds Resemble Water Flowing over a Rock," *Science* 231 (1986):1244–1245. Summarizes possible mechanisms responsible for destructive chinook winds.

MEYER, W. B. "Urban Heat Island and Urban Health: Early American Perspectives," *Professional Geographer* 43, No. 1 (1991):38–48. Demonstrates that the temperature contrast between a city and its surroundings was understood in America as far back as the mid-1700s.

PENNISI, E. "Dancing Dust," *Science News* 142 (1992):218–220. Summarizes recent research on wind erosion of soil and sediment.

REIFSNYDER, W. E. *Weathering the Wilderness.* San Francisco, CA: Sierra Club Books, 1980, 276 pp. Includes an informative chapter on mountain meteorology.

TUCKER, C. J., H. E. DREGNE, AND W. W. NEWCOMB. "Expansion and Contraction of the Sahara Desert from 1980 to 1990," *Science* 253 (1991):299–301. Describes how a satellite-derived vegetation index was used to plot the north/south shifts of the southern edge of the Sahara.

WEBSTER, P. J. "Monsoons," *Scientific American* 245, No. 2 (1981):109–118. Considers in some detail the energetics involved in monsoon circulations.

WOODCOCK, A. "Dust Devils in the Desert," *Weatherwise* 44, No. 4 (1991):39–41. Describes the development and characteristics of dust devils.

# 13 Thunderstorms

Blow, winds, and crack your
cheeks; rage, blow.
You cataracts and hurricanes.
spout
Till you have drench'd our
steeples, drown'd the cocks.
You sulph'rous and thought-
executing fires,
Vaunt-couriers of oak-cleaving
thunderbolts,
Singe my white head. And
thou, all-shaking thunder,
Strike flat the thick rotundity
o' th' world
Crack nature's moulds, all
germens spill at once,
That makes ingrateful man.

WILLIAM SHAKESPEARE
*King Lear*

Lightning is one of many hazards
associated with thunderstorms.
[Photograph by Arjen and Jerrine
Verkaik/SKYART]

296

M OST OF US are familiar with typical thunderstorm weather—the blackening sky and abrupt freshening of wind followed by bursts of torrential rain, flashes of lightning, and rumbles of thunder. Often the cool breezes and rains bring us welcome relief on hot, muggy summer afternoons. For a farmer whose crops are wilting under the summer sun, the rains may be an economic lifesaver. Some thunderstorms become violent, however, and wreak havoc. Lightning starts fires, strong winds level trees and buildings, heavy rains cause flooding, and, in some cases, thunderstorms spawn destructive hail or tornadoes.

At this moment, an estimated 2000 thunderstorms are in progress somewhere on Earth, mostly in tropical and subtropical latitudes. In this chapter, we discuss the life cycle and characteristics of thunderstorms and hazards posed by them. In the next chapter, we describe the most destructive of thunderstorm progeny, the tornado.

# Thunderstorm Life Cycle

A **thunderstorm** is a mesoscale weather system, that is, it affects a relatively small area and is short-lived. A thunderstorm is the product of vigorous convection that extends deep into the troposphere, sometimes reaching the tropopause or even higher. Upward-surging air currents are made visible by billowing cauliflower-shaped cumuliform clouds, as shown in Figure 13.1. A thunderstorm consists of one or more convection cells, each of which goes through a life cycle that is divided into three stages: cumulus, mature, and dissipating (Figure 13.2).

## CUMULUS STAGE

If atmospheric conditions are favorable, cumulus clouds build both vertically and laterally. This is the initial or **cumulus stage** of thunderstorm formation. Over a period of perhaps 15 minutes, the tops of cumulus clouds surge upward to altitudes of 8000 to 10,000 m (26,000 to 33,000 ft). At the same time, neighboring cumulus clouds merge so that by the end of the cumulus stage, the storm's lateral dimension may be 10 to 15 km (6 to 9 mi).

Recall from Chapter 3 that convection currents begin when heat is conducted from the relatively warm surface of the Earth to cooler air that overlies the surface. We can visualize the ascending branch of a convection current as a continuous stream of bubbles (or parcels) of warm, unsaturated air. The rising bubbles cool at the dry adiabatic lapse rate (10 Celsius degrees per 1000 m) until they reach the **convective condensation level (CCL),** where water vapor condenses and cumulus clouds start forming. The more humid the air is to begin with, the less the expansional cooling that is needed for the bubbles to achieve saturation and the lower is the base of cumulus clouds. Hence, the base of cumulus clouds usually is lower in Florida, where relative humidities are high, than in New Mexico, where relative humidities are low.

Latent heat that is released during condensation adds to the buoyancy of the saturated bubbles, and they surge upward while cooling at the moist adiabatic lapse

A          B

**FIGURE 13.1**

As convection currents surge upward within the troposphere, (A) cumulus clouds begin to show vertical development. In some instances, (B) cumulus clouds merge and build vertically into cumulus congestus and eventually cumulonimbus (thunderstorm) clouds. [Photograph by J. M. Moran]

rate (averaging 6 Celsius degrees per 1000 m). Convective bubbles continue ascending as long as they are warmer, and thus less dense, than the surrounding air, that is, as long as the ambient air is unstable (Chapter 6). Some of the saturated bubbles surge through the cloud top and evaporate in the relatively dry air above the cloud. As a consequence, the water vapor concentration of the air above the cloud increases. Because the air above the cloud is now more humid, subsequent bubbles are able to ascend higher before evaporating. As this process is repeated, the cumulus cloud billows upward. This bil-

lowing character of cumulus cloud growth is evident in Figure 13.1; at this stage of vertical development, the cloud is called **cumulus congestus.** If vertical growth continues, the cumulus congestus cloud builds into a **cumulonimbus** cloud, a thunderstorm cloud.

During the cumulus stage of the thunderstorm life cycle, saturated air streams upward throughout the cell as an *updraft*. The updraft is strong enough to keep water droplets and ice crystals suspended in the upper reaches of the cloud. For this reason, precipitation does not occur during the cumulus stage.

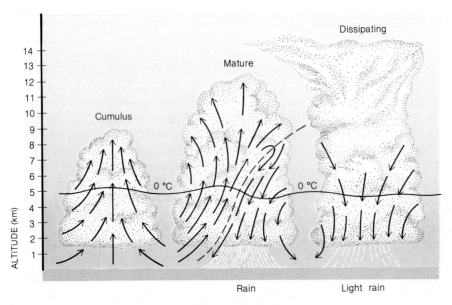

**FIGURE 13.2**
A thunderstorm's life cycle consists of cumulus, mature, and dissipating stages.

## MATURE STAGE

By convention, the cumulus stage ends and the **mature stage** begins when precipitation reaches the Earth's surface. Typically, this stage lasts for 15 to 30 minutes. The cumulative weight of water droplets and ice crystals eventually becomes so great that they are no longer supported by the updraft. Rain, ice pellets, and snow descend through the cloud and drag the adjacent air downward, creating a strong *downdraft* alongside the updraft. At the same time, unsaturated air at the edge of the cloud is drawn into the cloud, a process known as **entrainment.** (Actually, entrainment continues throughout the life cycle of a thunderstorm cell.) The entrained air mixes with the cloudy (saturated) air, causing some of the water droplets and ice crystals to vaporize. The consequent evaporative cooling weakens the buoyant uplift and strengthens the downdraft.

The downdraft exits the base of the cloud and spreads out along the ground, well in advance of the parent thunderstorm cell, as a mass of cool, gusty air. The downdraft is relatively cool because of evaporative cooling beneath the cloud base that offsets to some extent compressional warming of the downdraft. At the surface, the arc-shaped leading edge of downdraft air resembles a miniature cold front and is called a **gust front.** Convective clouds sometimes develop as a consequence of uplift along the gust front and may trigger formation of secondary thunderstorm cells tens of kilometers ahead of the parent cell.

Ominous-appearing low clouds are sometimes associated with thunderstorm gust fronts. A **roll cloud** is an elongated cylindrical dark cloud that appears to rotate slowly about its horizontal axis. The roll cloud occurs behind the gust front and beneath, but detached from, a cumulonimbus cloud. How this cloud forms is not fully understood. Although appearances might suggest otherwise, roll clouds are seldom accompanied by severe weather. This is *not* the case for shelf clouds.

A **shelf cloud** is a low, elongated cloud that is wedge-shaped with a flat base (Figure 13.3). This cloud appears at the edge of a gust front and beneath and attached to a cumulonimbus cloud. A shelf cloud is thought to develop as a consequence of uplift of stable warm and humid air along the gust front. Damaging

**FIGURE 13.3**
A shelf cloud such as this one may develop along a thunderstorm gust front. Often shelf clouds are accompanied by strong and gusty surface winds and may be associated with a severe thunderstorm. [Photograph by Arjen and Jerrine Verkaik/SKYART]

surface winds may occur under a shelf cloud, and sometimes this cloud is associated with a severe thunderstorm.

A thunderstorm cell attains its maximum intensity toward the end of the mature stage. This is when rain is heaviest, lightning bolts are most numerous, and hail, strong surface winds, and even tornadoes may develop. Cloud tops can exceed an altitude of 18,000 m (59,000 ft). Strong winds at such altitudes distort the cloud top into the anvil shape shown in Figure 13.4. The flat top of the anvil indicates that convection currents have reached the extremely stable air of the tropopause. Only in severe thunderstorms will convection currents overshoot this altitude, causing clouds to billow into the lower stratosphere before collapsing back into the troposphere. Temperatures within the upper portion of the cloud are so low that the anvil is composed exclusively of ice crystals, which give it a fibrous appearance.

Viewed from space by satellite, clusters of mature thunderstorm cells appear as bright white blotches, such as those over central Texas and northeast Oklahoma in Figure 13.5. The brightness of these clusters

**FIGURE 13.4**
When the billowing cumulonimbus cloud reaches the tropopause, it spreads out and forms a flat anvil top. [Photograph by Arjen and Jerrine Verkaik/SKYART]

is due to the high albedo (reflectivity) of the cloud tops (Chapter 2). Much of the solar radiation that penetrates cumulonimbus clouds is absorbed, so that the sunlight emerging at the cloud base is weakened considerably. For this reason, from our perspective on the Earth's surface, the daytime sky darkens as a thunderstorm approaches.

### DISSIPATING STAGE

As precipitation spreads throughout the thunderstorm cell, so does the downdraft, heralding the demise of the cell. During the **dissipating stage,** subsiding air replaces the updraft throughout the cloud. Adiabatic compression warms the subsiding air, the relative humidity drops, precipitation tapers off and ends, and convective clouds gradually vaporize.

Typically, a thunderstorm cell completes its life cycle in 30 minutes to an hour, but sometimes

**FIGURE 13.5**
In this visible satellite image, clusters of intense thunderstorm cells appear as bright white blotches over Texas and Northeast Oklahoma. [NOAA satellite photograph]

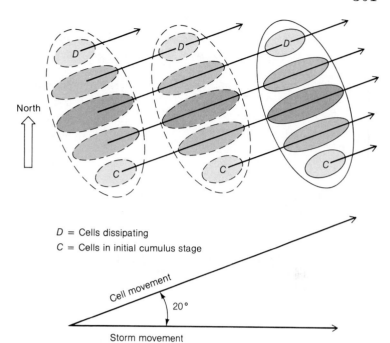

**FIGURE 13.6**
In this idealized situation, individual thunderstorm cells travel at about 20 degrees to the direction of movement of the multicellular thunderstorm. As they move, the individual cells progress through their life cycle. [After K. A. Browning and F. H. Ludlam, "Radar Analysis of a Hailstorm," *Technical Note* No. 5, Meteorology Department, Imperial College, London]

D = Cells dissipating
C = Cells in initial cumulus stage

lightning, thunder, and bursts of heavy rain persist for many hours. This is because a thunderstorm usually consists of a cluster of cells. Each cell may be at a different stage in its life cycle, and new cells form and old cells dissipate continuously. A succession of many cells is thus responsible for prolonged periods of thunderstorm weather. Although a locality may be in the direct path of a distant, intense cell, the relatively brief life span of an individual cell means that the severe weather may dissipate long before reaching that locality.

The multicellular characteristic of most thunderstorms complicates the motion of the weather system. A thunderstorm may track at some angle to the paths of its constituent cells. For example, as illustrated schematically in Figure 13.6, a thunderstorm tracks from west to east, whereas its five component cells head off toward the northeast. In this idealized case, new cells form in the southern sector of the storm, and old cells dissipate in the northern sector.

## Thunderstorm Genesis

Most thunderstorms develop within masses of warm, humid maritime tropical *(mT)* air when the air mass is destabilized. Uplift is the key to destabilizing *mT* air. Because maritime tropical air is usually conditionally stable, it becomes unstable (buoyant) only when lifted to the condensation level. Recall from Chapter 6 that conditional stability means that the ambient air is stable for unsaturated (clear) air parcels and unstable for saturated (cloudy) air parcels.

If the air is very humid to begin with, relatively little lifting (that is, expansional cooling) is needed for the *mT* air to reach saturation and for cumuliform clouds to develop. Recall that air is lifted (1) along fronts, (2) up the slopes of mountains, (3) via horizontal convergence of air at the surface, or (4) through intense solar heating of the ground. Any one or combination of these mechanisms may be sufficient to destabilize a mass of maritime tropical air. Furthermore, cold air advection aloft and/or warm air advection at the surface enhances the potential instability of *mT* air. Either of these processes increases the air temperature lapse rate, as shown in Figure 13.7, and reduces ambient air stability.

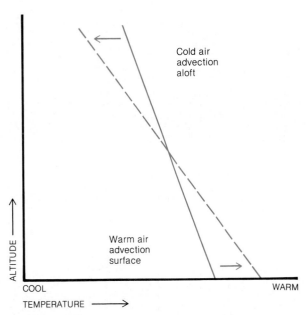

**FIGURE 13.7**
Cold air advection aloft, warm air advection near the surface, or a combination of the two increases the air temperature lapse rate and thereby destabilizes the air. Less stable
air, in turn, increases the likelihood of convection and thunderstorm development.

Outside of the tropics, thunderstorms are classified
as air mass thunderstorms, frontal thunderstorms, or
mesoscale convective complexes, depending on the
specific triggering mechanisms.

## AIR MASS AND FRONTAL TYPES

Both air mass and frontal thunderstorms are usually associated with the warm, humid air occupying
the southeast sector of a mature midlatitude cyclone
(Chapter 11). Convergence of surface winds and the
consequent upward motion of air associated with the
cyclone contribute to convection and thunderstorm
genesis.

An **air mass thunderstorm** seems to pop up
randomly almost anywhere within a mass of warm,
humid air.* Usually, they are relatively weak systems.

*In reality, air mass thunderstorms usually occur along some
subtle boundary within an air mass. The boundary, for example, may
represent the leading edge of the outflow from a distant thunderstorm.

Because they are caused by convection currents
driven by intense solar heating, most air mass thunderstorms develop in the afternoon, during the warmest hours of the day. A noteworthy exception to this
general rule is in the Missouri River Valley and adjacent portions of the upper Mississippi River Valley,
where air mass thunderstorms are more frequent at
night than during the day. Several explanations have
been proposed for this nocturnal thunderstorm maximum. One idea centers on the possible role of a low-
level jet stream of maritime tropical air that flows
northward up the Mississippi River Valley. This jet
stream strengthens at night and causes warm air advection at low levels that destabilizes the air and spurs the
buildup of cumuliform clouds.

As the name implies, a **frontal thunderstorm** is
associated with uplift along a frontal surface. Most
are triggered by vigorous uplift of maritime tropical air along or ahead of a well-defined cold front.
If the advecting warm air is unstable, however,
thunderstorms may break out in the overrunning
zone ahead of a surface warm front. In winter, in
northern regions, such thunderstorms sometimes produce snow.

Frontal thunderstorms are generally more energetic
than air mass thunderstorms because the synoptic-scale
circulation contributes more to their development. Usually, they parallel the surface front, and because frontal activity typically persists for many days, these thunderstorms may develop at any time of day or night. In
some instances, a line of thunderstorms forms perhaps
100 to 300 km (60 to 180 mi) in advance of a sharply
defined cold front. These thunderstorms are aligned
parallel to the cold front as a **squall line** and are often
severe.

## MESOSCALE CONVECTIVE COMPLEX

Satellite imagery reveals a third mode of thunderstorm occurrence in addition to the air mass and frontal types. A **mesoscale convective complex (MCC)** is
a nearly circular cluster of many interacting thunderstorms covering an area that may be a thousand times
larger than that of an individual air mass thunderstorm
(Figure 13.8). In fact, it is not unusual for a single MCC
to cover an area equal to that of the state of Iowa. These

**FIGURE 13.8**
An enhanced infrared satellite image taken on 20 July 1977 at 3:30 A.M. E.D.T. showing a
mesoscale convective complex (MCC) over Pennsylvania that resulted in a flash flood at Johns-
town. [Photograph courtesy of NOAA National Satellite, Data, and Information Service]

weather systems are primarily warm season (March through September) phenomena that generally develop at night and occur chiefly over the eastern two-thirds of the United States, where more than 50 may be expected in a single season.

A mesoscale convective complex is not associated with a front and usually develops under conditions of weak synoptic-scale flow. New thunderstorms develop continuously within an MCC so that the life expectancy of the system is at least 6 hours and often 12 to 24 hours. The longevity of an MCC coupled with its typically slow movement (15 to 30 km per hour) means that rainfall is widespread and substantial. MCCs account for perhaps 80% of growing season rainfall in the Great Plains and Midwest. An MCC also has the potential of producing severe weather. For example, NOAA reports that the 59 MCCs that occurred between February and October of 1985 produced 20 tornadoes and 16 flash floods that claimed 14 lives.

## Geographical Distribution

Convective activity is dependent on intense solar heating of the Earth's surface so that thunderstorms occur with the greatest frequency in the continental interiors of tropical latitudes. The steamy Amazon Basin of Brazil, the Congo Basin of equatorial Africa, and the islands of Indonesia have the greatest frequency of thunderstorms in the world. These localities are likely to experience thunderstorm activity on at least 100 days of the year. Because the surfaces of large bodies of water do not warm as much as adjacent land surfaces, thunderstorms are far less frequent over tropical oceans.

In some portions of subtropical and tropical latitudes, intense solar heating combines with converging surface winds to trigger thunderstorm development. As we saw in Chapter 10, this combination characterizes the ITCZ, which is actually a discontinuous line of thunderstorms that surges north and south seasonally with the sun.

Central Florida is the most frequent site of thunderstorms in North America (Figure 13.9). In some interior localities of that state, thunderstorms can be expected on more than 90 days of the year. The Florida thunderstorm maximum is due to the sea breeze circulation that characterizes both the east and west coasts of the Florida peninsula. Convergence of the sea breezes over the interior (Figure 13.10) induces upward motion of maritime tropical air and development of cumulonimbus clouds.

Portions of the Rocky Mountain Front Range rank second to Florida in thunderstorm frequency in North

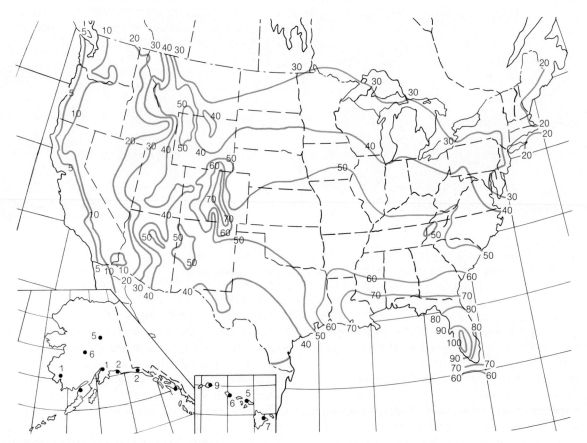

**FIGURE 13.9**
Thunderstorm frequency across the United States in average number of days per year. [NOAA data]

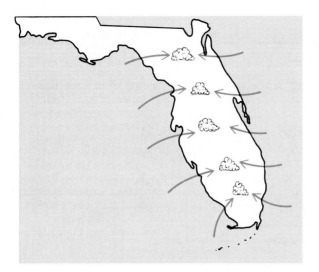

**FIGURE 13.10**
The relatively high frequency of thunderstorms in peninsular Florida is linked to the horizontal convergence of sea breezes blowing inland from both the Gulf and East coasts.

America. On average, more than 60 thunderstorm days occur per year in a band from southeastern Wyoming southward through central Colorado and into north-central New Mexico. This high thunderstorm frequency is linked to differences in heating arising from variations in topography.

Mountain slopes facing the sun absorb the direct rays of the sun and become relatively warm. The warm slopes, in turn, heat the air in immediate contact with the slopes and that air rises. At the same time, air at the same altitude, but located to the east of the mountains over the relatively flat terrain of the western Great Plains, is much cooler. The cooler air sinks and sweeps westward, replacing the warm air that rises over the mountain slopes. Updrafts over the mountain produce convective clouds that may billow upward to form thunderstorms. This process of thunderstorm development is enhanced whenever the synoptic-scale pressure pattern favors east winds over the western Great Plains.

The distinction is sometimes made between forced convection and free convection. In the case just described, topographic relief helps to force air upward to form convection currents. This situation is one of **forced convection.** On the other hand, convection that is triggered by intense solar heating of relatively flat terrain is known as **free convection.**

To this point we have considered those conditions conducive to deep convection and thunderstorm development. Under other circumstances, convection is inhibited and thunderstorms do not form. This situation usually occurs when air masses reside or travel over relatively cold surfaces and are thereby stabilized. Snow-covered ground cools and stabilizes the overlying air. Convection is thus suppressed, so thunderstorms are rare in middle and high latitudes during winter. Another reason for the rarity of winter thunderstorms is the relatively low water vapor concentration in cold air. As a general rule, over much of North America, thunderstorms are unlikely when the dew point is below 13 °C (55 °F).

Furthermore, thunderstorms are unusual over coastal areas that are situated downwind from relatively cold ocean waters. For example, thunderstorms are infrequent in coastal California, where prevailing winds are onshore and a shallow layer of maritime polar air often flows inland from off the relatively cold California current. Cool *mP* air at low levels suppresses deep convection and thunderstorm development.

## Severe Thunderstorms

By convention, a **severe thunderstorm** is accompanied by locally damaging winds, frequent lightning, or large hail.* As a rule, the greater the altitude of a thunderstorm top, the more likely that the system will produce severe weather. Why, then, do some thunderstorm cells surge to great altitudes and trigger severe weather, whereas others do not? One plausible explanation is that in a severe thunderstorm the updraft is tilted so that much of the precipitation falls alongside rather than against the updraft (Figure 13.11). Hence, the updraft maintains its strength and continues to build the cell to great altitudes. Apparently, strong vertical shear in the horizontal wind is responsible for tilting the updraft.

In the United States and Canada, most severe thunderstorms break out over the Great Plains and are as-

*The official National Weather Service criterion for designating a thunderstorm as *severe* includes any one or combination of the following: hailstones 1.9 cm (0.75 in.) or larger in diameter; tornadoes or funnel clouds; surface winds stronger than 93 km (58 mi) per hour.

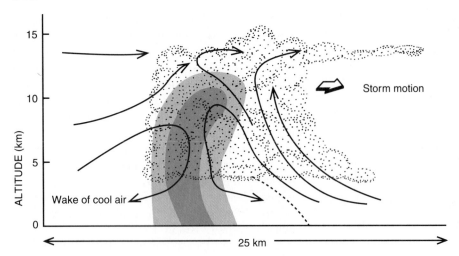

**FIGURE 13.11**
When a thunderstorm updraft is tilted, precipitation does not fall against (and thereby weaken) the updraft. Hence, the thunderstorm cell can billow upward to great altitudes and perhaps become severe. Precipitation is shown in green. [Reprinted with the permission of Macmillan College Publishing Company from *Meteorology,* 6th ed., by Richard A. Anthes. Copyright © 1992 by Macmillan College Publishing Company, Inc.]

sociated with mature synoptic-scale cyclones. Severe thunderstorm cells usually form along a squall line within the cyclone's warm sector, ahead of and parallel to a fast-moving, well-defined cold front. The squall line appears as an ominous, twisting mass of low, dark clouds (Figure 13.12), often hundreds of kilometers long. A squall line moves very rapidly; speeds may approach 80 km (50 mi) per hour.

The midlatitude jet stream is an important ingredient in the development of a severe thunderstorm cell,

sometimes called a *supercell.* The jet produces wind shear that tilts the updraft, thereby favoring great vertical development of the cell. In addition, the jet helps to produce a stratification of air that is especially favorable for formation of intense thunderstorm cells.

As we saw in Chapter 10, a midlatitude jet stream maximum induces both divergence and convergence of air aloft. Recall that diverging winds trigger cyclone development under the left front quadrant of the jet

**FIGURE 13.12**
An ominous-appearing, rolling, twisting mass of low black clouds marks a squall line. [Photograph by J. M. Moran]

stream maximum. Meanwhile, the air converges in the right front quadrant of the jet stream maximum, causing weak subsidence of air over the warm sector of the cyclone. The subsiding air is compressionally warmed, and its relative humidity decreases. The subsiding air is prevented from reaching the surface, however, by a shallow layer of maritime tropical air at the surface. The surge of *mT* air northward, from over the Gulf of Mexico, is particularly strong at about 3000 m (9800 ft) and is often described as a **low-level jet stream.** The warm, humid air is pumped northward by the circulation on the western flank of the Bermuda–Azores subtropical anticyclone.

The synoptic situation favorable to severe thunderstorm development is shown schematically in Figure 13.13. Note how the midlatitude (polar front) jet and the low-level *mT* jet cross to the southeast of the cyclone center.* Note also that the air ahead of the cold front and to the west of the warm, humid tongue is usually quite dry. For this reason, the western boundary of the maritime tropical air mass is called the **dry line.** West of the dry line, hot dry air flows east and downhill from the Southwest desert. A dry line usually occurs in Texas and northward on the Western Plains and is often the site of severe thunderstorm formation.

*The subtropical jet stream (Chapter 10) is sometimes also present at high levels in the troposphere, blowing from west to east over the warm sector of the cyclone. In that event, intense squall lines are likely to form between the midlatitude and subtropical jet streams.

As a consequence of compressional warming, the air subsiding from aloft becomes warmer than the underlying layer of maritime tropical air. A zone of transition develops between the two air masses as a temperature inversion (Figure 13.14). As we learned in Chapter 6, a temperature inversion is extremely stable, so the two air masses do not mix and convection is confined to the surface *mT* air layer. As long as this stratification persists, the contrast between the two air masses mounts. The subsiding air becomes drier, and the underlying air becomes more humid. The potential for severe weather continues to grow, and all that is needed is a trigger that will enable convection currents to penetrate the elevated temperature inversion. The needed upward impetus may be supplied by the intense solar heating of midafternoon or by the lifting of air associated with the approaching cold front. By either mechanism, convection currents eventually break through the temperature inversion, and cumulus clouds billow upward at speeds that may exceed 100 km (62 mi) per hour. Such explosive updrafts can even penetrate the tropopause and surge into the lower stratosphere. A severe thunderstorm is born.

Occasionally, ominous pouchlike mammatus clouds appear on the underside of the anvil top of a cumulonimbus cloud (Figure 13.15). Contrary to popular belief, **mammatus clouds** do not always indicate a severe thunderstorm. In fact, in the mountainous western United States, they often are associated with relatively

**FIGURE 13.13**
Synoptic situation most favorable for the development of severe thunderstorms.

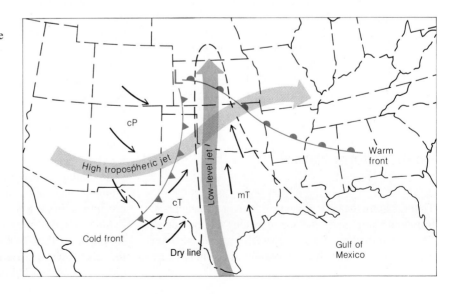

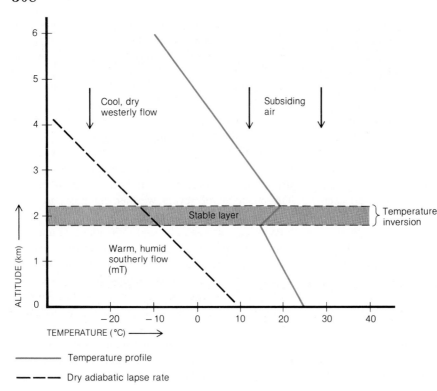

**FIGURE 13.14**
Vertical temperature profile that is most favorable for the sudden eruption of severe thunderstorm cells. An elevated temperature inversion separates subsiding dry air aloft from warm, humid air at the surface.

weak convective showers. Their unusual appearance is attributed to blobs of cold, cloudy air that descend from the anvil into the unsaturated clear air beneath the anvil. Ice crystals at the margin of a blob vaporize (sublimate), causing further cooling of the blob. This cooling offsets compressional warming of the descending air so that the blob sinks further into the dry air as a curious-looking protuberance. Ultimately, the size of the protuberance is limited by the increasing rate of vaporization of the constituent ice crystals.

## Thunderstorm Hazards

Thunderstorm hazards include lightning, downbursts, torrential rains, hail, and the spawning of tornadoes. Because tornadoes are an especially severe weather hazard, we consider them separately in the next chapter.

### *LIGHTNING*

By definition, a convective rain or snow shower is a thunderstorm if accompanied by lightning. Lightning is a weather phenomenon that is directly hazardous to human life, killing perhaps a hundred people in the United States and Canada each year. This death toll may seem surprisingly high, perhaps because fatal lightning bolts are typically isolated events that seldom make headlines. A single, fatal lightning bolt is not as newsworthy as a single disastrous tornado that takes many lives, but so many lightning bolts occur daily in much of North America that injuries and fatalities quickly add up. For some tips on lightning safety, refer to the Special Topic "Lightning Safety."

Lightning is feared not only because it can kill and injure people, but also because it ignites forest and brush fires. In the Rocky Mountain region, for example, lightning is the most common cause of forest fires, starting more than 9000 each year. One might think that the heavy rain associated with a thunderstorm would quickly quench a lightning-induced fire. In the west-

**FIGURE 13.15**
Pouchlike mammatus clouds occur on the underside of a thunderstorm anvil and sometimes indicate a severe storm. [Photograph by J. M. Moran]

ern basins, however, the base of a cumulonimbus cloud is usually so far above the ground, and the air below the cloud so dry, that much of the rain evaporates before reaching the fire.

Lightning is very costly to electrical utilities, each year causing tens of millions of dollars in damage to equipment (power lines, transformers) and in service restoration expenses. A new tool in utilities' efforts to cope with the lightning hazard is the **lightning detection network (LDN),** a system that provides real-time information on the location and severity of lightning strokes. Each monitoring station in a network consists of direction-finding equipment that can sense electromagnetic fields associated with cloud-to-ground lightning. A computer superimposes the locations of lightning strokes on a map of the region. A sample lightning stroke density map is shown in Figure 13.16.

As of this writing, an LDN consisting of 26 stations monitors the East Coast states as far west as a line from Erie, Pennsylvania, to Mobile, Alabama. Network stations send data to a processing facility at the State University of New York at Albany, where lightning locations are computed with an accuracy of a few kilometers. An electrical utility that subscribes to the LDN service accesses lightning data through a computer link. Such information enables the utility to mobilize repair crews more efficiently and speeds up res-

toration of disrupted service. As of this writing, a national lightning detection network is nearing completion.

What is lightning and why is it so dangerous? **Lightning** is a brilliant flash of light produced by an electrical discharge of about 100 million volts. The potential for an electrical discharge exists whenever a charge difference develops between two objects. A normally neutral object becomes negatively charged when it gains electrons (negatively charged subatomic particles) and positively charged when it loses electrons. That is, the object is ionized. When differences in electrical charge develop within a cloud or between a cloud and the ground, the stage is set for lightning.

On a clear day, the Earth's surface is negatively charged, and the upper troposphere is positively charged. This charge distribution changes as a cumulonimbus cloud develops. Within the cloud, charges separate so that the upper portion and a much smaller region near the cloud base become positively charged (Figure 13.17). In between, a pancake-shaped zone of negative charge forms that is a few hundred meters thick and several kilometers in diameter. At the same time, the developing cumulonimbus induces a positive charge on the ground directly under the cloud.

Air is a very good electrical insulator, and so, as a thunderstorm forms and electrical charges separate and

## SPECIAL TOPIC

# Lightning Safety

Lightning kills and injures, and a blinding flash of lightning followed by a crash of thunder frightens many people. In the United States between 1960 and 1985, an average of 96 people died each year as a result of lightning, and many times this number were injured. Although the danger of lightning cannot be ignored, some simple precautions will minimize the risk of injury when a thunderstorm threatens.

The odds of being struck and killed by lightning are actually very slim, about 350,000 to 1. By comparison, the odds of being struck and killed by an auto are 50 times greater. Although no place is absolutely safe, the risk can be reduced in even the most lightning-prone area of the nation, south and central Florida, where an estimated 10 lightning bolts strike each square kilometer of land yearly.

When a thunderstorm approaches, the best response is to seek shelter in a house or other building, avoiding contact with conductors of electricity that provide pathways for lightning. These include pipes (don't shower), stoves (don't cook), and wires (don't use the telephone). Electrical appliances pose no hazard if properly grounded, but why tempt fate by using them?

Some confusion surrounds the safety of motor vehicles during a lightning storm. A metal car or truck is a good shelter. Cars with cloth convertible tops and the backs of pickup trucks, on the other hand, are not. In Texas, in 1979, people riding in the back of a pickup truck were struck and killed by lightning while passengers in the cab of the truck were not harmed.

If a building or an auto is not accessible when a thunderstorm approaches, find shelter under a cliff, in a cave, or in a low area, such as a ravine, a valley, or even a roadside ditch. Avoid (1) tall, isolated objects, such as telephone poles and flagpoles, (2) metallic objects, such as wire fencing, rails, wire clotheslines, bicycles, and golf clubs, (3) high areas, such as hilltops and rooftops, and (4) bodies of water such as swimming pools and lakes. Stay off mowers and tractors. Individual trees in open spaces are hazardous, but a thick grove of small trees may offer safe haven. A group of people in the open should spread out, keeping several meters apart.

If your hair stands on end, lightning may be about to strike. In this unlikely event, immediately drop to your knees and, placing your hands on your knees, bend forward. In this way, you make your body the smallest possible target. One should not lie flat on the ground.

Contrary to popular belief, lightning can strike the same place more than once. New York City's Empire State Building typically is struck more than 20 times a year, and on one occasion was hit 15 times in only 15 minutes. Obviously, such sites should be avoided during a thunderstorm.

Fortunately, two of every three persons *struck* by lightning recover fully. Most survivors are jolted by a nearby lightning bolt and not hit directly. Immediate mouth-to-mouth resuscitation or cardiopulmonary resuscitation may revive victims who are not breathing. Those who appear merely stunned may require treatment for burns or shock. Victims of lightning bolts carry no electrical charge and can be handled safely.

build, a tremendous potential soon develops for an electrical discharge. When the thunderstorm enters its mature stage, the electrical resistance of air breaks down and lightning flows, thereby neutralizing the electrical charges. Lightning may forge a path between oppositely charged regions of a cloud, or between clouds, or between a cloud and the ground.

The cause of charge separation within cumulonimbus clouds is not well understood, but recent field measurements and laboratory simulations offer a promising explanation focused on the role of graupel and the convective circulation within thunderstorms. **Graupel** (German for "soft hail") consists of millimeter- to centimeter-sized ice pellets formed when supercooled water droplets collide and freeze on impact. Within the cloud, as falling graupel strikes smaller ice crystals in its path, opposite charges develop on the graupel and the ice crystals. For collisions that take place at temperatures below about $-15\,°C$ (5 °F), graupel becomes negatively charged and ice crystals acquire a positive

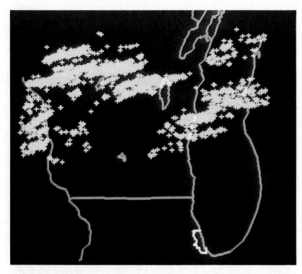

**FIGURE 13.16**
A sample terminal display of lightning strikes over Wisconsin and Lake Michigan obtained via a lightning detection network. [Photograph courtesy of R*Scan Corporation, Minneapolis, MN]

charge. Vigorous updrafts separate the particles, carrying the smaller positively charged ice crystals to the upper portion of the cloud, whereas the larger negatively charged graupel concentrate mostly in the lower portion of the cloud. This mechanism explains the posi-

tive charge of the upper cloud region and the negative charge of the pancake-shaped zone.

What accounts for the positive charge near the cloud base? Typically, temperatures near the cloud base are higher than −15 °C (5 °F). At those temperatures collisions between graupel and ice crystals induce a positive charge on graupel and a negative charge on ice crystals. The heavier graupel accumulate near the cloud base, giving that region a positive charge.

We are most concerned about lightning discharges between a cloud and the ground (Figure 13.18) because this path poses the greatest hazard. (Nonetheless, these discharges represent only about 20% of all lightning bolts.) Using high-speed photography, scientists have determined that a lightning flash consists of a regular sequence of events. Initially, streams of electrons surge from the cloud base toward the ground in discrete steps, each of which is about 20 to 100 m (66 to 328 ft) long. These so-called **stepped leaders** describe a branching path and produce a narrow ionized channel. When a branch of the stepped leaders comes within about 100 m (328 ft) of the ground, it is met by a positively charged **return stroke** from the ground. The return stroke follows the path of least resistance and often emanates from tall, pointed structures such as a metal flagpole or tower. Now an ionized channel only a few centimeters in diameter links the Earth's surface to the

**FIGURE 13.17**
Charge distribution within a thunderstorm at maturity.

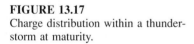

**FIGURE 13.18**
Cloud-to-ground lightning appears as bright streaks or bolts. [Photograph by Arjen and Jerrine Verkaik/SKYART]

cloud. Electrons flow, neutralization occurs, and the channel is illuminated.

Following this initial electrical discharge, subsequent surges of electrons from the cloud, called **dart leaders,** follow the same conducting path. Each dart leader is met by a return stroke (from the ground), and the conducting path is again illuminated. Typically, a single lightning discharge consists of two to four dart leaders plus return strokes. Sometimes, a dart leader is met by a return stroke that forges a new conducting path from the ground. The result is a forked lightning bolt that strikes the ground in more than one place.

The sequence just described is, by far, the most common occurrence of cloud-to-ground lightning. In less than 10% of cases, a positively charged leader emanates from the cloud and initiates a lightning discharge. Much more rarely, positive or negative stepped leaders propagate upward from the ground and meet a return stroke surging downward from a cloud. This ground-to-cloud lightning usually is initiated from mountaintops or tall structures such as antenna towers.

Electricity flows at the astonishing rate of nearly 50,000 km (31,000 mi) per second; hence, the entire lightning sequence takes place in less than two-tenths of a second. The human eye has difficulty separating the individual flashes of light that constitute a single lightning bolt, so that we perceive a lightning flash as a flickering light. **Sheet lightning** consists of bright flashes across the sky and indicates cloud-to-cloud discharges. **Heat lightning** is simply light reflected by clouds from distant thunderstorms that occur beyond the horizon.

Where there is lightning, there is thunder, although sometimes we see distant lightning but do not hear the thunder. Lightning heats the air along the narrow conducting path to temperatures that may exceed 25,000 °C (45,000 °F). For this reason, people can be burned severely by a lightning bolt that strikes nearby. Such intense heating expands the air violently and initiates a sound wave that we hear as **thunder.**

Because light travels about a million times faster than sound, we see a lightning bolt almost instantaneously, but we hear the thunder later. The closer we are to a thunderstorm cell, the shorter is the time interval between the lightning flash and thunder. As a rule, thunder takes about 3 seconds to travel 1 km (and 5

## WEATHER FACT

## The Rumble of Thunder

During a thunderstorm, we may be startled by a sudden, sharp clap of thunder that is followed by a low rumble that persists for tens of seconds. Why the rumble? As pointed out elsewhere in this chapter, the speed of light is so great that we see an entire lightning bolt instantly. Sound waves are generated all along the lightning path, but sound waves are considerably slower than light waves. This means that the first thunder we hear is from the part of the bolt that is closest to us (that is, where it strikes the ground). Subsequent sound waves reach us from portions of the bolt that are progressively further away from our ears (that is, at higher altitudes). The net result is a lingering rumble.

seconds to travel 1 mi). Thus, if you must wait 9 seconds between lightning flash and thunderclap, the thunderstorm cell is about 3 km (1.8 mi) away.

## DOWNBURSTS

Severe, and sometimes not so severe, thunderstorms can produce a **downburst,** an exceptionally strong downdraft that, upon striking the Earth's surface, diverges horizontally as a surge of potentially destructive winds.* T. Theodore Fujita of the University of Chicago is credited with the discovery of downbursts, and he coined the term. Fujita's discovery stemmed from his observation of a starburst pattern of wind damage while flying over an area near Beckley, West Virginia, in April 1974.

Downbursts occur with or without rain. If without rain, the downburst usually occurs under a curtain of virga (Chapter 8). Downbursts blow down trees (Figure 13.19), flatten crops, and wreck buildings. Often, downburst damage is erroneously attributed to an unseen tornado. Based on size, a downburst is classified as either a macroburst or a microburst.

A **macroburst** cuts a swath of destruction over a distance of more than 4.0 km (2.5 mi) and features surface winds that may top 210 km (130 mi) per hour. The leading edge of a macroburst may be marked by a gust front, and the system lasts up to 30 minutes. On 4 July 1977, an estimated 25 macrobursts struck several northern Wisconsin counties, felling trees and farm buildings over a 7200-km$^2$ (2800-mi$^2$) area. Based on extent of damage, surface winds probably peaked in the range of 180 to 250 km (112 to 155 mi) per hour.

A **microburst** is smaller and shorter-lived than a macroburst. By convention, its path of destruction is 4 km (2.5 mi) or less, top surface winds may be as high as 270 km (167 mi) per hour, and its life expectancy is less than 10 minutes. Microbursts are particularly dangerous at airports where their small size defies detection, and they can play havoc with the aerodynamics of aircraft. Microbursts trigger **wind shear,** a change in wind speed or direction with distance, that can

disturb the lift forces acting on an aircraft and may cause an abrupt change in altitude. An aircraft that flies into a microburst first encounters a strong headwind and then a strong tailwind. In only a matter of seconds, the aircraft's speed relative to air drops by more than 160 km (100 mi) per hour. If this takes place during takeoff or landing, the aircraft's speed may drop below the minimum required for flight.

Professor Fujita concluded that microbursts probably contributed to two commercial aircraft accidents in 1975. Apparently, in both cases, the pilots unwittingly flew their jet planes through the center of a microburst. One pilot did so on takeoff and the other while attempting to land. Both planes abruptly lost altitude and crashed. In retrospect, 30 or more civil airline ac-

**FIGURE 13.19**
Damage to a forest caused by a downburst. [Photograph courtesy of T. Theodore Fujita, Professor of Meteorology, The University of Chicago]

---

*We can simulate this flow by aiming a garden hose nozzle downward so that a stream of water strikes the ground at an angle and bursts outward. Depending on the angle at which a downburst strikes the ground, the resulting starburst pattern of destruction may be elongated in one direction.

cidents since 1964 have stemmed from encounters with microbursts.

Fujita and his colleagues conducted the first intensive field study of downbursts during the spring and summer of 1978 in the western suburbs of Chicago. The project was dubbed *NIMROD*, for *Northern Illinois Meteorological Research on Downbursts*. During the 42-day observational period, an array of weather instruments and radar units detected 50 microbursts. Another field study of microburst activity was conducted during the summer of 1982 over a 1,600-km$^2$ (615-mi$^2$) area near Denver's Stapleton International Airport. Using experimental aircraft and a dense grid of meteorological sensors, the *Joint Airport Weather Studies (JAWS)* project detected 186 microbursts during only 86 observational days. Subsequent field studies conducted during the 1980s found that an average of 25 microbursts occur at Stapleton airport each year, mostly in spring and summer.

One of the most important findings of *JAWS* is the apparent inadequacy of the *Low-Level Wind Shear Alert System (LLWSAS)*, an array of ground-level anemometers currently operated by the Federal Aviation Administration (FAA) at more than 100 airports. Many of Denver's microbursts were either too small or too far off the ground to be detected by *LLWSAS*. Indeed, microbursts may be so small as to affect only a small portion of one runway. Consequently, the FAA now requires airlines to install an approved microburst detection system on their aircraft by the end of 1995. As of this writing, the most promising systems are being tested aboard NASA's Boeing-737 research air-

**FIGURE 13.20**
Road sign in the Colorado Rockies warning people to climb to high ground in the event of a flash flood. [Photograph by J. M. Moran]

craft. Scientists are testing sensors that can detect wind shear using the same principle as Doppler radar (Chapter 14). The goal is to provide pilots with at least 20 to 40 seconds advanced warning of microburst-induced windshear. Such advanced warning is considered sufficient for pilots to take evasive action.

## FLASH FLOODS

A stationary or slow-moving intense thunderstorm can produce torrential rains that trigger a potentially disastrous **flash flood,** a sudden overflow of a river channel or other drainageway. Typically, a thunderstorm is stationary or slow moving because (1) it is imbedded in weak steering winds aloft and/or (2) the system is maintained by a persistent flow of humid air up a mountain slope. Both factors contributed to the flash flood that claimed 139 lives and caused $35.5 million in property damage in the Big Thompson Canyon of Colorado on 31 July 1976.

Big Thompson Canyon is located in the Colorado Front Range about 80 km (50 mi) northwest of Denver. During the late afternoon and evening of 31 July 1976, a persistent easterly flow of humid air up the mountain slopes triggered thunderstorm development and heavy rainfall. Thunderstorm activity remained nearly stationary because winds aloft were weak (less than 35 km per hour above 3000 m). Runoff poured down the steep mountain slopes and into the river that winds along the narrow canyon floor. The river rose abruptly and soon overflowed its banks as a flash flood. The National Weather Service estimates that in the river's headwaters, rainfall totaled 25 to 30 cm (10 to 12 in.) with perhaps 20 cm (8 in.) falling in only 2 hr. At one location along the river, the discharge (volume of water flowing per second) was more than 200 times normal. A wall of water almost 6 m (20 ft) high destroyed 418 houses and washed away 197 motor vehicles.

Flash flooding is especially hazardous in mountainous terrain, and motorists and campers are well advised to head for higher ground in the event of a flood warning (Figure 13.20). But even where the topography is relatively flat, prolonged periods of heavy rain (more than 7.6 mm, or 0.3 in., per hour) can greatly exceed the infiltration capacity of the ground. Simply put, the ground cannot absorb all the rainwater. Excess water

**FIGURE 13.21**
Motor vehicles and occupants may be trapped by the sudden rise of water accompanying flash floods. [NOAA photograph]

runs off to creeks, streams, rivers, or sewers or collects in other low-lying areas. If a drainage system cannot accommodate the sudden input of huge quantities of water, flash flooding is the consequence.

A post–World War II upswing in flood fatalities in the United States is largely due to more people visiting remote areas prone to flash flooding. During the 1970s, floods took an average of 200 lives annually, twice the flood fatalities of the 1960s and three times those of the 1940s. Furthermore in the mid-1980s, R. E. Hallgren, director of the National Weather Service, estimated that, in an average year, floods cause up to $3 billion in property damage.

Because of their design and composition, urban areas are prone to flash floods during intense downpours. Concrete and asphalt render the surfaces of a city virtually impervious to water, so elaborate storm sewer systems are required to transport runoff to nearby natural drainageways. Storm sewer systems have a lim-

ited capacity for water, however, and may be unable to handle the excess water produced during a torrential rainfall. Water backs up and collects under viaducts and in other low-lying areas. Sometimes water levels rise so fast in these areas that motorists are trapped in their vehicles (Figure 13.21). For example, on the evening of 1 August 1985, a slow-moving severe thunderstorm drenched Cheyenne, Wyoming, with 15.4 cm (6.06 in.) of rain in less than 4 hr. Flash flooding filled city streets with up to 2 m (6.6 ft) of water, 12 people lost their lives, and estimates of property damage exceeded $65 million.

Severe thunderstorms are not the only culprits in triggering flash flooding. Breaching of a dam or levee, or the sudden release of water during breakup of a river ice jam, can also cause an abrupt rise in water level. In any event, when in flood-prone areas, we are well advised to heed the precautions listed in Table 13.1.

**TABLE 13.1**
**Some Flash Flood Safety Tips[a]**

1. Avoid driving on flooded roadways or bridges. Flood waters only a half meter deep can carry away most autos. Futhermore, flooded roads may be undermined.
2. If your vehicle stalls in high water, abandon it and seek high ground.
3. On foot, never try to cross a stream if the water level is above your knees.
4. Keep children away from drainage ditches and culverts.
5. Find out the elevation of your property and the flood history of the area. If there is a risk of flooding, plan an evacuation route.
6. Be especially cautious when camping in remote areas. Avoid camping in mountain valleys or near dry stream beds. Take a battery-operated radio along to monitor changing weather conditions.

[a]Based on NOAA recommendations.

## HAIL

**Hail** is precipitation in the form of balls or lumps of ice, usually called *hailstones*. Hail falls from intense thunderstorm cells that are characterized by strong updrafts, great vertical development, and an abundant supply of supercooled water droplets. Hailstones range from pea size to the size of a golfball or even larger. In the United States, the largest hailstone on record was collected at Coffeyville, Kansas, on 3 September 1970. It weighed 758 g (1.67 lb) and measured 44.5 cm (17.5 in.) in circumference and 14.2 cm (5.5 in.) in diameter—about the size of a softball. Canada's largest authenticated hailstone weighed 290 g (10.23 oz) and measured 10.2 cm (4 in.) in diameter. It fell at Cedoux, Saskatchewan, on 27 August 1973.

A hailstone forms when an ice pellet is transported through portions of a cumulonimbus cloud containing varying concentrations of supercooled water droplets. The ice pellet may descend slowly through the entire cloud, or it may follow a more complex pattern of ascent and descent as it is caught alternately in updrafts and downdrafts. In the process, the ice pellet grows by accretion (addition) of freezing water droplets. In general, the stronger the updraft, the larger will the ice pellet grow. Eventually, when it becomes too large and heavy for updrafts to support it, the ice pellet descends and falls out of the cloud base. If the ice does not melt completely during its journey through the above-

freezing air beneath the cloud, it reaches the ground as a hailstone.

When an ice pellet enters portions of the cloud containing a relatively high concentration of supercooled water droplets, water collects on the ice pellet as a liquid film, which freezes slowly to form a transparent layer, or **glaze.** When the ice pellet travels through portions of the cloud where the concentration of supercooled water droplets is relatively low, droplets freeze immediately on contact with the ice pellet. As droplets freeze, many tiny air bubbles are trapped within the ice, producing an opaque whitish layer of granular ice, or **rime.** The result is alternating laminae of clear (glaze) and opaque (rime) ice, which, in cross section, resemble the internal structure of an onion (refer back to Figure 8.7). As many as 25 layers have been counted in a single large hailstone.

Often hailstones cover the landscape in a long, narrow stripe known as a **hailstreak.** A typical hailstreak may be 2 km (1.2 mi) wide and 10 km (6.2 mi) long, and a single large thunderstorm may produce several hailstreaks. From a study of numerous hailstorms and hailstreaks, scientists at the Illinois State Water Survey devised a model to explain hailstreak development. As shown in Figure 13.22, hail forms within the upper portion of a cumulonimbus cloud. After several minutes of hail formation, the updraft weakens, allowing the hail to descend within the cloud; about 4 minutes later, the first hailstones reach the ground. In

**FIGURE 13.22**
A model of hailstreak development. [From S. A. Changnon and J. L. Ivens, *Hail in Illinois.* Public Information Brochure 13, Illinois State Water Survey, 1987, p. 5]

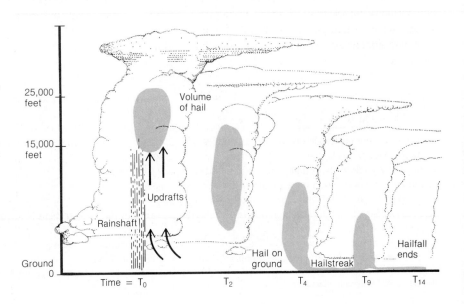

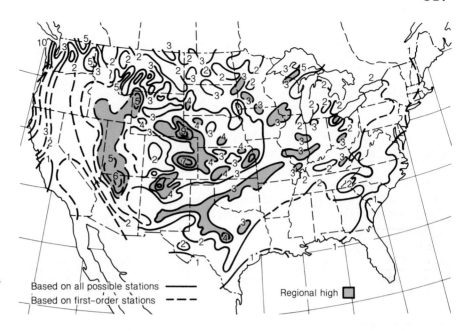

**FIGURE 13.23**
Hail frequency across the United States in average number of days per year. [Courtesy of the Illinois State Water Survey]

Based on all possible stations ——
Based on first-order stations ‑ ‑ ‑
Regional high ▨

the ensuing 10 minutes, as the thunderstorm continues to move along, the entire volume of hail is deposited along the ground as a hailstreak.

On rare occasions, the fall of hail is so great that snowplows must be called out to clear highways. This happened in Milwaukee, Wisconsin, on 4 September 1988, when pea-size hail formed drifts to 46 cm (18 in.) in some north-side neighborhoods. On 6 August 1980 at Orient, Iowa, drifts of hail were reported to be 1.8 m (6 ft) deep. On the afternoon of 13 June 1984, hailstones as large as golfballs fell for up to 1.5 hours in the western suburbs of Denver, producing an accumulation of 25 cm (10 in.), with drifts greater than a meter.

Hailstones may be large enough to smash windows and dent automobiles, but the most costly damage is to crops. Hail usually falls during the growing season, and, in only a matter of minutes, large hail can wipe out the fruits of a farmer's year of labor. On 11 July 1990, the most costly hailstorm in U.S. history struck an area just east of the Colorado Front Range from Estes Park south to Colorado Springs. Golfball- to baseball-sized hailstones injured 60 people who were caught out in the open during the storm. Total property damage was estimated at $625 million.

Traditionally, farmers cope with the hail hazard by purchasing insurance. Illinois farmers, for example, lead the United States in crop-hail insurance, purchasing an average annual liability coverage that tops $600 million. It is not surprising, then, that efforts have been made to suppress or prevent hail. To date, such efforts have not been fruitful. A brief historical sketch of these activities is presented in the Special Topic "Hail Suppression."

Perhaps surprisingly, hail frequency is not necessarily related to thunderstorm frequency. Although Florida experiences the greatest frequency of thunderstorms in the United States, hail is unusual in that state. In North America, hail is most likely on the High Plains just east of the Rockies, where it can be expected to fall from about 10% of all thunderstorms. Figure 13.23 is a map of hail frequency in the United States.

## Hail Suppression*

Efforts to suppress hail have deep historical roots. Indeed, in 14th-century Europe, church bells were rung and cannons fired in the belief that the attendant noise would somehow ward off hail. A period of particularly intense hail suppression activity took place in the grape-growing regions of Austria, France, and Italy during the late 19th century. M. Albert Stiger, a wine grower and burgomaster of Windisch-Feistritz, Austria, designed and built a special funnel-shaped hail suppression cannon (Figure 1). Stiger believed that smoke particles in the cannon fire would inhibit hailstone development. Amazingly, in experimental firings in 1896–97 at Windisch-Feistritz, Stiger reported no hail, although severe hail damage occurred in neighboring areas.

Word of Stiger's apparent success spread throughout the vineyard regions of Europe, and hail cannons soon became commonplace. There were so many cannons that accidental shooting of people became a serious problem in some localities. After Stiger's much heralded success was not duplicated elsewhere,

*This discussion is based on S. A. Changnon, Jr., and J. L. Ivens. "History Repeated: The Forgotten Hail Cannons of Europe," *Bulletin of the American Meteorological Society* 62 (1981): 368–375.

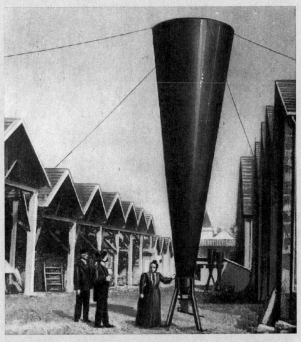

**FIGURE 1**
A hail suppression cannon popular in the grape-growing region of Austria, France, and Italy during the late nineteenth century. [Photograph courtesy of S. A. Changnon and J. L. Ivens, Illinois State Water Survey; © American Meteorological Society]

## Conclusions

Thunderstorms are the products of convection currents that surge to great altitudes within the troposphere. As such, thunderstorms channel excess heat at the Earth's surface into the atmosphere via a combination of sensible and latent heating. Thunderstorms are thus most frequent in the warmest regions of the Earth, that is, over the continental interiors of tropical latitudes.

Most thunderstorms consist of more than one cell, each of which completes its life cycle in an hour or less. A thunderstorm cell reaches its maximum intensity late in its mature stage. During this stage, thunderstorms produce lightning, heavy rain, and strong gusty surface winds. Potential hazards associated with thunderstorms are downbursts, hail, and tornadoes.

Although fewer than 1% of all thunderstorms produce tornadoes, in some areas of North America, the possibility of a tornado is the principal reason why people fear thunderstorms. Tornadoes are spawned by certain severe thunderstorm cells (supercells) that are usually part of a squall line. We focus on the characteristics and genesis of tornadoes in the next chapter.

however, interest in hail cannons waned rapidly, and by 1905 this early attempt at weather modification ended.

The modern era of hail suppression experimentation began after World War II. Although founded on a much better, albeit not yet complete, understanding of cloud physics, the new techniques shared some similarities with earlier efforts. For example, until the practice was outlawed in the early 1970s, Italian farmers regularly fired explosive rockets into threatening clouds in an attempt to shatter developing hailstones. In the former Soviet Union, scientists fire silver iodide (AgI) crystals, a cloud-seeding agent (Chapter 8), into thunderclouds. They theorize that silver iodide crystals stimulate the formation of large numbers of small hailstones, which will melt long before they reach the ground, instead of the normal development of small numbers of larger hailstones.

In the United States, annual agricultural losses due to hail exceed $700 million (Figure 2). It is not surprising, then, that scientists in this country set out to test the Soviet hypothesis of hail suppression and to learn more about hail-producing thunderstorms. To these ends the *National Hail Research Experiment* was launched over northeastern Colorado in 1972. Although much was learned about hailstorms, three years of seeding potential hailstorms failed to confirm the Soviet hypothesis. Some scientists questioned the experimental

**FIGURE 2**
In only minutes, a hail storm devastated this field of corn. [National Center for Atmospheric Research/University Corporation for Atmospheric Research/National Science Foundation]

design and whether it was a viable test of the Soviet technique. In an interesting parallel with events in 19th-century Europe, declining public confidence in the effectiveness of modern hail suppression efforts brought an end to federal funding of hail suppression research in 1979.

## Key Terms

| | | | |
|---|---|---|---|
| thunderstorm | frontal thunderstorm | graupel | macroburst |
| cumulus stage | squall line | stepped leaders | microburst |
| convective condensation level (CCL) | mesoscale convective complex (MCC) | return stroke | wind shear |
| cumulus congestus | forced convection | dart leaders | flash flood |
| cumulonimbus | free convection | sheet lightning | hail |
| mature stage | severe thunderstorm | heat lightning | glaze |
| entrainment | low-level jet stream | thunder | rime |
| gust front | dry line | downburst | hailstreak |
| roll cloud | mammatus clouds | | |
| shelf cloud | lightning detection network (LDN) | | |
| dissipating stage | lightning | | |
| air mass thunderstorm | | | |

## Summary Statements

☐ A thunderstorm is a mesoscale weather system produced by strong convection currents that surge high into the troposphere. The life cycle of a thunderstorm cell consists of a three-stage sequence: cumulus, mature, and dissipating.

☐ During the cumulus stage, cumulus clouds build vertically and laterally, updrafts characterize the entire system, and there is no precipitation. The mature stage begins when precipitation reaches the Earth's surface. At that stage, updrafts occur alongside downdrafts and the system reaches maximum intensity. During the dissipating stage, subsiding air spreads through the entire cell and clouds vaporize.

☐ A thunderstorm usually consists of a cluster of cells, each of which may be at a different stage of its life cycle. A cluster of thunderstorm cells may track in a different direction than the individual cells.

☐ Most thunderstorms develop in maritime tropical air as a consequence of uplift (1) along fronts, (2) on mountain slopes, (3) via convergence of surface winds, or (4) through intense solar heating of the Earth's surface.

☐ Thunderstorms occur along or ahead of fronts, within a mass of maritime tropical air, or as a mesoscale convective complex (MCC). Frontal thunderstorms are usually more energetic than air mass thunderstorms. MCCs account for a substantial portion of growing season rainfall over the Great Plains and Midwest.

☐ Worldwide, thunderstorms are most common over the continental interiors of tropical latitudes. In North America, sea breeze convergence makes central Florida the most frequent site of thunderstorms. Convection is inhibited and thunderstorms fail to develop where air masses reside or travel over relatively cold surfaces and are thereby stabilized.

☐ Severe thunderstorm cells typically form along a squall line ahead of a fast-moving, well-defined cold front associated with a mature midlatitude cyclone. The midlatitude jet stream causes dry air to subside over a surface layer of maritime tropical air. This produces a layering of air that can lead to explosive convection and the development of severe thunderstorms.

☐ Lightning is a brilliant flash of light produced by the discharge of electricity within a cloud, between clouds, or between a cloud and the ground. Cloud-to-ground lightning consists of a very rapid sequence of events involving stepped leaders, return strokes, and dart leaders.

☐ The reason for electrical charge separation in a cumulonimbus cloud is not well understood, but collisions between graupel and ice crystals plus updrafts and downdrafts within the cloud likely play important roles.

☐ Some thunderstorm cells produce downbursts—very intense downdrafts that spread out (diverge) at the Earth's surface as potentially destructive winds. Based on size, a downburst is categorized as either a macroburst or microburst.

☐ Flash flooding is a special hazard in mountainous terrain, where steep slopes channel excess runoff into narrow stream and river valleys, and in urban areas, where impervious surfaces cause excess water to collect in low-lying areas.

☐ Hail develops in intense thunderstorm cells characterized by strong updrafts, great vertical development, and an abundant supply of supercooled water droplets.

## Review Questions

1. Describe the characteristics of each stage in the life cycle of a thunderstorm cell.
2. What is a gust front? Explain how a gust front can spur development of secondary thunderstorm cells.
3. What is the significance of the relatively short life span of a thunderstorm cell for local weather forecasting?
4. Distinguish between air mass thunderstorms and frontal thunderstorms. Which is more likely to be severe and why?
5. Why do air mass thunderstorms usually develop in the afternoon, during the warmest hours of the day?
6. What is a squall line?
7. Describe how thunderstorms develop over the interior of the Florida peninsula.
8. Why are thunderstorms most frequent in the continental interiors of tropical latitudes?
9. Under what conditions is a thunderstorm considered to be severe?
10. In what sector of a mature cyclone are thunderstorms most likely to develop? Explain why.
11. Describe the synoptic situation in middle latitudes that is most favorable for the formation of severe thunderstorms.
12. What causes lightning? Why is it dangerous?
13. What causes thunder?
14. What is a microburst, and how might it pose a hazard for aircraft?
15. Speculate on why the damage produced by a downburst is sometimes mistaken for tornado damage.

**16.** How might the flash flood hazard vary with the season in middle and high latitudes?

**17.** Why is flash flooding a particular hazard in mountainous terrain? In urban areas?

**18.** What type of thunderstorm may produce hail?

**19.** Explain the internal concentric layering of hailstones.

**20.** Today, little research is directed at hail suppression. Explain why.

## Questions for Critical Thinking

**1.** Compare and contrast a sea breeze front with a gust front.

**2.** What causes the anvil shape of a thunderstorm top?

**3.** Thunderstorms may track at an angle to the paths of their constituent cells. What is the significance of this behavior for thunderstorm forecasting?

**4.** Explain the diurnal and seasonal variations in thunderstorm occurence. Also explain why thunderstorms are rare along the coast of southern California and over snow-covered terrain.

**5.** How does a tilted updraft affect the likelihood of hail formation?

## Selected Readings

BLANCHARD, D. O. "Mesoscale Convective Patterns of the Southern High Plains," *Bulletin of the American Meteorological Society* 71 (1990):994–1005. Classifies mesoscale convective systems into three basic patterns.

CHANGNON, S. A., JR., AND J. L. IVENS. "History Repeated: The Forgotten Hail Cannons of Europe," *Bulletin of the American Meteorological Society* 62 (1981):368–375. Provides a fascinating discussion of the parallels between modern and early weather modification efforts.

FEW, A. A. "Thunder," *Scientific American* 233, No. 1 (1975):80–90. Discusses the relationship between thunder and lightning flash.

FUJITA, T. T. *The Downburst*. Chicago: Department of the Geophysical Sciences, University of Chicago, 1985. 122 pp. Presents a well-illustrated account of macrobursts and microbursts; features many case studies.

FUJITA, T. T., AND F. CARACENA. "An Analysis of Three Weather-Related Aircraft Accidents," *Bulletin of the American Meteorological Society* 58 (1977):1164–1181. Focuses on the possible role of microbursts in aircraft accidents.

HUGHES, P., AND R. WOOD. "Hail: The White Plague," *Weatherwise* 46, No. 2 (1993):16–21. Elaborates on the impacts of hail.

MADDOX, R. A., AND J. M. FRITSCH. "A New Understanding of Thunderstorms: The Mesoscale Convective Complex," *Weatherwise* 37, No. 3 (1984):128–135. Discusses the discovery of large clusters of interacting thunderstorms.

RHODES, S. L. "Mesoscale Weather and Aviation Safety: The Case of Denver International Airport," *Bulletin of the American Meteorological Society* 73 (1992):441–447. Reviews concerns raised about potential mesoscale weather hazards at the new Denver airport.

WILLIAMS, E. R. "The Electrification of Thunderstorms," *Scientific American* 259 (1988):88–99. Provides a well-illustrated discussion of the causes of charge separation within thunderstorms.

# 14 Tornadoes

*Toto, I've a feeling we're not in Kansas anymore.*

WIZARD OF OZ
*MGM, 1939, based on the novel
by L. Frank Baum*

A tornado is a small-scale weather system that has the potential of taking lives and producing considerable property damage. [Photograph by Arjen and Jerrine Verkaik/ SKYART]

T
ORNADOES ARE by far the most violent of weather systems. Fortunately, they are small and short-lived and often strike sparsely populated regions. Occasionally, however, a major tornado outbreak causes incredible devastation, death, and injury. During a 16-hr period on 3–4 April 1974, 148 tornadoes left 307 people dead and 5,484 people injured and more than $600 million in property damage in 13 states (Figure 14.1). In this chapter, we discuss the characteristics and genesis of tornadoes. We also describe weather radar, a valuable tool for tracking severe thunderstorms and detecting tornado development.

## Tornado Characteristics

A **tornado** is a small mass of air that whirls rapidly about a nearly vertical axis. It is made visible by clouds and by dust and debris sucked into the system. Torna-
does are approximately funnel shaped, although a variety of forms have been observed, ranging from cylindrical masses of roughly uniform lateral dimension (Figure 14.2A) to long, slender, ropelike pendants (Figure 14.2B). By convention, when the circulation remains aloft, the system is termed a **funnel cloud,** but when it touches down on the ground, it is called a tornado.

A weak tornado's path on the ground typically is less than 1.5 km (1 mi) long and 100 m (328 ft) wide, and the system has a life expectancy of only 1 to 3 minutes. Wind speeds are under 182 km (113 mi) per hour. At the other extreme, an intense tornado can produce damage along a path more than 160 km (100 mi) long and hundreds of meters wide, and the lifetime of the system may be more than 2 hours. Estimated wind speeds in violent tornadoes range up to 513 km (318 mi) per hour. The most deadly tornado in North American history was the Tri-State tornado of 18 March 1925. Traveling at an average speed of 118 km (73 mi) per hour, the system lasted 3.5 hr and produced a 353-km (219-mi) path of devastation stretching from southeastern Missouri through the southern tip of Illinois and into southwest Indiana. Along the tornado's path, fatalities totaled 695, injuries topped 2000, and 11,000 people were made homeless.

Most tornadoes are spawned by and travel with severe thunderstorm cells (*supercells*). Tornadoes and their parent thunderstorm cells usually (about 90% of the time) travel from southwest to northeast. Trajectories are often erratic, however, with many tornadoes exhibiting a hopscotch pattern of destruction as they alternately touch down and lift off the ground (Figure 14.3). Tornadoes have been known to move in circles and to describe a figure eight. The average forward speed is around 55 km (34 mi) per hour, although there are reports of tornadoes racing along at speeds approaching 240 km (149 mi) per hour.

An extremely steep horizontal air pressure gradient between the tornado center and outer edge is the force ultimately responsible for a tornado's violence. The air pressure drop over a distance of only 100 m (328 ft) or so may be equivalent to the normal air pressure drop between sea level and an altitude of 1 km, that is, about a 10% reduction. The Coriolis effect is also present, but the system is so small that its influence is negligible. Thus, winds in a tornado may rotate in either a clockwise or a counterclockwise direction, although the lat-

323

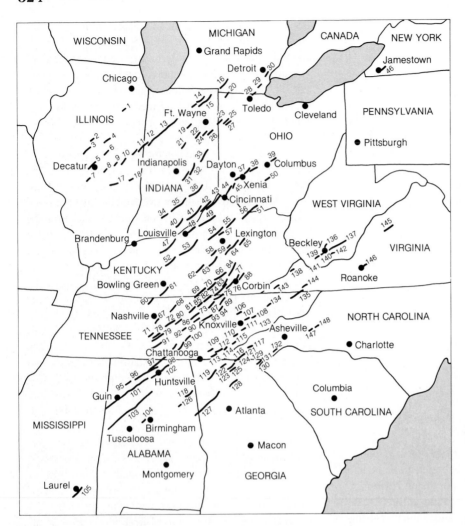

**FIGURE 14.1**
Tracks of the 148 tornadoes that struck a 13-state area on 3–4 April 1974. [Courtesy of T. Theodore Fujita, Professor of Meteorology, The University of Chicago]

ter dominates by far in Northern Hemisphere tornadoes. This counterclockwise bias may be inherited from the parent thunderstorm in which the Coriolis effect has a stronger influence.

## Distribution of Tornadoes

In Chapter 13, we described the special synoptic-scale circulation pattern that favors development of severe thunderstorms. To this pattern we must add another requirement for tornado development: Tornadoes are most likely to form over relatively flat and dry terrain.

They are rare in areas of great topographic relief such as the Rocky Mountain region. Flat terrain offers a minimum of frictional resistance, and dry conditions mean that most of the absorbed radiation is channeled into sensible heating, thereby spurring deep convection.

The central United States is one of only a few places in the world where synoptic weather conditions and terrain are ideal for tornado development; interior Australia is another. Although tornadoes have been reported in all 50 states and throughout southern Canada, most occur in **tornado alley,** a corridor stretching from eastern Texas and the Texas Panhandle northward through Oklahoma, Kansas, and portions of Nebraska. Central Oklahoma has the highest annual incidence of tornadoes, whereas local tornado frequency maxima occur

**FIGURE 14.2**
A tornado may appear as (A) a cylindrical mass of relatively uniform lateral dimensions or (B) a long, slender ropelike pendant. [Photographs by Arjen and Jerrine Verkaik/SKYART]

in central Illinois and Indiana and in southern Mississippi (Figure 14.4).

Each year, the United States can anticipate between 700 and 1100 tornadoes. The annual average for 1961 to 1990 is 803. Slightly more than half of all tornadoes develop during the warmest hours of the day (10 A.M. to 6 P.M.), and almost three-quarters of tornadoes in the United States occur from March to July. The months of peak tornado activity are April (13%), May (22%), and June (21%). During that time of year, weather conditions are optimal for spurring vigorous convection and the severe thunderstorms that spawn tornadoes.

One factor that contributes to the spring peak in tornado frequency is the relative instability of the lower atmosphere at that time of year. During the transition from winter to summer, days lengthen and insolation increases, thereby warming the ground. Heat is transported from the ground into the troposphere (Chapter 4), but it takes time for the entire troposphere to adjust to heating from below. The upper troposphere, in fact, usually retains its winterlike cold well into spring. The result is a relatively great air temperature lapse rate that favors supercell development.

Another factor that contributes to the spring tornado maximum is the greater likelihood that ideal synoptic weather conditions will occur at that time of year. Recall from Chapter 13 that severe thunderstorms typically develop in the warm southeast sector of a strong midlatitude cyclone. Such cyclones achieve their greatest intensity when sharp temperature contrasts develop across the nation, that is, in spring when the polar front is well defined.

During spring there is a general northward progres-

**FIGURE 14.3**
The pattern of destruction caused by a tornado is often erratic, as shown by this aerial photograph of a residential area in Birmingham, Alabama, hit by a tornado on 4 April 1977. Note how some houses are completely destroyed while neighboring houses are still standing. [NOAA photograph]

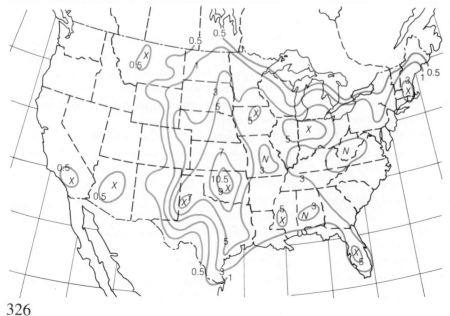

**FIGURE 14.4**
Tornado frequency in number per year within areas defined by 91-km (56.5-mi) radius circles, based on 29 years of data collected since 1 January 1950. An *X* indicates a relative maximum, and an *N* denotes a local minimum. [From NSSFC, "Tornadoes," *Weatherwise* 33, No. 2 (1980):54]

**TABLE 14.1**
**What to Do If a Tornado Is Approaching[a]**

1. Seek shelter in a tornado cellar, an underground excavation, or a steel-framed or substantial reinforced concrete building.
2. Avoid auditoriums, gymnasiums, supermarkets, or other structures that have wide, free-span roofs.
3. In an office building or school, go to an interior hallway on the lowest floor or a designated shelter area. Lie flat on the floor with your head covered.
4. At home, go to the basement. If there is no basement, go to a small room (closet, bathroom, or interior hallway) in the center of the house on the lowest floor. Seek shelter under a mattress or a sturdy piece of furniture.
5. Stay clear of all windows and outside walls.
6. In open country, do not try to outrun a tornado in an auto. Tornadoes often move too fast and their paths are too erratic to avoid even if you drive at right angles to the apparent track of the storm. Instead, stop and seek shelter indoors, and if this is not possible, lie flat in a ravine, creek bed, or open ditch.
7. Do not seek shelter in mobile homes or motor vehicles.

[a]Modified slightly from NOAA recommendations.

sion of tornado occurrences. In effect, the center of maximum tornado frequency follows the sun, as do the midlatitude jet stream, the principal storm tracks, and northward incursions of maritime tropical air. In late winter, maximum tornado frequency, on average, is along the Gulf Coast states. By May, the maximum frequency shifts to the southern Great Plains, and by June, the highest tornado incidence is usually in the northern plains and the Prairie Provinces east of the Rockies.

What are your chances of experiencing a tornado? Very slim. Fewer than 1% of all thunderstorms produce tornadoes, and even in the most tornado-prone regions of North America, a tornado is likely to strike only once every 250 years. There are, of course, exceptions to the rule. Tornadoes have hit Oklahoma City no less than 26 times since 1892. To be prepared in the unlikely event that a tornado should strike your area, study the recommendations in Table 14.1. These precautions will reduce the hazard posed by tornadoes.

km (100 mi) per hour, strong enough to lift a house off its foundation. The most intense tornadoes usually include subsidiary vortices, small whirlwinds that swirl about within the tornado. The strongest winds in the system may be associated with these whirlwinds.

It was once widely believed that a tornado causes buildings to explode. The air pressure within the building supposedly could not adjust rapidly enough to the abrupt pressure drop associated with the tornado. In the event of a tornado sighting, people were advised to open windows to help equalize the internal and external air pressure. However, most buildings have sufficient air leaks so that a potentially explosive pressure differential never really develops. The destruction of buildings is due, instead, to very strong currents of air that stream over roofs and cause the structure to lift, much as air induces lifting as it flows over the curved upper surface of an airplane wing. As the roof is lifted, the walls collapse or are blown in, and the building disintegrates.

## Hazards of Tornadoes

Hazards of tornadoes are (1) extremely high winds, (2) strong updraft, (3) subsidiary vortices, and (4) an abrupt drop in air pressure. Winds blow down trees, power poles, buildings, and other structures. Flying debris causes much of the death and injury associated with tornadoes as broken glass, lumber, and even vehicles become lethal projectiles. In violent tornadoes, the updraft near the center of the storm's funnel may top 160

## The F-Scale

Professor T. Theodore Fujita of the University of Chicago devised a six-point intensity scale for evaluating tornado strength and damage to structures. Called the **F-scale,** and presented as Table 14.2, it is based on wind speeds estimated from property damage and categorizes tornadoes as *weak* (F0, F1), *strong* (F2, F3), or *violent* (F4, F5).

## Tornado Oddities

In addition to reports of the terrible devastation caused by tornadoes, strange and even bizarre events associated with tornadoes are sometimes recounted. Reports of straws driven deeply into trees are readily explained. As a tree is bent by the wind, its grain opens, thus allowing a straw to enter. When the wind slackens, the grain closes, and the straw is trapped. Rains of fish or frogs are probably the result of a tornado updraft pulling the water and its inhabitants out of a nearby pond or lake. Among the oddest reports is the deplumation of chickens and other fowl during a tornado. This phenomenon has variously been attributed to the pressure drop within the tornado, strong wind (perhaps aided by natural molting), or the relaxation of the feather follicles, a reaction brought on by nervous stress.

An F0 tornado on Fujita's scale produces minor damage, snapping twigs and small branches and breaking some windows. F1 and F2 tornadoes can cause moderate to considerable property damage and even take lives. An F1 tornado can down trees and shift mobile homes off their foundations, and an F2 tornado can rip roofs off frame houses, demolish mobile homes, and uproot large trees. An F3 tornado can partially destroy even well-constructed buildings and lift motor vehicles off the ground. At the violent end of Fujita's scale, destruction is described as devastating to incredible with the potential for many fatalities. An F4 tornado can level sturdy buildings and other structures and toss automobiles about like toys. In an F5 tornado, sturdy frame houses are lifted and transported some distance before disintegrating.

Fortunately, F5 tornadoes are rare. Of the average 803 tornadoes that strike the United States each year, perhaps only one will be rated an F5. When one does occur, however, the results are catastrophic. In about 1 minute, an F5 tornado leveled the village of Barneveld, Wisconsin, on 8 June 1984; 100 homes were totally destroyed and nine lives were lost. The Xenia, Ohio, tornado, which was part of the massive tornado outbreak of 3–4 April 1974, rated F5 over a portion of its 51-km (32-mi) path and claimed 34 lives.

The American Meteorological Society reports that, in a typical year, 79% of all tornadoes are weak, 20% are strong, and only about 1% are violent. The few violent systems are responsible for the majority of all fatalities, however. Between 1961 and 1990, the average annual fatalities from tornadoes was 82 with 42% in April and 17% in May.

## The Tornado-Thunderstorm Connection

The most intense tornadoes usually appear on the rain-free rear portion of a severe thunderstorm. They develop in the thunderstorm's strong updraft (Figure 14.5), although the precise relationship between the two systems is not completely understood.

A tornadic circulation apparently stems from an interaction between the updraft in the thunderstorm and the larger scale horizontal wind. The horizontal wind must exhibit strong vertical shear in both speed and direction. That is, wind speed must increase with altitude, and wind direction must veer (turn clockwise) with altitude—from southeast at the surface to southwest or west aloft. The shear in wind speed causes air to rotate about a horizontal axis. When this rotation interacts with the updraft, the region of rotating air is tilted to a

**TABLE 14.2**
**The Fujita Tornado Intensity Scale**

| F-Scale | Category | Estimated wind speed, km/hr | (mi/hr) |
|---------|----------|------|---------|
| 0 | Weak | 65–118 | 40–73 |
| 1 | | 119–181 | 74–112 |
| 2 | Strong | 182–253 | 113–157 |
| 3 | | 254–332 | 158–206 |
| 4 | Violent | 333–419 | 207–260 |
| 5 | | 420–513 | 261–318 |

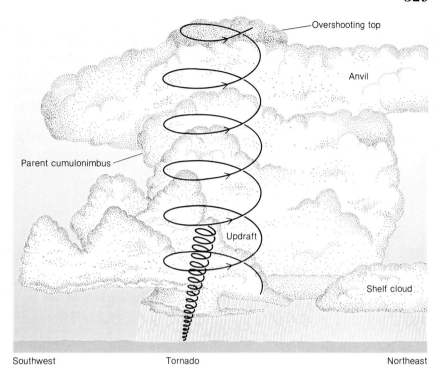

**FIGURE 14.5**
Features of a severe thunderstorm that spawns a tornado. [From J. T. Snow, "The Tornado," *Scientific American* 250, No. 4 (1984):91. George V. Kelvin Science Graphics; copyright © 1984 by Scientific American, Inc. All rights reserved.]

vertical position. The shear in the horizontal wind direction also adds to this rotation of air about a vertical axis. As a consequence, the entire updraft spins as a cylinder 10 to 20 km (6.2 to 12.4 mi) in diameter. This circulation system is known as a **mesocyclone,** and odds are about 1 in 2 that a mesocyclone will evolve into a tornado.

The mesocyclone circulation actually begins in the midtroposphere and, from there, builds upward and downward. Meanwhile, the updraft strengthens as more air converges toward the base of the thunderstorm. The updraft may eventually become so strong that it overshoots the top of the thunderstorm and produces a cloud bulge on top of the thunderstorm anvil (Figure 14.6).

For reasons again not well understood, the mesocyclone in a tornadic thunderstorm narrows and spirals downward toward the ground as a funnel. As the spinning mass of air narrows, its speed increases, perhaps to violent levels. This increase in speed is analogous to that of an ice skater performing a spin: The skater's rate of spin increases as she brings her arms closer to her body.

When a mesocyclone spawns a tornado, the intensity can be anywhere on the F-scale, from weak to devastating. On the other hand, tornadoes sometimes spin

off the gust front associated with a severe thunderstorm; these tornadoes are almost always weak.

Perhaps 80% of all North American tornadoes are linked to thunderstorms associated with midlatitude

**FIGURE 14.6**
The bulging top of a thunderstorm anvil indicates a strong updraft and a potentially severe system. [Photograph by Arjen and Jerrine Verkaik/SKYART]

cyclones. Most of the others are the product of convective instability triggered by hurricanes. In fact, most hurricanes that strike the southeastern United States are accompanied by tornadoes. Tornadoes often develop on the northeast flank of a hurricane, after the system has curved toward the north and northeast (Chapter 15).

Our description of tornado genesis is tentative because direct monitoring of tornado characteristics is not feasible; tornadic winds destroy traditional weather instruments. Between 1981 and 1983, *storm chasers* from the University of Oklahoma tried deploying a 180-kg (400-lb) tornado-resistant instrument package in the path of tornadoes without much success. Today, storm chasers rely primarily on photography, balloon-borne instruments that monitor conditions surrounding severe storms, plus portable Doppler radar, a new type of radar system that can resolve the circulation within thunderstorms.

## Tornado Look-Alikes

A **waterspout** is a tornadolike disturbance that occurs over the ocean or over a large inland lake (Figure 14.7). It is so named because it consists of a whirling mass of water that appears to stream out of the base of its parent cloud, which can be either a cumulus congestus or cumulonimbus. A waterspout is usually considerably less energetic, smaller, and shorter-lived than a tornado. The rare intense waterspout may well be a tornado that formed over land and then traveled out over a body of water. In any event, boaters should stay clear of waterspouts.

A tornado has other look-alikes. Distant **virga** (rain or snow that vaporizes before reaching the ground) may be mistaken for a tornado because of its cylindrical or funnel shape (Chapter 8). Absence of any rotary motion, however, distinguishes virga from a funnel cloud. In deserts, or wherever exposed soil dries out, intense solar heating often gives rise to a swirling mass of dust, called a **dust devil,** which resembles a tornado but forms at ground level, is not associated with any clouds, and causes little if any damage (Chapter 12).

## Weather Radar

Weather **radar*** is a valuable tool for detecting and tracking severe weather systems. Thunderstorm cells are so small that often they escape detection by the widely spaced network of weather stations. Weather radar, on the other hand, scans a wide area continuously and can locate small and isolated pockets of precipitation. As of this writing, conventional weather radar, which has served weather observers for more than 30 years, is being replaced by much more sophisticated Doppler radar systems.

### CONVENTIONAL RADAR

A conventional weather radar unit operated by the National Weather Service emits short pulses of microwaves having wavelengths of 5 or 10 cm. Radar waves are scattered by precipitation but are not scattered by

*The acronym for RAdio Detection And Ranging.

**FIGURE 14.7**
Waterspouts, such as these off the Grand Bahama Islands, are weak tornado like systems that develop over large bodies of water. [NOAA photograph]

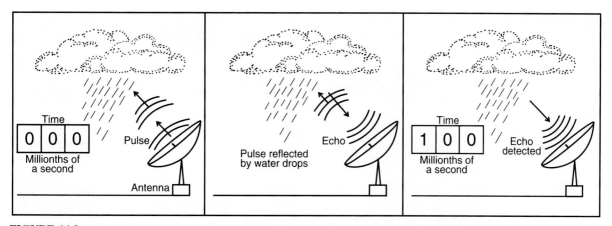

**FIGURE 14.8**
A radar system continuously sends out pulses of energy. Precipitation scatters radar signals back
to a receiving unit that displays echos on a television-type screen.

the very small droplets or ice crystals that compose
clouds. Thus, weather radar detects *(sees)* rain or snow
but not the parent clouds. When precipitation is targeted, a portion of the radar waves is scattered back to
a receiving unit, which displays the return signal, called
a **radar echo,** as electrical pulses on a cathode ray tube
similar to a television screen (Figure 14.8). Radar signals are sent out and received hundreds of times each
second as the radar continuously scans a 360-degree
circle (Figure 14.9). The product is a map of the precipitation pattern surrounding the radar unit. Because
the speed of the radar pulse is known, the time interval
between emission and reception of the radar signal can
be calibrated to give the distance to the precipitation.

The intensity of a radar echo depends on the reflectivity of the targeted precipitation and is greatest for
large drops and hailstones. To a lesser extent the concentration of raindrops in the path of the radar beam
also influences echo intensity. Hence, echo intensity is
used as an index of rainfall rate and to identify severe
thunderstorm cells, which often contain large hail.
Some radar units are equipped with electronic devices
that display echo intensity on a color scale, so, for example, a patch of red indicates very heavy rain and, at
the other end of the scale, green indicates very light
rain (Figure 14.10).

Weather radar monitors the development and dissipation of thunderstorm cells, their direction and speed
of forward motion, and the spiral bands of rainfall associated with hurricanes (Figure 14.11). Conventional
radar cannot detect a tornado directly, but in some cases

**FIGURE 14.9**
This dome houses a radar dish that continually scans a 360-
degree circle. [Photograph by J. M. Moran]

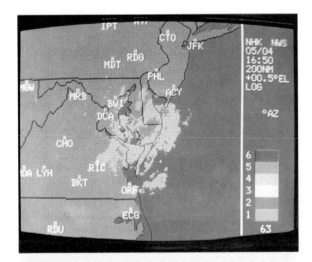

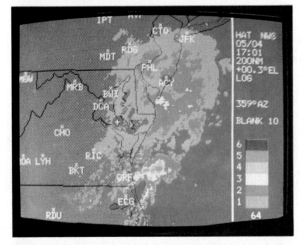

**FIGURE 14.10**
Conventional weather radar displays in which echo intensity is graduated by color. The most intense echo (caused by heavy rain) is indicated by dark red, and the weakest echo (light rain) by light green. Radar image (A) was taken 11 minutes before radar image (B); this indicates an intensification of rain showers over the mid-Atlantic coast. [Courtesy of Alden Electronics]

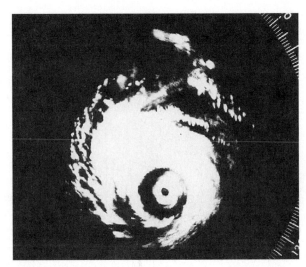

**FIGURE 14.11**
Conventional radar display of spiral bands of heavy rainfall associated with a hurricane. [NOAA photograph]

when a mesocyclone is present, a hook-shaped echo appears on a radar screen (Figure 14.12). If present, the hook usually appears on the south side of a severe thunderstorm cell. Apparently, a **hook echo** indicates rainfall being drawn around the mesocyclone within a severe thunderstorm.

Appearance of ground clutter on a radar screen can be misleading. **Ground clutter** usually refers to echos from nearby objects on the ground such as buildings or smokestacks. The ground clutter pattern is unique

to the radar site, appears all the time, and hence is readily distinguished from precipitation echos. On the other hand, special atmospheric conditions sometimes develop that give rise to ground clutter that may be mistaken for precipitation. For example, a strong temperature inversion may cause the outgoing radar signal to bend downward and intercept the ground. Such ground clutter differs from precipitation echos in that it is stationary, stronger, and more grainy in appearance.

To this point, we have described the *plan-position indicator (PPI)* radar, in which a microwave beam sweeps out an almost horizontal circle. The observing circle is limited in size by the curvature of the Earth and may have a radius of up to 400 km (248 mi). A second type of radar display, the *range-height indicator (RHI),* scans up and down rather than horizontally and is used to approximate the altitude of thunderstorm tops by detecting precipitation in the upper reaches of clouds. As noted in Chapter 13, the altitude of cloud top gives an indication of thunderstorm intensity: the higher the echo, the more intense is the storm. Conventional weather radar units can operate in either the *PPI* or the *RHI* mode.

The first reported use of radar for meteorological purposes was on 20 February 1941, when a radar unit on the south coast of England tracked a thunderstorm a distance of 11 km (7 mi). More widespread application of radar for weather analysis began shortly after World War II. The early weather radars were short-

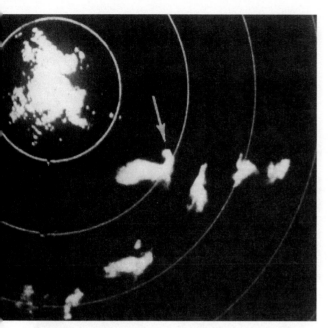

**FIGURE 14.12**
Conventional radar display of a hook echo, which may indicate tornadic circulation of air. [NOAA photograph]

range surplus military units. Not until the mid-1950s, following major tornado and hurricane disasters, did the U.S. Congress allocate funds for purchase of new long-range radar units designed specifically for meteorological applications. In 1959, these units, called *WSR-57* radars, went into service for hurricane detection along the Atlantic and Gulf coasts and for tornado and severe thunderstorm monitoring in the central United States. In the ensuing years, area coverage by weather radar expanded steadily; by the late 1960s, radar had become a routine component of televised weathercasts.

## DOPPLER RADAR

Over the past two decades, much research and field testing have focused on a new type of weather radar that utilizes the **Doppler effect,** named for Johann Christian Doppler, the Austrian physicist who first explained the phenomenon in 1842. This principle is also the basis for the police radar used to monitor traffic movement and the *gun* that measures the speed of a pitched ball. The Doppler effect refers to the shift in frequency* of sound waves or electromagnetic waves emanating from a moving source. For example, the pitch (frequency) of a train whistle is higher as a train approaches you and drops off as the train moves away. As shown schematically in Figure 14.13A, the crests of sound waves generated by a stationary source form evenly spaced concentric circles about the source, and the wave frequency is uniform. If the source is moving, however, as in Figure 14.13B, the wave crests become more closely spaced in the direction in which the source is moving so that the frequency is greater ahead of the source and less behind the source. You can experience this same change in pitch by moving a buzzing electric razor past your ear.

**Doppler radar** is a conventional radar that has the added capability of determining the detailed motion of the targeted precipitation toward or away from the radar unit. As a raindrop moves away from or toward a radar unit, the frequency of the radar signal (microwaves) shifts slightly between emission and the return echo. This frequency shift, the Doppler effect, is calibrated in terms of the motion of the target precipitation. With this capability, it is possible to measure the

*Frequency is the number of wave crests that pass through a point in a specified period of time.

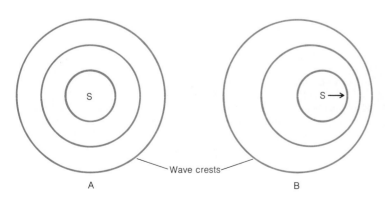

A          B

Wave crests

**FIGURE 14.13**
The Doppler effect is the shift in frequency of sound or electromagnetic waves that accompanies the relative motion of the wave source or wave receiver. (A) A sound-wave source (a train whistle, for example) is stationary, and wave frequency is uniform everywhere. (B) The wave source is in motion so that wave frequency is greater ahead of the source than behind the source.

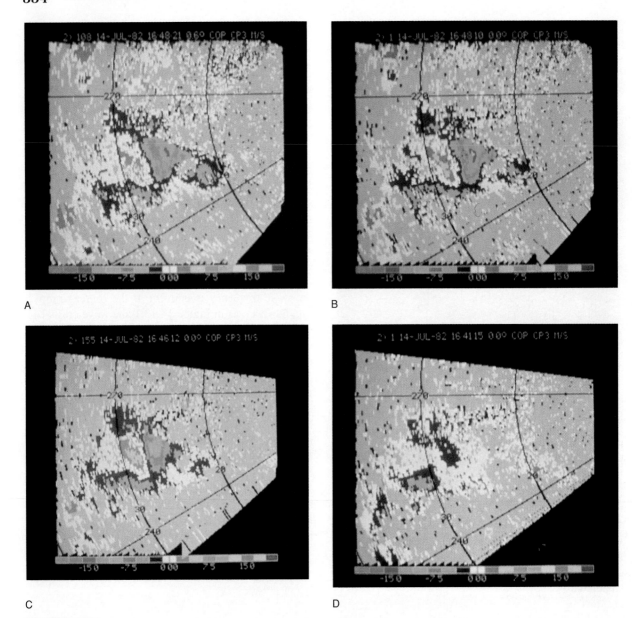

**FIGURE 14.14**

The life cycle of a microburst as documented in a sequence of four Doppler radar images. The radar unit is located to the northeast, and range rings are drawn at 5-mile intervals. Orange and yellow (warm colors) indicate air motion away from the radar; blue and green (cold colors) indicate air motion toward the radar. A starburst circulation pattern—characteristic of a microburst—appears at the center of the first three images. A microburst develops in the first two images (A & B), reaches maximum intensity in the third image (C), and rapidly dissipates in the fourth image (D). The time span between the first and final images is about 7 minutes. [National Center for Atmospheric Research/University Corporation for Atmospheric Research/National Science Foundation]

air circulation pattern within a thunderstorm or other storm cloud. Multiple Doppler radar units viewing the same storm simultaneously create a three-dimensional image of air circulation.

The principal advantage of Doppler radar over conventional radar is that it can determine the detailed circulation *within* a weather system rather than just the general displacement of an area of precipitation. Thus, Doppler radar sees inside clouds and severe thunderstorms and detects mesocyclones, developing tornadoes, strong wind shears, and gust fronts. Doppler radar displays, such as the sequence in Figure 14.14, are color-coded so that greens and blues (cold colors) indicate motion toward the radar and reds and yellows (warm colors) indicate motion away from the radar.

Because Doppler radar monitors a tornado as it develops within a mesocyclone and before it begins descending to the surface, Doppler provides more advance warning of severe weather than does conventional radar. Field tests of Doppler radar indicate that a warning of up to 20 minutes is quite feasible. Conventional radar, on the other hand, provides an average warning time of only about 2 minutes because spotters on the ground must confirm a suspected tornado signature on radar. The greater advanced warning time provided by Doppler radar likely will save many more lives.

Current weather radars are scheduled to be replaced by a new multimillion dollar network of Doppler radars during the mid-1990s. *NEXRAD,* for *NEXt Generation Weather RADar,* is a joint effort of the National Weather Service, the Air Weather Service of the U.S. Air Force, and the Federal Aviation Administration (FAA). It is intended to provide data for civil, military, and aviation needs. Plans call for a network of 113 10-cm microwave units, each capable of completing one 360-degree sweep every 5 minutes, detecting target precipitation at a range of 460 km (285 mi), and measuring velocity of particles as distant as 230 km (143 mi). A computer will process data from each NEXRAD unit into a variety of guidance products (such as Figure 14.14). Through a telephone link, users may access these products at 157 centers nationwide.

A more recent application of Doppler technology is to monitor winds aloft. For more on this subject, see the Special Topic "Wind Profilers."

# Conclusions

A tornado is a short-lived, small-scale weather system that is usually spawned by a severe thunderstorm. Tornado development requires a special combination of atmospheric and terrain conditions so tornadoes are most frequent in spring over the central United States. Hazards of tornadoes are extremely high winds, strong updraft, subsidiary vortices, and an abrupt drop in air pressure. The intensity of a tornado is rated from F0 (weak) to F5 (violent) based on wind speed estimated from property damage. The most intense tornadoes develop in the strong updraft of supercells.

Weather radar sends out pulses of microwave radiation that detect areas of precipitation. The new Doppler radar has the advantage over conventional radar of monitoring the detailed circulation of air within a thunderstorm. In this way, Doppler radar can provide early warning of tornado development.

In the next chapter, we continue our discussion of weather systems with a focus on hurricanes.

## *Key Terms*

| | |
|---|---|
| tornado | dust devil |
| funnel cloud | radar |
| tornado alley | radar echo |
| F-scale | hook echo |
| mesocyclone | ground clutter |
| waterspout | Doppler effect |
| virga | Doppler radar |

## *Summary Statements*

☐ A tornado is a small mass of air that whirls rapidly about a nearly vertical axis and is made visible by clouds, dust, and debris sucked into the system.

☐ An exceptionally steep horizontal air pressure gradient between the tornado center and outer edge is the force ultimately responsible for the violence of a tornado. A tornado is too small to be influenced by the Coriolis effect.

☐ Most tornadoes occur during the spring in a corridor stretching from Texas northward to Nebraska and eastward to Illinois and Indiana.

## SPECIAL TOPIC

# Wind Profilers

A recent application of Doppler radar promises to provide meteorologists with much closer surveillance of winds aloft than is now possible by rawinsondes (Chapter 1). A *wind profiler* consists of a wire-mesh antenna about half the size of a tennis court that sends radar signals (wavelengths of 33 cm to 6 m) straight up into the atmosphere. The signal detects changes in atmospheric density caused by the turbulent mixing of volumes of air that differ slightly in temperature and humidity. Fluctuations in the index of refraction are used as a tracer of the mean horizontal wind.

In May 1992, the National Atmospheric and Oceanic Administration (NOAA) announced that the nation's first wind profiler network was up and operating. The network consists of 29 wind profilers located in 15 midwestern states (Figure 1). Each profiler monitors wind speed and direction at 72 different levels up to an altitude of 16,000 m (52,500 ft). Data are automatically collected every 6 minutes and transmitted to a quality-control hub in Boulder, Colorado. Wind data are then averaged over 1-hr periods and made available to National Weather Service offices throughout the United States.

The principal advantage of a wind profiler over a rawinsonde is the much greater frequency of wind observations. As noted in Chapter 1, rawindsondes are launched once every 12 hr. Significant changes can take place in the upper-level steering currents over much shorter periods. Such shifts could, for example, alter a cyclone track so that a city receives snow instead of rain. More detailed monitoring of winds aloft will also permit airlines to better plan flight routes that avoid strong headwinds and take advantage of tailwinds.

Wind profilers are unlikely to replace rawindsondes, however. Most profilers cannot obtain temperature soundings, and none can monitor the variation of humidity with altitude.

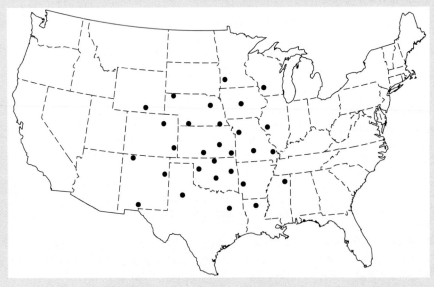

**FIGURE 1**
Locations of stations in the NOAA wind profiler network.

☐ Synoptic weather conditions favorable for the outbreak of tornadoes progress northward (with the sun) from the Gulf Coast in early spring to southern Canada by early summer.

☐ When a tornado strikes, damage is caused by very high winds, a strong updraft, subsidiary vortices, and an abrupt air pressure drop.

☐ Based on wind speed estimated from damage to structures, tornadoes are classified as weak, strong, or violent on the F-scale. Most tornadoes are weak, and most fatalities are caused by rare violent tornadoes.

☐ Most tornadoes develop out of a mesocyclone that forms in the strong updraft of a severe thunderstorm (supercell) by a process that is not yet well understood. The circulation in a tornado is apparently the consequence of an interaction between a thunderstorm updraft and strong shear in the horizontal wind.

☐ Waterspouts, virga, and dust devils resemble tornadoes in appearance only.

☐ Weather radar determines the location and movement of areas of precipitation. Echo strength increases with precipitation intensity.

☐ Currently, conventional radar is being replaced by Doppler radar, which can determine the detailed movement of targeted precipitation toward or away from the radar unit based upon a frequency shift. Hence, Doppler radar monitors the circulation within a severe thunderstorm and thus can provide advance warning of tornado development.

## Review Questions

1. What is a tornado?
2. Distinguish between a tornado and a funnel cloud.
3. Tornadoes usually move in what direction? Speculate on the reason why.
4. What is the principal force operating within a tornado? Is the Coriolis effect important?
5. Why are tornadoes more likely to form over flat and dry terrain than over mountainous or wet terrain?
6. Where is *tornado alley?*
7. Why is the troposphere less stable in spring than in fall?
8. The location of principal tornadic activity shifts northward during the spring. Why?
9. Is it a good idea to open windows when a tornado approaches in order to help equalize air pressure between the interior and exterior of a building? Explain your response.
10. What is the basis for the F-scale of tornado intensity? Where do most U.S. tornadoes rank on this scale?
11. Describe what happens as a mesocyclone evolves into a tornado.
12. Why is our understanding of tornado genesis far from complete?
13. How do tornadoes compare in appearance and intensity with waterspouts?
14. Compare and contrast a tornado with a dust devil.
15. How can virga be distinguished from a funnel cloud?

16. What type of electromagnetic radiation is utilized in a weather radar?
17. Conventional radar cannot detect clouds. Explain why.
18. What is the relationship between radar echo intensity and the severity of a thunderstorm?
19. What causes a hook echo? What is ground clutter?
20. What is the Doppler effect, and what advantages does Doppler radar offer over conventional radar?

## Questions for Critical Thinking

1. Does the centripetal force operate in a tornado? Explain your response.
2. Comment on whether conventional radar can be used to determine the precise altitude of a thunderstorm top.
3. Why are tornadoes unusual in southern California?
4. Compare and contrast a tornado and a downburst. How do the patterns of property damage differ?
5. A cloud bulge appears on the anvil top of a thunderstorm. Why might this suggest a severe thunderstorm?

## Selected Readings

AMERICAN METEOROLOGICAL SOCIETY. "Tornado Forecasting and Warning," *Bulletin of the American Meteorological Society* 72 (1991):1270–1272. Presents a policy statement on the tornado warning system in the United States.

FUJITA, T. T. "Rugged Rocky Mountain Tornado," *Weatherwise* 41, No. 2 (1988):80–83. Reports on a rare tornado in the mountainous Teton Wilderness and Yellowstone National Park.

GALWAY, J. G. "Ten Famous Tornado Outbreaks," *Weatherwise* 34, No. 3 (1981):100–109. Describes great tornado outbreaks during the period 1870 to 1979.

MARSHALL, T. "Dryline Magic," *Weatherwise* 45, No. 2 (1992):25–28. Discusses conditions that are favorable for severe thunderstorm development along a dryline on the Western Plains.

MILLER, P. "Tornado!" *National Geographic* 171 (1986):690–715. Presents a well-illustrated summary of the causes and effects of nature's most violent weather system.

MILNER, S. "NEXRAD, The Coming Revolution in Radar Storm Detection and Warning," *Weatherwise* 39, No. 2 (1986):72–85. Provides an excellent synopsis of the history of development of Doppler radar systems.

SNOW, J. T. "The Tornado," *Scientific American* 250, No. 4 (1984):86–96. Gives a well-illustrated review of tornado characteristics and genesis.

# 15 Hurricanes

*Wheeling, the careening winds
arrive with lariats
and tambourines of rain.
Torn-to-pieces, mud-dark
flounces of Caribbean
cumulus keep passing,
keep passing. By afternoon
rinsed transparencies begin
to open overhead.
Mediterranean
windowpanes of clearness.*

AMY CLAMPITT
*The Kingfisher, "The Edge of the
Hurricane"*

Visible satellite image of Hurricane Hugo as its eye slammed ashore near Charleston, South Carolina, about midnight on 21 September 1989. This violent storm took 21 lives and caused an estimated $7 billion in property damage on the United States mainland. [NOAA National Environmental Satellite, Data, and Information Service]

I N THE three hours it took Hurricane Andrew* to cross extreme southern Florida on the morning of 24 August 1992, its winds, gusting in excess of 264 km (165 mi) per hour, contributed to the deaths of 25 people and left 180,000 homeless (Figure 15.1). With property damage estimated at $30 billion, Hurricane Andrew was the most costly natural disaster in U. S. history. To make matters worse, Andrew continued to push west and north-westward across the Gulf of Mexico and two days later struck the Louisiana coast where it claimed 4 more lives and caused $400 million in property damage.

In this chapter, our focus is on hurricanes. We describe their characteristics, geographical and seasonal distribution, associated hazards, and life cycle. We also consider in separate sections, the hurricane threat to the southeastern United States and efforts to modify hurricanes.

*Hurricane names alternate each year between male and female names in alphabetical order.

# Hurricane Characteristics

A **hurricane,** named for *Huracan*, a Carib god of evil, is a violent cyclone that originates over tropical oceans, usually in late summer or early fall. By definition, a hurricane has a maximum sustained wind speed greater

**FIGURE 15.1**
The path of Hurricane Andrew as its eye tracked across southern Florida on 24 August 1992 (A). Some of the extensive property damage caused by the winds of Hurricane Andrew (B). [Diagrams from J. Williams, "Tracking Andrew from Ground Zero," *Weatherwise* 45, No. 6 (1993):11; photograph © *Orlando Sentinel*/Gamma Liaison]

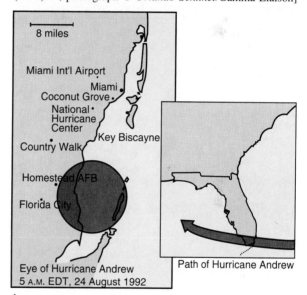

A

B

# Naming Hurricanes

When sustained winds in a tropical depression reach 65 km (40 mi) per hour, the intensifying system is designated a tropical storm and assigned a name. Between 1953 and 1978, meteorologists used an alphabetical sequence of female names for tropical storms and hurricanes in the Atlantic, Caribbean, and Gulf of Mexico.

Since 1979, female names have alternated with male names. A six-year cycle of names is now used, although the names of very destructive hurricanes (e.g., Gilbert, 1988; Hugo, 1989; Bob, 1991; Andrew, 1992) are removed from the list. Tropical storm and hurricane names for 1994–1996 are as follows.

| 1994 | 1995 | 1996 |
| --- | --- | --- |
| Alberto | Allison | Arthur |
| Beryl | Barry | Bertha |
| Chris | Chantal | Cesar |
| Debby | Dean | Diana |
| Ernesto | Erin | Edouard |
| Florence | Felix | Fran |
| Gordon | Gabrielle | Gustav |
| Helen | Humberto | Hortense |
| Isaac | Iris | Isidore |
| Joyce | Jerry | Josephine |
| Keith | Karen | Klaus |
| Leslie | Luis | Lili |
| Michael | Marilyn | Marco |
| Nadine | Noel | Nana |
| Oscar | Opal | Omar |
| Patty | Pable | Paloma |
| Rafael | Roxanne | Rene |
| Sandy | Sebastien | Sally |
| Tony | Tanya | Teddy |
| Valerie | Van | Vicky |
| William | Wendy | Wilfred |

than 119 km (74 mi) per hour, although winds in very intense hurricanes may top 250 km (155 mi) per hour.

Perhaps the most convenient way to describe a hurricane is to contrast it with the midlatitude, or *extratropical*, cyclone, we examined in Chapter 11. Hurricanes develop in a uniform mass of very warm and humid air, so they have no associated fronts or frontal weather. Air pressure is distributed symmetrically about the system center, and thus, as shown in Figure 15.2, isobars form a pattern of closely spaced concentric circles. Typically, the central pressure is considerably lower and the horizontal air pressure gradient much steeper in a hurricane than in an extratropical cyclone.

Reconnaissance aircraft extrapolated a surface air pressure of 888 mb (26.22 in.) at the center of Hurricane Gilbert while the storm was over the northwest Caribbean Sea on 13 September 1988. This is the lowest sea-level air pressure ever observed in the Western Hemisphere. In addition, a hurricane is usually much smaller, averaging a third of the diameter of a typical midlatitude cyclone. Rarely do hurricane-force winds extend much more than 120 km (75 mi) from the storm center (Figure 15.3).

Structurally, a mature hurricane is a warm-core, low-pressure system that weakens rapidly with altitude, especially above 3000 m (9800 ft). In the upper tropo-

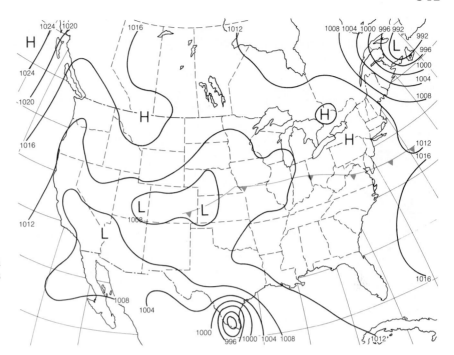

**FIGURE 15.2**
Surface isobar pattern on the morning of 10 August 1980 as Hurricane Allen tracked northwestward from the Gulf of Mexico and into extreme southeast Texas. Isobars (in mb) form closely spaced concentric circles about the hurricane center. See Figure 15.5 for the corresponding visible satellite image. [Redrawn from NOAA surface weather map]

sphere, at an altitude of about 15,000 m (49,000 ft), it is usual to find anticyclonic airflow above the hurricane. At the hurricane center is an area of almost cloudless skies, subsiding air, and light winds (less than 25 km per hour), called the **eye** (Figure 15.4). The eye generally ranges from 20 to 65 km (12 to 40 mi) across, shrinking in diameter as the hurricane intensifies and winds strengthen.

At a hurricane's typical rate of forward motion, it is not unusual for the eye to take up to an hour to pass over a locality. Hence, people may be deceived into thinking the storm has ended when clearing skies and slackening winds follow a hurricane's initial blow. They may well be experiencing the hurricane's eye— heavy rains and ferocious winds will soon resume but from a different direction.

Bordering the eye of a mature hurricane is the **eye wall,** a ring of cumulonimbus clouds that produce

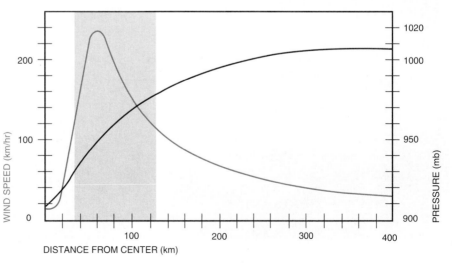

**FIGURE 15.3**
Vertical cross-section through a typical hurricane showing the variation of wind speed and sea-level pressure with distance from the storm center. Shaded area indicates hurricane-force winds. [Reprinted with the permission of Macmillan College Publishing Company from *Meteorology,* 6th ed., by Richard A. Anthes. Copyright © 1992 by Macmillan College Publishing Company, Inc.]

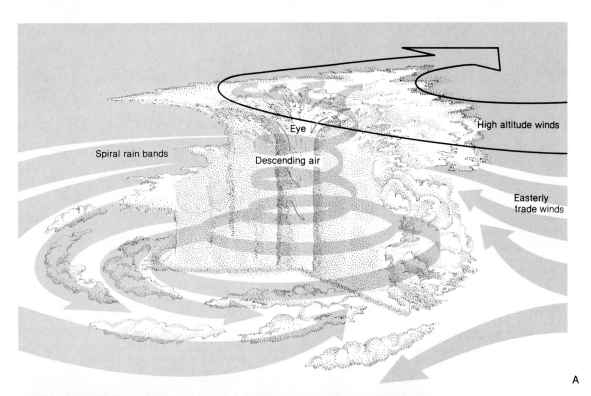

A

B

**FIGURE 15.4**
(A) The internal structure of a hurricane as determined by radar and satellite monitoring. In this artist's conception, the vertical dimension is greatly exaggerated. Actual hurricanes are less than 15 km in altitude and have diameters of several hundred kilometers.
(B) Hurricane Elena over the Gulf of Mexico as photographed from the Space Shuttle Discovery on 2 September 1985. Note the spiral cloud bands surrounding the eye of the storm. The Earth's curvature is visible in the background. [Diagram from NOAA, *Hurricane.* Washington, DC: Superintendent of Documents, 1977, p. 11; NASA photograph courtesy of the Johnson Space Center, Houston, Texas]

heavy rains and very strong winds. The most dangerous and destructive part of the hurricane is near the eye on the side where the wind blows in the same direction as the storm's forward motion. On that side, hurricane winds are added to the steering current producing the storm's stongest surface winds. Cloud bands accompanied by hurricane-force winds and heavy convective showers spiral inward to the eye wall. All this is surrounded by an outer region of high clouds (cirrus or cirrostratus) and cyclonic winds. The satellite photograph in Figure 15.5 shows the typical cloud pattern associated with a hurricane.

## Distribution of Hurricanes

Two conditions must be met before a hurricane can develop. One of these is very warm surface ocean water. Hurricane formation requires a sea-surface temperature of at least 26.5 °C (80 °F) through a depth of 60 m (200 ft) or more. Such exceptionally warm water sustains the hurricane circulation by the latent heat released when water evaporates from the ocean surface and subsequently condenses within the storm. Temperature governs the rate of evaporation of water, so the higher the sea-surface temperature, the greater is the supply of latent heat in the storm system. As a hurricane moves over colder water or land, however, it loses its warm-water energy source and weakens.

The second prerequisite for hurricane formation is a significant Coriolis effect. The influence of the Earth's rotation must be strong enough to sustain a cyclonic circulation. As noted in Chapter 9, the Coriolis effect weakens toward lower latitudes and becomes zero at the equator. The minimum latitude where the Coriolis effect is strong enough for hurricane formation is about 4 degrees.

The required combination of sufficient Coriolis effect and relatively high sea-surface temperature occurs only over certain portions of the world's oceans, identified in Figure 15.6. Most hurricanes form in the 5- to 20-degree latitude belt. Major hurricane breeding grounds are (1) the western tropical North Pacific, where a hurricane is called a **typhoon**,* (2) the South Indian Ocean east of Madagascar, (3) the North Indian Ocean (including the Arabian Sea and the Bay of Bengal), (4) tropical waters adjacent to Australia, (5) the Pacific Ocean west of Mexico, and (6) the tropical North Atlantic west of the bulging west coast of Africa (including the Caribbean Sea and Gulf of Mexico). Absence of hurricanes off either South American coast is noteworthy; it is due to relatively low sea-surface temperatures.

Only hurricanes spawned over the tropical Atlantic, Caribbean Sea, and Gulf of Mexico pose a serious threat to coastal North America. During an average hurricane season, six hurricanes and four tropical storms (precursors to hurricanes) form over these waters. And, on average, two hurricanes strike the United States coast each year. The Florida Keys is the most hurricane-prone region of the United States, having recorded 43 hurricanes in the past 100 years. For your use, we have included a hurricane tracking chart in Appendix III.

The West Coast of North America is rarely a target

*By convention, tropical storms over the Pacific Ocean are called *typhoons* when they occur west of 180 degrees longitude, and *hurricanes* when they occur east of 180 degrees longitude.

**FIGURE 15.5**
Visible satellite image of the cloud pattern associated with Hurricane Allen over southeast Texas on 10 August 1980. [NOAA photograph]

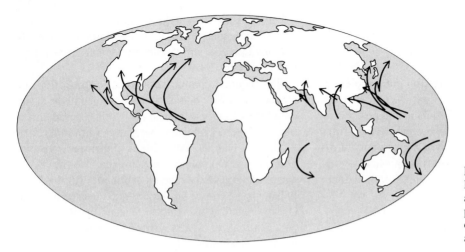

**FIGURE 15.6**
Hurricane breeding grounds are located only over certain portions of the world's oceans. Arrows indicate average hurricane trajectories.

of hurricanes for two reasons: (1) the prevailing winds (northeast trades) are offshore and, hence, steer tropical storms that form west of Central America away from the coast, and (2) surface waters just off the southern California (and Baja California) coast normally are too cold to sustain hurricanes. During unusual circulation regimes, however, tropical storms have struck coastal southern California and even traveled over the desert Southwest. For example, on 9–10 September 1976, Hurricane Kathleen crossed Baja California and tracked near Yuma, Arizona, bringing torrential rains and floods to the Imperial and Lower Colorado River valleys.

The requirement of relatively high sea-surface temperature also makes hurricane occurrence distinctly seasonal. Because of the great thermal stability of ocean water, sea-surface temperatures reach a seasonal maximum long after the time of peak solar radiation. Consequently, most Atlantic hurricanes develop in late summer and early autumn; the official hurricane season runs from 1 June to 30 November.*

## Hazards of Hurricanes

Hazards of hurricanes are (1) strong winds, (2) associated tornadoes, (3) heavy rains, and (4) storm surge. About 90% of hurricane-related fatalities are caused by coastal and inland floodwaters.

*Over the North Pacific Ocean, the hurricane season is 15 May to 30 November.

Meteorologists have assumed that strong wind gusts are responsible for the most serious hurricane damage. After detailed analysis of the aftermath of Hurricane Andrew, however, T. Theodore Fujita argued that small but powerful whirlwinds embedded in the hurricane circulation were responsible for the most severe damage. The whirlwinds combined with hurricane winds to produce winds estimated at 320 km (200 mi) per hour. Based on mode of origin, Fujita describes these whirlwinds as *spin-up vortices.* Apparently, small eddies spin off the eye wall and are stretched vertically by powerful convection currents. Stretching accelerates the eddy circulation.

Hurricane winds diminish rapidly once the storm makes landfall so that most wind damage is confined to within about 200 km (124 mi) of the coastline. Two factors account for the abrupt drop in wind speed once a hurricane makes landfall. As noted earlier, over land a hurricane is no longer in contact with its energy source, warm ocean water. In addition, the increased surface roughness over land weakens the system. The land surface is rougher than the sea surface so that when a hurricane moves over land its surface winds are slowed and blow at a greater angle across isobars and toward the storm center (Chapter 9). This wind shift causes the storm to begin to fill, that is, the central pressure rises, the horizontal pressure gradient weakens, and winds slacken.

As noted in Chapter 14, it is common for tornadoes to accompany a hurricane once it makes landfall. Tornadoes are most probable after the hurricane enters the westerly steering current and curves toward the north and northeast. Various studies have found that tornadoes are most common to the northeast of the storm center and often develop outside of the region of hurricane-force winds.

Hurricanes also produce torrential rains with amounts typically in the range of 13 to 25 cm (5 to 10 in.). Even if the storm moves well inland, heavy rains often persist and may trigger flooding. Such was the impact of Hurricane Agnes in June 1972. Although Agnes was a weak hurricane, its heavy rains produced flooding that accounted for much of the storm's $6.3 billion in property damage.

Agnes tracked up the Eastern Seaboard from southwest of the Florida Keys to North Carolina and then northward into Pennsylvania. The storm brought beneficial rains to the Southeast,* which had been experiencing a dry spell, but heavy rains in the mid-Atlantic region were anything but beneficial. During the week prior to Agnes's arrival, frontal showers and thunderstorms brought soaking rains from Virginia to New England. Rainfall averaged 3 to 8 cm (1 to 3 in.) and locally topped 15 cm (6 in.). Agnes's torrential rains falling on already saturated soils produced runoff that triggered many record-breaking river crests and devastating floods.

Heaviest rains fell in a band 350 km (217 mi) wide from western North Carolina northeastward into central New York State. In this hilly terrain, rainfall totals were quite variable but generally ranged from 20 to 30 cm (8 to 12 in.) for the period 18–25 June. Some localities reported total rainfall in excess of 46 cm (18 in.) with more than 25 cm (10 in.) falling in only 24 hours.

Central Pennsylvania was particularly hard hit by flooding. Rising waters forced more than 250,000 people from their homes. Crests on the Susquehanna River averaged 4 to 5.5 m (13 to 18 ft) above flood stage, and at Wilkes-Barre the river crested at more than 5.5 m (18 ft) above flood stage (Figure 15.7). Of the 122 fatalities attributed to Agnes, 50 were in Pennsylvania, and flood damage in that state accounted for almost two-thirds of total storm damage.

---

*Although much can be said about the destructive aspects of hurricanes, their rains can be beneficial. For example, in the southeastern United States, hurricanes and tropical storms account for an average of 10% to 15% of June through October rainfall.

## *STORM SURGE*

More than strong winds and heavy rains, potentially the most devastating feature of hurricanes that strike coastal areas is the **storm surge,** a flood of ocean water that accompanies the storm. Low air pressure in a hurricane causes sea level to rise (about 0.5 m for every 50 mb drop in air pressure), and hurricane-force winds drive ocean waters over low-lying coastal areas, ravaging property and sometimes taking many lives (Figure 15.8).

A storm surge is superimposed on normal ocean tidal oscillations. Hence, the greatest potential for flooding and shoreline erosion occurs when a storm surge coincides with high tide in localities where the tidal range (difference in water level between high and low tides) is great. Average tidal range is relatively small along the Gulf Coast (less than 1 m) and greater along the Eastern Seaboard (typically several meters).

A storm surge of 1 to 2 m (3 to 6.5 ft) can be expected with a weak hurricane, whereas the storm surge accompanying an intense hurricane may top 5 m (16.4 ft). In the most deadly natural disaster in United States history, an estimated 6000 people perished, mostly by drowning, when a hurricane storm surge flooded Galveston, Texas, on 8 September 1900. Hurricane Camille, an exceptionally intense storm with winds to 300 km (186 mi) per hour, produced a maximum storm surge of 7.3 m (24 ft) at Pass Christian, Mississippi, on 17 August 1969. One of history's most disastrous

**FIGURE 15.7**
Waters of the Susquehanna River raged through downtown Wilkes-Barre, Pennsylvania, following heavy rainfall associated with Hurricane Agnes in June 1972. [U.S. Coast Guard photograph]

**TABLE 15.1**
**Saffir–Simpson Hurricane Intensity Scale**

| Scale number (category) | Central pressure | | Wind speed | | Storm surge | | Damage |
|---|---|---|---|---|---|---|---|
| | mb | in. | mi/hr | km/hr | ft | m | |
| 1 | ≥980 | ≥28.94 | 74-95 | 119-154 | 4-5 | 1-2 | Minimal |
| 2 | 965-979 | 28.50-28.91 | 96-110 | 155-178 | 6-8 | 2-3 | Moderate |
| 3 | 945-964 | 27.91-28.47 | 111-130 | 179-210 | 9-12 | 3-4 | Extensive |
| 4 | 920-944 | 27.17-27.88 | 131-155 | 211-250 | 13-18 | 4-6 | Extreme |
| 5 | <920 | <27.17 | >155 | >250 | >18 | >6 | Catastrophic |

storm surges hit the Bay of Bengal coast of Bangladesh on 13 November 1973. A storm surge of nearly 7 m (23 ft) flooded a vast coastal plain, claiming an estimated 300,000 lives by drowning.

## SAFFIR–SIMPSON INTENSITY SCALE

Like tornadoes, hurricanes have an intensity scale, known as the **Saffir–Simpson Hurricane Intensity Scale** after its designers, H. S. Saffir, a consulting engineer, and R. H. Simpson, former director of the National Hurricane Center. The scale rates hurricanes from 1 to 5 corresponding to increasing intensity. As shown in Table 15.1, each intensity category specifies (1) a range of central air pressure, (2) a range of wind speed, (3) storm surge potential, and (4) the potential for property damage. Of the 126 tropical storms or hurricanes that struck the U. S. Gulf or Atlantic coasts between 1949 and 1990, 25 (19.8%) were classified as

**FIGURE 15.8**
A hurricane storm surge produced by strong winds and low pressure associated with a hurricane approaching a coastal area. [After Michael D. Morgan et al., *Environmental Science: Managing Biological and Physical Resources.* Copyright © 1993 Wm. C. Brown Communications, Inc., Dubuque, Iowa.]

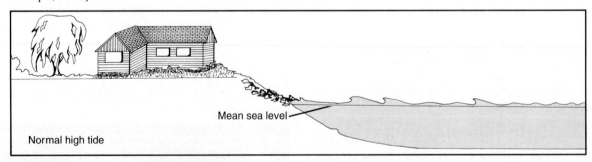

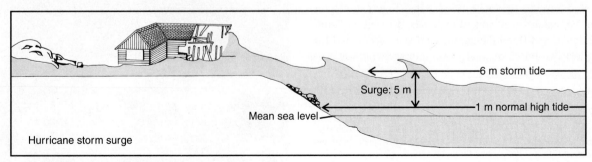

**TABLE 15.2**
**Ten Deadliest Hurricanes to Strike the United States Since 1900**

| Hurricane | Year | Category | Deaths |
|---|---|---|---|
| 1 Texas (Galveston) | 1900 | 4 | 6000 |
| 2 Florida (Lake Okeechobee) | 1928 | 4 | 1836 |
| 3 Florida (Keys and S. Texas) | 1919 | 4 | 600-900[a] |
| 4 New England | 1938 | 3 | 600 |
| 5 Florida (Keys) | 1935 | 5 | 408 |
| 6 Audrey (Louisiana and Texas) | 1957 | 4 | 390 |
| 7 Northeast U.S. | 1944 | 3 | 390[b] |
| 8 Louisiana (Grand Isle) | 1909 | 4 | 350 |
| 9 Louisiana (New Orleans) | 1915 | 4 | 275 |
| 10 Texas (Galveston) | 1915 | 4 | 275 |

*Source:* P. J. Hebert and G. Taylor. "The Deadliest, Costliest, and Most Intense United States Hurricanes of This Century (and Other Frequently Requested Hurricane Facts)." NOAA Technical Memorandum, NWS, NHC 18, 1983.

[a]Over 500 of these lost on ships at sea.

[b]Some 344 of these lost on ships at sea.

**TABLE 15.3**
**Ten Costliest Hurricanes to Strike the United States Since 1900**

| Hurricane | Year | Category | Damage (millions of $)[a] |
|---|---|---|---|
| 1 Andrew (FL, LA) | 1992 | 4 | 30,000 |
| 2 Hugo (South Carolina) | 1989 | 4 | 7,000 |
| 3 Betsy (FL to LA) | 1965 | 3 | 6,321 |
| 4 Agnes (FL to northeast U.S.) | 1972 | 1 | 6,279 |
| 5 Camille (MI and LA) | 1969 | 5 | 5,128 |
| 6 Diane (northeast U.S.) | 1955 | 1 | 4,108 |
| 7 New England | 1938 | 3 | 3,515 |
| 8 Frederic (AL & MS) | 1979 | 3 | 3,427 |
| 9 Alicia | 1983 | 3 | 2,340 |
| 10 Carol (northeast U.S.) | 1954 | 3 | 2,318 |

*Source:* NOAA, National Weather Service.

[a]Adjusted to 1989 dollars.

*major;* that is, they rated 3 or higher on the Saffir–Simpson scale. The 10 deadliest and 10 costliest U. S. hurricanes since 1900 are listed in Tables 15.2 and 15.3, respectively.

Property damage potential rises rapidly with ranking on the Saffir–Simpson Scale. In fact, destruction from a category 4 or 5 hurricane is 100 to 300 times greater than that caused by a category 1 hurricane. The 25 major hurricanes that made landfall on the Gulf or Atlantic coast between 1949 and 1990 accounted for 76% of all property damage from all landfalling tropical storms and hurricanes during the same period.

The last category 5 hurricane to make landfall on the Atlantic Coast was Gilbert in September 1988. (Camille was the previous one.) Gilbert originated as a tropical wave off the West African coast on 3 September. The wave followed a steady west–northwest track and, seven days later, strengthened to a hurricane while south of Puerto Rico. Gilbert then passed over the length of Jamaica on the 12th, intensified over the northwest Caribbean, and on 14 September made landfall as a category 5 hurricane near Cozumel on

Mexico's Yucatan Peninsula. At landfall sustained winds were estimated at 275 km (171 mi) per hour. The hurricane then continued northwestward into the Gulf of Mexico and made its final landfall (as category 3) on the coast of Mexico about 200 km (124 mi) south of the Texas border. Gilbert brought much death and destruction to both Mexico and Jamaica. The death toll was 202 in Mexico and 45 in Jamaica, and property damage in the two countries topped $4 billion.

## Life Cycle of Tropical Storms

The first sign that a hurricane may be in the making is the appearance of an organized cluster of thunderclouds over tropical seas. This region of convective activity is labeled a **tropical disturbance** if a center of low pressure is detectable at the surface. Chances are that the tropical disturbance was triggered by the ITCZ, by a trough in the westerlies intruding into the tropics from midlatitudes, or by a wave (or ripple) in the easterly trade winds (an *easterly wave*).

The most intense hurricanes that threaten North America usually develop out of convective cloud clusters associated with easterly waves that continually travel westward off the West African coast. Convergence on the east side of a wave helps to organize convective activity into a developing system (Figure 15.9). Storms originating in this way are referred to as *Cape Verde–type* hurricanes.

A very small percentage of convective cloud clusters in the tropical Atlantic actually evolve into full-blown hurricanes for several reasons.

1. Strong subsidence of air on the eastern flank of the Bermuda–Azores anticyclone and the associated *trade-wind inversion* inhibit deep convection.

2. The vertical wind shear (change of horizontal wind with altitude) over the tropical Atlantic is usually too great for hurricane formation. Strong winds aloft disrupt convection.

3. The middle troposphere is usually too dry; low vapor pressure at these levels inhibits intensification of the system.

If conditions favorable to hurricane genesis persist, a cyclonic circulation develops and the surface air pressure begins to fall. Water vapor condenses within the storm, releasing latent heat of vaporization, and the heated air rises. Expansional cooling of the rising air triggers more condensation, release of even more latent heat, and a further increase in buoyancy. Rising temperatures in the core of the storm, coupled with an anticyclonic outflow of air aloft, cause a sharp drop in air pressure, which, in turn, induces more rapid convergence of air at the surface. The consequent uplift surrounding the developing eye leads to additional condensation and release of latent heat, an example of positive feedback.

Through these processes, the tropical disturbance intensifies and winds strengthen. When sustained wind speeds top 37 km (23 mi) per hour, the developing storm is called a **tropical depression.** When wind speeds reach 65 km (40 mi) per hour, the system is classified as a **tropical storm.** If winds exceed 119 km (74 mi) per hour, the storm is officially designated a hurricane. As a hurricane decays, the storm is downgraded by reversing this classification sequence.

Hurricanes that threaten the Atlantic and Gulf coasts of the United States usually drift slowly westward with the trade winds across the tropical North Atlantic and into the Caribbean.* At this stage in the storm's trajectory, it is not unusual for the storm to travel at a mere 10 to 20 km (6 to 12 mi) per hour and take a week to cross the Atlantic. Once over the western Atlantic, however, the storm usually speeds up and begins to curve northward, and eventually northeastward, as it is caught in the midlatitude westerlies. Precisely where this curvature takes place determines whether the hurricane enters the Gulf of Mexico and tracks up the Mississippi River Valley, moves up the Eastern Seaboard, or curves back out to sea.

By the time a coastal hurricane reaches a latitude of about 30 degrees N, it may begin to acquire extratropical (midlatitude) characteristics, as colder air is drawn into the circulation of the system and fronts develop. From then on, the storm follows a life cycle similar to that of any other midlatitude cyclone and often ends by occluding over the North Atlantic.

Many hurricanes depart significantly from the track just described. The hurricane tracks shown in Figure 15.10, for example, are quite erratic. Sometimes a hurricane describes a complete circle or reverses direction. In addition, some hurricanes maintain their tropical

**FIGURE 15.9**
An easterly wave moving westward off the west coast of North Africa. Such waves sometimes develop into Cape Verde–type hurricanes.

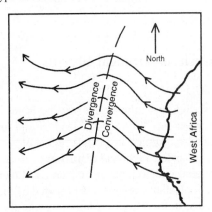

---

*Some hurricanes also originate over the warm waters of the Gulf of Mexico or Caribbean Sea.

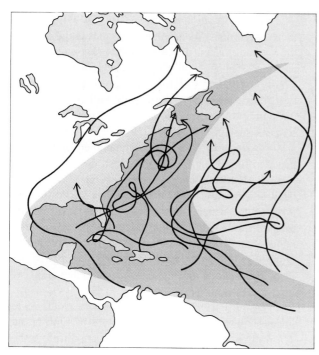

**FIGURE 15.10**
Hurricane trajectories are often erratic, as shown by these sample tracks. As indicated by the shaded area, however, hurricanes initially describe a generally westerly drift and eventually curve toward the northeast when they reach the western Atlantic. [From NOAA, *Hurricane*. Washington, DC: Superintendent of Documents, 1977, p. 13.]

characteristics even after traveling far north along the Atlantic Coast where they are fueled by warm Gulf Stream waters. The eye of Hurricane Hazel, for example, was still discernible when the storm passed over Toronto in October 1954.

New England, located more than 25 degrees of latitude north of the usual hurricane breeding grounds, has been the target of many full-blown hurricanes. The most deadly of these was the unnamed hurricane of 21 September 1938 (category 3). A storm surge ravaged the New England coast. Winds gusting over 200 km (124 mi) per hour severely damaged forests, and torrential rains caused flash flooding by rivers and streams. Fatalities were estimated at 600.

A more recent New England hurricane was Bob (category 3) in August 1991. Hurricane Bob moved swiftly up the East Coast from its origin over the northern Bahamas. The storm center passed just east of Cape Hatteras, North Carolina, and then remained

offshore until striking Rhode Island's south coast. The eye then passed just east of Boston, Massachusetts, and Portland, Maine. Top wind speed was 185 km (115 mi) per hour, and property damage totaled $1.5 billion.

## Hurricane Threat to the Southeast

Today, many atmospheric scientists and others are concerned about the hurricane threat to coastal areas of the southeastern United States. The rarity of major hurricanes during the 1970s and most of the 1980s lulled many residents of the coastal Southeast into a false sense of security and encouraged rapid population growth. (In fact, fewer hurricanes affected the United States in the 1970s than in any prior decade of the century.) More and more resort hotels, condominiums, and homes were constructed perilously close to the shoreline and even among coastal sand dunes (Figure 15.11). Population along the Gulf Coast from Florida west to Texas climbed from 5.2 million in 1960 to 10.1 million in 1990. Along the Atlantic Coast from Florida north to Virginia, the population more than doubled, from 4.4 million in 1960 to 9.2 million in 1990. Today, most Atlantic and Gulf Coast residents have never experienced a major hurricane.

The hurricane danger is particularly acute for residents of the nearly 300 barrier islands that fringe por-

**FIGURE 15.11**
The relative infrequency of tropical storms and hurricanes in recent decades has lulled some people in coastal areas of the Southeast into a false sense of security and inspired construction of hotels and condominiums within a few meters of high-tide level. [Photograph by J. M. Moran]

**FIGURE 15.12**
Miami Beach, Florida, is built on a barrier island and is thus vulnerable to a hurricane storm surge. [Photograph © *Southern Living*/Photo Researchers]

tions of the Atlantic and Gulf coasts. Barrier islands protect coastal beaches and wetlands by bearing the brunt of powerful storm-driven sea waves. The energy of storm waves is partially dissipated in shifting the sands that compose barrier islands. However, for decades, many barrier islands have undergone rapid development particularly for resorts and cottages, and some coastal cities, including Miami Beach, are built entirely on barrier islands (Figure 15.12). The exposed location of these islands makes them vulnerable to a hurricane storm surge. Thus, evacuation of people from barrier islands as well as other low-lying coastal areas is the most prudent strategy in the event of a major hurricane threat (Figure 15.13).

The effectiveness of coastal evacuation plans was tested in the late summer of 1985 when Hurricane Elena menaced the Gulf of Mexico coast (Figure 15.14). For four days, Elena followed an erratic path over the Gulf, first taking aim at southern Louisiana, then the Florida panhandle, and later central Florida, before reversing direction and finally coming ashore near Biloxi, Mississippi, on 2 September. Nearly a million people from Sarasota, Florida, to New Orleans, Louisiana, were forced to leave beachfront communities and flee to inland shelters. Some returned home only to evacuate again as Elena changed course. Although property damage was considerable ($1.25 billion) because of extensive flooding and winds that exceeded 160 km (100 mi) per hour, only four fatalities were attributed to Elena and none in the area of landfall.

Timely evacuation of residents of low-lying coastal areas saved many lives when hurricanes Hugo and Andrew struck. About midnight on 21 September 1989, Hurricane Hugo (category 4) moved ashore near Charleston, South Carolina. Hugo was a Cape Verde–type hurricane that several days earlier had ripped through the islands of the northeast Caribbean. Driven by peak sustained winds of 215 km (134 mi) per hour, a storm surge of 5 m (16.4 ft) slammed onto barrier islands off the South Carolina coast, destroying many beachfront homes and structures. Hugo's winds also dealt a severe blow to South Carolina's forests. Over 36% of the state's forested area, winds twisted and toppled more than half the trees—a timber loss estimated at more than $1 billion. Until Hurricane Andrew, Hugo was the most costly hurricane ever to strike the U. S. mainland, with total damage estimated at $7 billion.

In August 1992, when Hurricane Andrew (category 4) threatened, 1 million people in Florida and 1.7 million in Louisiana and Mississippi were either asked or ordered to evacuate their homes. The death tolls of 21 from Hugo and 29 from Andrew no doubt would have been much higher without evacuation.

**FIGURE 15.13**
In Florida, hurricane evacuation routes are marked by special road signs. [Photograph by J. M. Moran]

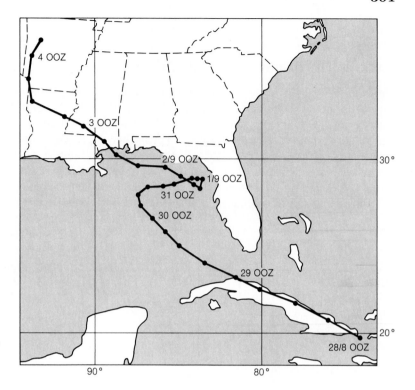

**FIGURE 15.14**
The track of Hurricane Elena from 28 August 1985 through 4 September 1985. The hurricane's position is reported every six hours. [From the National Hurricane Center, Coral Gables, Florida]

Successful evacuation, however, hinges on sufficient advance warning of a hurricane's approach. Unfortunately, as noted earlier, hurricanes often follow erratic paths and are notorious for sudden changes in direction and forward speed. This characteristic of hurricanes is especially troublesome for congested cities and isolated localities where the required evacuation time may be lengthy. For example, estimated evacuation time is 37 hours for the Florida Keys and 50 hours for New Orleans.

Other strategies adopted or advocated by public safety and other government officials in order to minimize loss of life and property to hurricanes are (1) stringent building codes, (2) preservation of mangrove swamps, and (3) elimination of federal floodplain insurance. Building codes may call for new homes to be more wind resistant by requiring bolts that anchor the floor to the foundation and steel brackets that attach the roof to the walls. In addition, in many areas, all new buildings must be elevated above the once-in-a-century flood level. Unfortunately, many structures were built prior to these tough codes, and, as the aftermath of Hurricane Andrew

demonstrated, in some areas, stringent codes were not always enforced.

Mangrove swamps, the final line of natural defense against hurricanes, help dissipate the energy of a storm surge. In too many cases, however, shoreline development has destroyed these swamps—often because the trees blocked the ocean view for residents of seaside dwellings.

A controversial strategy is to eliminate federal flood insurance for flood-prone coastal areas, as has been done already on undeveloped portions of barrier islands. Some people argue that by providing policies at low cost, federal flood insurance programs encourage development in flood-prone areas. Furthermore, this insurance enables homeowners to rebuild structures destroyed by a storm surge in the same hazardous location.

The need for better safeguards against the hurricane threat to the Southeast is underscored by recent indications that the frequency of major hurricanes in the tropical Atlantic may be on the rise. For more on this, see the Special Topic "Atlantic Hurricanes and West African Rainfall."

SPECIAL TOPIC

## Atlantic Hurricanes and West African Rainfall

Professor William M. Gray of Colorado State University noted an interesting relationship between the frequency of major hurricanes in the tropical Atlantic and rainfall in West Africa. Specifically, major Atlantic hurricanes (categories 3–5 on the Saffir–Simpson Scale) are much more numerous when West Africa is relatively wet than when West Africa is dry. This finding may prove valuable to long-range weather forecasters

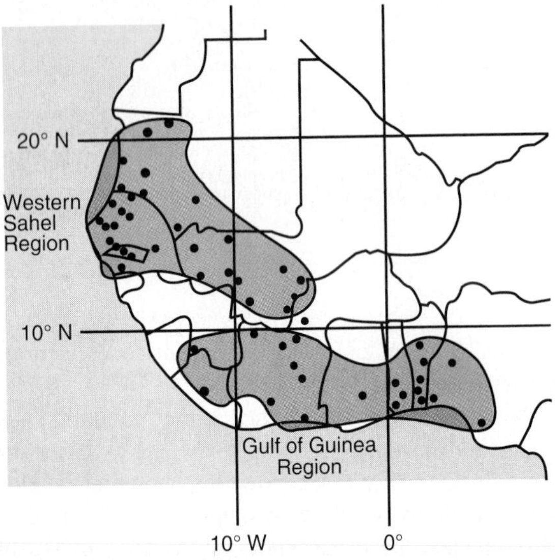

**FIGURE 1**
Locations of climate stations that supplied rainfall data for W. M. Gray's study of the relationship between West African rainfall and hurricane activity in the western Atlantic. [From W. M. Gray and C. W. Landsea, "African Rainfall as a Precursor of Hurricane-Related Destruction on the U.S. East Coast," *Bulletin of the American Meteorological Society* 73 (1992):1355]

and, based on historical trends in West African rainfall, suggests that Atlantic hurricane activity may be on the upswing.

Gray analyzed rainfall records from 38 climate stations in the Western Sahel (Senegal, southern Mauritania, western Mali, Gambia, and Guinea Bissau) and 24 stations in the Gulf of Guinea (Figure 1) for the 42-year period 1949–1990. Recall from Chapter 12 that the climate of this region is dominated by a monsoon circulation such that summer is the rainy season and winter is dry. Recall also that this region is subject to prolonged drought.

In West Africa, the 1950s and 1960s were relatively wet. During that period, 13 major hurricanes made landfall along the U. S. East Coast. On the other hand, during the Sahelian drought of 1970–1987, only one major hurricane (Gloria, 1985, category 3) struck the East Coast. In fact, Gray found that when rainfall in West Africa is above average prior to August 1, the likelihood of a major East Coast hurricane after August 1 is 10 to 20 times greater than when West Africa experiences below average rainfall prior to August 1 (Figure 2). On the other hand, Gray found no clear relationship between West African rainfall and hurricane intensity along the Gulf Coast (including the Florida panhandle).

What is the reason for the apparent link between Atlantic hurricane intensity and West African rainfall? As noted elsewhere in this chapter, the convergence associated with easterly waves that travel westward from West Africa is a key to hurricane development. According to Gray, such waves are stronger and better organized when rainfall in West Africa is relatively

## Hurricane Modification

Spurred by a succession of six destructive hurricanes that lashed the East Coast during the mid-1950s, atmospheric scientists stepped up efforts to learn more

about hurricanes and to devise hurricane modification techniques. To these ends, *Project STORM-FURY* was initiated in 1962. Over the following two decades, project scientists advanced our understanding of the formation, structure, and dynamics of hurricanes; improved hurricane forecasting; and up-

abundant. As long as all other atmospheric and sea-surface conditions are favorable, stronger and better organized waves are more likely to develop into hurricanes.

Experts on the climate of the African Sahel, such as Professor Sharon Nicholson of Florida State University, point out that for at least several hundred years the Sahel has experienced alternating multidecadal episodes of dry and wet conditions. Beginning in the late 1980s, signs appeared that perhaps the latest Sahelian drought was coming to an end. If this is the case and if past weather is any indicator of the future, then West Africa may be entering an extended wet episode. If Gray's correlation holds, an upswing in intense hurricane activity in the Atlantic may also be in the offing.

**FIGURE 2**

Contrast in hurricane activity between a 3-year period of higher than normal rainfall in West Africa (top diagram) and a similar period of lower than normal rainfall (bottom diagram). [From W. M. Gray and C. W. Landsea, "African Rainfall as a Precursor of Hurricane-Related Destruction on the U.S. East Coast," *Bulletin of the American Meteorological Society* 73 (1993):1363]

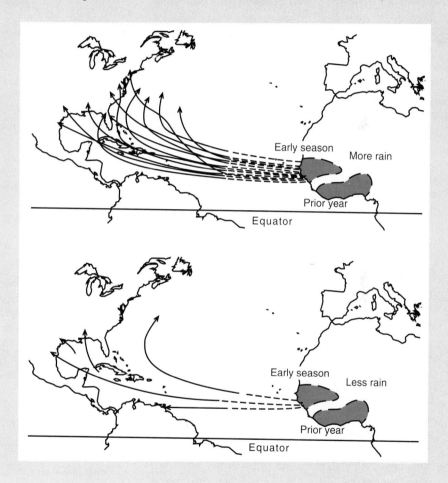

graded hurricane reconnaissance methods. The original goal of hurricane modification, however, was never realized.

The working hypothesis of *STORMFURY* was that seeding hurricanes with silver iodide (AgI) crystals would reduce wind strength. The argument went as follows: Seeding the band of convective clouds just beyond the eye wall would trigger additional latent heat release (as supercooled water droplets converted to ice crystals) that would enhance convection. This artificially invigorated convection would then dominate the convection in the eye wall, and a new eye wall would form at a greater distance from the storm center. Thereby, the band of strongest winds would be displaced farther from the eye center. Hence, the hurricane's circulation would also weaken just as a skater's spin rate slows when she extends her arms.

In the 1960s and 1970s, there were few opportunities for hurricane seeding experiments primarily because few hurricanes met the necessary criteria of being intense, having a well-defined eye, and being situated far from land. Some success was reported, however, in reducing winds (between 10% and 30%) in four hurricanes that were seeded. Later, these apparent successes were dismissed when new data called into question the original *STORMFURY* hypothesis. For one, it was discovered that convective clouds in hurricanes contain too little supercooled water for seeding to be effective. For another, recent monitoring of unmodified hurricanes found that changes in eye-wall diameter occur as part of the storm's natural evolution. It is likely, then, that *STORMFURY's* apparent successes at hurricane modification were by chance rather than by seeding.

## Conclusions

Hurricanes are long-lived, synoptic-scale weather systems that develop over tropical oceans. The requirements of high sea-surface temperature plus a significant Coriolis effect restrict hurricane formation to certain regions of the globe. Strong winds and torrential rains associated with hurricanes can take lives and cause considerable property damage. Based on intensity and potential for storm surge and property damage, a hurricane is assigned a rating on the Saffir–Simpson Scale. Most major Atlantic hurricanes drift slowly westward across the tropical North Atlantic and then curve north and northeastward as the system encounters the prevailing westerlies of midlatitudes.

With this chapter, we have completed our description of the genesis, life cycle, and characteristics of the various atmospheric circulation systems that affect midlatitude weather. With this background, we are now ready to apply principles of meteorology to weather forecasting.

## Key Terms

hurricane
eye (of a hurricane)
eye wall
typhoon
storm surge

Saffir–Simpson
Hurricane Intensity
Scale
tropical disturbance
tropical depression
tropical storm

## Summary Statements

☐ A hurricane is a violent cyclone that originates over tropical oceans, usually in late summer or early fall. It develops in a mass of warm, humid air, has no fronts, and is about one-third the diameter of an average extratropical cyclone.

☐ For tropical storms and hurricanes to develop, sea-surface temperatures must be 26.5 °C (80 °F) or higher through a depth of at least 60 m (200 ft), and the Coriolis effect must be of significant magnitude (latitude of at least 4 degrees north or south of the equator) to maintain a cyclonic circulation.

☐ The hazards of hurricanes include strong winds, associated tornadoes, torrential rains, and storm surge.

☐ Once a hurricane makes landfall, it loses its warm-water energy source and experiences greater surface roughness. Hence, its circulation weakens rapidly so that most wind damage is confined to the coast. Heavy rains often continue well inland, however, and can cause severe flooding.

☐ A storm surge is a flood of ocean water driven by hurricane winds. The greatest potential for coastal flooding and erosion occurs when a storm surge coincides with high tide and where the tidal range is relatively great.

☐ Based on intensity and potential for property damage, a hurricane is rated 1 (weak) to 5 (very intense) on the Saffir–Simpson Scale.

☐ Most major hurricanes that affect North America originate as convective cloud clusters associated with waves in the trade winds. For reasons not totally understood, some convective cloud clusters develop into tropical disturbances that evolve into tropical depressions, then tropical storms, and finally hurricanes.

☐ Rapid population growth in coastal areas coupled with the lull in intense hurricane activity during the 1970s and 1980s has heightened the hurricane hazard in the southeastern United States.

☐ Attempts to modify hurricanes through cloud seeding have been unsuccessful. However, out of such efforts has come a greater understanding of the dynamics of hurricanes and tropical storms.

## Review Questions

1. Compare and contrast a hurricane with a mature, mid-latitude cyclone.
2. Why are no fronts associated with a hurricane?
3. Describe the isobar pattern about the center of a hurricane.
4. Describe the weather within the eye of a hurricane.
5. What conditions are required for hurricane development?
6. Why do hurricanes fail to develop off the east and west coasts of South America?
7. Why is it rare for hurricanes to strike southern California?
8. Why are hurricanes unusual in winter and spring?
9. Why do hurricane winds weaken once the system makes landfall?
10. What is the most destructive impact of a hurricane on a low-lying coastal area?
11. What is the greatest hazard posed by a hurricane that tracks well inland?
12. What is the basis of the Saffir–Simpson Scale?
13. Describe the life cycle of a *Cape Verde–type* hurricane.
14. While in tropical latitudes, a hurricane or other tropical storm usually drifts slowly toward the west. Why west?
15. Describe what happens to a hurricane when it reaches middle latitudes.
16. Distinguish among (a) tropical disturbance, (b) tropical depression, (c) tropical storm, and (d) hurricane.
17. What factors make the southeastern United States particularly vulnerable to a hurricane disaster?
18. Why are barrier islands susceptible to a hurricane storm surge?
19. What was the working hypothesis in Project *STORM-FURY?* Why were efforts to modify hurricanes unsuccessful?
20. Are there any beneficial aspects of tropical storms and hurricanes? Elaborate on your response.

## Questions for Critical Thinking

1. Why is sea-surface temperature a critical factor in the development of tropical storms and hurricanes?
2. Suppose that efforts to modify hurricanes were success-ful. Speculate on what this might mean for poleward heat transport (Chapter 4).
3. Speculate on why the ten deadliest hurricanes occurred prior to 1958.
4. In view of the requirements for hurricane development, how is it possible that hurricanes have struck New England?
5. Speculate on the relationship between the circulation of the Bermuda–Azores anticyclone and the usual path taken by *Cape Verde–type* hurricanes.

## Selected Readings

AMERICAN METEOROLOGICAL SOCIETY. "Is the United States Headed for Hurricane Disaster?" *Bulletin of the American Meteorological Society* 67 (1986):537–538. Includes a statement of concern issued by the AMS.

COBB, H. "The Siege of New England," *Weatherwise* 42, No. 5 (1989):262–266. Describes two hurricanes, Carol and Edna, that struck New England within 10 days during the fall of 1954.

DOLAN, R., AND H. LINS. "Beaches and Barrier Islands," *Scientific American* 257, No. 1 (1987):68-77. Presents a detailed summary of the vulnerability of barrier islands to coastal storms, especially hurricanes.

GRAY, W. M., AND C. W. LANDSEA. "African Rainfall as a Precursor of Hurricane-Related Destruction on the U.S. East Coast," *Bulletin of the American Meteorological Society* 73 (1992):1352–1364. Discusses the strong correlation between rainfall in West Africa and the frequency of major hurricanes along the Atlantic seaboard.

HUGHES, P. "The Great Galveston Hurricane," *Weatherwise* 32, No. 4 (1979):148–156. Describes the deadliest natural disaster in U.S. history.

LUDLUM, D. M. "The Great Hurricane of 1938," *Weatherwise* 41, No. 4 (1988):214–216. Details one of the most intense hurricane to strike New England.

WILLIAMS, J. ET AL. "Hurricane Andrew in Florida," *Weatherwise* 45, No. 6 (1992/93):7–17. Presents three different perspectives on the landfall of Hurricane Andrew.

WILLOUGHBY, H. E., ET AL. "Project STORMFURY: A Scientific Chronicle 1962–1983," *Bulletin of the American Meteorological Society* 66 (1985):505–514. Traces the historical roots of efforts to modify hurricanes via seeding techniques.

# 16 Weather Analysis and Forecasting

*Probable nor'east to sou'west wind, varying to the souhard and westard and eastard and points between; high and low barometer, sweeping round from place to place; probable areas of rain, snow, hail, and drought, succeeded or preceded by earthquakes with thunder and lightning.*

MARK TWAIN
*New England Weather*

Monitoring of the atmosphere is an essential first step in the preparation of a weather forecast. [Photograph by J. M. Moran]

M OST PEOPLE readily recall occasions when an erroneous weather forecast upset their plans. It may have been an unexpected thundershower that brought an abrupt end to a softball game, or a raging blizzard that appeared instead of the anticipated clearing skies, or the promised springlike weekend that turned out to be anything but springlike. People seem to remember missed weather forecasts all too clearly and conveniently overlook the vast majority of times when the forecast was on target.

Actually, when viewed with the objectivity of statistical analysis, short-range weather forecasting is surprisingly accurate. For example, the United States **National Weather Service (NWS),** an agency of the **National Oceanic and Atmospheric Administration (NOAA),** issues 24-hour weather forecasts that are correct nearly 85% of the time. The popular notion that weather forecasting is seldom accurate probably stems from the simple fact that we notice and remember a missed forecast more readily than an accurate one because of the inconvenience we experience.

How are weather forecasts made? What are the limits of forecast accuracy? On the basis of what you have learned so far, how can you make your own weather forecasts? These are some of the questions that we consider in this chapter.

# World Meteorological Organization

Because the atmosphere is a continuous fluid that envelops the globe, weather observation, analysis, and forecasting require international cooperation. To this end, the *International Meteorological Organization (IMO)* was created in 1878. In 1947, the IMO changed its name to the **World Meteorological Organization (WMO)** and became an agency of the United Nations. Today, the WMO, headquartered in Geneva, Switzerland, coordinates the efforts of more than 145 member nations in a standardized global weather-monitoring network called **World Weather Watch.**

At standard observation times, the state of the atmosphere worldwide is monitored daily by almost 4000 land stations, more than 7000 ships at sea, almost 1000 radiosondes, and reconnaissance aircraft and satellites. These data are transmitted to the three World Meteorological Centers near Washington, D.C., Moscow, Russia, and Melbourne, Australia, where maps and charts are drawn up representing the current state of the atmosphere. From analyses of this information, generalized weather forecasts are prepared. Maps, charts, and forecasts are then sent to National Meteorological Centers (NMCs) located in WMO member nations as well as to 26 Regional Meteorological Centers (RMCs). At NMCs and RMCS, weather information and forecasts are generated and interpreted for each center's area of responsibility and distributed to local weather service offices and then to the public. The U.S. NMC is located at Camp Springs, Maryland; the Canadian NMC is in Toronto, Ontario.

From the above description we see that weather forecasting entails (1) acquisition of present weather data, (2) depiction of data on weather maps, (3) analysis of data and prediction, and (4) dissemination of weather information and forecasts to users.

# Acquisition of Weather Data

Since invention of the first weather instruments in the seventeenth century, weather observation has undergone considerable refinement. Denser monitoring networks, more sophisticated instruments and communications systems, and better trained weather observers have produced an increasingly detailed, reliable, and representative record of weather and climate.

## *SURFACE WEATHER OBSERVATIONS*

Across the United States, nearly 1000 stations routinely monitor surface weather. These stations may be operated by any of the following: (1) National Weather Service personnel, (2) the staff of other government agencies, including the Federal Aviation Administration (FAA), or (3) private citizens or businesses in cooperation with the NWS. At sea and on the Great Lakes, more than 2000 ships also voluntarily gather surface weather data.

The NWS also maintains networks of automated weather stations in locations where manned observations are not feasible. These include, for example, more than 110 automated weather stations operated by the National Data Buoy Center (NDBC). Some NDBC stations are attached to moored buoys in offshore areas (including the Great Lakes), and others are located at lighthouses, fishing piers, and offshore oil platforms. NDBC stations provide data on storm intensity and track and are considered an essential part of the hurricane warning system. Satellites relay data from the buoy networks to users.

Weather stations gather data for (1) preparation of weather maps and forecasts, (2) exchange with other nations, and (3) use by aviation. Stations report cloud type and sky cover, wind speed and direction, visibility, precipitation, air temperature, dew point, and air pressure. Also, for aviation purposes, airport weather stations report cloud height and altimeter setting.

Weather observations must be synchronous around the globe. To this end, all weather stations observe **Greenwich Mean Time (GMT),** which is the time at the Old Royal Observatory, Greenwich, England. Greenwich is located at 0 degrees longitude, the prime meridian. For reference, at 0600 GMT, it is midnight

Central Standard Time (CST) in Chicago, and 10 P.M. Pacific Standard Time (PST) in Vancouver, British Columbia. Greenwich Mean Time is commonly called *Universal Coordinated Time (UCT).*

As of this writing, weather-observing facilities of the National Weather Service are undergoing extensive restructuring and modernization with the goal of upgrading the quality and reliability of weather observations and forecasts. New **Weather Forecast Offices (WFOs)** are planned for 115 locations across the United States. In addition, automated weather stations are replacing the old manual system of hourly observations. This **Automated Surface Observing System (ASOS)** consists of modern sensors (Table 16.1), computers, and fully automated communications ports. When completed by the late 1990s, a network of about 1700 ASOS units will operate continuously 24 hr a day and feed data to NWS offices and local airport control towers.

Also as part of its modernization program, the National Weather Service is phasing out conventional weather radar and installing Doppler radars at 121 sites nationwide. The FAA and the Department of Defense will operate another 39 Doppler units. As noted in Chapter 14, Doppler radar offers significant (possibly life saving) advantages over conventional radar in providing more advance warning of the development of a severe weather system.

Besides the numerous land-based weather stations that provide information of potential use in weather forecasting and aviation, another 10,600 cooperative weather stations are scattered across the United States. These stations are cooperative in that volunteers supply their time and labor to monitor instruments and the NWS provides the equipment and data management. The principal function of the **NWS Cooperative Observer Network** is to record daily precipitation and

**TABLE 16.1**
**Sensors in the Automated Surface Observing System (ASOS) of the National Weather Service**

- Cloud height indicator
- Visibility sensor
- Precipitation identification sensor
- Freezing rain sensor
- Ambient temperature/dew point sensor
- Anemometer
- Rainfall accumulation sensor

maximum/minimum temperatures for hydrologic, agricultural, and climatic purposes.

## UPPER-AIR WEATHER OBSERVATIONS

As noted in Chapter 1, meteorologists monitor the upper atmosphere with radiosondes. A **radiosonde** is a radio-equipped instrument package borne aloft by a balloon. The instrument transmits to a ground station vertical profiles of air temperature, pressure, and relative humidity up to an altitude of about 30 km (19 mi). In addition, winds at various levels are computed by tracking the balloons with a radio direction-finding antenna (a rawinsonde observation).* Worldwide readings are made twice each day, at 0000 GMT and 1200 GMT.

In the United States, there are 126 radiosonde stations, 25 of which use special high-altitude balloons that provide data from altitudes above 30 km (19 mi). Meteorological rockets probe much higher altitudes (to 100 km), but the data obtained from these probes are used primarily for research. Upper-air weather data are also obtained by aircraft, dropwindsondes, radar, and satellites. In some urban areas, low-level soundings (up to 3000 m) monitor weather conditions to assess air pollution potential.

Canada's Atmospheric Environment Service operates similar networks of surface and upper-air weather observation stations. The Canadian service has 270 first-order weather stations and more than 3000 climatological stations nationwide.

# Meteorology by Satellite

The foremost advantage of weather satellites is that they provide a continuous picture of the state of the atmosphere. Furthermore, networks of surface and upper-air observation stations detect weather conditions at discrete points only, some of which may be hundreds or thousands of kilometers apart. Satellite imagery fills in the gaps. Hence, satellites *see* subsynoptic weather systems, such as severe thunderstorms, that might es-

*As noted in the Special Topic in Chapter 14, NOAA now operates a 29-station wind profiler network in the central United States that monitors upper-air winds.

cape detection by weather stations. Satellites also enable forecasters to locate and track tropical storms over oceans where weather observations are few and far between.

A variety of weather satellites are either currently in orbit or planned for the future. The United States, the former Soviet Union, Japan, and the 11-nation European Space Agency have launched weather satellites. Most are in near-polar orbits at altitudes of 800 to 1000 km (500 to 620 mi), that is, orbits that take the satellites on longitudinal (meridional) trajectories over the polar regions. For example, the U.S. weather satellites NOAA-7 and NOAA-8 observe a strip 1800 km (1100 mi) wide from pole to pole and back in 102 minutes. The satellites then observe the next adjacent meridional strip. With 14 orbits each day, a **polar-orbiting satellite** *sees* a particular locality twice every 24-hr day.

Some weather satellites are in geosynchronous orbits over the equator; that is, they orbit at a rate that matches the Earth's rotation, so they are always above the same spot and scan the same region. From an altitude of about 36,000 km (22,320 mi), each **geosynchronous satellite** monitors almost one-third of the Earth's surface and has a *full-disk* view of the Earth from pole to pole. Because polar-orbiting satellites fly at much lower altitudes, they provide greater resolution (more detail) but over a smaller area and with a longer time interval between images. For more on the orbital characteristics of both polar and geosynchronous satellites, refer to the Mathematical Note at the end of this chapter.

Sensors aboard satellites intercept two types of radiation that emanate from the Earth-atmosphere system: reflected solar radiation (visible) and emitted infrared radiation. Visible satellite imagery is like a black and white photo of Earth taken from space. As a rule, satellite sensors can discern any object on the Earth's surface larger than 1.0 km (0.6 mi) across. From cloud patterns, we can locate storm centers, fronts, fog banks, and thunderstorms. The cloud pattern of a hurricane tells meteorologists something about the storm's size, spiral structure, and eye characteristics, from which it is possible to estimate the stage of the hurricane's development and its strength. Sensors also detect dust storms and the extent of snow cover over land and ice cover at sea, as well as geographical features such as the Black Hills of South Dakota and the Great Lakes.

Surveillance of weather systems is made possible by viewing successive satellite images of the same area as a movie loop. Geosynchronous satellites obtain images every 30 minutes, although, under special circumstances, the National Weather Service can direct the satellite to scan only a portion of the *full-disk* and thus obtain images as frequently as every 3 to 5 minutes. Sequential images are therefore used to monitor the development of thunderstorms and squall lines and to predict the track of tropical storms by simple extrapolation.

Satellite infrared (IR) sensors measure the heat emitted by land and sea surfaces and the tops of clouds. In Chapter 2, we discussed how emitted radiation depends on the temperature of the radiating body. Based on measured differences in IR emission, sensors can distinguish clouds at low levels, which are relatively warm, from clouds at high levels, which are relatively cold. These measurements can be calibrated on a gray scale in which low clouds appear dark gray and high clouds appear white (Figure 16.1). IR sensing of cloud-top temperature is also a means of gauging the depth of convection and the intensity of thunderstorms. As noted in Chapter 13, the altitude of a cloud-top is an indicator of thunderstorm intensity. The higher the top, the lower is the cloud-top temperature, and the more intense is the thunderstorm.

Horizontal winds are inferred from satellite observations of cloud displacement. Clouds move with the

**FIGURE 16.1**
GOES full-disk infrared image showing clouds on a grey scale. High clouds are bright white, and low clouds are grey. [NOAA National Environmental Satellite, Data, and Information Service]

wind so that wind speed and direction can be estimated from the drift of clouds in successive satellite images. To determine the actual altitude of those winds, the satellite's IR sensor measures the cloud's radiation temperature, which is then matched to the nearest radiosonde sounding to give cloud (and wind) altitude. Meteorologists routinely monitor winds in the troposphere by this cloud-track method four times daily and transmit those data to forecast centers.

Vertical winds can also be inferred from satellite imagery even in the absence of convective clouds. IR measurements can distinguish regions of relatively high water vapor concentration from regions of relatively low water vapor concentration. In the satellite image of Figure 16.2, the lighter regions are more humid than the darker regions. We can assume that upward motion characterizes humid regions and subsiding motion characterizes dry regions.

By instrument analysis of the spectrum of radiation emitted by the atmosphere, satellites can obtain soundings.* This technique of obtaining vertical profiles of temperature and humidity is expected to be a routine function of geostationary satellites by the late 1990s.

*The specific method whereby soundings are obtained by satellite sensors is beyond the scope of this book. For more information, interested readers are urged to consult the article by Smith et al. cited at the end of this chapter.

**FIGURE 16.2**
GOES full-disk infrared image displaying measurements in the water vapor channel. On this grey scale, the lighter the shade, the more humid is the air. [NOAA National Environmental Satellite, Data, and Information Service]

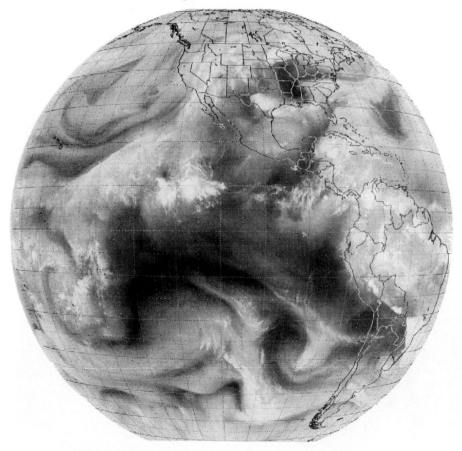

Normally, two **Geostationary Operational Environmental Satellites (GOES)** monitor atmospheric conditions over the western hemisphere, at or near longitudes 75 degrees W and 135 degrees W. Together, these two satellites provide continuous coverage from the central Pacific eastward to the Atlantic hurricane breeding grounds. Failure of one of these satellites in 1989 forced NOAA to borrow a European weather satellite (Meteosat-3) to fill in the gap. Unfortunately, the other GOES is now on its last legs, having exceeded its 5-year life expectancy in February 1992. Meanwhile, technical problems have thwarted NOAA's plan to orbit a series of five new-generation geostationary satellites *(GOES-NEXT)*. Launch of the first satellite in this series was scheduled for 1989 but as of this writing still has not taken place.

Besides the satellites used for weather observation, some meteorological satellites, such as those in the Nimbus series, are designed mainly for research purposes. Nimbus satellites are in polar orbits and monitor concentrations of water vapor, ozone ($O_3$), and other trace atmospheric gases.

Remote sensing by satellite as well as other advances in monitoring technology are generating a deluge of real-time weather data. This has necessitated de-velopment of computerized data management systems. Perhaps foremost among these systems is **McIDAS** *(Man-computer Interactive Data Access System),* designed by scientists at the Space Science and Engineering Center at the University of Wisconsin–Madison. McIDAS receives satellite imagery and soundings, radar displays, and surface and upper-air weather observations. McIDAS integrates and organizes those data into guidance products for potential users. Products include two- and three-dimensional composite images such as the one shown in Figure 16.3. Users access McIDAS on a computer terminal, select some satellite image or other data display, and may choose, for example, to overlay different fields of data on a satellite image.

## Data Depiction on Weather Maps

Computers at the National Meteorological Centers use the special symbols listed in Appendix III to plot weather observations on synoptic and hemispheric weather maps. The weather for each observation station is depicted on a map by following a conventional

**FIGURE 16.3**
Cloud image generated from GOES observations over the Gulf of Mexico, Louisiana, Mississippi, and Alabama at 1800 GMT, 10 September 1985. This technique of animated display was developed in conjunction with McIDAS. [Photograph courtesy of W. L. Hibbard, Space Science and Engineering Center, University of Wisconsin–Madison, and the American Meteorological Society]

**station model.** The station model in Figure 16.4 shows symbols for surface weather conditions. By international agreement, the same station model and symbols are used throughout the world.

Because weather systems are three-dimensional, both surface and upper-air weather maps are needed. A very different approach is used for the two types of maps, however. Surface weather data are plotted on a constant-altitude (usually sea-level) surface, and upper-air weather data are plotted on constant-pressure (isobaric) surfaces.

## SURFACE WEATHER MAPS

It is standard practice for meteorologists to reduce surface air pressure readings to sea-level (Chapter 5). This procedure enables meteorologists to compare surface air pressure readings at weather stations that are situated at different elevations above sea level. The adjusted air pressure readings are plotted on surface weather maps. Isobars, lines of equal air pressure, are then constructed at 4-mb intervals to reveal such features as anticyclones and cyclones,

troughs and ridges, and horizontal air pressure gradients.

The major weather features on the sample surface weather map in Figure 16.5 are a cold anticyclone centered over western Minnesota (central pressure, 1034 mb) and an intense cyclone centered over the southeastern United States. On surface weather maps, a storm center, where air pressure is lowest, is indicated by the symbol *L* or *LOW*. Closely spaced isobars surrounding the storm center indicate a steep horizontal pressure gradient and strong surface winds. Note how fronts originate at the storm center and how isobars bend *(kink)* as they cross fronts. Thus, fronts occupy troughs. Because we can infer surface wind direction from the isobar pattern, the bending of isobars at fronts tells us that the wind direction shifts as we cross a front. Recall from our discussion in Chapter 11 that a wind shift is one characteristic of a frontal passage. Note also how relatively weak air pressure gradients (widely spaced isobars) are associated with the anticyclones (mapped as *H* or *HIGH*).

Surface synoptic weather maps are drawn every 3 hr for North America and at 6-hr intervals for the

**FIGURE 16.4**
This conventional weather station model shows the symbols that are used to depict weather observations on surface weather maps.

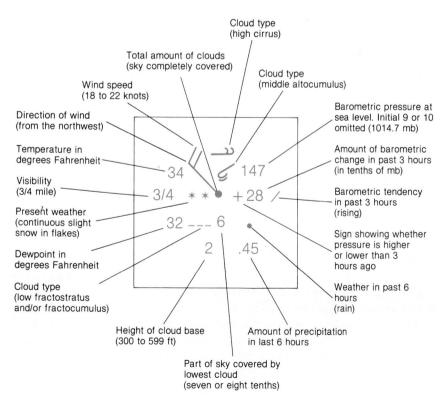

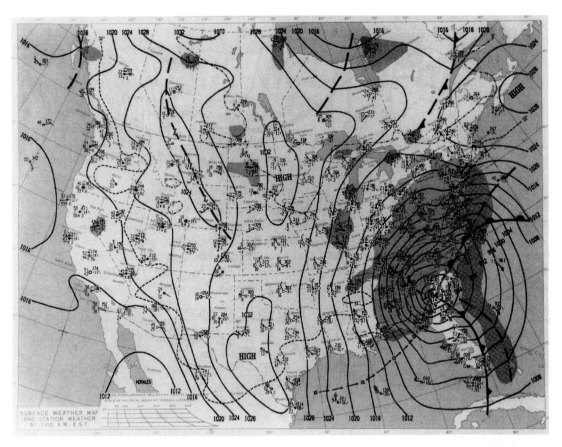

**FIGURE 16.5**
A sample surface weather map (March 13, 1993). [From NOAA]

Northern Hemisphere. Special charts are also constructed that summarize a variety of weather elements
including, for example, (1) maximum and minimum
temperatures for 24-hr periods, (2) precipitation
amounts for 6 hr and for 24 hr, and (3) observed snow
cover. A radar summary chart is also issued hourly (Figure 16.6).

## UPPER-AIR WEATHER MAPS

Upper-air weather data acquired by rawinsondes are
plotted on constant-pressure surfaces. Applying the basic laws of atmospheric physics, meteorologists compute the actual altitudes corresponding to these measurements. For example, we can determine the altitude
of the 500-mb surface, that is, the altitude at which the

air pressure drops to about one-half its average sea-
level value. By plotting altitudes of the 500-mb level
as monitored simultaneously by all rawinsonde stations
across North America, meteorologists can construct a
map representing the *topography* of the 500-mb surface. They simply draw contours through localities
where the 500-mb level is at the same altitude, a procedure that requires some interpolation between stations.

The altitude of a pressure surface (such as the
500-mb surface) varies from one place to another, primarily because of differences in temperature of the air
below the pressure surface. The drop of air pressure
with altitude is more rapid in cold air masses than in
warm air masses. Because raising the temperature of
air reduces its density, greater altitudes are required for
warm air to exhibit the same drop in pressure as cold

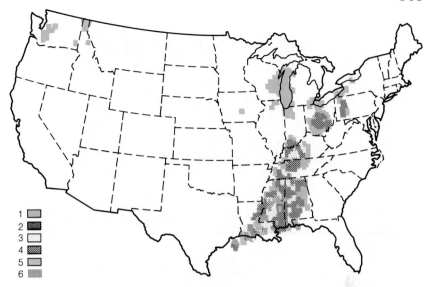

**FIGURE 16.6**
A sample national weather radar summary on a color scale. Red represents the most intense echoes (heaviest rain) and light green the weakest echoes (light rain).

air. This means, for example, that the 500-mb level is at a low altitude when the air below is relatively cold and at a high altitude when the air below is relatively warm. The contours of an isobaric surface therefore always show a gradual slope downward from the warm tropics to the colder polar latitudes.

Upper-air observations indicate that horizontal winds parallel height contour lines at the 500-mb level. Where contours are closely spaced (a steep height gradient), winds are strong, and where contours are far apart (a weak height gradient), winds are light. Why are winds associated with a gradient in the height of a pressure surface? A height gradient develops in response to a horizontal air temperature gradient, and where there is a horizontal air temperature gradient, there are also horizontal gradients in air density and air pressure. As we saw in Chapter 9, a horizontal air pressure gradient generates wind.

The 500-mb surface is so far above the friction layer that the planetary- and synoptic-scale winds are essentially geostrophic where contours are straight, and gradient where contours are curved (Chapter 9). The large-scale wind at the 500-mb level is thus the product of interactions among a horizontal height gradient, the Coriolis effect, and the centripetal force.

On upper-air weather maps, contours exhibit both cyclonic (counterclockwise) and anticyclonic (clockwise) curvature. These are the ridges and troughs in the westerlies that we described in Chapter 10. Contours may also define a series of nearly concentric circles, perhaps indicating a cutoff or blocking circulation pattern. At the center of a ridge, the air column is relatively warm and contour heights are high, so we label the ridge with an *H*. An upper-air ridge may be linked to a warm-core anticyclone at the surface. In contrast, near the center of a trough, the air column is relatively cold and contour heights are low, so the trough is labeled with an *L*. An upper-air trough may be linked to a cold-core cyclone at the surface.

From the preceding discussion, it follows that warm-core cyclones (thermal lows) and cold-core anticyclones (polar and arctic highs) do not appear on 500-mb weather maps. These pressure systems are simply too shallow to influence the air circulation pattern at the 500-mb level.

Figure 16.7 is the 500-mb analysis that corresponds to the surface weather map in Figure 16.5. The 500-mb circulation is particularly useful because winds at that level closely approximate the upper-level steering winds and the trough and ridge patterns responsible for the development and displacement of surface weather systems. Solid lines are height contours of the 500-mb surface labeled in meters above sea level; the contour interval (difference between successive contours) is 60m. Dashed lines are isotherms labeled in degrees Celsius. Flags indicate wind direction and speed at the 500-mb level (Appendix III). Winds paralleling contour lines describe cyclonic flow in troughs and anticyclonic flow in ridges.

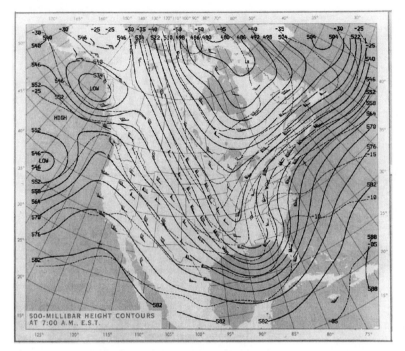

**FIGURE 16.7**
A sample 500-mb analysis (March 13, 1993). Contour lines are in meters with the final zero omitted. For example, 564 = 5640 m. [From NOAA]

Winds flowing across isotherms produce cold or warm air advection (Chapter 4). **Cold air advection** takes place when winds blow from colder localities toward warmer localities. **Warm air advection** occurs when winds blow in the opposite sense. Warm air advection causes the 500-mb surface to rise, and cold air advection causes the 500-mb surface to lower. Cold air advection at 500-mb thus deepens troughs and weakens ridges, whereas warm air advection strengthens ridges and weakens troughs.

When cold air advects into troughs at the same time that warm air advects into ridges, the circulation becomes more meridional. On the other hand, if warm air advects into troughs while cold air advects into ridges, the circulation becomes more zonal. As we saw in Chapter 10, shifts in the upper-air circulation pattern between meridional and zonal have important implications for air mass exchange and storm development.

National Meteorological Centers (NMCs) issue 500-mb maps twice each day based on upper-air observations at 0000 GMT and 1200 GMT. One set of maps covers North America and another the Northern Hemisphere. Although we have focused our discussion of upper-air weather maps on the 500-mb level, similar analyses are routinely constructed twice daily for the 850-, 700-, 300-, 250-, 200-, and 100-mb levels. The 300- and 250-mb analyses are important because they are near the strongest winds of the midlatitude jet stream.

## Weather Prediction

Meteorologists at National Meteorological Centers analyze satellite observations and weather data plotted on surface and upper-air weather maps. From these analyses, meteorologists prepare weather forecasts. Daily, the National Weather Service issues 12-, 24-, 36-, and 48-hr forecasts plus 3- to 5-day extended outlooks. And, three times weekly, the NWS releases 6- to 10-day weather outlooks. In addition, the Long-Range Prediction Branch of the Climate Analysis Center generates monthly (30-day) and seasonal (90-day) outlooks.

Weather forecasting is extremely challenging, primarily because it involves many variables and a vast quantity of weather data. For these reasons, computerized numerical models of the atmosphere have been developed to assist forecasters.

## NUMERICAL WEATHER FORECASTING

In the early 1950s, John von Neuman and Jule Charney pioneered the use of electronic computers for weather forecasting at the Institute for Advanced Study at Princeton University. A primitive computer, less powerful than today's personal computer, successfully forecast the horizontal air pressure pattern at 5000-m (16,400-ft) altitude over North America 24 hours in advance. By 1955, computers were routinely generating weather forecasts from surface and upper-air weather observations.

An electronic computer is programmed with a numerical model of the atmosphere, that is, a model consisting of mathematical equations that relate winds, temperature, pressure, and water vapor concentration. (Many of these basic equations are presented in the Mathematical Notes of this book.) Beginning with present (real-time) weather data, numerical models predict altitudes of pressure-surfaces for some future time, say, 10 minutes from now. With these predicted conditions as a new starting point, another forecast is then computed for, say, the subsequent 10 minutes. The computer repeats this process again and again until a weather map is generated for the next 12, 24, 36, and 48 hr. In this process, tens of millions of computations must be performed each second on a vast array of observational data; hence, a high-speed electronic computer that can accommodate huge quantities of data is necessary.

NMC computers apply different numerical models to specific portions of the atmosphere. As of this writing the *LFM (Limited-area Fine-mesh Model)* has been replaced by the *Nested Grid Model (NGM),* which was developed in 1984 by staff scientists at the National Meteorological Center. Already the NGM has proved its value in providing better predictions of East Coast winter storms. Table 16.2 is a sample of the guidance materials produced by numerical models.

Some computerized numerical models of the atmosphere are designed to operate over different spatial scales depending on the forecast range. For medium-range forecasts (up to 10 days), observational data are fed into the computer from all over the globe, since within that forecast range a weather system may travel long distances. On the other hand, for short-range forecasts (up to 3 days), the model utilizes data drawn from a more restricted region of the globe. Compared to a

**TABLE 16.2**
**Sample Computer-Generated Guidance Maps Used in Weather Forecasting for Periods of 12, 24, 36, and 48 Hours**

- Surface air pressure
- 1000-mb to 500-mb thickness (a measure of temperature)
- 700-mb vertical velocity and 12-hour precipitation totals
- 700-mb height contours
- Surface to 500-mb average relative humidity
- 500-mb height contours
- 500-mb vorticity
- 850-mb temperature
- 850-mb height contours

global model, a regional model offers the advantage of greater resolution of data over a smaller area of interest.

## SPECIAL FORECAST CENTERS

Not all weather forecasts are prepared by the National Meteorological Center. Responsibility for forecasting tropical storms and hurricanes is divided between two centers: the **National Hurricane Center (NHC)** in Coral Gables, Florida, and the Central Pacific Hurricane Center in Honolulu, Hawaii. Local and regional weather service offices transmit information from these centers to the public as advisories, warnings, or special weather statements. The goal is to provide at least 12 hr of daylight warning so that coastal residents may prepare for a hurricane.

Specially instrumented reconnaissance aircraft complement satellite and radar surveillance of hurricanes and tropical storms. Aircraft fly directly into and through the storms and determine the precise location of the eye, measure wind speeds, and obtain soundings by dropwindsonde. By extrapolating from air pressure readings at flight altitude (usually 1500 to 3000 m), scientists also determine the sea-level air pressure at the storm center. These data are immediately radioed back to the NHC.

The principal challenge to forecasters at the National Hurricane Center is prediction of the track of a hurricane. Normally, such forecasts are issued every 6 hr and cover periods up to 72 hr in the future. The basis for hurricane track forecasts is a blend of climatology (records of tracks of similar hurricanes in the past), numerical models, and the experience of the forecaster. Although the advent of satellite monitoring of hurricanes (and their precursors) led to significant upgrading in the accuracy of track forecasts during the 1960s, track forecasting skill has improved very little since then. Beyond 24 hr, accuracy declines rapidly and approaches zero for 72-hr forecasts. Apparently, the major problem is the poor quality of observational data initially fed into the numerical model.

If, on the other hand, the hurricane track forecast is on target, another numerical model can predict quite accurately the location and height of the storm surge along coastal areas (Chapter 15). Forecasters report success with the NHC numerical model *SLOSH (Sea, Lake, and Overland Surges from Hurricanes)*. By consulting local topographic maps, officials can use SLOSH predictions to identify areas that are likely to be inundated by floodwaters and plan evacuation routes accordingly.

Since the 1983 hurricane season, the NHC has included a probability forecast as part of its advisory statements. Probability is defined as the percent chance that the center of a hurricane or tropical storm will pass within 105 km (65 mi) of any of 46 designated Gulf and East Coast communities from Brownsville, Texas, to Eastport, Maine. The first probability forecast is usually issued 72 hr in advance of the storm's anticipated landfall. At that time, by convention, the probability is set no higher than 10% for any community. Probabilities increase to 13 to 18% at 48 hr, 20 to 25% at 36 hr, 35 to 50% at 24 hr, and 60 to 80% at 12 hr, before the storm's expected landfall.

Meteorologists at the **National Severe Storms Forecast Center (NSSFC)** at Kansas City, Missouri, monitor atmospheric conditions for the potential development of severe local storms (including winter blizzards) and issue watches for severe thunderstorms and tornadoes. A watch goes out about 1 hr prior to the beginning of the watch period, which typically lasts for 6 hr. Watches usually cover an area of about 65,000 square km (25,000 square mi). Local or regional weather service offices are responsible for issuing severe weather warnings. In addition, the NSSFC generates *Convective Outlooks* that identify regions of possible severe weather 1 or 2 days in advance.

## FORECAST SKILL

Just how accurate are today's computer-guided weather forecasts? This question can be answered by comparing the skill of modern weather forecasting with that of predictions based on persistence (forecasting no change in present weather) or climatology (forecasts derived from past weather records). This comparison is summarized in Table 16.3. Forecasting skill declines rapidly for periods longer than 48 hr and is minimal beyond 10 days.

Several factors contribute to the drop in forecasting skill. Computerized forecasts are only as accurate as the input (observational) data and predictive equations (numerical models) allow them to be. There is nothing magical about a computer. Errors are introduced by (1) missing or inaccurate observations, (2) failure of weather stations to detect all mesoscale and microscale circulation systems, and (3) imprecise predictive equations that include assumptions and first approximations. Unfortunately, the adverse effects of these errors grow as the forecast period lengthens. In today's numerical models of the atmosphere, the impact of even small errors doubles about every 2 to 2.5 days over the fore-

**TABLE 16.3**
**Weather Forecasting Skill**[a]

| Forecast period | Skill |
|---|---|
| 0–12 hr | Considerable for general weather conditions and trends |
| | Much less for onset and location of severe local storms |
| 12–24 hr | Moderate for cyclone tracks and associated weather, and areas of likely thunderstorm activity |
| 3–5 days | Moderate for large-scale circulation events (e.g., cold waves, major cyclones) |
| | Fair to good for temperature and fair to marginal for precipitation |
| 6–10 days | Some skill for mean temperatures and mean precipitation |
| Month (30 days) | Slight |
| Seasonal (90 days) | Slight |

*Source:* AMS. "Weather Forecasting," *Bulletin of the American Meteorological Society* 72 (1991):1273–1276.

[a]Applies to middle latitudes of the Northern Hemisphere; skill based on comparison to persistance and climatology.

cast period. When computerized numerical models are run out to 10 days, some elements of daily forecasts are useful for some localities to about 6 or 7 days. Beyond that, chaos sets in.

Nonetheless, the accuracy of numerical weather forecasting, particularly in the 1- to 5-day range, has shown slow but steady improvement over the past few decades, and prospects for continued progress are good. This trend is largely due to (1) better understanding of atmospheric processes that makes possible the development of more realistic numerical models of the atmosphere, (2) larger and faster computers, (3) more reliable and sophisticated observational tools, including Doppler radar and remote soundings by satellite, and (4) denser observational networks worldwide.

The computer is never likely to replace human weather forecasters, however. The products of computerized numerical models function as guidance materials for the forecaster. He or she must understand the characteristic errors and biases of the various numerical models as well as atmospheric processes and weather-observing systems. As much as numerical weather prediction has advanced in recent years, the best forecasters still rely heavily on their personal experience and intuition, tempered by their knowledge of how the atmosphere works. A good forecaster does not start with the model output in preparing a forecast, but rather he or she begins with previous and current observations, forming a view of what the atmosphere has been doing and ideas on what is likely to happen in the future. Furthermore, special local or regional conditions may call for significant modification of a computer-generated weather forecast. Forecasters must analyze and interpret computerized predictions and adapt those forecasts as necessary to regional and local circumstances. Consider an example.

Suppose that an intense early winter storm tracks northeastward through the Midwest, across central Illinois, and into eastern lower Michigan, as shown in Figure 16.8. Muskegon, Michigan, is on the cold, snowy side of the storm's path, so residents experience strong northeast winds accompanied by blowing snow. After the storm passes, winds shift to the north and northwest, and cold air advection begins. A computerized numerical model might predict clearing skies for Muskegon, but local conditions dictate a different result. Strong winds from the northwest advect cold, dry air across the relatively warm waters of Lake Michigan,

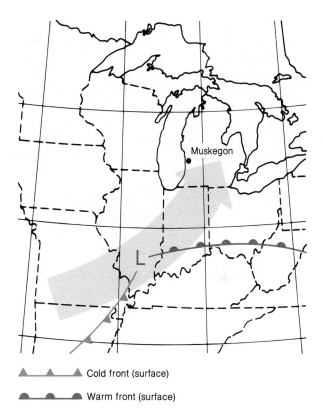

**Cold front (surface)**

**Warm front (surface)**

**FIGURE 16.8**
An early winter storm tracks northeastward through the Midwest and causes heavy snowfall at Muskegon, Michigan.

giving rise to lake-effect snows on the lake's southeastern shore (Chapter 12). Because of this local effect, northwest winds may bring more snow to Muskegon after the storm has passed than northeast winds did when the storm was nearby.

## LONG-RANGE FORECASTING

The Long Range Prediction Branch of the Climate Analysis Center in Camp Springs, Maryland, prepares 30-day (monthly) and 90-day (seasonal) generalized weather *outlooks* that identify areas of expected positive and negative **anomalies** (departures from long-term averages) in temperature and precipitation. An example is shown in Figure 16.9.

Forecasting the prevailing circulation pattern at the 700-mb level is the first step in forecasting monthly

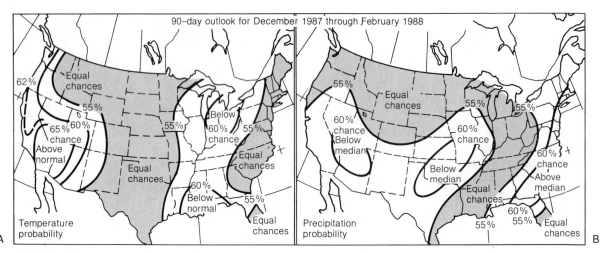

**FIGURE 16.9**
Winter weather outlook for 1987–88 as issued in late fall 1987 by the National Weather Service.
(A) The probability of the mean temperature being above or below the long-term average. (B)
The probability of the total precipitation being above or below the long-term median. Winter
encompasses the 90-day period of December 1987 through February 1988. [From the Long
Range Prediction Branch, Climate Analysis Center, NOAA]

temperature and precipitation anomalies at the surface.
The present circulation pattern is extrapolated into the
future, although an effort is also made, based on historical data, to identify features of the present pattern that
are most likely to persist. The prevailing westerly flow
at the 700-mb level permits identification of areas of
strong warm and cold air advection as well as principal
storm tracks. Predictions of surface temperature and
precipitation anomalies are derived from these data.

A somewhat different approach is utilized for 90-
day outlooks. Forecasters rely more on long-term
trends and on recurring events and attempt to isolate
persistent circulation features from prior months and
seasons. In the analog technique, for example, a computer searches its 40-year archive (memory) of past seasonal weather patterns for the closest match with the
present season's weather pattern. The 90-day outlook
is then based on whatever followed the best historical
match.

Statistics on seasonal forecast skill can be a bit confusing. For more than two decades, the National
Weather Service has reported average skills of only 8%
for temperature outlooks and 4% for precipitation outlooks. To put these statistics in perspective, forecast
skill would be 0% if based on chance alone and 100%
for a perfect forecast. The greatest success so far has

been with winter temperature outlooks, with a forecast
skill of 16% to 18%.

## TELECONNECTIONS

A promising area of research in long-range weather
forecasting relies on identification of teleconnections.
A **teleconnection** is a linkage between changes in atmospheric circulation occurring in widely separated regions of the globe, often many thousands of kilometers apart. An example is the linkage between an ENSO
or La Niña with weather extremes in the tropics and
midlatitudes (Chapter 10).

Recall that an extreme El Niño **(ENSO)** begins
when trade winds over the tropical Pacific weaken.
Weaker trades allow a pool of relatively warm surface
water in the western equatorial Pacific to gradually
spread eastward. Anomalously high sea-surface temperatures develop in the eastern tropical Pacific and
spur thunderstorm activity, thereby enhancing the flow
of latent heat into the atmosphere. This alters atmospheric circulation patterns and causes weather extremes in many regions of the world.

Essentially, **La Niña** is the opposite of ENSO. Unusually strong trade winds cause vigorous upwelling off

the northwest coast of South America and anomalously low sea-surface temperatures over the eastern tropical Pacific. The resulting alteration of atmospheric circulation patterns also causes weather extremes in various parts of the world, but typically opposite in sign to those accompanying an ENSO. La Niña alternates with ENSO about every 3 to 7 years.

Scientists report success with numerical models designed to predict the onset of an ENSO. Two of the best known models were developed at Scripps Institution of Oceanography, La Jolla, California, and at Columbia University's Lamont-Doherty Earth Observatory, Palisades, New York. The Lamont-Doherty model, developed by S. Zebiak and M. Cane, is a coupled atmosphere-ocean simulation that successfully predicted the onset of both the 1986–87 and 1991–92 ENSOs a year ahead of time. The Scripps model is statistically based and also predicted the 1991–92 ENSO a year in advance.

Sea-surface temperature anomaly patterns are not the same during all ENSO and La Niña events. Hence, associated weather extremes may differ, although some are quite consistent from one ENSO to the next. For example, an ENSO is usually accompanied by drought in northern Australia and above average winter rainfall in the southeastern United States. Hence, for the first time, in 1991, the National Weather Service incorporated the latter association in developing its 90-day forecast for the winter of 1991–92. Based on model predictions for continuation of a moderate ENSO, the Long Range Prediction Branch of the Climate Analysis Center called for above average rainfall over the Southeast, a forecast that was verified.

While numerical models have succeeded in predicting the onset of an ENSO, they are not so skillful in forecasting its demise. Both the Lamont-Doherty and Scripps models called for a return to lower sea-surface temperatures over the eastern tropical Pacific by late 1992 or early 1993 and an end to the ENSO. In fact, an ENSO persisted through 1992 and well into 1993.

In another example of a teleconnection, Jerome Namias and his colleagues at the Scripps Oceanographic Institution argue that a coupling between ocean and atmosphere triggered the 1988 drought over the North Central United States. Anomalously cold water in the central and eastern tropical Pacific (La Niña) plus unusually high sea-surface temperatures to the north (near Hawaii) interacted with the atmosphere to produce a long-wave pattern that was more meridional than usual. Recall from Chapter 10 that this circulation pattern featured a warm anticyclone over the central contiguous United States that persisted through much of the spring and summer of 1988.

## SINGLE-STATION FORECASTING

Short-term weather forecasts based on weather observations at one location, known as **single-station forecasts,** may be derived from the principles of weather behavior discussed in the previous chapters of this book. Because such forecasts are based on rules applied at only one location, they tend to be quite generalized and tentative, and complications often crop up as local conditions are modified by changes elsewhere. Table 16.4 is a sample list of rules of thumb applicable to midlatitude weather. You may wish to add to this

**TABLE 16.4**
**Rules of Thumb for Single-Station Weather Forecasting**

- At night, air temperatures will be lower if the sky is clear than if the sky is cloud covered.
- Clear skies, light winds, and a fresh snow cover favor extreme nocturnal radiational cooling and very low air temperature by dawn.
- Falling air pressure may indicate the approach of stormy weather, whereas rising air pressure suggests that fair weather is in the offing.
- The appearance of cirrus, cirrostratus, and altostratus clouds, in that order, indicates overrunning ahead of a warm front and the possibility of precipitation.
- A counterclockwise wind shift from northeast to north to northwest (called *backing*) is usually accompanied by clearing skies and cold air advection.
- A clockwise wind shift from east to southeast to south (called *veering*) is usually accompanied by clearing skies and warm air advection.
- A wind shift from northwest to west to southwest is usually accompanied by warm air advection.
- If radiation fog lifts by late morning, a fair afternoon is likely.
- With west or northwest winds, a steady or rising barometer, and scattered cumulus clouds, fair weather is likely to persist.
- Towering cumulus clouds by midmorning may indicate afternoon showers or thunderstorms.

list. Sometimes, but not usually, weather proverbs are useful in forecasting. This is the subject of the Special Topic "Weather Proverbs: Fact or Fiction?"

Analysis of records of past weather events may aid single-station weather forecasting. Such records exhibit a **fair-weather bias.** That is, fair-weather days outnumber stormy days almost everywhere. In fact, if we boldly predict that all days will be fair, we probably will be correct more than half the time. The only merit of this exercise would be to establish a baseline for evaluating the skill of more sophisticated weather forecasting techniques (discussed earlier). That is, we would expect traditional forecasting methods to score higher than forecasting based solely on fair-weather bias.

Another characteristic behavior of weather is **persistence,** that is, the tendency for weather episodes to persist for some period of time. For example, if the weather has been cold and stormy for several days, the weather may well continue that way for many more days. Weather records show, however, that an episode of one weather type typically gives way to another weather type very abruptly, usually in a day or less. Weather forecasts based on persistence alone are therefore prone to serious error.

A third approach is based on climatology; that is,

we prepare a weather forecast for a particular day based on the type of weather that occurred on the same day in years past. Suppose, for example, that climate records of your area indicate that it has rained on 7 August only 12 times in the past 100 years. Accordingly, you predict that the probability of rain next 7 August is only 12%, and you confidently plan a picnic. The problem with this approach is that, statistics aside, there is no guarantee that it will not rain next 7 August. Another example of climate-derived weather forecasting is statistics on the likelihood of a white Christmas, as shown in Figure 16.10.

## PRIVATE FORECASTING

In our description of weather forecasting, we have focused mainly on the role of government agencies. In addition, numerous private weather forecasters and forecast services analyze weather maps and other guidance materials supplied by the National Meteorological Centers and tailor them for the private sector's special needs. For example, a private weather forecaster retained by an appliance store chain might alert store executives to a pending heat wave so that stores might be stocked with an adequate supply of fans and air con-

**FIGURE 16.10**
Percentage of Christmas mornings with snow on the ground over the 30-year period 1956–1985. [From D. R. Cook, "Dreaming of a White Christmas," *Weatherwise* 39, No. 6 (1986):310]

ditioners. Another private forecaster might advise an electric power company of expected summer temperatures so the energy supplier could better anticipate customer demand for air conditioning. In this way, private forecasters supplement the efforts of government weather forecasting.

## Communication and Dissemination

Weather maps, charts, forecasts, and outlooks issued by the U.S. National Meteorological Center are transmitted to regional Weather Forecast Offices (WFOs).

When hazardous weather appears either possible or probable, the National Weather Service issues weather watches and weather warnings covering the affected geographical area for a specified time period. In general, a **weather watch** is indicated when hazardous weather is considered possible on the basis of current or anticipated atmospheric conditions. People in the designated area need not interrupt their normal activities ex-

cept to remain alert for threatening weather and to keep the television or radio on for further advisories. A **weather warning** is issued when hazardous weather is occurring somewhere in the region. People are then advised to take all necessary safety precautions.

Watches and warnings are issued for tornadoes, severe thunderstorms, hurricanes, floods, and winter storms. A tornado warning is issued only after detection of a thunderstorm that is known or likely to include a tornado. If conventional radar is used, a suspected tornado signature (a hook echo, for example) must be confirmed by storm-spotters, individuals trained for this purpose. On the other hand, if the more reliable Doppler radar is used (Chapter 14), ground confirmation may not be required before a tornado warning is issued. The warning bulletin specifies the location of the tornado, its anticipated path, and the time when the tornado is expected in the warning area. Warning areas are typically much smaller than watch areas.

Winter storm warnings may specify heavy snow, blizzard conditions, or an ice storm (Figure 16.11). Usually, a *heavy snow warning* is issued if snowfall is

**FIGURE 16.11**
Winter storm warnings are issued when heavy accumulations of snow or ice are anticipated. [Photograph by Arjen and Jerrine Verkaik/SKYART]

# Weather Proverbs: Fact or Fiction?

Over the centuries, people have developed weather proverbs to serve as general guides for predicting weather. These proverbs are based on casual observations, usually of a weather event or a change in animal behavior, that are linked to a subsequent weather event. When weather proverbs are subjected to close scientific scrutiny, a few prove to be substantially correct, but most turn out to be merely picturesque myths with little or no scientific basis.

The Zuni tribe of New Mexico has a weather proverb that often proves to be true.

*When the sun is in his house, it will rain soon.*

The phrase "in his house" refers to a halo surrounding the sun. Recall from Chapter 8 that a halo forms when the sun's rays are refracted by ice crystals that compose high, thin cirrus clouds. Recall also from Chapter 11 that cirrus clouds signal the beginnings of overrunning as warm air is advected aloft. The appearance of a halo around the sun thus signals the approach of a storm, but a halo does not ensure that precipitation will follow at your location. Because cirrus clouds may spread as far as 1000 km (620 mi) ahead of the storm center, the storm may well change direction or die out before reaching your area.

Another scientifically sound proverb uses rainbows to predict the day's weather.

*A rainbow in the morning*
*Is the sailor's warning.*
*A rainbow at night*
*Is the sailor's delight.*

The basis of this proverb's validity (as described in Chapter 8) is that (1) in midlatitudes, weather systems usually progress from west to east and (2) an observer

of a rainbow must face the rainshower with the sun at her back.

Several other proverbs also have scientific merit. For example:

*As the days lengthen,*
*the cold strengthens.*

This saying correctly describes the lag time that takes place in the adjustment of midlatitude temperatures to changing insolation (Chapter 4). That is, although the minutes of daylight begin to increase after the winter solstice, the coldest part of the year typically occurs about a month later.

The saying *"The more snow, the more healthy the seasons"* illustrates the beneficial effects of a deep snow cover on overwintering vegetation. Because snow is a good insulator (Chapter 3), it moderates fluctuations in soil temperature even though major changes in air temperature are common as air masses from different source regions replace one another. Without the snow cover, however, soil is subject to varying periods of freezing and thawing. As a consequence, water in the soil expands (with freezing) and contracts (with thawing). Accompanying soil movements cause frost heaving, a phenomenon that damages plant roots and may kill plants. Frost heaving can severely damage overwintering hay fields and winter wheat crops. A deep snow cover, however, also has a negative side. While protecting the underlying plants, the insulating cover also protects overwintering insects from temperature extremes. As a consequence, insect damage to crops may be more severe the following growing season.

Another sound proverb reads

*No weather is ill*
*if the wind is still.*

expected to total at least 15 cm (6 in.) over 12 hr (or less) or 20 cm (8 in.) over 24 hr (or less). A **blizzard warning** means that falling or blowing snow is expected to be accompanied by winds in excess of 56 km

(35 mi) per hour, reducing visibility to 140 m (450 ft) or less. A severe blizzard produces winds stronger than about 73 km (45 mi) per hour, visibility near zero, air temperature below $-12$ °C (10 °F), and dangerously

We can validate this proverb by recalling from Chapter 9 that winds about the center of a high pressure system are light or calm. Because the arrival of a high pressure system commonly ushers in a period of fair weather, the proverb is often correct. However, the proverb does not account for the realities of life in today's industrial societies. In Chapter 17, we will see that air pollution episodes are commonly associated with stagnant high-pressure systems. Thus, still winds favor the accumulation of air pollutants that can have ill effects on human health.

Although a few weather proverbs have merit, most have no validity. For example, many cultures have proverbs that deal with the ability of animals to forecast weather changes. Probably the most famous of these is the myth that a groundhog can determine on the second day of February whether winter is about to end. If the sun is shining and the groundhog can see its shadow, winter will supposedly last another six weeks. If it is cloudy, winter will soon end. Interestingly, the proverb originated in medieval Europe, and Groundhog Day was known as Candlemas Day. At that time, no definite period was assigned for winter to stay. Then the proverb read

> If Candlemas Day be fair and bright
> Winter will have another fight;
> But if Candlemas Day brings clouds and rain
> Winter is gone and won't come again.

Common sense shows this proverb to be false. Many parts of the United States and Canada are usually snow covered on the second of February and it is unlikely that the groundhog even makes an appearance. What if the day dawns clear and then turns cloudy? Would the weather for the next six weeks depend on whether or not the groundhog was an early riser? Ridiculous, isn't it? Moreover, no evidence exists that the presence or absence of sunshine on one particular day is an indicator of the weather for weeks to come. Weather is too variable and the atmosphere too complex to support such a simple cause/effect relationship.

Other weather proverbs deal with the ability of animals such as squirrels, deer, rabbits, and caterpillars to predict the severity of the upcoming winter. For example:

> When squirrels lay in a large store of nuts, expect a cold winter.

However, no evidence of any linkage exists between the two. Squirrels are more likely to store more nuts if more nuts are available! The squirrels are simply responding to a good acorn-growing season, rather than exhibiting an innate ability to predict future weather.

Another familiar, but false, weather proverb concerns lightning:

> Lightning never strikes twice in the same place.

Recall our discussion of cloud-to-ground lightning in Chapter 13. An electrical discharge usually follows the shortest path between the ground and a cloud, so that lightning often strikes a tree, an electrical pole, a tower, or other tall object. As long as this object is not destroyed by the first lightning bolt, there is no physical reason why lightning cannot strike the object again.

This concluding verse, attributed to E. V. Lucas (1863–1938), illustrates one universal truth about weather—no matter what the weather, not everybody will be happy.

> The Duke of Rutland urged the folk to pray
> For rain; the rain came down the following day,
> The pious marvelled, the skeptics murmured fluke,
> The farmers, late with hay, said "Damn the Duke."

low windchill equivalent temperatures. Ice storm warnings mean that potentially dangerous accumulations of freezing rain or sleet are expected on the ground and other exposed surfaces.

Because weather is changeable, weather observations and guidance information such as weather maps, forecasts, watches, and warnings must be communicated as rapidly as possible both nationally and inter-

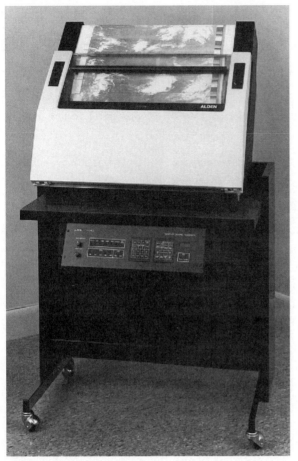

**FIGURE 16.12**
A conventional weather facsimile machine used to relay
and reproduce weather maps, satellite imagery, and other
weather guidance materials. [Courtesy of Alden Electronics]

nationally. For this reason, the World Meteorological
Organization and its member nations maintain elaborate communications networks consisting of a variety
of systems. Weather information is relayed by satellite,
teletypewriter, radio, and facsimile systems like the one
shown in Figure 16.12, which reproduce maps, charts,
and satellite images.

Modernization of National Weather Service facilities is making the weather communications network
more efficient and reliable. Earlier in this chapter, we
described the Automated Surface Observing System
(ASOS). In addition, new workstations have been designed to receive and organize ASOS data plus analysis and guidance products from the National Meteorological Center (Figure 16.13). This **Advanced Weather**

**Interactive Processing System (AWIPS)** enables the
meteorologist to display, process, and overlay pictures,
graphics, and other data.

The public receives regular weather reports and
forecasts via commercial and public radio, the NOAA
weather radio, commercial television and cable-TV
weather channels, and newspapers. In many cities,
weather information is also available via recorded telephone announcement systems.

In addition, during hurricane season, the National
Hurricane Center (in cooperation with NOAA, NBC
News, and *USA Today*) operates a 24-hour telephone
hotline. A recorded message provides late information
on an approaching hurricane or tropical storm. The telephone number (in the United States) is (900) 410-
NOAA; there is a small per-minute fee.

Pilots obtain weather briefings prior to and during
flight and are advised on any weather conditions that
may pose a hazard to flight. For a summary description of these hazards, refer to the Special Topic "Aviation Weather Hazards."

## Conclusions

Weather forecasting is a complex and challenging science that depends on the efficient interplay of weather
observation, data analysis by meteorologists and computers, and rapid communication systems. Meteorologists have achieved a very respectable level of skill for
short-range weather forecasting. Further improvement
is expected with denser surface and upper-air observational networks, more precise numerical models of the
atmosphere, larger and faster computers, and more sophisticated techniques of remote sensing by satellite. If
these advances are to be realized, however, continued
international cooperation is essential, for the atmosphere is a continuous fluid that knows no political
boundaries.

What, then, are the prospects for climate forecasting? So far, the accuracy of long-range forecasting has
been minimal, and the likelihood of significant improvement in the near future is slim. The reasons for
this will become evident in Chapters 18, 19, and 20,
where we discuss principles of climate, climatic trends,
and possible causes of climatic variability. Before that,
however, we turn our attention to the topic of air pollution meteorology.

**FIGURE 16.13**
As part of the modernization of the National Weather Service, new workstations are designed to enable meteorologists to display and analyze different fields of data on the same computer screen. This is the prototype AWIPS facility at Sterling, Virginia. [Photograph by J. M. Moran]

## MATHEMATICAL NOTE

## Some Orbital Characteristics of Weather Satellites

Both types of weather satellites, geosynchronous and polar-orbiting, move through space at constant speeds in orbits that are very nearly circular. Because friction is negligibly small, the only forces acting on an orbiting satellite are gravitation and the centripetal force.

Gravitation $(F)$, the attractive force between two objects, is directly proportional to the product of the masses of the two objects $(m_1, m_2)$ and inversely proportional to the square of the distance $(r)$ between the two objects. That is,

$$F \propto \frac{m_1 \times m_2}{r^2}$$

The constant of proportionality is $G$, the gravitational constant. Hence,

$$F = \frac{G(m_1 \times m_2)}{r^2}$$

Assume that a weather satellite has a unit mass and that the distance between the satellite and the Earth's center of mass is equal to the Earth's radius $(R)$ plus the satellite's altitude $(h)$. Then the gravitational attraction between the Earth and the satellite is given by

$$F = \frac{Gm_E}{(R + h)^2}$$

where $m_E$ is the mass of the Earth.

The centripetal force $(C_e)$ acting on a satellite of unit mass is computed as

$$C_e = \frac{V^2}{R + h}$$

where $V$ is the orbital speed of the satellite.

For computational purposes, we can set the gravitational force equal to the centripetal force since the satellite's acceleration is zero. But recall from earlier discussions (Chapter 9) that this is *not* a case of balanced forces because the satellite continually changes direction as it describes a circular orbit. The centripetal force is due to the gravitational force. Hence,

$$F = C_e$$

or

$$\frac{Gm_E}{(R + h)^2} = \frac{V^2}{R + h}$$

Solving for *V* and simplifying gives

$$V = \left[ \frac{Gm_E}{R + h} \right]^{1/2}$$

From this equation, the orbital speed *(V)* of a satellite is inversely proportional to its altitude *(h)*. Thus, in order to remain in orbit, a satellite at low altitudes must move faster than a satellite at high altitudes. For a geosynchronous satellite, an orbital altitude is se-lected so that the corresponding orbital speed means that the satellite always views the same portion of the Earth. For today's polar-orbiting weather satellites, on the other hand, an orbital altitude is selected so that the corresponding orbital speed takes the satellite over all points on the Earth's surface twice daily at the same local time, at 7:30 A.M. and 2:30 P.M. Such an orbit is described as *sun-synchronous*.

A geostationary satellite at an altitude of 36,000 km (22,320 mi) has an orbital speed of about 11,000 km (6820 mi) per hour and completes one orbit in 24 hours. A polar-orbiting satellite at an altitude of 855 km (530 mi) has an orbital speed of almost 27,000 km (16,740 mi) per hour and completes on orbit every 102 minutes.

## SPECIAL TOPIC

## Aviation Weather Hazards

The *Federal Aviation Administration (FAA)* estimates that 50% of all aircraft accidents are weather related. Between 1975 and 1986 weather was the chief causal factor in 26% of fatal accidents involving U.S. commercial jetliners. Three major aviation weather hazards are obstructed visibility, turbulence, and icing. In Chapter 13 we described a fourth hazard, microbursts, which can accompany even relatively weak thunderstorms.

In the interest of aircraft safety, pilots obtain preflight and in-flight weather briefings tailored to their individual flight plans. These briefings report current and anticipated weather conditions, including any potential weather hazards. The briefings are given in a special code, which is presented in Appendix III.

*Visibility* is the maximum horizontal distance at which prominent objects can be seen and identified. The *ceiling,* the altitude of the base of the lowest cloud layer covering more than half the sky, is also important because visibility can be restricted by low stratus clouds. Other visibility-restricting conditions include fog, polluted air, haze, precipitation, and blowing snow, sand, or dust. In general, visibility is good and ceilings are high in anticyclones (except for radiation fog); con-versely, visibility can be poor, and ceilings low, in cyclones.

Depending on visibility, pilots fly under different flight rules. When visibility is good, *Visual Flight Rules (VFR)* apply, and a pilot relies on vision to spot landmarks and other aircraft. When visibility is poor, *Instrument Flight Rules (IFR)* apply, and navigation must be aided by instruments.

Because low visibility is the chief reason for flight delays and cancellations, airports should be located where local climate favors good visibility. The ideal site is a moderately elevated area; a low area that is subject to cold air drainage and frequent radiation fogs is especially unsuitable. The airport should also be upwind of industrial areas to avoid the adverse effects of air pollution on visibility.

As we saw in Chapter 9, *turbulence* is an irregular flow of air analogous to the white water in the rapids of a swiftly flowing stream. Eddies of varying size, generated by thermal or mechanical influences, develop in the wind. As the mean wind speed increases, eddies generally become more energetic and can adversely affect aircraft in flight.

Wind shear—that is, a change in wind speed over a relatively short distance—most often generates turbu-

# Key Terms

National Weather Service
(NWS)
National Oceanic and
Atmospheric
Administration (NOAA)
World Meteorological
Organization (WMO)
World Weather Watch
Greenwich Mean Time
(GMT)
Weather Forecast Offices
(WFOs)
Automated Surface
Observing System
(ASOS)

NWS Cooperative
Observer Network
radiosonde
polar-orbiting satellite
geosynchronous satellite
Geostationary
Operational
Environmental
Satellites (GOES)
McIDAS
station model
cold air advection
warm air advection
National Hurricane
Center (NHC)

National Severe Storms
Forecast Center
(NSSFC)
anomalies
teleconnection
ENSO
La Niña
single-station forecasts

fair-weather bias
persistence
weather watch
weather warning
blizzard warning
Advanced Weather
Interactive Processing
System (AWIPS)

# Summary Statements

☐ The fluid nature of the atmosphere means that international cooperation is required in the gathering and interpretation of surface and upper-air weather data. To this end, the World Meteorological Organization (WMO) co-

lence. The greater the wind shear, the more severe is the turbulence. Downbursts, convection currents, the jet stream, and fronts (especially cold fronts) all produce wind shear and turbulence. In convection currents, for example, the wind shear between updrafts and downdrafts produces turbulence that aircraft passengers feel as bumpiness, especially on takeoff and landing. Cumulus clouds are visible evidence of convective turbulence: air is ascending where the sky is cloudy and descending where the sky is clear. Convective turbulence usually poses no problem for aircraft unless the currents surge to high altitudes and trigger thunderstorms and severe wind shear. The likelihood of extreme turbulence and the possibility of damaging hail mean that aircraft should never attempt to fly through a thunderstorm.

Mountain waves can also generate turbulence. Recall from Chapter 12 that a lofty mountain range disturbs large-scale horizontal winds so that *standing waves* develop to the lee of the range. Vigorous eddies form below mountain waves and sometimes penetrate to the ground as a rotor-like circulation. Low-flying aircraft can be caught in the rotor and be forced to the ground or into the mountain side. Turbulence may also be encountered at altitudes above the mountain waves, perhaps up to the tropopause.

Often turbulence aloft can be detected by merely observing clouds. In addition to cumulus clouds, clouds

exhibiting a wave pattern (stratocumulus, altocumulus, and cirrocumulus) are associated with turbulent airflow. In other instances, turbulent air occurs without clouds. *Clear air turbulence (CAT)* is difficult to detect except by sophisticated airborne instruments currently under development. Normally, however, CAT is confined to relatively thin layers of the atmosphere, and pilots readily escape this turbulence by changing altitude.

The usual turbulence encountered by commercial airliners is so light as to be merely annoying. On rare occasions when an aircraft flies into an area of severe turbulence, abrupt changes in altitude, attitude, or both occur. For this reason, passengers are advised to keep seat belts buckled throughout a flight.

Aircraft flown through freezing rain or clouds composed of supercooled water droplets are prone to *icing,* the accumulation of ice on the leading edges of the aircraft. Supercooled raindrops or cloud droplets freeze on contact with the surface of the wings and fuselage. If the aircraft does not have deicing or antiicing equipment, ice buildup adds weight and interferes with the aircraft's aerodynamics. Icing is most often a hazard for aircraft flying below about 6000 m (19,700 ft) and at air speeds of less than 400 knots. Above 6000 m, air temperatures are so low that clouds are composed almost exclusively of ice crystals, which do not accumulate on the aircraft.

ordinates an international effort of weather observation, analysis, and forecasting.

☐ Surface weather is monitored at land stations and by automated weather stations and ships at sea. Upper-air weather data are profiled by rawinsonde measurements, and additional weather observations are supplied by radar, aircraft, and satellites.

☐ The 1990s is a decade of transition for the National Weather Service with the establishment of new Weather Forecast Offices (WFOs) and the Automated Surface Observing System (ASOS).

☐ Weather satellites have the advantage of continuously monitoring a broad field of view. Polar-orbiting satellites follow relatively low meridional orbits. Geosynchronous satellites are positioned over the equator and orbit at the Earth's rotational rate.

☐ Weather satellites monitor solar radiation reflected by the Earth-atmosphere system plus infrared radiation emitted by the planet.

☐ Surface weather data are plotted on a constant-altitude (sea-level) surface, and upper-air weather data are plotted on constant-pressure (isobaric) surfaces.

☐ Because many variables are involved in the atmospheric system and because huge quantities of weather observations are generated, weather data are analyzed and forecasts are prepared by high-speed electronic computers using numerical models of the atmosphere.

☐ Hurricane track forecasts are based on climatology (record of past storm tracks), numerical models, and the experience of the forecaster.

☐ Although the skill of short- and medium-range weather forecasting has improved steadily in recent decades, forecasting skill declines rapidly for periods longer than 48 hr and is minimal beyond 10 days.

☐ An important tool in long-range weather forecasting is identification of teleconnections, linkages between changes in atmospheric circulation occurring in widely separated regions of the globe.

☐ Single-station weather forecasting is based on principles of meteorology, fair-weather bias, climatic records, or persistence of weather episodes.

☐ Weather information and forecasts are communicated to users via teletypewriters, radio, facsimile systems, television, and newspapers.

## Review Questions

1. Why does weather forecasting require international cooperation?

2. Describe the principal steps involved in the preparation of weather forecasts.

3. Why do airport weather stations report altimeter settings for aircraft?

4. What is Greenwich Mean Time (GMT)?

5. What is your local time at 0100 GMT?

6. What is the chief purpose of the NWS Cooperative Observer Network? In what sense is the network cooperative?

7. What advantages do satellites bring to weather observation?

8. Distinguish between a satellite in polar orbit and a satellite in geosynchronous orbit.

9. Why do polar-orbiting satellites provide greater spatial resolution than geosynchronous satellites?

10. What is the basic difference between surface weather maps and upper-air weather maps?

11. What is meant by a numerical model of the atmosphere?

12. Why does the reliability of numerical weather prediction deteriorate for forecast periods beyond 48 hours?

13. Why are computerized weather forecasts not likely to replace human weather forecasters? Provide some illustrations.

14. Describe the basis for long-range (monthly, seasonal) weather forecasts.

15. What is a teleconnection? Describe how teleconnections aid long-range weather forecasting. Provide a specific example.

16. Speculate on how long-term weather records could be used to formulate seasonal weather forecasts.

17. Explain why there is a linkage between sea-surface temperatures and planetary-scale atmospheric circulation patterns.

18. An episode of one weather type usually gives way to another weather type very abruptly—in a single day or less. Explain why.

19. What is meant by *fair-weather bias?*

20. Distinguish between weather watches and weather warnings.

## Questions for Critical Thinking

1. Explain how weather satellites are able to provide night-time observations of cloud cover.
2. Explain why warm air advection causes the 500-mb surface to rise and why cold air advection causes the 500-mb surface to lower.
3. Explain why cold-core anticyclones and warm-core cyclones do not appear on 500-mb maps.
4. Why are winds associated with contour gradients on upper-air weather maps?
5. On average in midlatitudes in summer, a 500-mb ridge develops over the continents and a 500-mb trough occurs over the oceans. In winter, it's just the reverse: A 500-mb trough develops over the continents and a 500-mb ridge occurs over the oceans. Please explain.

## Selected Readings

AMERICAN METEOROLOGICAL SOCIETY. "Tornado Forecasting and Warning," *Bulletin of the American Meteorological Society* 72 (1991):1270–1272. A policy statement regarding the forecasting and detection of tornadic storms.

AMERICAN METEOROLOGICAL SOCIETY. "Weather Forecasting," *Bulletin of the American Meteorological Society* 72 (1991):1273–1276. A policy statement that evaluates the skill of weather forecasting.

FAA AND NOAA. *Aviation Weather Services.* Washington, DC: U.S. Government Printing Office, 1979, 123 pp. Describes the weather maps and forecast charts available for pilots.

FLEMING, J. R. *Meteorology in America, 1800–1870.* Baltimore, MD: The Johns Hopkins University Press, 1990, 264 pp. Recounts the principal people and concepts during the early days of meteorology.

GILMAN, D. "Predicting the Weather for the Long Term," *Weatherwise* 36, No. 6 (1983):290–297. Includes a discussion on how long-range weather outlooks are prepared.

HILL, J. *Weather from Above.* Washington, DC: Smithsonian Institution Press, 1991, 89 pp. Provides an historical perspective on the development of weather satellites.

HUGHES, P. "American Weather Services," *Weatherwise* 33, No. 3 (1980):100–111. Discusses the principal people and events in the history of the National Weather Service.

KIEREIN, T. "The Hi-Tech World of TV Weathercasting," *Weatherwise* 41, No. 2 (1988):150–154. Gives an illustrated tour of the making of a television weathercast.

MEINDL, E. A. AND G. D. HAMILTON, "Programs of the National Data Buoy Center," *Bulletin of the American Meteorological Society* 73 (1992):985–993. Describes the networks of coastal and offshore automated weather stations.

POOL, R. "Is Something Strange About the Weather?" *Science* 243 (1989):1290–1293. Summarizes the recent interest in applying chaos theory to long-range weather forecasting.

SMITH, W. L., ET AL. "The Meteorological Satellite: Overview of 25 Years of Operation," *Science* 231 (1986):455–462. Thoroughly reviews the capabilities of today's weather satellites along with projected developments for the future.

SPENCER, R. W., AND J. R. CHRISTY. "Precise Monitoring of Global Temperature Trends from Satellites," *Science* 247 (1990):1558–1562. Describes how satellites are used to monitor Earth's surface temperature patterns.

TRIBBIA, J. J., AND R. A. ANTHES. "Scientific Basis of Modern Weather Prediction," *Science* 237 (1987):493–499. Provides a relatively sophisticated summary of the historical roots and future directions of numerical weather prediction.

# 17

# Air Pollution Meteorology

*This goodly frame, the earth, seems to me a sterile promontory; this most excellent canopy the air, look you, this brave o'erhanging firmament, this majestical roof fretted with golden fire—why, it appeareth no other thing to me than a foul and pestilent congregation of vapours.*

WILLIAM SHAKESPEARE
*Hamlet*

Wind speed and atmospheric stability govern the dispersal of air pollutants. [Photograph by Mike Brisson]

ALTHOUGH WE usually think of air pollution as an undesirable byproduct of modern industrialism, it is at least as old as civilization. The first air pollution episode probably took place when some of the first humans made a fire in a poorly ventilated cave. Reference to polluted air appears as early as Genesis (19:28): "Abraham beheld the smoke of the country go up as the smoke of a furnace." Hippocrates noted the pollution of city air in about 400 B.C., and in 1170, Maimonides, referring to Rome, wrote: "The relation between city air and country air may be compared to the relation between grossly contaminated, filthy air, and its clear, lucid counterpart."

The Industrial Revolution was the single greatest contributor to air pollution as a chronic problem in Europe and North America. In the United States, in post–Civil War days, cities swelled with new industries and new immigrants to work in them. By the turn of the century, the urban environment was increasingly fouled by the fumes of foundries and steel mills. In those days, a city took pride in its smokestacks, which were considered a sign of a prosperous economy (Figure 17.1). Efforts to regulate air quality were meager, and little was known about the effects of air pollution on human health. In an attempt to placate the wheezing and coughing citizenry, some physicians even argued that polluted air had medicinal value.

Even today, concern over polluted air does not stem from disenchantment with the fruits of industrialism, but many people are troubled by reports that polluted air adversely affects our health, agricultural productivity, and the weather. In this chapter, we examine two questions related to these concerns: (1) How do weather conditions influence air pollution levels? (2) What is the impact of air pollution on weather? In Chapter 20, we consider the possible influence of air pollution on global climate.

## Air Pollutants

As we pointed out in Chapter 1, many gases and aerosols that can be air pollutants are actually normal constituents of the atmosphere. These substances become pollutants when their concentrations increase to levels that threaten the well-being of living things or disrupt physical or biological processes. In the Special Topic "Principal Air Pollutants," we survey the major air pollutants, noting their natural cycling within the environment, and the contributions to pollution made by human activity.

## Air Pollution Episodes

On the morning of 26 October 1948, a fog blanket that reeked of pungent sulfur dioxide ($SO_2$) fumes spread over the town of Donora in Pennsylvania's Monongahela Valley. Before the fog lifted five days later, almost half of the area's 14,000 inhabitants had fallen ill and 20 had died. That killer fog resulted from a combination of mountainous topography and weather conditions that trapped and concentrated deadly effluents from the community's steel mill, zinc smelter, and sulfuric acid plant.

383

# Principal Air Pollutants

Here we summarize the sources, cycling, and some of the impacts of the major air pollutants. Usually, we express the concentration of a gaseous pollutant in parts per million (ppm), that is, the number of pollutant molecules per million molecules of air. Aerosol concentrations, on the other hand, are normally given in mass of pollutant per volume of air (micrograms per cubic meter).

**Oxides of Carbon.** Through cellular respiration, organisms release carbon dioxide ($CO_2$) to the atmosphere, and through photosynthesis, plants take up carbon dioxide. Other natural sources of $CO_2$ include forest and brush fires and volcanic activity. Combustion of fossil fuels (coal, oil, and natural gas) for electric power generation and space heating releases carbon dioxide into the atmosphere. About 55% of this carbon dioxide remains in the atmosphere; most of the remainder dissolves in ocean water. Today the concentration of carbon dioxide in the atmosphere is about 350 parts per million (ppm) and is rising at a rate of about 17 ppm per decade. Because carbon dioxide is an important greenhouse gas, this increasing concentration may affect global climate (Chapters 2 and 20).

By far the most important natural source of atmospheric carbon monoxide (CO) is the combination of oxygen with methane and other volatile organic compounds (described later). Carbon monoxide is removed from the atmosphere by the activity of certain soil microorganisms and by chemical reactions that convert CO to $CO_2$. The net result is a harmless Northern Hemisphere average concentration of less than 0.15 ppm.

In the more developed nations of the Northern Hemisphere, the principal source of CO derived from human activity is incomplete combustion of fossil fuels, especially by motor vehicles. Recently, scientists discovered that the major source of CO in the Southern Hemisphere and the tropics is the burning of forests and savannas to clear land. Instrument measurements from aircraft and the space shuttle indicate that this contribution may compare in magnitude to global fossil fuel combustion.

Carbon monoxide is a colorless, odorless, and tasteless gas that defies direct human detection. CO is an asphyxiating agent that constitutes a serious health hazard, especially where the concentration is high, as it can be in highway tunnels and underground parking garages. Breathing CO causes drowsiness, slows reflexes, and impairs judgment; at high concentrations death ensues.

**Volatile Organic Compounds.** Volatile organic compounds (VOCs), commonly called *hydrocarbons,* include a wide variety of chemicals made up of only hydrogen and carbon atoms. Of the hydrocarbons that occur naturally in the atmosphere, methane ($CH_4$) has the highest concentration (1.74 ppm, on average). Methane is produced when organic material decays in the absence of oxygen, for example, in swamps, rice paddies, and the rumen of cattle. At normal background levels, methane is nonreactive; that is, it does not chemically interact with other substances and is not harmful. However, methane is the chief component of natural gas, and air with a methane concentration of 5% or greater is explosive. Methane is also a greenhouse gas, and its mean global concentration has been rising at a rate of 0.009 ppm per year. This increase may also contribute to climatic change (Chapter 20).

All vegetation emits various hydrocarbons. Among them are terpenes, which occur in concentrations of less than 0.1 ppm. They are chemically reactive and responsible for the aromas of pine, sandalwood, and eucalyptus trees. Recent studies suggest that such natural hydrocarbons (other than methane) may play an important role in the formation of photochemical smog.

Reactive hydrocarbons are also emitted during the incomplete combustion of gasoline by motor vehicles. This source is responsible for hundreds of different hydrocarbons. Because gasoline is very volatile, some hydrocarbons (perhaps as much as 15% of the total in some cities) escape to the air during gasoline delivery and refueling operations at service stations. Hydrocarbons are also emitted by solvents used in a variety of industrial and commercial processes, from painting to cleaning heavy equipment. Chemical manufacturing and petroleum refining also emit hydrocarbons to the atmosphere.

Although our understanding of the natural cycling of hydrocarbons in the atmosphere is still incomplete, we do know that the typically low concentrations of most hydrocarbons found in city air likely pose little direct environmental threat. However, serious health hazards arise from products of complex chemical reactions involving hydrocarbons and other air pollutants (particularly nitrogen oxides) in the presence of sunlight. These reactions produce photochemical smog. Some hydrocarbons, such as benzene ($C_6H_6$), an industrial solvent, and benzopyrene, a product of fossil fuel and tobacco combustion, are also carcinogenic.

**Oxides of Nitrogen.** The action of soil bacteria is responsible for most of the nitric oxide (NO) that is produced naturally and released to the atmosphere. Within the atmosphere, NO combines readily with oxygen to form nitrogen dioxide ($NO_2$). Together, these two oxides of nitrogen are usually referred to as $NO_x$.

Although human activities contribute only about 10% of the atmosphere's total load of $NO_x$, our contributions tend to be much more concentrated than the natural atmospheric average. $NO_x$ forms when high combustion temperatures, such as those inside an automobile engine, cause nitrogen ($N_2$) and oxygen ($O_2$) in the air to combine. In addition, $NO_x$ forms when nitrogen compounds in a fuel, such as coal, are oxidized. ($N_2$ in the fuel combines with $O_2$ in the air.) For both modes of formation, NO is generated initially, and when it is vented and cooled, some of the NO converts to $NO_2$. About half of $NO_x$ comes from stationary sources (power plants, primarily), and the other half comes from motor vehicles.

Nitrogen dioxide ($NO_2$) is a much more serious air pollutant than its precursor, NO; the toxicity of $NO_2$ is about four times that of NO. Nitrogen dioxide at high levels is believed to contribute to heart, lung, liver, and kidney damage, and it is linked to the incidence of bronchitis and pneumonia. Moreover, because nitrogen dioxide occurs as a brownish haze, it reduces visibility. Oxides of nitrogen are precursors of photochemical smog. In addition, $NO_2$ combines with moisture in the air to form nitric acid ($HNO_3$), a corrosive substance and acid rain precursor.

**Compounds of Sulfur.** Sulfur enters the atmosphere naturally as sulfur dioxide ($SO_2$) from volcanic eruptions, as sulfate particles from sea spray, and as hydrogen sulfide ($H_2S$) produced when organic matter decays in the absence of oxygen. These sulfur compounds are washed from the air by precipitation and are taken up by soil, vegetation, and surface waters.

Human activities also release sulfur compounds into the atmosphere, at about one-third the rate of emission by natural sources. Most of our contribution comes from fossil fuels (chiefly coal and oil) that contain sulfur as an impurity and emit sulfur dioxide when burned. The principal U.S. source of sulfur dioxide is coal burning for electric utilities (65.7%). Industrial processes contribute 16.4%, transportation, 4.4%; and nonutility fuel combustion from stationary sources, 13.5%.

Within the atmosphere, sulfur dioxide is converted to sulfur trioxide ($SO_3$) and sulfate-containing aerosols. Sulfate aerosols restrict visibility and, in the presence of water, form droplets of sulfuric acid ($H_2SO_4$), a highly corrosive substance that also lowers visibility. Both $SO_2$ and $SO_3$ irritate respiratory passages and can aggravate asthma, emphysema, and bronchitis. Sulfate aerosols and sulfuric acid droplets are thought to increase human vulnerability to respiratory infection.

Certain industrial activities, including paper and pulp processing, emit hydrogen sulfide ($H_2S$) as well as a family of organic sulfur-containing gases called mercaptans. Even at extremely low concentrations, these compounds are foul smelling. Hydrogen sulfide tarnishes silverware and copper facings and blackens lead-based paints.

**Photochemical Smog.** On sunny days, when vehicular traffic is congested (during morning and evening rush hours, for example), *photochemical smog* is likely to form. Oxides of nitrogen in motor vehicle exhaust and hydrocarbons (from various anthropogenic and biogenic sources) react in the presence of sunlight to produce a noxious, hazy mixture of aerosols and gases. Products include ozone ($O_3$), formaldehyde ($CH_2O$), ketones, and PAN (peroxyacetyl nitrates), substances that irritate the eyes and damage the respiratory system. Photochemical smog is most common in urban areas, but winds can transport auto exhaust into suburban and rural areas, where the sun's rays also trigger smog development.

While levels of ozone at the Earth's surface average only about 0.02 ppm, ozone concentrations may exceed 0.5 ppm in thick photochemical smog. Exposure to these relatively high concentrations for an hour or two can cause healthy people to experience coughing, painful breathing, and temporary loss of some lung function when they engage in vigorous physical activity. Some medical experts fear that chronic exposure to ozone may cause structural changes in the lungs, perhaps leading to lung disease. It also degrades rubber and fabrics, retards tree growth, and damages some crops.

**Suspended Particulates.** So far, our survey of the major air pollutants has focused primarily on gases. Now we turn our attention to the multitude of tiny solid particles and liquid droplets that are suspended in the atmosphere. Sea-salt spray, soil erosion, volcanic activity, and various industrial emissions account for about one half the atmosphere's total aerosol load. The other half is largely the consequence of atmospheric reactions among various gases.

Perhaps the most common particulates are dust and soot. Most dust is produced when wind erodes soil;

agricultural activity often accelerates such erosion. Soot, tiny solid particles of carbon, is emitted during the incomplete combustion of coal, oil, and refuse. In urban-industrial air, suspended aerosols may include a wide variety of materials, depending on the specific types of manufacturing. Urban-industrial aerosols usually include a diverse array of trace metals such as lead, nickel, iron, zinc, copper, magnesium, and cadmium. These aerosols pose a significant health hazard because their small size allows them to be inhaled deeply into the lungs. In addition, air may contain asbestos fibers, pesticides, and fertilizer dust.

Air normally also contains fungal spores and pollen. Disturbance of the land by wildfires, farming, or construction promotes the abundant growth of ragweed and other weeds whose pollen evokes allergic reactions, such as hay fever, in roughly one out of every 20 people.

**FIGURE 17.1**
To many people, belching industrial smokestacks in 1906 Pittsburgh signaled a prosperous economy. [Carnegie Library of Pittsburgh]

VIEW N.W. FROM MIDDLE POINT OF UNION STATION

Air pollutants are especially dangerous when atmospheric conditions reduce their rate of dilution. Once pollutants enter the atmosphere through smokestacks or exhaust pipes, their concentrations usually begin to decline as they mix with cleaner air. The more thorough the mixing, the more rapid is the rate of dilution. When conditions in the atmosphere favor rapid dilution, the impact of air pollution is usually minor. On other occasions, called **air pollution episodes,** conditions in the atmosphere minimize dilution, and the impact can then be severe, especially on human health. The two weather conditions that most influence the rate of dilution are wind speed and atmospheric stability.

## WIND SPEED

Intuitively, we know that air is likely to mix more thoroughly on a windy day than on a calm one. When it is windy, turbulent eddies mix polluted air and cleaner air and thereby accelerate dilution. But when the wind is calm, dilution is by the much slower process of **molecular diffusion,** that is, dispersal at the molecular level. As a general rule, a doubling of wind speed cuts the concentration of air pollutants in half (Figure 17.2).

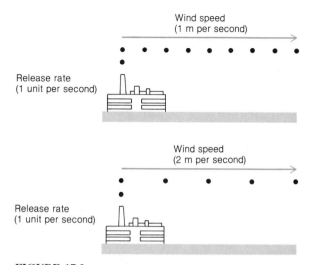

Wind speed
(1 m per second)

Release rate
(1 unit per second)

Wind speed
(2 m per second)

Release rate
(1 unit per second)

**FIGURE 17.2**
Doubling the wind speed from 1 to 2 m per second increases the spacing between puffs of smoke by a factor of two, thereby reducing pollution concentrations by one half.

Certain weather patterns favor light winds and thus inhibit dispersal of contaminants. Over a broad region at the center of an anticyclone, for example, the horizontal air pressure gradient is weak. Consequently, winds are very light or calm, and pollutants do not disperse readily. On the other hand, within a cyclone, a steeper air pressure gradient means stronger winds and more rapid dilution of air pollutants. In addition, the rain or snow associated with a cyclone cleans the air by washing pollutants to the ground.

Wind speed is influenced not only by horizontal air pressure gradients but also by surface roughness (friction). In a city, the canyonlike topography of tall buildings and narrow streets creates a rough surface that slows the wind; average near-surface wind speeds may be 25% lower in a city than in the surrounding countryside. When regional (synoptic-scale) winds are light (less than 15 km per hour), the contrast between city and country winds is even more pronounced, amounting to a wind speed reduction of up to 30%. Dilution of air pollutants by wind is thus impeded in urban localities, the very places where most contaminants are generated.

Another consequence of the frictional interaction of the wind with obstacles on the Earth's surface is the formation of zones of light and irregular winds that can trap air pollutants. As we saw in Chapter 9, the horizontal wind breaks into turbulent eddies to the lee of obstacles such as trees or a building. Immediately downwind from the obstacle, the irregular eddy motion forms a closed circulation, known as a **wake.** If a smoke plume enters the wake of a building, as in Figure 17.3A, the smoke is trapped and may circulate into the building's ventilation system. Hence, smokestacks must be constructed to a height such that the plume clears the wake of nearby buildings (Figure 17.3B). This stack height is generally taken to be 2.5 times the height of the nearest obstacle.

## *ATMOSPHERIC STABILITY*

As we saw in Chapter 6, stability affects vertical motion within the atmosphere. Convection and turbulence are enhanced when the air is unstable and inhibited when the air is stable. The stability of air thus influences the rate at which polluted air mixes with clean air. A parcel of polluted air (a puff of smoke, for example) emitted into unstable air undergoes more mixing than does the same parcel of polluted air emitted into stable air. Stable air inhibits the upward transport of air pollutants, and a layer of stable air aloft may act as a lid over the lower troposphere and thus traps air pollutants. Continual emission of contaminants into stable air results in the accumulation of pollutants.

**Mixing depth** is the vertical distance between the Earth's surface and the altitude to which convection currents (that is, mixing) reach. When mixing depths are great (many kilometers, for example), the relative abundance of clean air allows pollutants to readily mix and dilution is enhanced. When mixing depths are shallow (less than 1000 m, for example), air pollutants are restricted to a smaller volume of air, and concentrations may build to unhealthy levels. When air is stable, convection and turbulence are suppressed and mixing depths are shallow. When air is unstable, convection and turbulence are enhanced and mixing depths increase. Because solar heating triggers convection, mixing depths tend to be greater in the afternoon than in the morning, greater during the day than at night, and greater in summer than in winter.

We can sometimes estimate the stability of air layers by observing the behavior of a plume of smoke belching from a smokestack. If the smoke enters an unstable air layer, the plume undulates, as in Figure 17.4.

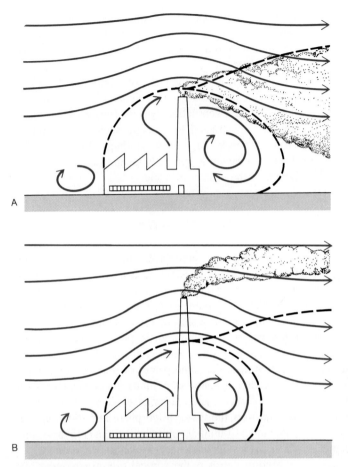

**FIGURE 17.3**
(A) If a smokestack is too low, effluents may be trapped within the wake of nearby buildings or the chimney itself. (B) If a smokestack is constructed to the height of good engineering practice (2.5 times the height of the nearest obstacle), effluents clear the wake and downwash and trapping are avoided.

In general, this plume behavior indicates that polluted air is mixing readily with the surrounding cleaner air and being diluted. The net effect is improved air quality—except where the plume loops to the ground. On the other hand, a plume of smoke that flattens and spreads slowly downwind, as shown in Figure 17.5, indicates very stable conditions and minimal dilution.

In summary, air stability influences the rate at which

**FIGURE 17.4**
Smoke plumes entering unstable air exhibit looping that is indicative of strong mixing and dilution of air pollutants.

- - - - Dry adiabatic lapse rate
———— Temperature profile

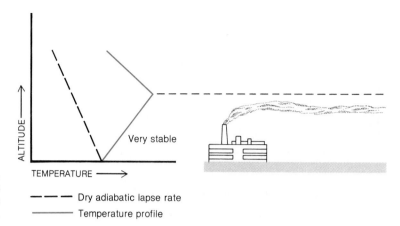

**FIGURE 17.5**
A temperature inversion within a surface air layer indicates very stable conditions. A smoke plume entering such stable air forms a thin ribbon extending downwind from the stack, and dilution of pollutants is minimal.

polluted air and cleaner air mix. If air is stable, dilution is inhibited, but if air is unstable, dilution is enhanced.

## TEMPERATURE INVERSIONS

An air pollution episode is most likely when a persistent **temperature inversion** develops. Within an air layer characterized by a temperature inversion, air temperature increases with altitude; that is, warmer, lighter air overlies cooler, denser air. (Note that a temperature inversion is the *inverse* of the usual temperature profile of the troposphere.) This is an extremely stable stratification that strongly inhibits mixing and dilution of pollutants. A temperature inversion can form by (1) subsidence of air, (2) extreme radiational cooling, or (3) advection of air masses. The resulting inversion may occur aloft or at the surface.

A **subsidence temperature inversion** forms a lid over a broad area, often encompassing several states or provinces at once. It develops during a period of fair weather when the planetary-scale circulation pattern causes a warm anticyclone to stall. An anticyclone is characterized by air that subsides and is thereby warmed by adiabatic compression. Subsiding warm air is prevented from reaching the Earth's surface by the **mixing layer,** in which air is thoroughly mixed by convection (Figure 17.6). The air temperature within the mixing layer declines with altitude, but air just above the mixing layer, having been warmed by adiabatic compression, is warmer than air at the top of the mixing layer. A temperature inversion thus separates the

mixing layer from the compressionally warmed air above. Under these conditions, air pollutants are distributed throughout the mixing layer, but no higher than the temperature inversion. Pollutants are thus confined to a relatively small volume of air, and continual emissions will elevate concentrations. This situation is sometimes referred to as **fumigation.**

A **radiational temperature inversion** is perhaps more common and often more localized than a subsidence temperature inversion. Under clear night skies and light winds, radiational cooling chills the ground, which in turn chills the air in contact with it. Because the air near the surface is colder than the air aloft, a low-level temperature inversion develops. Such temperature inversions usually disappear within a few hours after sunrise because the sun heats the ground, which heats the overlying air and eventually reestablishes a normal sounding in which air temperature decreases with increasing altitude. In winter, however, where the sun's rays are weak and the ground is covered by a highly reflective layer of snow, a radiational temperature inversion may persist for days, or even weeks.

Advecting air masses can also give rise to temperature inversions. This sometimes happens at the base of the Rocky Mountains. As shown in Figure 17.7, a westerly airflow is compressionally warmed as it is drawn down the leeward slopes of the mountain range. Along the foot of the mountains, however, surface winds advect cold air southward. Hence, a temperature inversion aloft separates the warm air above from the surface layer of cold air. Although temperature inversions also characterize warm and cold fronts, these inversions have little adverse effect on air quality because the

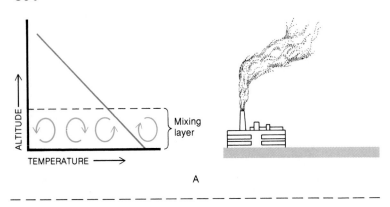

A

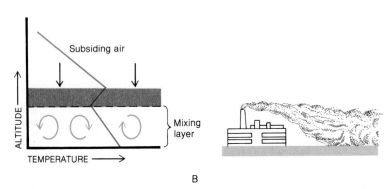

B

**FIGURE 17.6**
A temperature inversion can develop aloft through subsidence of air. A sounding prior to subsidence (A) is compared with a sounding during subsidence (B). The temperature inversion acts as a lid over the lower atmosphere, trapping air pollutants.

fronts are moving and the accompanying precipitation washes pollutants from the air.

## AIR POLLUTION POTENTIAL

Weather conditions that favor shallow mixing depths, air stagnation, and air pollution episodes occur with varying frequency in different places and during different seasons of the year. Areas with particularly high potential for air pollution include southern and coastal California, portions of the Rocky Mountain states, and Appalachian Mountain valleys. In general, in much of the West, air quality is lowest in winter; in the East, air quality is lowest in autumn. In southern California, air pollution potential is highest in summer. The usual seasonal shifts in atmospheric circulation patterns are responsible for seasonal changes in air pollution potential.

Many regions with high air pollution potential are

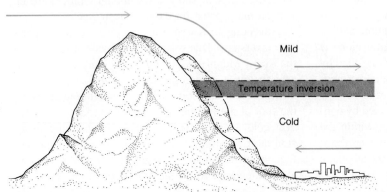

**FIGURE 17.7**
A temperature inversion develops aloft on the leeward slopes of the Rocky Mountains. Air descending the leeward slopes is warmed by compression and overlies colder air advected on northerly surface winds. This situation creates a high air pollution potential for the city of Denver.

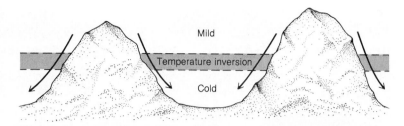

**FIGURE 17.8**
Under the influence of gravity, cold dense air drains downslope and strengthens the temperature inversion in the valleys.

also locales of great topographic relief, because hills and mountain ranges can block horizontal winds that would disperse polluted air. In addition, a radiational temperature inversion that forms in a lowland, such as a river valley, is often strengthened by drainage of cold, dense air from nearby highlands. As illustrated in Figure 17.8, the result is a persistent stable stratification of mild, lighter air over cooler, denser air.

Los Angeles is particularly susceptible to air pollution episodes because of frequent temperature inversions, topographic setting, and high concentration of pollutant sources (9 million motor vehicles). Figure 17.9 shows the air circulation and topographic features that influence air quality in that city. Weather in the Los Angeles area, like the weather throughout much of California, is strongly influenced by the eastern edge of the semiperimanent Pacific anticyclone. This subtropical high is responsible not only for California's famous fair weather but also for air that gently subsides over Los Angeles. Subsiding air is prevented from reaching the ground, however, by a shallow onshore flow of cool marine air from over the adjacent ocean waters. Because of adiabatic

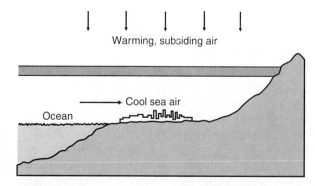

**FIGURE 17.9**
Prevailing atmospheric circulation patterns and topographic features combine to give Los Angeles an unusually high air pollution potential.

compression, the subsiding air aloft is warmer than the underlying marine air layer. Consequently, a subsidence temperature inversion typically develops at about 700 m (2300 ft) over Los Angeles, and convection (mixing and dilution of pollutants) is restricted to the shallow marine layer. This weather pattern contributes to a relatively high air pollution potential and occurs on perhaps two out of every three days of the year.

The exceptionally high incidence of temperature inversions over Los Angeles is aggravated by topographic barriers. The city is situated in a basin that opens to the Pacific and is rimmed on three sides by mountains. Cool breezes that sweep inland from the ocean are unable to flush pollutants out of the city. Mountains and a temperature inversion aloft thus encase the city in its own fumes, and within this crucible a complex photochemistry takes place that produces photochemical smog (Figure 17.10).

### NATURAL CLEANSING PROCESSES

Conditions that favor accumulation and concentration of pollutants in the air are countered to some extent by natural removal (cleansing) mechanisms. Some aerosols are removed from the air when they strike and adhere to buildings and other structures, a process called **impaction.** Aerosols with radii greater than 0.1 micrometer are also subject to **gravitational settling** (or *sedimentation*). Heavier (and larger) aerosols have greater terminal velocities and settle more rapidly than do smaller ones (Chapter 8). For this reason, larger aerosols tend to settle nearer their source, whereas the wind may transport smaller ones many kilometers and to great altitudes before they finally settle to the Earth's surface. The combined process of impaction and gravitational settling is sometimes referred to as **dry deposition.**

**FIGURE 17.10**
Restricted visibility caused by photochemical smog in Los Angeles. [Photograph by J. M. Moran]

The most effective natural pollution-removing mechanism is **scavenging** by rain and snow.* From personal experience, we know that the air seems cleaner just after a rain shower. In fact, in regions that experience moderate percipitation, scavenging is responsible for perhaps 90% of aerosol removal. Although gaseous pollutants are somewhat less susceptible to scavenging than aerosols, they dissolve to some extent in raindrops and cloud droplets. While scavenging improves air quality, it degrades the quality of rain and snow—sometimes to the point of polluting surface water and harming aquatic life. We discuss this problem later in the chapter.

## Air Pollution's Impact on Weather

We have seen how weather conditions influence air pollution potential. Air pollution also impacts weather. Air pollution affects the amount of cloudiness and the quantity and quality of precipitation, especially downwind from large urban-industrial areas.

*Related to this process, some aerosols function as condensation or deposition nuclei and thereby are incorporated into cloud droplets or ice crystals that may or may not grow into raindrops or snowflakes.

### URBAN WEATHER

The influence of the urban environment on cloudiness and precipitation is illustrated by typical climatic contrasts between urban and rural areas. Winter fogs are about twice as frequent in cities as in the surrounding countryside. Downwind from cities, rainfall may be enhanced by 5% to 10%. The greater contrast tends to occur on weekdays, when urban-industrial activity is at its peak, suggesting that increased precipitation is at least partially due to urban-industrial air pollution. Furthermore, studies show that the size of the urban area influences both the magnitude and the areal extent of the climatic impact.

Data from the *Metropolitan Meteorological Experiment (METROMEX)* indicate significantly more precipitation enhancement downwind of St. Louis, Missouri. METROMEX scientists analyzed weather observations during an intensive field study over a six-year period during the 1970s and concluded that average summer rainfall was 25% greater downwind of St. Louis than upwind of the city.

Analysis of findings from METROMEX and other field studies indicates that increased cloudiness and precipitation downwind of cities is due to a combination of factors. Urban-industrial areas are sources of (1) condensation nuclei (many of which are hygroscopic) that spur cloud development, (2) water vapor that increases the relative humidity, and (3) heat (urban heat island) that adds to the buoyancy of air and favors ascent of air. In addition, irregularities of urban topography induce convergence of horizontal winds and uplift.

Because precipitation, fog, and cloudiness in urban areas often adversely impact both surface and air transportation, any artificial increase in these conditions is potentially troublesome. Reduced visibility, for example, slows surface traffic, curtails air travel, and contributes to auto accidents. In the last two decades, local urban visibilities have improved significantly, apparently due to enforcement of strict air-quality regulations.

Jet aircraft traffic is modifying the cloud cover, especially along heavily traveled air corridors between major cities. The visible jet contrails etching the sky, like those shown in Figure 17.11, are composed of ice crystals that are traceable to the water vapor and condensation nuclei produced by jet engines as combustion products (refer back to the Special Topic "Clouds by Mixing" in Chapter 6). Contrails may dissipate rapidly or spread laterally and form a thin cirrus overcast. Increased cloudiness, in turn, reduces sunshine penetration and may enhance local precipitation by serving as a source of ice crystal nuclei for lower clouds. In fact, S. A. Changnon of the Illinois State Water Survey reports that in portions of the Midwest where jet air traffic is heavy, cloud cover increased 20% between 1960 and 1980.

**FIGURE 17.11**
Contrails produced by jet aircraft rapidly spread laterally to form cirrus clouds. [Photograph by Arjen and Jerrine Verkaik/SKYART]

## *ACID DEPOSITION*

As we saw in Chapter 6, the atmospheric subcycle of the global hydrologic cycle purifies water. That is, when water evaporates (or ice sublimates), dissolved and suspended substances are left behind. As raindrops and snowflakes fall from clouds to the Earth's surface, however, they wash pollutants from the air, and the chemistry of the precipitation changes.

Rain and snow are normally slightly acidic because they dissolve some atmospheric carbon dioxide, producing a weak acidic solution. Where air is polluted with oxides of sulfur and oxides of nitrogen, these gases interact with moisture in the atmosphere to produce tiny droplets of sulfuric acid ($H_2SO_4$) and nitric

acid ($HNO_3$). These substances dissolve in precipitation and increase its acidity. Precipitation that falls through such contaminated air may become 200 times more acidic than normal. Furthermore, in the absence of precipitation, sulfuric acid droplets convert to acidic aerosols that reduce visibility and may cause human health problems when inhaled. Eventually, acidic (and other) aerosols settle to the ground as dry deposition (Figure 17.12). The combination of acid precipitation and dry deposition is often referred to as **acid deposition.**

The range of acidity and alkalinity, called the **pH scale,** is shown in Figure 17.13, which compares the normal acidity of rainwater with the pH values of some other familiar substances. The normal pH of rainwater is 5.6; rain that is more acidic than normal is called

**FIGURE 17.12**
An instrument for monitoring precipitation quality. Separate containers collect dry and wet deposition for chemical analysis. One container is open only during fair weather and collects dry deposition. Rainfall activates a switch that covers the dry deposition container and opens the other to collect precipitation. [Photograph by J. M. Moran]

acid rain.* Note that the pH scale is logarithmic; that is, each unit increment corresponds to a tenfold change in acidity. Hence, a drop in pH from 5.6 to 3.6 represents a hundredfold (10 × 10) increase in acidity.

Gene E. Likens, past director of the Institute for Ecosystem Studies of the New York Botanical Garden, was one of the first to sound the alarm regarding acid rain. Likens and his colleagues reported an increase in the acidity of rainfall over the eastern United States between 1955 and 1973. Their findings were later confirmed and updated by measurements made by the National Atmospheric Deposition Program in the United States and by the Canadian Network for Sampling Precipitation (Figure 17.14). In general, the mean annual pH of precipitation is under 5 east of the Mississippi and over 5 west of the Mississippi. Rain and snow are most acidic in the northeastern United States and adjacent portions of Canada, where mean annual pH in 1991 ranged from 3.0 to 5.5 and individual storms produced rainfall pH as low as 2 to 3.

*Distilled water that is saturated with carbon dioxide has a pH of 5.6. This is why a pH of 5.6 is taken as the threshold for acid precipitation; that is, rain or snow having a pH under 5.6 is described as *acidic*. However, recent studies demonstrate that small amounts of naturally occurring acids (other than carbonic acid) lower the normal pH of precipitation closer to 5.0. This argues for a revision of the current criterion for acid rain and snow.

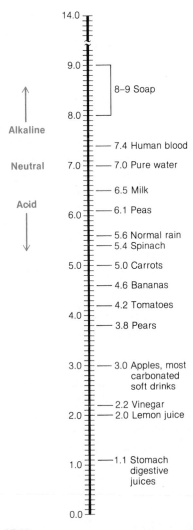

**FIGURE 17.13**
The scale of acidity and alkalinity, the pH scale, plus pH values for some common substances. pH is a measure of hydrogen ion concentration.

Acid precipitation is attributed to gaseous precursors emitted as byproducts of fuel combustion for electric power, industry, and motor vehicles. Coal burning for electric power generation is the principal source of sulfur oxides, while high-temperature industrial processes and internal combustion engines (in motor vehicles) produce nitrogen oxides.

Where soils are thin and the bedrock is noncarbonate (neither limestone nor dolomite), acid rains and snowmelt are not usually neutralized and may adversely impact waterways (Figure 17.15). Acid drain-

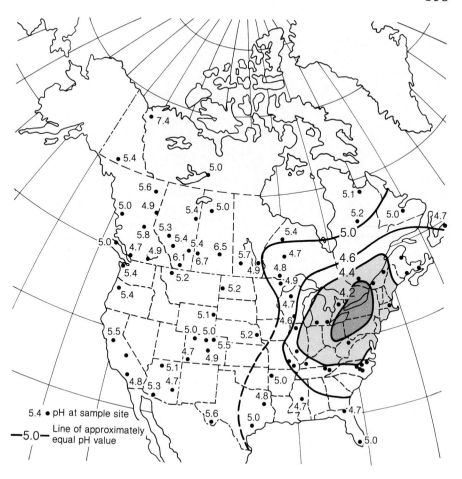

**FIGURE 17.14**
Acidity of precipitation over North America. Figures are average annual pH values of rain in 1982. [Data supplied by the United States National Atmospheric Deposition Program and the Canadian Network for Sampling of Precipitation]

5.4 • pH at sample site

—5.0— Line of approximately equal pH value

age lowers the pH of lakes and streams to the point that the reproductive cycles of fish are disrupted. Furthermore, acid rains leach metals (such as aluminum) from the soil, washing them into lakes and streams where they may harm fish, microorganisms, and aquatic plants. A 1991 Environmental Protection Agency (EPA) survey of 1180 lakes and 4670 streams in acid-sensitive regions of the United States found that of the waterways with excessive acidity, acid deposition was the principal cause in 75% of the lakes and 47% of the streams.* Increased acidity has caused the decline or elimination of fish populations in some lakes and streams in Norway, Sweden, eastern Canada, and the northeastern United States.

*Acid mine drainage was responsible for 26% of the acidified streams, and natural organic sources were implicated in 25% of the lakes and streams having below normal pH.

As part of the EPA's National Acid Precipitation Assessment Program, scientists artificially acidified a portion of a small lake in northern Wisconsin to determine the impact on the lake's aquatic life under controlled conditions. The 18-hectare (45-acre) Little Rock Lake is typical of the thousands of small, acid-sensitive lakes that dot northern Wisconsin, Minnesota, and Upper Michigan. Scientists anticipate that the experiment will enable them to predict the probable long-term response of similar lakes to acid deposition.

Little Rock Lake is shaped like an hourglass: two basins of roughly equal area are joined by a narrows. Following an intensive field study of the lake's physical and biological properties, scientists installed a 71-m (230-ft) plastic barrier to separate the two basins at the narrows (Figure 17.16). Over a six year period (1985 to 1990), 1500 liters (400 gallons) of sulfuric acid were added to one of the lake basins so that its pH gradu-

**FIGURE 17.15**
Surface waters (lakes and rivers) in many areas of North America are sensitive to acidification. These areas typically lack natural buffers (substances that neutralize acids) in the soil or bedrock. [From J. M. Galoway and E. B. Cowling, "The Effects of Precipitation on Aquatic and Terrestrial Ecosystems: A Proposed Precipitation Chemistry Network," *Journal of Air Pollution Control Association* 28 (1978):233]

Sensitive areas

ally declined from its initial 6.1 value to 4.7. The other lake basin was not disturbed and served as the control for comparison with the acidified basin. Scientists discovered that the acidic waters spured algae blooms, reduced the reproductive activity of certain species of fish, including rock and large-mouth bass, and caused some microorganisms to die and others to thrive. As of this writing, the acidified lake basin is slowly recovering without human intervention.

In addition to threatening aquatic systems, acid rains and fogs have been implicated along with other air pollutants (especially ozone) in the decline and dieback of coniferous forests in West Germany and in the Appalachian Mountains from North Carolina to New England. Another costly impact of acid precipitation is accelerated weathering of building materials, especially limestone, marble, and concrete. Metals, too, corrode faster than normal when exposed to acidic moisture.

Winds aloft can transport oxides of sulfur and nitrogen many thousands of kilometers from tall stack sources, so acid rain is becoming a global problem. Acid rains have been reported from such isolated localities as the Hawaiian Islands and the central Indian Ocean. Long-range transport of acid rain precursors has even strained the traditionally amiable relationship between the United States and Canada.

Acid rains threaten Canada's primary industries, lumber and fisheries. An estimated 300,000 lakes in eastern Canada are susceptible to acidic deposition; the number in the eastern United States is about 11,000. In 1985, the United States exported more than three times as much sulfur dioxide ($SO_2$), the chief acid rain precursor, to Canada than Canada exported to the United States. Perhaps half of the acid rainfall in Canada can be traced to emissions from Ohio Valley coal-fired power plants and industries. Canadian government of-

**FIGURE 17.16**
Little Rock Lake in northern Wisconsin was the subject of an experiment on the effects of acidified waters on aquatic life. The basin to the left of the barrier was artificially acidified, while the basin to the right was unaltered and served as a standard for comparison with the acidified basin. [Photograph by J. M. Moran]

ficials have pressed their Washington counterparts to enact stricter controls on polluting power plants and industries in order to ease the problem, which some Canadians feel will become severe within the next decade. In partial response to this international issue, the 1990 amendments to the U.S. Clean Air Act call for a sharp reduction in emissions of sulfur dioxide.

## The Ozone Shield

Erosion of the planet's **ozone shield** and the potential impacts on human health are among today's major environmental concerns. Recall from Chapter 2 that certain chemicals react with and destroy stratospheric ozone ($O_3$), allowing more intense solar ultraviolet radiation to reach the Earth's surface. More intense UV is likely to have serious consequences for human health, including a greater incidence of skin cancer and cataracts of the eye. A portion of the UV radiation that reaches the Earth's surface is responsible for sunburn and causes or contributes to skin cancer. This biologically active radiation, designated **UVB,** spans the wavelengths from 0.28 to 0.32 micrometer (Figure 17.17).

### *CHLOROFLUOROCARBONS (CFCs)*

The most serious threat to the ozone shield is posed by a group of chemicals, collectively known as CFCs (for chlorofluorocarbons), which were first synthesized in 1928 (Table 17.1). CFCs have been used widely as chilling (heat-transfer) agents in refrigerators and air conditioners, for cleaning electronic circuit boards, and in the manufacture of foams used for insulation. Their use in common household aerosol sprays such as deodorants, hairsprays, and furniture polish was banned in the United States and Canada in 1979. Less impor-

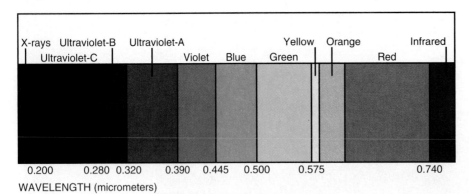

**FIGURE 17.17**
Ultraviolet and visible portions of the electromagnetic spectrum.

tant ozone-depleting chemicals include halons (used in fire extinguishers) and the industrial chemicals carbon tetrachloride and methyl chloroform.

F. S. Rowland and M. J. Molina of the University of California at Irvine first warned of the threat of CFCs to the ozone shield in 1974. They proposed the following disturbing scenario: Certain CFCs are inert (chemically nonreactive) in the troposphere, where they have been accumulating since production began more than 60 years ago. The inert CFCs gradually migrate upward into the stratosphere. At altitudes above about 25 km (16 mi), intense UV radiation breaks down the CFCs, causing them to release chlorine (Cl), a gas that readily reacts with and destroys ozone (Figure 17.18). Through a chain reaction, each chlorine atom can destroy tens of thousands of ozone molecules.

Although 20 years have passed since scientists first warned of threats to stratospheric ozone, evidence that the ozone shield is thinning appeared only recently. The first indications of a problem came from Antarctica.

**TABLE 17.1 Chronology of Events Related to Stratospheric Ozone Depletion**

| | |
|---|---|
| 1881 | Discovery of the stratospheric ozone layer by W. N. Hartley and A. Cornu. |
| 1928 | Synthesis of CFCs by T. Midgley as a nontoxic, nonflammable alternative to hazardous chemicals (e.g., ammonia) then used in home refrigerators. |
| 1950s | Rapid growth in use of CFCs in refrigeration, air conditioning, aerosol spray cans, foams, and solvents. |
| 1974 | Description of chemical reactions by M. J. Molina and F. S. Rowland whereby CFCs trigger depletion of stratospheric ozone. |
| 1979 | Ban on the use of CFCs in nearly all aerosol spray cans in the United States, Canada, Norway, and Sweden. |
| 1980s | Growth in world wide use of CFCs after a leveling off in the mid to late 1970s. |
| 1985 | Discovery of the Antarctic ozone hole. |
| 1987 | Adoption of the *Montreal Protocol on Substances That Deplete the Ozone Layer,* calling for a 50% cut in annual use of CFCs by 1998. |
| 1988–89 | Confirmation of declines in stratospheric ozone in midlatitudes of the Northern Hemisphere and in the Arctic. |
| 1990 | Amendment of the Montreal Protocol to require a complete phase out of CFCs and other major ozone-depleting substances by the year 2000. |
| 1990 | Requirement by U.S. Clean Air Act Amendments for recycling CFCs and periodic reports on the status of the ozone shield. |
| 1991 | Announcement of the phase out of production of CFCs by the end of 1996 and halons by the end of 1994 by Du Pont, the world's largest producer of CFCs. |
| 1992 | Discovery of relatively high levels of ClO over middle and high latitudes of the Northern Hemisphere. This prompts a U.S. Senate resolution calling for phase out of CFC production by 1995. |
| 1992 | Amendment of the Montreal Protocol on November 25 at Copenhagen meeting, calling for phase out of CFCs and carbon tetrachloride by 1 January 1996 and halons by 1994. |

*Source*: EarthQuest, Fall 1991, Office for Interdisciplinary Earth Studies, University Corporation for Atmospheric Research, Boulder, Colorado.

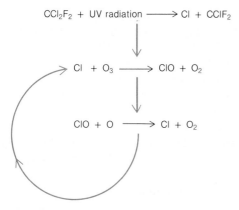

$$CCl_2F_2 + UV\ radiation \longrightarrow Cl + CClF_2$$

$$Cl + O_3 \longrightarrow ClO + O_2$$

$$ClO + O \longrightarrow Cl + O_2$$

**FIGURE 17.18**
A chain reaction involving $CCl_2F_2$, a chlorofluorocarbon (CFC) that is widely used as a refrigerant. Through this reaction, chlorine repeatedly reacts with and destroys ozone molecules.

## THE ANTARCTIC OZONE HOLE

As noted in Chapter 2, the ozone layer in the Antarctic stratosphere thins drastically during the Southern Hemisphere spring (mainly in September and October). Every November the ozone level recovers. The area of Antarctic ozone depletion, dubbed an **ozone hole,** is at least the size of the continental United States (Figure 17.19).

What causes the Antarctic ozone hole, and why does it fill in by November? During the long, dark Antarctic winter, extreme radiational cooling causes temperatures in the stratosphere to plunge below $-85\ °C$ ($-121\ °F$). At such frigid temperatures, what little water vapor there is in the stratosphere deposits as ice crystal clouds. Formation of those ice crystals is key to ozone depletion; they provide surfaces on which chlorine compounds that are inert toward ozone are converted to active forms that destroy ozone. Once the sun reappears in spring, solar radiation supplies the energy that causes active forms of chlorine to begin destroying ozone.

**FIGURE 17.19**
The Antarctic ozone hole (violet color) is much more evident in later 1989–1992 sequence than in the 1979–1982 maps. These images were produced at NASA's Goddard Space Flight Center using data obtained by the TOMS (Total Ozone Mapping Spectrometer) aboard the Numbus-7 satellite. Ozone concentrations are expressed in Dobson units (see Figure 2.17). [Courtesy of NASA]

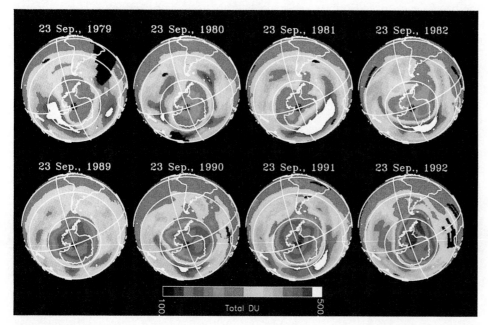

Ozone depletion takes place while the Antarctic atmosphere is essentially cut off from the rest of the global atmospheric circulation by the **circumpolar vortex,** a band of strong winds that encircles the outer margin of the Antarctic continent. A month or so into spring, however, the circumpolar vortex begins to weaken and allows warmer, ozone-rich air from lower latitudes to invade the Antarctic stratosphere. Ice crystal clouds vaporize, and the stratospheric ozone concentration returns to normal levels, that is, the ozone hole fills in.

Scientists investigating stratospheric chemistry in the Arctic in early 1989 discovered ozone-destroying chlorine compounds and a slight thinning of ozone. An Arctic ozone hole comparable in magnitude to that of Antarctica is considered unlikely for two reasons: (1) in winter, the Arctic stratosphere averages about 10 Celsius degrees (18 Fahrenheit degrees) warmer than the Antarctic stratosphere, making development of stratospheric ice crystal clouds unlikely; (2) the circumpolar vortex that surrounds the Arctic weakens earlier than its Antarctic counterpart. An exceptionally cold Arctic winter coupled with an unusually persistent circumpolar vortex could, however, translate into considerable ozone depletion in the Arctic, an area not very far from population centers.

Discovery of the Antarctic ozone hole and speculation about a possible link to CFCs spurred questions about trends in stratospheric ozone levels over North America. This concern led to a reanalysis of global ozone measurements made by ground-based and satellite instruments between 1969 and 1986. In March 1988, NASA's Ozone Trends Panel reported that average ozone levels were declining by about 1% to 3% per decade in the latitude belt roughly bounded by New Orleans and Seattle. An update of this study released in April 1991 by the EPA reports ozone losses in the same latitude belt at about twice the previous rate: 4% to 5% per decade. The EPA also noted that a thinner ozone shield persists into spring and summer when more people are outdoors and exposed to more solar radiation.

A new threat to stratospheric ozone followed the spectacular eruption of Mount Pinatubo in the Philippines in mid-June 1991. This, the century's most violent volcanic eruption, injected huge quantities of sulfur dioxide ($SO_2$) into the stratosphere. Within a month or so, $SO_2$ converted to tiny sulfuric acid droplets and sulfate aerosols. These particles are so small that they remain suspended in the lower stratosphere for perhaps 1 to 3 years. In the meantime, they spread as a sulfur haze around the globe (Chapter 20).

Apparently, volcanic sulfate aerosols function like ice crystals in the Antarctic stratosphere. That is, they help convert chlorine compounds from inactive to active molecules that can destroy ozone. This may at least partially explain the record low levels of stratospheric ozone in 1992–93 over most of the planet. In the belt between 30 and 60 degrees N, for example, ozone concentration was 9% below normal and the lowest since satellite measurements began in 1979.

## INTERNATIONAL RESPONSE

The first international response to threats to the stratospheric ozone shield was the so-called *Montreal Protocol* of September 1987. Under United Nations auspices, representatives of 23 nations meeting in Montreal negotiated a treaty to cut by 50% (from 1986 levels) the global production and consumption of CFCs by June 1998. By June 1990, however, ominous signs of ozone depletion in midlatitudes and the Arctic prompted a revision of the Montreal Protocol calling for a complete phase out of CFC production and use by the year 2000 for more developed nations and by 2010 for less developed nations. All this spurred the chemical industry to strengthen efforts to search for environmentally safe, nontoxic alternatives to CFCs.

In 1992, the U.S. Senate passed a resolution calling for a complete phase out of CFC production by 1995. This change in timetable was in response to the January 1992 discovery of surprisingly high levels of ozone-destroying chlorine monoxide (ClO) in the stratosphere of middle and high latitudes of the Northern Hemisphere. This discovery was made by high-altitude NASA aircraft surveys.

In November 1992, representatives of more than half the world's nations gathered in Copenhagen, Denmark, with the goal of revising again the Montreal Protocol. They agreed to phase out CFCs and carbon tetrachloride by 1 January 1996, and halons by 1994.

As of this writing, CFC emission rates have dropped by about 50% since 1988. Even if production and use of CFCs and other ozone-depleting chemicals cease within the next several years, the atmospheric content of chlorine will continue to grow because of the long residence times of CFCs in the atmosphere (refer back to Table 2.1). CFCs may remain for a century or more before being washed from the atmosphere. Consequently, the stratospheric chlorine concentration, now at 3.4 parts per billion (ppb), may top out at 4.1 ppb by the turn of the century.

## Conclusions

Weather influences, and is influenced by, air quality. Atmospheric stability and circulation patterns affect the rate of dilution of air pollutants. As a rule of thumb, every doubling of wind speed cuts air pollution concentrations in half. Dispersal of air pollutants is inhibited by stable air and enhanced by unstable air. The highest air pollution potential occurs where temperature inversions are frequent and topographic barriers block the horizontal wind.

The urban-industrial environment affects cloud and precipitation development downwind. Condensation nuclei, heat, and moisture emitted by urban sources are important contributing factors. In addition, sulfur oxides and nitrogen oxides from various industrial sources lower the pH of rain and snow. Where excessively acidic precipitation is not neutralized, aquatic systems may be threatened.

Thinning of the ozone shield is one of today's most pressing environmental concerns. CFCs have been implicated in the formation of the Antarctic ozone hole and may be involved in the thinning of stratospheric ozone elsewhere. All this has spurred international agreements to phase out production of CFCs and other ozone-destroying chemicals.

In Chapter 20 we consider the possible impact of air pollution on trends in global climate, but first we turn to the nature of climate and the climatic record.

## Key Terms

air pollution episodes
molecular diffusion
wake
mixing depth
temperature inversion
subsidence temperature
  inversion
mixing layer

fumigation
radiational temperature
  inversion
impaction
gravitational settling
dry deposition
scavenging
acid deposition
pH scale
acid rain
ozone shield
UVB
ozone hole
circumpolar vortex

## Summary Statements

☐ Most air pollutants are cycled naturally within the atmosphere, but concentrations may reach levels that threaten human health or disrupt physical and biological processes.

☐ Strong winds and unstable air enhance the rate of dilution of air pollutants, whereas weak winds and stable air suppress dilution.

☐ A temperature inversion consists of an extremely stable stratification of light, mild air over denser, cooler air. An inversion may develop through subsidence of air, extreme radiational cooling, or air mass advection. Temperature inversions greatly inhibit the dispersal of air pollutants.

☐ Many regions with great topographic relief have high air pollution potential because hills and mountain ranges can block horizontal winds that disperse polluted air. Furthermore, cold air drainage into river valleys and other lowlands can strengthen radiational temperature inversions.

☐ Conditions that favor the accumulation and concentration of pollutants in the atmosphere are countered to some extent by natural cleansing processes, that is, dry deposition (impaction plus gravitational settling) and scavenging by rain and snow.

☐ Urban heat and air pollutants affect precipitation, cloudiness, and fog development within and downwind from large metropolitan areas. Some urban air pollutants are hygroscopic and thereby function as efficient cloud condensation nuclei.

☐ Normally, precipitation is slightly acidic. In localities where the air is polluted by oxides of sulfur or nitrogen, precipitation becomes strongly acidic. Acid deposition threatens aquatic life and corrodes structures.

☐ Formation of ozone in the stratosphere protects life on Earth by filtering out harmful intensities of solar ultraviolet radiation. This ozone shield is threatened by chlorofluorocarbons (CFCs) accumulating in the atmosphere. CFCs are inert in the troposphere, but they drift into the stratosphere where intense UV causes them to break down and release chlorine, a gas that reacts with and destroys ozone.

☐ The deepening of the Antarctic ozone hole is likely due to CFCs and variations in Antarctic atmospheric circulation.

## Review Questions

1. Under what conditions are natural aerosol or gaseous components of the atmosphere considered to be air pollutants?
2. As a general rule, how does wind speed affect the concentration of air pollutants? Explain why.
3. Why is dilution of air pollutants particularly impeded in urban areas?
4. Define mixing depth. Explain how and why the mixing depth varies with (a) time of day and (b) season.
5. How does the stability of air influence the rate of dilution of air pollutants?
6. Describe the behavior of smoke plumes in (a) stable air and (b) unstable air.
7. Why do subsidence temperature inversions generally cover a larger geographical area than radiational temperature inversions?
8. Why are radiational temperature inversions often short-lived? Under what conditions might a radiational temperature inversion persist for many days?
9. In areas of great topographic relief, cold air drainage strengthens radiational temperature inversions. Explain this statement.
10. What factors contribute to the relatively high air pollution potential of Los Angeles?
11. Compare the natural atmospheric cleansing processes for the troposphere and the stratosphere.
12. What is the most effective natural mechanism for removing air pollutants?
13. How do air polllutants influence urban weather?
14. Describe the potential climatic influence of jet plane contrails.
15. Why is rainfall normally slightly acidic?
16. What air pollutants are acid rain precursors? Identify their sources.
17. What is the ozone shield?
18. What is thought to be the most serious threat to the ozone shield? How might these chemicals affect stratospheric ozone?
19. How might thinning of the stratospheric ozone shield affect the amount of biologically active ultraviolet radiation (UVB) that reaches the Earth's surface? How might this impact human health?
20. Speculate on the major sources and types of air pollutants in nonindustrialized nations.

## Questions for Critical Thinking

1. Midlatitude cyclones are beneficial for air quality. Explain this statement.
2. Some electric power utilities have reduced ground-level concentrations of air pollutants by building taller smokestacks. How do taller smokestacks contribute to the acid rain problem?
3. Why is a temperature inversion a case of *extreme* atmospheric stability?
4. What types of weather patterns favor (a) low air quality and (b) high air quality?
5. It is possible for temperature inversions to form simultaneously at the surface and aloft in the atmosphere. Speculate on the sequence of weather events that would give rise to such a situation.

## Selected Readings

BAKER, L. A. ET AL. "Acidic Lakes and Streams in the United States: The Role of Acid Deposition," *Science* 252 (1991):1151–1153. Reports on the results of an EPA survey of lakes and streams in acid-sensitive areas of the United States.

CHANGNON, S. A. "Inadvertent Weather Modification in Urban Areas: Lessons for Global Climate Change," *Bulletin of the American Meteorological Society* 73 (1992): 619–627. Discusses parallels between urban climatology and global climate change.

CHANGNON, S. A. "More on the LaPorte Anomaly: A Review," *Bulletin of the American Meteorological Society* 61 (1980):702–711. Concludes that anomalous precipitation at LaPorte, Indiana, between the late 1930s and the 1960s was linked to urban-industrial air pollution.

KERR, R. A. "New Assaults Seen on Earth's Ozone Shield," *Science* 255 (1992):797–798. Reports on the discovery of unusually high levels of ozone-depleting ClO in the stratosphere as far south as Cuba.

LIKENS, G. E., ET AL. "Acid Rain," *Scientific American* 241, No. 4 (1979):43–51. Discusses trends in acid deposition over North America and Western Europe.

MORGAN, M. D., J. M. MORAN, AND J. H. WIERSMA. *Environmental Science, Managing Biological and Physical Resources*. Dubuque, IA: Wm. C. Brown Publishers, 1993, 526 pp. Chapter 16 includes strategies to improve air quality.

POSTEL, S. "Air Pollution, Acid Rain, and the Future of Forests," *Worldwatch Paper,* No. 58 (1984):1–54. Reviews possible link between acid deposition and the dieback of coniferous forests in portions of Europe and northeastern North America.

SCHINDLER, D. W. "Effects of Acid Rain on Freshwater Ecosystems," *Science* 239 (1988):149–156. Presents an excellent summary of what is understood about the impacts of acidic precipitation.

SHAW, R. W. "Air Pollution by Particles," *Scientific American* 258, No. 2 (1987):96–103. Discusses the role of acidic sulfate particles in the development of haze.

STOLARSKI, R. S. "The Antarctic Ozone Hole," *Scientific American* 258, No. 1 (1988):30–36. Reviews the thinning of ozone during the Antarctic spring and possible causes of the recent acceleration of that thinning.

TOON, O. B. AND R. P. TURCO. "Polar Stratospheric Clouds and Ozone Depletion," *Scientific American* 264, No. 6 (1991):68–74. Describes the role of stratospheric cloud particles in the formation of the Antarctic ozone hole.

# 18 World Climates

*Antiphanes said merrily, that in a certain city the cold was so intense that words were congealed as soon as spoken, but that after some time they thawed and became audible; that the words spoken in winter were articulated next summer.*

PLUTARCH
*Of Man's Progress in Virtue*

The mountain range in the distance produces a rain shadow for the area in the foreground. Topography is one of many controls of climate. [Photograph by J. M. Moran]

# H

AVING EXAMINED the properties of the atmosphere, the governing principles of weather, and the various atmospheric circulation systems, we now turn to climate. In this chapter we examine some of the basics of climatology, the classification of climate, and global patterns of climate. Climatology is a complex discipline, however, and space does not permit more than a cursory summary of major concepts. The interested reader is referred to the Selected Readings at the close of this chapter for more on the subject. In the next chapter, we review the climatic record and how climate changes. In the final chapter, we summarize the possible causes of climatic variability.

## Describing Climate

**Climate** is defined as the weather of some locality averaged over some time period plus the extremes in weather behavior. Climate must be specified for a place and time period because, like weather, climate varies both spatially and temporally. Thus, for example, the climate of Minneapolis is different from that of Miami, and winters in Minneapolis were somewhat milder in the 1940s and 1950s than in the 1840s and 1850s.

Extremes in weather are important aspects of the climatic record. Hence, daily weather reports usually include the highest and lowest temperatures ever recorded for that date. Climatic summaries typically identify such extremes as the coldest, warmest, driest, wettest, snowiest, or cloudiest month or year on record.

Knowledge of climatic extremes has many practical applications. Farmers, for example, are interested in knowing not only the long-term average rainfall during the growing season but also the frequency of drought. Electric utilities are interested in the coldest winters and hottest summers on record so that they might assess the likely variability in residential energy demand for space heating and cooling.

Climate is usually described in terms of normals, means, and extremes of a variety of weather elements including temperature, precipitation, and wind. Climatic summaries are available in tabular form for climatic divisions of each of the states and Canadian provinces as well as for major cities. A narrative description of local or regional climate usually accompanies these data. Weather extremes for the world and for North America are listed in Appendix IV, and climatic data for selected U.S. and Canadian cities are presented in Appendix V.

In the United States, the National Weather Service is responsible for gathering the basic weather data used in climatological summaries. Data are processed, entered into archives, and made available to the public at the **National Climate Data Center (NCDC)** in Asheville, North Carolina. The NCDC also disseminates climatic information and prepares analyses for specialized uses. The Atmospheric Environment Service, headquartered at Downsview, Ontario, provides comparable services in Canada.

### THE CLIMATIC NORM

Traditionally, the climatic *norm* or *normal* is equated to the average value of some climatic element such as temperature or snowfall. Such an equation sometimes fosters misconceptions. For one, *normal* may be taken to imply that climate is static when, in fact, climate is inherently variable with time (Chapter 19). Furthermore, normal may imply a Gaussian (bell-shaped) probability distribution, although many climatic elements are non-Gaussian.

For our purposes, we can think of the climatic norm as encompassing the total variation in the climatic record, that is, both averages and extremes. Hence, for example, the occurrence of an exceptionally cold winter may not be *abnormal* because it may fall well within the expected range of variability of winter temperature. Even the cold summer of 1816, described in the Special Topic "1816, The Year Without a Summer?," was a normal, albeit extreme, event.

By international convention, climatic norms are computed from averages of weather elements compiled over a 30-year period.* Current climatic summaries are based on weather records from 1961 to 1990. The average July rainfall, for example, is the simple average of the total rainfall during each of 30 consecutive Julys from 1961 through 1990. The 30-year period is adjusted every 10 years to add the latest decade and drop the earliest one.

In the United States, 30-year averages are computed for temperature, precipitation, and air pressure only. Averages of other climatic elements such as wind speed and cloudiness are derived from the entire period of record. Extremes such as the highest temperature or driest month are also drawn from the period of record.

Selection of a 30-year period for averaging weather data may be inappropriate for many applications because climate varies over all time scales and can change significantly in periods much shorter than 30 years. For other purposes, a 30-year period provides a short-sighted view of the climatic record. Compared with the longer term climatic record, the current 1961–1990 *norm,* for example, was an unusually mild period over the eastern two-thirds of the United States. Nonetheless, people use the averages and extremes of past climate as a general guide to future expectations.

Many people assume that the mean value of some climatic element is the same as the median (middle value). That is, 50% of all cases are above the mean and 50% of all cases are below the mean. This is a rea-

*The Climatology Commission of the International Meteorological Organization adopted the 30-year period as a compromise at a meeting in Warsaw, Poland, in 1933.

sonable assumption for those climatic elements, such as temperature, that approximate a simple Gaussian-type distribution. Hence, for example, we can expect that about half the Januaries will be warmer and half the Januaries will be colder than the 30-year mean January temperature.

On the other hand, the distribution of some climatic elements, such as precipitation, is distinctly non-Gaussian, and the mean does not correspond to the median. For example, in a dry climate subject to infrequent deluges of rain during the summer, considerably fewer than half the Julys are wetter than the mean and much more than half of Julys are drier than the mean. In fact, for many purposes, the median value of precipitation may be more representative of climate than the mean value.

## CLIMATIC ANOMALIES

Climatologists often compare the average weather of a specific week, month, or year with the past climatic record. Such comparisons carried out over wide geographical areas show that departures from long-term climatic averages, called **anomalies,** never occur with the same sign or magnitude everywhere. As an illustration, consider Figure 18.1, which shows the temperature anomaly pattern for the winter of 1980–81 (December through February) across the United States, excluding Alaska and Hawaii. Note that average winter temperatures were above long-term averages *(positive anomalies)* in the western two-thirds of the nation and below long-term averages *(negative anomalies)* in the eastern third of the nation. Furthermore, the magnitude of the anomaly, positive or negative, varied from one place to another. For example, the winter of 1980–81 was up to 4 Fahrenheit degrees (2 Celsius degrees) milder than the long-term average over most of Texas and 2 to 4 Fahrenheit degrees (1 to 2 Celsius degrees) colder than the long-term average over the bordering state of Louisiana.

The geographic nonuniformity of climatic anomalies is linked to the prevailing westerly wave pattern that ultimately governs air mass advection and storm tracks. Hence, during the winter of 1980–81, the prevailing westerly wave pattern favored more than the usual cold air advection into the East and anomalous warm air advection into the West. The airflow aloft (the westerlies)

# 1816, The Year Without a Summer?

The summer of 1816 is famous for its anomalous cold. In fact, 1816 is sometimes described as "the year without a summer." Unseasonable snowfall and freezes damaged many crops in the then agrarian northeastern United States and adjacent portions of Canada. More than 90% of the corn crop, the prime food staple, was lost. Some superficial accounts of events of that summer give the impression that unusual cold persisted throughout the summer and affected the entire civilized world. Such was not the case.

In the Northeast, the summer of 1816 was punctuated by several outbreaks of unusually cold weather. Killing frosts struck northern and interior southern New England as well as Quebec in early June and July and again in late August. The June cold snap was accompanied by moderate to heavy snowfalls in the highlands. No sooner had farmers replanted their frozen crops than a killing freeze would strike again. The miserably short and disastrous growing season came to an end with a general hard freeze on 27 September.

These brief but unusually cold episodes interrupted longer spells of seasonably warm weather. Mean June and July temperatures were 1.6 to 3.3 Celsius degrees (3 to 6 Fahrenheit degrees) below average. Individually, these monthly anomalies fall within the expected range of climatic variability. Some previous and subsequent Junes, Julys, and Augusts have been as cold as the summer months of 1816, if not colder. What is most notable about the summer of 1816 is the persistence of a strong negative temperature anomaly through all three summer months. Nevertheless, the summer in the Northeast as a whole fell within the range of expected climatic variability.

What about the weather elsewhere during the summer of 1816? We know that it was probably not the same everywhere as in the Northeast, for climatic anomalies are typically geographically nonuniform in both magnitude and sign. Unfortunately, little weather information is available from the central and western United States and Canada because few settlements existed there at the time. France, Germany, and Great Britain did experience an unusually cold summer, but this is not surprising. The westerly long-wave pattern responsible for advecting unseasonably cold air into the Northeast would also favor western Europe with the same anomalous weather. The two localities are typically one wavelength apart. In contrast, available data indicate that east-central and eastern Europe experienced a warmer than usual summer.

The purpose of this discussion is not to diminish the climatic significance of the summer of 1816 or the hardships that people suffered then. It is, rather, to show that weather extremes typically fall within the expected range of climatic variability. We have more to say about the summer of 1816 in Chapter 20, where we discuss the impact of volcanic eruptions on climate.

determines the location of weather extremes such as drought or very cold temperatures. In view of the number of westerly waves that typically encircle the hemisphere (Chapter 10), a single weather extreme never occurs over an area as large as the United States or Canada; that is, severe cold or drought never grips the entire nation at the same time.

Rainfall typically forms considerably more complex anomaly mosaics than does temperature, as illustrated by Figure 18.2. This is due to the greater spatial differences in rainfall arising from the variability of storm tracks and the almost random distribution of convective showers. For these reasons, in spring in midlatitudes, for instance, even adjoining counties may experience opposite rainfall anomalies (one having above-average rainfall and the other below-average rainfall).

From an agricultural perspective, the geographic nonuniformity of climatic anomalies may be advantageous in that some compensation is implied. That is, poor growing weather and consequent low yields in one area may be compensated for to some extent by better growing weather and increased yields elsewhere. This is known as **agroclimatic compensation** and is discussed further in the Special Topic "Agroclimatic Compensation: The Benefits and Limitations."

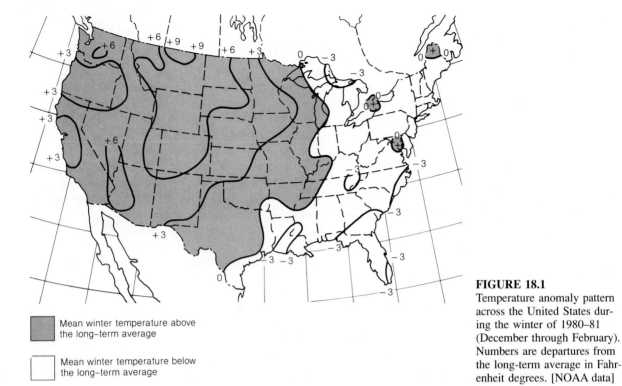

**FIGURE 18.1**
Temperature anomaly pattern across the United States during the winter of 1980–81 (December through February). Numbers are departures from the long-term average in Fahrenheit degrees. [NOAA data]

▨ Mean winter temperature above the long–term average

☐ Mean winter temperature below the long–term average

Geographic nonuniformity also characterizes trends in climate. A trend in the average annual temperature of the Northern Hemisphere is therefore not necessarily representative of all localities within the hemisphere. Over the same period, some locations experience cooling trends, whereas others experience warming trends, regardless of the direction of the hemispheric temperature trend. Not only is it misleading to assume that the direction of large-scale climatic trends applies to all localities, but it is also erroneous to assume that the magnitude of climatic trends is the same everywhere. A small change in the average hemispheric temperature typically translates into a much larger change in some areas and into little or no change in other areas.

## AIR MASS CLIMATOLOGY

Although the climate of a locality is traditionally described by normals, means, and extremes of various weather elements, an alternate approach, known as **air mass climatology**, has some interesting implications. In this approach, the frequency with which various types of air masses develop over a locality or are ad-

vected into a locality is used to describe the climate of that locality. For example, during January a northern U.S. city on average might have cold, dry air 60% of the time; mild, humid air 30% of the time; and mild, dry air 10% of the time.

Earlier we identified the major source regions for North American air masses (Chapter 11). Air masses actually originate in the clockwise and diverging flow that characterizes surface winds in anticyclones. As air streams away from an anticyclone, the temperature and humidity of the air modify to some extent, depending on the nature of the surface over which the air travels *(air mass modification)*. In this way, a single anticyclone can be the source of many different types of air masses, depending on the specific modifications that take place. Recall, for example, that winds spiraling outward from subtropical anticyclones are dry on the eastern flank and humid on the western flank. Air masses to the east of subtropical highs are therefore dry, whereas those to the west are humid.

In a study of air mass climatology, W. M. Wendland of the Illinois State Water Survey and R. A. Bryson of the University of Wisconsin–Madison identified the anticyclonic sources for Northern Hemisphere air

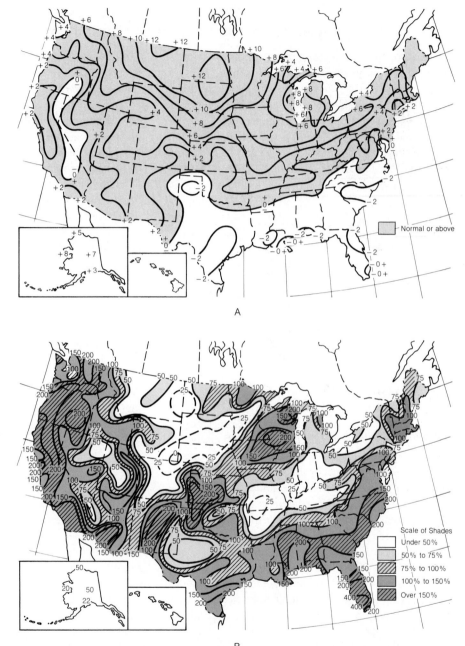

A

B

**FIGURE 18.2**
A precipitation anomaly pattern typically is much more complex than a temperature anomaly pattern. (A) Temperature anomalies across the United States for February 1983 are departures from 30-year averages and are expressed in Fahrenheit degrees. (B) Precipitation anomalies for the same month are expressed as percentages of long-term averages. [NOAA data]

masses. They did this by determining the average surface streamline pattern across the hemisphere for each month of the year. Average **streamlines** represent the mean paths of air moving horizontally. Figure 18.3 is an example of the Wendland–Bryson analysis. Anticyclonic flow is indicated by streamlines that turn into a clockwise and outward spiral. Of the 19 anticyclonic source regions, some were over the oceans, some were over the continents, and a few were in the Southern Hemisphere. Five sources persisted through the entire year, and three persisted for about 11 months. The others were prominent for 1 to 9 months.

SPECIAL TOPIC

## Agroclimatic Compensation, The Benefits and Limitations

Because of the geographic nonuniformity of climatic anomalies, crop yields in any one season are likely to vary from one place to another. Consequently, lower crop yields in one area may be compensated for to some extent by higher crop yields in another area. For example, in the spring of 1982, wet fields delayed corn planting for nearly a month in Iowa and Kansas. As a consequence, per-hectare yields in these states were 5% and 9% lower, respectively, than in the prior year. Meanwhile, eastern corn belt states such as Ohio and Indiana experienced quite favorable weather throughout the growing season, and their per-hectare yields jumped 22% and 18%, respectively, from 1981 levels. Interestingly, these yields resulted in a record high average yield per hectare for the corn belt as a whole.

The impact of a weather extreme on crop production typically hinges on its time of occurrence during the growing season. For example, by the time drought and excess heat developed in the Midwest and Plains states in 1983, much of the wheat crop had already matured. Although yields per hectare of the later maturing corn and soybeans declined by more than 28% and 19%, respectively, from the previous year, the 1983 wheat harvest broke a record for yield per hectare.

The impact of weather extremes on crop production also varies on a larger geographical scale. Although the United States experienced very favorable growing conditions for corn from May to September 1982, the Republic of South Africa received less than half its normal summer rainfall (December 1981 through March 1982). The United States enjoyed a record corn harvest, whereas the Republic of South Africa's corn yield plunged nearly 30%. Yields declined over much of Africa and South America that year but increased in North America and Europe. On a global scale in 1982, gains more than compensated for declines, and the worldwide average yield per hectare increased by 3%.

Agroclimatic compensation has its limitations. More favorable weather conditions in a region where low soil fertility already limits crop yields would probably not compensate for a yield reduction caused by unfavorable weather in a region with highly fertile soils. For example, many of the soils in Illinois and Iowa developed from loess, a dusty, wind-blown sediment that was deposited during the Ice Age. As a rule, the thicker the deposit of loess, the more fertile is the soil. Even with heavy applications of fertilizer, corn yields on the shallower loess soils are often only half those of deeper loess soils. Hence, favorable corn-growing weather in areas with thin loess soils would probably not compensate for the decline in yields resulting from poor growing weather in areas with deep loess soils.

Another limitation of agroclimatic compensation is that intensive cultivation of most vegetable and fruit crops generally is confined to specific regions. In the United States, nearly half the vegetables for fresh market and food processing are grown in California. In March and April 1983, cold winds and heavy rainfall delayed the planting and harvesting of vegetables, thereby reducing significantly the nation's supply of these vital foodstuffs. Although the return of good weather spurred vegetable production, the inclement episode contributed to a 4% production decline for that year and a nearly 6.5% increase in market price, a rate twice that of the inflation rate.

For nations as large as the United States and Canada, which have diverse climatic regimes that support a variety of crops, agroclimatic compensation provides some flexibility in feeding the population, but it may mean switching to more abundant foodstuffs and paying more for food. For smaller nations, agroclimatic compensation may not be possible within their borders, and weather extremes may require the importation of food, perhaps at elevated prices.

Where streamlines from different anticyclones meet, a zone of confluence forms. A **confluence zone** is a boundary between air masses and is equivalent to an average frontal position. Confluence zones are indicated by dashed lines in Figure 18.3. As a rule, these zones separate distinctly different types of climate. Climate is quite uniform within airstreams but changes significantly across confluence zones.

**FIGURE 18.3**
In this streamline analysis of mean January winds over the Northern Hemisphere, the anticyclonic spirals are air mass source regions. Dashed lines mark confluence zones that are climatic air mass boundaries. [From W. M. Wendland and R. A. Bryson, "Northern Hemisphere Airstream Regions," *Monthly Weather Review* 109 (1981):257. Graphics by Raymond Steventon, Center for Climatic Research, University of Wisconsin–Madison. *Monthly Weather Review* is a publication of the American Meteorological Society.]

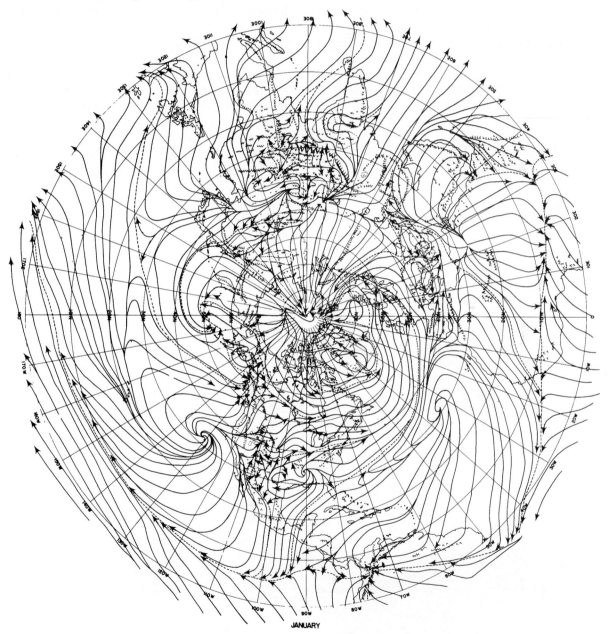

JANUARY

Using air mass frequency to describe climate appears to be a valid approach in light of the apparent air mass control of the location of certain vegetational communities. For example, R. A. Bryson has demonstrated a close correspondence between the region dominated by cold, dry arctic air and the location of the coniferous boreal forest of Canada. The southern boundary of the boreal forest nearly coincides with the average position of the leading edge of arctic air (the arctic front) during winter, and the northern border of the forest closely corresponds to the average position of the leading edge of arctic air during summer (Figure 18.4).

## Climate Controls

Many factors, working together, shape the climate of any locality. Controls of climate consist of (1) latitude, (2) elevation, (3) topography, (4) proximity to large bodies of water, and (5) prevailing atmospheric circulation. For all practical purposes, the first four controls are fixed and so exert a regular and predictable influence on climate. In Chapter 2, we saw that seasonal changes in incomming solar radiation as well as length of day vary with latitude. Recall from Chapter 4 our discussion of how air temperature responds to these regular variations of solar energy input. Elevation also influences air temperature and whether precipitation falls in the form of rain or snow. Topography can affect the distribution of cloud and precipitation patterns. For example, in Chapter 6, we learned why the windward slopes of high mountain barriers are wetter than the leeward slopes. The great thermal stability of large bodies of water moderates the temperature of downwind localities, reducing the seasonal temperature contrast and lengthening the growing season.

The fifth climatic control, atmospheric circulation, is considerably less regular and, consequently, less predictable than the others. This control encompasses the combined influence of all the weather systems that we described in Chapters 10 through 15. This variability is especially characteristic of synoptic- and subsynoptic-scale weather systems. Planetary-scale circulation systems, such as the prevailing wind belts, the subtropical anticyclones, and the intertropical convergence zone

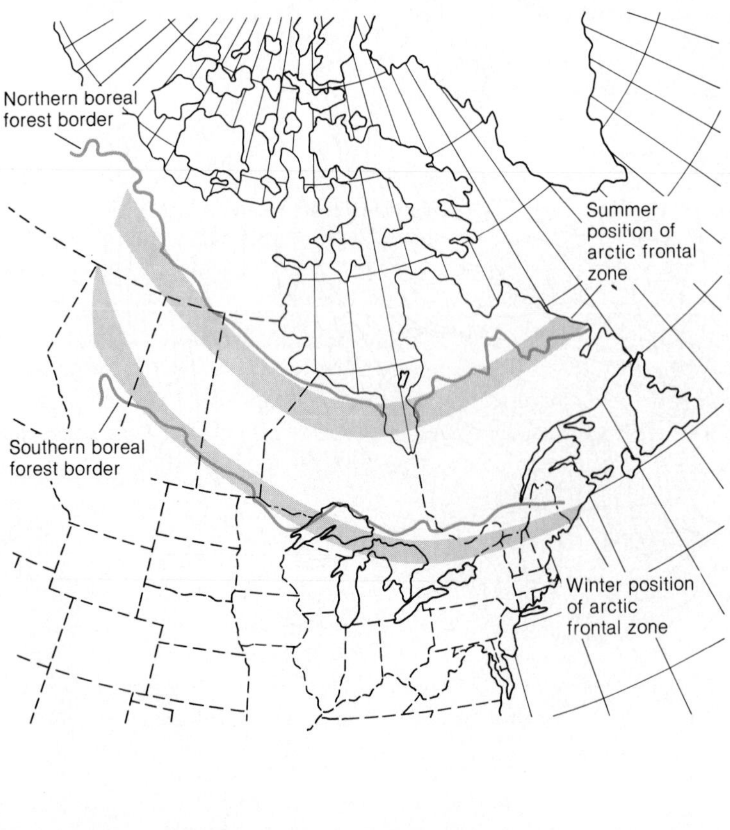

Northern boreal forest border

Southern boreal forest border

Summer position of arctic frontal zone

Winter position of arctic frontal zone

**FIGURE 18.4**
Northern and southern borders of Canada's boreal forest correspond closely to the average positions of the leading edge of arctic air in summer and winter, respectively. [From R. A. Bryson. "Air Masses, Streamlines and the Boreal Forest." *Geographical Bulletin* 8 (1966):266. Reproduced by permission of the Minister of Supply and Services, Canada.]

(ITCZ), exert a somewhat more systematic impact on climate. How these and other controls interact to shape the climates of the continents will become clearer as we describe the Earth's major climate groups.

# Global Patterns of Climate

Viewed globally, climate exhibits some regular patterns. As we describe these patterns, keep in mind that they may be significantly altered by local and regional climate controls.

## *TEMPERATURE*

If we ignore the influence of mountainous terrain on air temperature, then mean annual isotherms roughly parallel latitude circles, underscoring the importance of solar radiation and solar altitude as climate controls. Interestingly, the latitude of highest mean annual temperature, the so-called **heat equator,** is located about 10 degrees north of the geographical equator. Mean annual isotherms are symmetrical with respect to the heat equator and decline in magnitude toward the poles.

The heat equator is in the Northern Hemisphere because, overall, that hemisphere is warmer than the Southern Hemisphere. Several factors contribute to this hemispheric temperature contrast. For one, the polar regions of the two hemispheres have different radiational characteristics. Most of the Antarctic continent is submerged under a massive glacial ice sheet, so the surface has a very high albedo for solar radiation and is the site of intense radiational cooling, especially during the long polar night. In contrast, the Northern Hemisphere polar region is mostly ocean. Although the Arctic Ocean is usually ice covered, patches of open water develop in summer and lower the overall surface albedo. Hence, differences in radiational characteristics mean that the Arctic is warmer than the Antarctic.

A second factor contributing to the relative warmth of the Northern Hemisphere is that hemisphere's greater fraction of land in tropical latitudes. Because land surfaces warm up more than water surfaces in response to the same insolation, tropical latitudes are warmer in the Northern Hemisphere than in the South-

ern Hemisphere. A third contributing factor is ocean circulation. Apparently, ocean currents transport more warm water to the Northern Hemisphere than to the Southern Hemisphere.

Systematic patterns also appear when we consider the worldwide distribution of mean January temperature (Figure 18.5) and mean July temperature (Figure 18.6); January and July are usually the coldest and warmest months of the year. If we neglect the influence of topography on air temperature, isotherms tend to parallel latitude circles. However, monthly isotherms exhibit some notable north-to-south bends primarily because of land/sea contrasts and the influence of ocean currents.

As the year progresses from January to July and to January again, isotherms in both hemispheres shift north and south in tandem. The latitudinal (north-south) shift in isotherms is greater over the continents than over the oceans. That is, the annual range in air temperature is greater over land than over the sea (Figure 18.7). Furthermore, the meridional temperature gradient is steeper in the winter hemisphere than in the summer hemisphere. From earlier discussions (Chapter 10), we know that this steeper temperature gradient means a more vigorous circulation and stormier weather in the winter hemisphere.

## *PRECIPITATION*

The global pattern of mean annual precipitation (rain plus melted snow) exhibits great spatial variability (Figure 18.8). Some of this variability can be attributed to topography and the distribution of land and sea, but the planetary-scale circulation is also important. The ITCZ, subtropical anticyclones, and the prevailing wind belts impose a roughly zonal pattern on precipitation distribution. In addition, the regular shifts of these circulation features through the year cause the seasonality of precipitation that is typical of many localities.

At low latitudes near the equator, convective activity associated with the trade wind convergence triggers abundant rainfall year-round. In the adjacent belt poleward to about 20 degrees latitude, rainfall depends on seasonal shifts of the ITCZ and the subtropical anticyclones. Poleward shifts of the ITCZ cause summer rains, whereas the equatorward shift of the subtropical

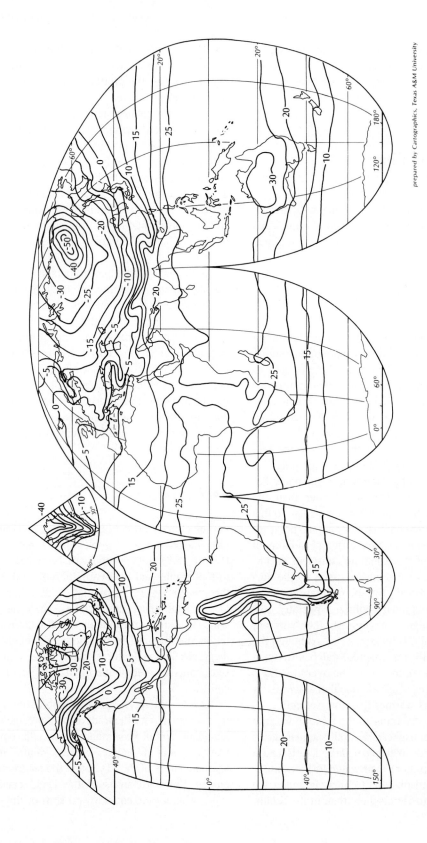

**FIGURE 18.5**
Mean sea-level temperatures for January in °C. [From Hidore, J. J. and J. E. Oliver. *Climatology, An Atmospheric Science,* New York: Macmillan Publishing Company, 1993, p. 63]

*prepared by Cartographics, Texas A&M University*

**FIGURE 18.6**
Mean sea-level temperature for July in °C. [From Hidore, J. J. and J. E. Oliver. *Climatology, An Atmospheric Science*, New York: Macmillan Publishing Company, 1993, p. 62]

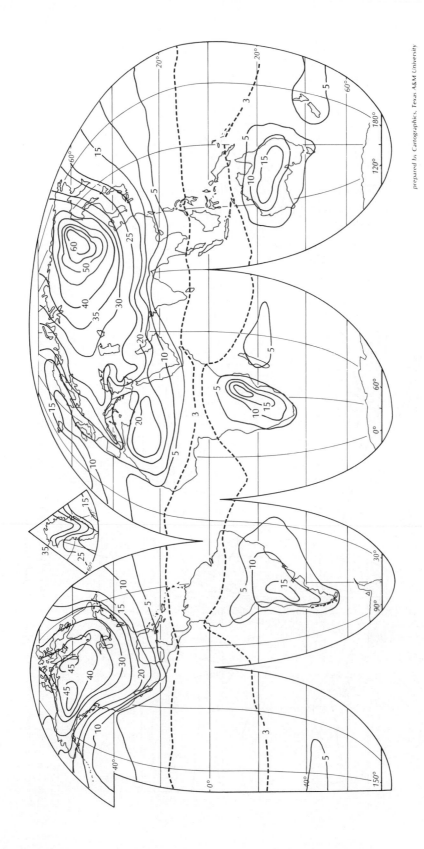

**FIGURE 18.7**
Mean annual range in temperature in Celsius degrees. [From Hidore, J. J. and J. E. Oliver. *Climatology, An Atmospheric Science*, New York: Macmillan Publishing Company, 1993, p. 64]

prepared by Cartographics, Texas A&M University

416

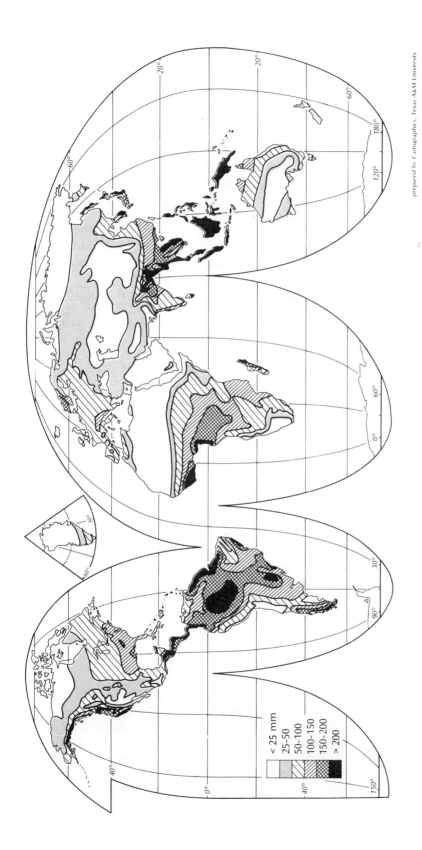

**FIGURE 18.8**
Mean annual precipitation (rain plus melted snow) in millimeters. [From Hidore, J. J. and J. E. Oliver. *Climatology, An Atmospheric Science*, New York: Macmillan Publishing Company, 1993, p. 118]

highs brings winter drought. This is the belt of tropical monsoon circulation described in Chapter 12.

Poleward of this belt, from about 20 to 35 degrees N and S, subtropical anticyclones dominate the climate all year. Subsiding dry air on the anticyclones' eastern flanks is responsible for the Earth's major subtropical deserts. On the other hand, unstable humid air that characterizes the western flanks of subtropical anticyclones causes relatively moist conditions.

Between about 35 and 40 degrees latitude, precipitation is governed by the prevailing westerlies and subtropical anticyclones. Typically, on the western side of continents, winter cyclones migrating with the westerlies bring moist weather, but in summer, westerlies shift poleward and the area lies under the dry eastern flank of a subtropical anticyclone. Hence, summers are dry. At the same latitudes, but on the eastern side of continents, the climate is dominated by westerlies in winter and the moist airflow on the western flank of the subtropical anticyclone in summer. Thus, rainfall is triggered by cyclonic activity in winter and by convection in summer and shows little seasonal variability.

Poleward of about 40 degrees latitude, precipitation generally declines as lower temperatures reduce the mean saturation vapor pressure (Chapter 6). Although precipitation is generally not seasonal, the tendency in the continental interiors is for more precipitation in summer. This is a result of higher air temperatures and more vigorous convection in summer and lower air temperatures and more frequent anticyclones in winter.

Our description of annual global precipitation is somewhat idealistic and requires some qualification. Land/sea distribution and topography add complexity to the zonal distribution of precipitation. More rain falls over the oceans than over the continents, and north-to-south-oriented mountain ranges induce wet windward slopes and extensive leeward rain shadows. Furthermore, annual precipitation totals fail to convey some other important aspects of precipitation, including the average amount of rainfall per day and the season-to-season and year-to-year reliability of precipitation. As a rule, rainfall is most reliable in maritime climates, less reliable in continental localities, and least reliable in arid regions. However, drought is possible anywhere, even in maritime climates.

# Climate Classification

In response to the interaction of many controls, the world's climates form a complex mosaic. Climatologists have attempted to simplify and organize the myriad of climate types by devising classification schemes that group together climates with common characteristics. Classification schemes typically group climates according to (1) the meteorological basis of climate or (2) the environmental effects of climate.

The first is a *genetic* climate classification that asks why climate types occur where they do. The second is an *empirical* climate classification that infers the type of climate from such climatic impacts as the distribution of vegetation types or the degree of weathering of exposed bedrock. In addition, the advent of electronic computers and databases has made possible *numerical* climate classification schemes that utilize sophisticated statistical techniques.

One of the most widely used climate classification systems, designed by Wladimir Köppen, combines the genetic and empirical approaches. Recognizing that vegetation indigenous to a region is a natural indicator of regional climate, Köppen delineated climatic boundaries throughout the world based on the limits of vegetational communities plus monthly mean temperature and precipitation and mean annual temperature. His climate classification scheme uses letters to symbolize five main groups of world climates: (**A**) tropical rainy; (**B**) dry; (**C**) midlatitude rainy, mild winter; (**D**) midlatitude rainy, cold winter; and (**E**) polar. Each of these climates is further subdivided into principal climate types, designated by the addition of another letter symbol to form a pair. Even further qualification of the climate may be specified by adding a third letter symbol. Hence, a combination of only two or three letters describes the general climate characteristics of a region.

Since its introduction in 1918, Köppen's climate classification has undergone numerous and substantial revisions by Köppen himself and by other climatologists. One version of the Köppen-based climate classification is presented by G. T. Trewartha and L. H. Horn in their text, *An Introduction to Climate* (1980). Trewartha and Horn identify seven main climate groups (Table 18.1); five are based on temperature, one is based on precipitation, and one applies to mountain-

**TABLE 18.1**
**Climate Classification**

| Climate groups | Climate types | Precipitation |
|---|---|---|
| **A** Tropical humid | **Ar,** tropical wet | Not over 2 dry months |
|  | **Aw,** tropical wet-and-dry | High-sun wet |
|  |  | low-sun dry |
| **B** Dry | **BS,** semiarid (steppe) |  |
|  | **BSh** (hot), tropical–subtropical | Short moist season |
|  | **BSk** (cold), temperate–boreal | Meager rainfall, mostly in summer |
|  | **BW,** arid (desert) |  |
|  | **BWh** (hot), tropical–subtropical | Constantly dry |
|  | **BWk** (cold), temperate–boreal | Constantly dry |
| **C** Subtropical | **Cs,** subtropical dry summer | Summer drought, winter rain |
|  | **Cf,** subtropical humid | Rain in all seasons |
| **D** Temperate | **Do,** oceanic | Rain in all seasons |
|  | **Dc,** continental | Rain in all seasons, accent on summer; winter snow cover |
| **E** Boreal | **E,** boreal | Meager precipitation throughout year |
| **F** Polar | **Ft,** tundra | Meager precipitation throughout year |
|  | **Fi,** ice cap | Meager precipitation throughout year |
| **H** Highland | **H,** variable | Variable |

*Source:* G. T. Trewartha and L. H. Horn. *An Introduction to Climate,* 5th ed. New York: McGraw-Hill, 1980, p. 227.

ous regions. The worldwide distribution of these climate groups is shown in Figure 18.9.

Several criticisms have been leveled at Köppen-based climate classification systems. For one, although climate is the principal control of worldwide biomes,* it is not the only control. Soil type and drainage conditions, for example, also influence the distribution of vegetation. Furthermore, in the Köppen scheme, climate and vegetation boundaries do not always coincide. Nonetheless, Köppen classification schemes are quite useful. In the remainder of this section, we summarize the climate of each of the major groupings.

### TROPICAL HUMID CLIMATES (A)

Tropical humid climates constitute a discontinuous belt straddling the equator and extending poleward to near the Tropic of Cancer in the Northern Hemisphere and the Tropic of Capricorn in the Southern Hemisphere. Mean monthly temperatures are high and show

---

*A *biome* is a community of plants and animals that usually occupies an extensive area. Biomes are named for the types of plants that dominate the landscape. Examples include temperate grasslands, savanna, and boreal forest.

little variability throughout the year. The coolest month averages no lower than 18 °C (64 °F), and there is no frost. The temperature contrast between the warmest and coolest month is typically less than 10 Celsius degrees (18 Fahrenheit degrees). In fact, the diurnal temperature range generally exceeds the annual temperature range. This monotonous air temperature regime is the consequence of consistently intense insolation and nearly uniform day length throughout the year.

Although tropical humid climate types are not readily distinguishable on the basis of temperature, there are important differences in precipitation regimes. Tropical humid climates are therefore subdivided into tropical wet climates (**Ar**) and tropical wet-and-dry climates (**Aw**). Although both climate types feature abundant annual rainfall, more than 100 cm (39 in.), their rainy seasons differ in length.

In tropical wet climates, the yearly rainfall of 175 to 250 cm (69 to 98 in.) supports the world's most luxuriant vegetation (Figure 18.10). Tropical rainforests occupy the Amazon Basin of Brazil, the Congo Basin of Africa, and the islands of Micronesia. For the most part, rainfall is distributed uniformly throughout the year, although in some areas there is a brief (1 to 2 months) relatively dry season. Rainfall occurs as heavy down-

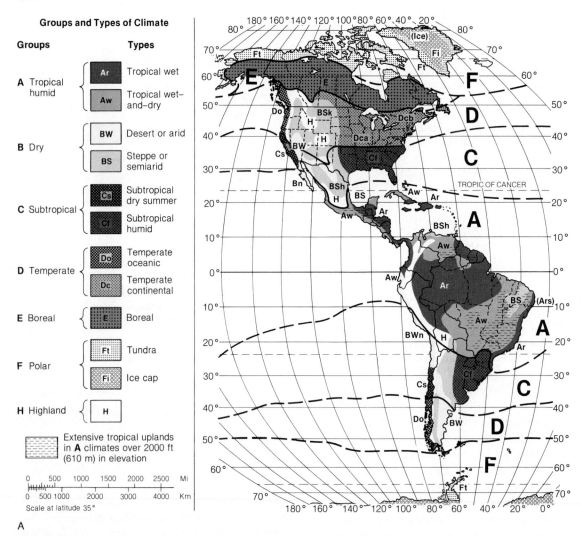

**Groups and Types of Climate**

| Groups | Types | |
|---|---|---|
| **A** Tropical humid | Ar | Tropical wet |
| | Aw | Tropical wet-and-dry |
| **B** Dry | BW | Desert or arid |
| | BS | Steppe or semiarid |
| **C** Subtropical | Cs | Subtropical dry summer |
| | Cf | Subtropical humid |
| **D** Temperate | Do | Temperate oceanic |
| | Dc | Temperate continental |
| **E** Boreal | E | Boreal |
| **F** Polar | Ft | Tundra |
| | Fi | Ice cap |
| **H** Highland | H | |

Extensive tropical uplands in **A** climates over 2000 ft (610 m) in elevation

Scale at latitude 35°

A

**FIGURE 18.9**
Distribution of major climate groups across the globe. [Modified after G. T. Trewartha and L. H. Horn. *An Introduction to Climate,* 5th ed. New York: McGraw-Hill, 1980. Used with permission.]

pours in frequent thunderstorms triggered by local convection and by surges of the ITCZ. Convective rainfall, controlled by insolation, typically peaks in midafternoon, the warmest time of day. Because water vapor concentrations are very high, even the slightest cooling during the early morning hours results in dew or fog, which gives such a region a sultry, steamy appearance.

For the most part, tropical wet-and-dry climates (**Aw**) border tropical wet climates and are transitional poleward to the subtropical dry climates. **Aw** cli-

mates support the savanna, tropical grasslands with scattered deciduous trees (Figure 18.11). Summers are wet and winters are dry, with the dry season lengthening poleward. This marked seasonality of rainfall is linked to shifts of the ITCZ and subtropical anticyclones, which follow the seasonal excursions of the sun. In summer, surges of the ITCZ trigger convective rainfall; in winter, the weather is dominated by the dry eastern flank of the subtropical anticyclones.

Annual mean temperatures in **Aw** climates are only

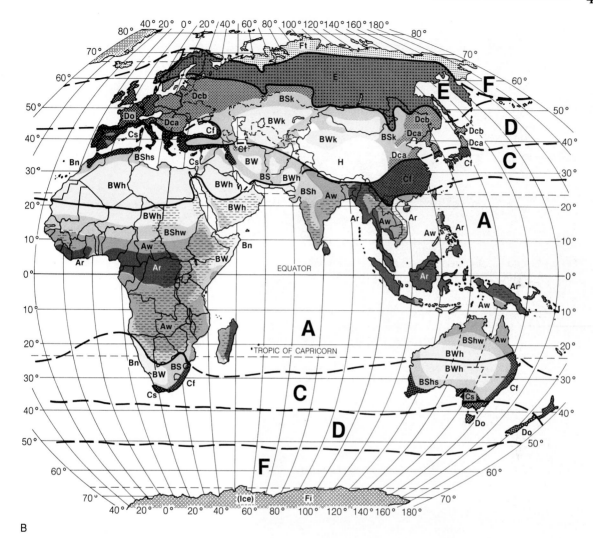

B

**FIGURE 18.9 (continued)**

slightly lower, and the seasonal temperature range is only slightly greater, than in the tropical wet climates (**Ar**). Diurnal temperature variations are noticeably greater, however. In summer, frequent cloudy skies and high humidities suppress the diurnal temperature range by reducing solar heating during the day and radiational cooling at night. In winter, on the other hand, persistent fair skies have the opposite effect and increase the diurnal temperature contrast. Cloudy, rainy summers plus dry winters also mean that the year's highest temperatures typically occur toward the close of the dry season in late spring.

## DRY CLIMATES (B)

Dry climates characterize those regions where annual potential evaporation exceeds annual precipitation. Because evaporation is controlled by air temperature, it is not possible to specify some rainfall amount as the criterion for dry climates. Rainfall is not only limited in **B** climates but is also highly variable and unreliable. As a general rule, the lower the mean annual rainfall, the greater is its variability from one year to the next.

The world's dry climates encompass a larger land area than any other single climate grouping. Perhaps

**FIGURE 18.10**
Luxuriant tropical rainforests grow where
rainfall is abundant and temperatures are
relatively high throughout the year. [Photo-
graph by J. M. Moran]

30% of the Earth's land surface, stretching from the
tropics to midlatitudes, is characterized by a moisture
deficit of varying degree. These are the climates of the
world's deserts and steppes where vegetation is sparse
and equipped with special adaptations that permit survival under conditions of severe moisture stress (Figure 18.12). Dryness is the consequence of either subtropical anticyclones or the rain shadow effect of high
mountain barriers. Mean annual temperatures are latitude dependent, as is the variation of mean monthly
temperatures through the year.

On the basis of degree of dryness, we distinguish
two climate types: steppe or semiarid (**BS**) and arid or
desert (**BW**). The steppe or semiarid climates are transitional between more humid climates and arid or desert
climates. We further distinguish warm, dry climates of
tropical latitudes (**BSh** and **BWh**) from cold, dry climates of higher latitudes (**BSk** and **BWk**).

**FIGURE 18.11**
Grassland savanna. [Photograph © Darrell Gulin/ Allstock]

Subsiding stable air on the eastern flanks of subtropical anticyclones gives rise to tropical dry climates **(BSh and BWh).** These huge semipermanent pressure cells dominate the weather year-round near the Tropics of Cancer and Capricorn. Consequently, dry climates characterize North Africa eastward to northwest India, the southwestern United States and northern Mexico, coastal Chile and Peru, southwest Africa, and much of the interior of Australia.

**FIGURE 18.12**
Sparsely vegetated desert of the American Southwest. [Photograph by Stephen J. Krasemann/DRK Photo]

Although persistent and abundant sunshine is the general rule in dry tropical climates, there are some important exceptions. In coastal deserts bordered by cold ocean waters, a shallow layer of stable marine air drifts inland. The desert air thus features high relative humidity, persistent low stratus clouds and fog, and considerable dew formation. Examples are the Atacama Desert of Peru and Chile, the Namib Desert of southwest Africa, and portions of the coastal Sonoran Desert of Baja California and stretches of the coastal Sahara Desert of northwest Africa. These anomalous desert climates are designated **Bn** or **BWn.**

Cold, dry climates of higher latitudes (**BWk** and **BSk**) lie in the rain shadows of great mountain ranges. They occur primarily in the Northern Hemisphere, to the lee of the Sierra Nevada and Cascade ranges in North America and the Himalayan chain in Asia. Because these dry climates are at higher latitudes than their tropical counterparts, mean annual temperatures are lower and the seasonal temperature contrast is greater. Anticyclones dominate winter weather, resulting in cold, dry conditions, whereas summers are hot and generally dry. The meager precipitation is produced by scattered convective showers mostly in summer (Figure 18.13).

## SUBTROPICAL CLIMATES (C)

Subtropical climates are situated just poleward of the Tropics of Cancer and Capricorn and are dominated

**FIGURE 18.13**
Most of the time this riverbed in southwestern Utah is dry. (A dry streambed is known as an arroyo.) But an occasional thunderstorm in its watershed will produce a surge of water that flows down the streambed. [Photograph by J. M. Moran]

by seasonal shifts of subtropical anticyclones. There are two basic climate types: subtropical dry summer (or *Mediterranean*) climates **(Cs)** and subtropical humid climates **(Cf).**

Mediterranean climates occur on the west side of continents between about 30 and 45 degrees latitude. In North America, mountain ranges confine this climate type to a narrow coastal strip of California. Elsewhere, **Cs** climates rim the Mediterranean Sea and occur in portions of extreme southern Australia. Summers are dry because at that time of year **Cs** regions are under the influence of stable subsiding air on the eastern flanks of subtropical highs. The equatorward shift of subtropical highs in autumn allows oceanic cyclones to migrate inland, bringing moderate winter rainfall. Annual precipitation totals are in the range of 40 to 80 cm (16 to 31 in.).

Although Mediterranean climates exhibit a pronounced seasonality in precipitation (dry summers, wet winters), the temperature regime is quite variable. In coastal areas, cool onshore breezes prevail, lowering mean annual temperatures and reducing seasonal temperature contrasts. Well inland, however, away from the ocean's moderating influence, summers are considerably warmer, so that mean annual temperatures are higher and seasonal temperature contrasts are greater than in coastal **Cs** localities. Climatic records of coastal San Francisco and inland Sacramento, California, illustrate the contrast in temperature regime within **Cs** re-

gions (Figure 18.14). Although the two cities are separated by only about 145 km (90 mi), the climate of Sacramento is much more continental than that of San Francisco.

Subtropical humid climates **(Cf)** occur on the eastern sides of continents between about 25 and 40 degrees latitude. **Cf** climates are situated primarily in the southeastern United States, a portion of southeastern South America, eastern China, southern Japan, on the extreme southeastern coast of South Africa, and along much of the east coast of Australia. These climates feature abundant precipitation (76 to 165 cm, or 30 to 65 in.), which is distributed throughout the year. In summer, **Cf** regions are dominated by a flow of sultry maritime tropical air on the western flanks of the subtropical anticyclones. Consequently, summers are hot and humid with frequent thunderstorms. As noted in Chapter 15, hurricanes and tropical storms contribute significant rainfall to some North American and Asian **Cf** regions. In winter, after the subtropical highs shift toward the equator, Cf regions come under the influence of migrating midlatitude cyclones and anticyclones.

A noteworthy exception to the nonseasonality of precipitation in **Cf** regions is the interior of southern China, where winters are relatively dry. This is because prevailing winter winds are from the cold and dry north and northwest. This anomalous **Cf** climate is designated **Cw.**

**FIGURE 18.14**
Note the contrast in the march of monthly mean temperatures between Sacramento, California, and more maritime San Francisco. [NOAA data]

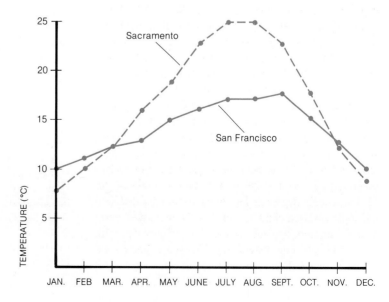

In **Cf** localities, summers are hot and winters are mild. Mean temperatures of the warmest month are typically in the range of 24 to 27 °C (75 to 81 °F). For the coolest month, average temperatures are typically in the 4 to 13 °C (39 to 55 °F) range. Subfreezing temperatures and snowfalls are infrequent.

## TEMPERATE CLIMATES (D)

The primary distinction among temperate climates is between temperate oceanic (**Do**) and temperate continental (**Dc**). As the names imply, the first features a strong maritime influence, whereas the second is highly continental. These are midlatitude climates.

Temperate oceanic climates occur on the windward side of continents, mainly poleward of 40 degrees latitude. These climates characterize a narrow coastal strip in the Pacific Northwest from northern California northwestward into Alaska, along the southwestern coastal plain of South America, throughout most of western Europe, in Tasmania (Australia), and over most of the South Island of New Zealand. In these regions, a strong maritime influence prevails year-round. Consequently, temperatures are relatively mild for the latitude, and seasonal and diurnal temperature contrasts are reduced. Cold waves and heat waves are rare, and the growing season is long (up to 210 freeze-free days).

The precipitation regime in **Do** climates also reflects the maritime influence. Maritime polar air (mP) dominates so that there are persistent episodes of low clouds and light to moderate rainfall, with rainfall amounts dependent on topography. Droughts are infrequent. The mP air is stable, so that thunderstorms are rare.

Temperate continental climates (**Dc**) occur only in the Northern Hemisphere from about 40 to 50 degrees N. These **Dc** climates are inland and on the leeward side of continents: Eurasia, the northeastern third of the United States, southern Canada, and extreme eastern Asia. Continentality increases inland with maximum temperature contrasts between the coldest and warmest months as great as 25 to 35 Celsius degrees (45 to 63 Fahrenheit degrees). **Dc** climates are divided into the southerly **Dca** climates with cool winters and warm to hot summers and the northerly **Dcb** climates with cold winters and mild summers. The freeze-free period varies in length from 7 months in the south to only 3 months in the north.

The weather in **Dc** regions is very changeable and dynamic because these areas are swept by cyclones and anticyclones and by surges of contrasting air masses. Polar front cyclones dominate winter, bringing episodes of light to moderate frontal precipitation. These storms alternate with incursions of dry polar and arctic air masses. In summer, cyclones are weak and infrequent because the principal storm track shifts poleward. Summer rainfall is mostly convective, and locally amounts can be very heavy in severe thunderstorms and mesoscale convective complexes (Chapter 13). Although precipitation is distributed rather uniformly throughout the year, most places experience a summer maximum.

In the northern portions of **Dc** climates, winter snowfall becomes an important factor. The amount of snow and the persistence of snow cover increase northward (Table 18.2). Because of its high albedo for solar radiation and its excellent emission of infrared, a snow cover chills and stabilizes the overlying air. For these reasons, a snow cover is self-sustaining; once established in early winter, an extensive snow cover tends to persist.

**TABLE 18.2**
**The 20 Snowiest Major Cities in the United States Based on Cumulative Totals for the Period 1975–1985**

|  | cm (in.) |
| --- | --- |
| Buffalo | 2710.9 (1067.3) |
| Rochester, NY | 2635.8 (1037.7) |
| Salt Lake City | 1622.6 (638.8) |
| Minneapolis–St. Paul | 1605.5 (632.1) |
| Albany | 1604.3 (631.6) |
| Cleveland | 1592.6 (627.0) |
| Denver | 1483.4 (584.0) |
| Milwaukee | 1400.3 (551.3) |
| Detroit | 1176.8 (463.3) |
| Chicago | 1114.8 (438.9) |
| Pittsburgh | 1069.6 (421.1) |
| Hartford, CT | 1062.0 (418.1) |
| Boston | 1058.7 (416.8) |
| Providence | 872.7 (343.6) |
| Dayton, OH | 849.9 (334.6) |
| Indianapolis | 824.2 (324.5) |
| Columbus, OH | 804.9 (316.9) |
| St. Louis | 700.3 (275.7) |
| Cincinnati | 685.5 (269.9) |
| Philadelphia | 678.2 (267.0) |

*Source:* P. R. Chaston, "1975–1985 Snowiest Major Cities," *Weatherwise* 39, No. 1 (1986):44–45.

## *BOREAL CLIMATE (E)*

The boreal climate occurs only in the Northern Hemisphere as an east-west band between 50 or 55 degrees N and 65 degrees N latitude. It is a region of extreme continentality and very low mean annual temperature. Summers are short and cool, and winters are long and bitterly cold. Both continental polar *(cP)* and arctic *(A)* air masses originate here, and this area is the site of an extensive coniferous forest. Because midsummer freezes are possible, the growing season is precariously short.

Weak cyclonic activity occurs throughout the year and yields meager annual precipitation (typically less than 50 cm, or 20 in.). Convective activity is rare. A summer precipitation maximum is due to the winter dominance of cold, dry air masses. Snow cover persists throughout the winter.

## *POLAR CLIMATES (F)*

Polar climates occur poleward of the Arctic and Antarctic circles. These boundaries correspond roughly to localities where the mean temperature for the warmest month is 10 °C (50 °F). These limits also approximate the tree line, the poleward limit of tree growth. Poleward are tundra (Figure 18.15) and the Greenland and Antarctic ice sheets. A distinction is made between tundra (**Ft**) and ice cap (**Fi**) climates, with the dividing criterion being 0 °C (32 °F) for the mean temperature of the warmest month. Vegetation is sparse in **Ft** regions and almost nonexistent in **Fi** areas.

Polar (**F**) climates are characterized by extreme cold and slight precipitation, which is mostly in the form of snow (less than 25 cm, or 10 in., melted, per year). Although summers are cold, the winters are so extremely cold that **F** climates feature a marked seasonal temperature range. Mean annual temperatures are the lowest in the world.

## *HIGHLAND CLIMATES (H)*

Highland climates encompass a wide variety of climate types that characterize mountainous terrain (Figure 18.16). Altitude, latitude, and exposure are among the factors that shape a complexity of climate types. Climate-ecological zones are telescoped in mountain-

**FIGURE 18.15**
Tundra in the Magellanic Forest Reserve near Punta Arenas, Chilean Patagonia. [Photograph © Galen Rowell]

ous areas. That is, in ascending several thousand meters of altitude, we encounter the same biotic-climatic zones that we would experience in traveling several thousand kilometers of latitude. As a general rule, every 300 m (980 ft) of elevation corresponds roughly to a northward advance of 500 km (310 mi).

# Conclusions

Climate encompasses both averages and extremes in weather for some locality over some time period (either 30 years or the entire period of record). The cli-

**FIGURE 18.16**
Climate and ecological zones are telescoped in mountainous areas. [Photograph by M. D. Morgan]

matic norm or normal is usually interpreted as the average value of some climatic element such as temperature or precipitation. While average values of many climatic elements are useful, for some, such as precipitation, the median value is more representative of climate.

The world is a mosaic of many climate types that are shaped by a variety of many interacting controls. Genetic, empirical, and numerical schemes have been devised to classify climates. Among the most popular and widely used are classifications based on the Köppen system.

Climate varies not only spatially but also with time. The climatic record and what it tells us about climatic behavior are subjects of the next chapter.

## Key Terms

climate
National Climate Data
    Center (NCDC)
anomalies
agroclimatic
    compensation

air mass climatology
streamlines
confluence zone
heat equator

## Summary Statements

☐ Climate is defined as weather conditions averaged over some time period plus extremes in weather. Climate is usually specified for a particular location and for some time period.

☐ The climatic norm or normal encompasses both means and extremes of weather elements. By convention, average values of temperature, precipitation, and air pressure are computed for a 30-year period (currently, 1961–1990). Extremes, on the other hand, are derived from the entire period of record.

☐ Anomalies in climate, departures from long-term averages, are geographically nonuniform in both sign (positive or negative) and magnitude. The same geographic nonuniformity also characterizes large-scale trends in climate.

☐ An alternate way of describing the climate of a locality is by frequency of occurrence of the various types of air masses that develop over or regularly invade that locality.

☐ Climatic controls include the various weather systems operating at all spatial and temporal scales, along with insolation, land/water distribution, and topography.

☐ The globe is a mosaic of many different types of climate that may be grouped and classified by meteorological causes (genetic) or environmental effects (empirical). The Köppen climate classification system combines these two approaches and has undergone considerable refinement since its introduction.

☐ The globe's major climate groups are tropical humid, dry, subtropical, temperate, boreal, polar, and highland.

## *Review Questions*

1. Define climate.
2. What is meant by the climatic norm?
3. By international convention, what period of time is used as the basis for computation of climatic averages? Is this a reasonable time frame?
4. What is a climatic anomaly?
5. How is air mass frequency used to describe the climate of a locality?
6. Explain how a single anticyclone can be the source of several different types of air masses.
7. What is the purpose of a streamline analysis?
8. How does arctic air mass frequency correspond to the location of Canada's boreal forest?
9. List and describe briefly the principal controls of climate.
10. What are the bases for climate classification?
11. How does Köppen's classification scheme group global climates?
12. In tropical humid climates, how does the diurnal temperature range compare with the annual temperature range?
13. On the continents, which climate grouping covers the largest area?
14. How does the variability (and hence, reliability) of rainfall change as annual rainfall totals decline?
15. How is a desert climate defined? What are the principal controls of desert climates?
16. How do seasonal shifts of the subtropical anticyclones induce seasonal changes in rainfall in subtropical latitudes?
17. Explain the seasonality of rainfall in (a) India and (b) coastal California.
18. Contrast temperate oceanic climates with temperate continental climates.
19. How does a seasonal snow cover influence climate?
20. Why is convective activity rare in boreal climates and polar climates?

## *Questions for Critical Thinking*

1. Explain why a 30-year period provides a short-sighted view of the climatic record.
2. Why are climatic anomalies geographically nonuniform in sign (direction) and magnitude?
3. Explain why a precipitation anomaly map typically is considerably more complex than a temperature anomaly map.
4. What is the significance of the geographic nonuniformity of climatic anomalies for (a) agricultural productivity and (b) home heating and cooling demands?

5. Climatic controls are really not independent of one another. Provide some examples of linkages.

## *Selected Readings*

BRYSON, R. A., AND F. K. HARE (eds.). *World Survey of Climatology 11, Climates of North America.* New York: Elsevier Scientific, 1974, 420 pp. Includes data summaries and reviews of the major features of the climates of the United States, Canada, and Mexico.

GUTTMAN, N. B. "Statistical Descriptors of Climate," *Bulletin of the American Meteorological Society* 70 (1989): 602–607. Discusses how climatic data may be made more useful for predictive purposes.

HAMILTON, K. "Early Canadian Weather Observers and the 'Year Without a Summer'," *Bulletin of the American Meteorological Society* 67 (1986):524–532. Presents a description of an exceptionally cold summer based on newspaper accounts and records of amateur weather observers.

HIDORE, J. J. AND J. E. OLIVER. *Climatology, An Atmospheric Science.* New York: Macmillan Publishing Company, 1993, 423 pp. Surveys the basic principles of climatology and describes the climates of the continents.

HUGHES, P. "1816—The Year Without a Summer," *Weatherwise* 32, No. 3 (1979):108–111. Speculates on possible causes of the anomalously cold summer.

KUNKEL, K. E., AND A. COURT. "Climatic Means and Normals—A Statement of the American Association of State Climatologists," *Bulletin of the American Meteorological Society* 71 (1990):201–204. Discusses the limitations of 30-year climate averages.

OLIVER, J. E., AND R. W. FAIRBRIDGE (eds.). *The Encyclopedia of Climatology.* New York: Van Nostrand Reinhold, 1987, 986 pp. Provides a comprehensive and detailed treatise on climatology.

PHILLIPS, D. *The Climates of Canada.* Ottawa: Minister of Supply and Services Canada, 1990, 176 pp. Describes the climates of Canada by province plus the Yukon and Northwest territories.

TREWARTHA, G. T., AND L. H. HORN. *An Introduction to Climate,* 5th ed. New York: McGraw-Hill, 1980, 416 pp. Serves as a basic textbook on the meteorological basis of climate and describes primary climate belts.

WENDLAND, W. M., AND R. A. BRYSON. "Northern Hemisphere Airstream Regions," *Monthly Weather Review* 109 (1981):255–270. Determines air mass source regions on the basis of streamline analysis.

# 19

# The Climatic Record

*Now from the smooth deep ocean-*
*stream the sun*
*Began to climb the heavens, and*
*with new rays*
*Smote the surrounding fields*

HOMER
*Iliad*

Change is one of the basic character-istics of climate. [Photograph by Tranquality/Philip Chaudoir]

C LIMATE IS INHERENTLY VARIABLE. As we saw in the previous chapter, climate differs from one region to another; climate also varies with time. In this chapter, we describe what is understood about past variations in climate. We also examine what the climatic record tells us about the behavior of climate and its impact on society.

# The Climatic Past

In most places, the reliable instrument-based record of past weather and climate is limited to only 100 years or so. For information on earlier variations in climate, we must rely on indirect evidence. In this chapter's first Special Topic, "Reconstructing Past Climates," we describe some of the methods that have enabled scientists to extend the climatic record back in time. In this

section, we summarize what is currently understood about the Earth's climatic record.

## GEOLOGIC TIME

As we journey back through the millions and millions of years that constitute **geologic time,** information on climate becomes extremely fragmented and unreliable. Scientists must infer climatic information from geologic evidence such as the fossil remains of ancient plants and animals and the physical properties of sediments and sedimentary rocks. Lengthy gaps appear in the record, and time control and correlation of events become more uncertain. Consequently, with increasing time before present, climate is described in increasingly generalized terms (Table 19.1).

Dealing with time frames of hundreds of millions of years is complicated by geologic processes such as mountain building and continental drift. Today we consider topography and land/water distribution to be fixed controls of climate, but in the perspective of geologic time, they are variable. Mountain ranges have risen and eroded away; seas have invaded and withdrawn from the land, continually altering the shapes of continents; and land masses have drifted slowly across the face of the globe.

The solid outer 100 km (62 mi) of the planet is divided into a dozen gigantic, rigid plates that drift slowly (typically less than 20 cm per year) over the face of the globe. The continents are part of some plates so that as those plates move, the continents drift. Although the concept of **continental drift** was first formally proposed in 1912 by Alfred Wegener, a German geophysicist, it was not widely accepted until the 1960s as mounting geologic evidence became convincing.

Geologic evidence suggests that about 200 million years ago there was just one supercontinent, called *Pangaea* (for "all land"). Subsequently, Pangaea broke up, and its constituent land masses, the continents we know today, drifted apart, eventually reaching their present locations (Figure 19.1). What preceded Pangaea is unknown and the subject of much speculation.

Continental drift explains such seemingly anomalous finds as 200-million-year-old glacial deposits in the Sahara Desert, fossil tropical plants in Greenland, and fossil coral reefs in Wisconsin (Figure 19.2). Such

# Reconstructing Past Climates

For times and places where there is no instrument record of weather, it may be possible to infer past climatic information from various sensors that substitute for actual weather instruments. These climatic sensors, called *proxy climate data sources,* include historical records, pollen, tree growth rings, deep-sea sediment cores, and glacial ice cores.

Under cautious scrutiny certain historical documents can yield a wealth of information on past climates. Almanacs, personal diaries, old newspapers, and ships' logs may contain qualitative and some quantitative references to weather and climate. Other types of documents refer only indirectly to weather and climate but can be useful nonetheless. Records of success of grain harvests, quality of wine, or various phenological events (such as dates of blooming of plants) provide indirect indications of growing-season weather. For example, in his fascinating book, *Times of Feast, Times of Famine* (1971), Emmanuel Le Roy Ladurie relies heavily on vineyard records to reconstruct the climate of western Europe in the Middle Ages.

We must be careful in interpreting climatic information from historical documents chiefly because many factors other than weather and climate usually influence such records. For example, in addition to growing-season weather, vineyard harvest dates are affected by fluctuations in the wine market as well as by human judgment as to when the grapes are ripe and ready for harvest. It is also well to bear in mind that people have always applied ingenuity to ameliorate the impact of climate—especially extremes in climate. Hence, climatic information derived from records of human activity is not always reliable, and corroborative data from other sources are needed to support any climatic inferences.

Ponds, peat bogs, and swamps are favorable sites for the accumulation and preservation of wind-borne pollen, the tiny fertilizing component of a seed plant. Pollen grains mix with sediments (clay, silt, and other organic debris) that also settle and accumulate in these low-lying depositional areas. Upward of 20,000 pollen grains may be mixed in a cubic centimeter of mud. If we assume that accumulated pollen is the product of nearby vegetation and that climate largely governs vegetation types, then climate may be inferred from pollen. When the climate changes significantly, the vegetation changes and so does the pollen. Thus, changes in abundance of pollen of different species at various depths within accumulated sediment may reflect changes in climate. Scientists extract sediment cores from swamps and pond bottoms, separate pollen from its host sediment, and reconstruct the sequence of past vegetational changes. From the climatic requirements of the reconstructed vegetation (based on modern species distribution and modern climate), scientists decipher the sequence of past climatic shifts.

Much of the pollen/climate research conducted in North America has focused on vegetation and climate variations over the past 15,000 years, that is, since the last glacial maximum. Using sophisticated statistical techniques, scientists have reconstructed remarkably detailed quantitative climatic data. For example, the pollen record from Kirchner Marsh near Minneapolis, Minnesota, yielded a record of variations in July mean temperature and annual precipitation back to 12,000 years ago. Unfortunately, very few sites favor the accumulation and preservation of continuous long-term pollen/climate records. Most such records come from a few geographical areas, including the Great Lakes region, interior New England, and western mountain valleys.

Analysis of variations in the thickness of annual growth rings of certain trees also can yield detailed information on past climates. The study of tree growth rings for climatic data is known as *dendroclimatology.* Andrew E. Douglass, who was a solar astronomer at the University of Arizona, pioneered this work in the American Southwest early in this century. Today, his successors at the University of Arizona's Laboratory of Tree-Ring Research are attempting to reconstruct past climates using special statistical approaches and computers. At the onset of each growing season in spring, plant tissue that is located immediately beneath tree bark produces relatively large thin-walled wood cells that give the wood a relatively light appearance. Wood cells produced in summer, however, are thick-walled and give the wood a darker appearance. A year's growth of spring wood plus summer wood constitutes an annual growth ring. Hence, counting the number of growth rings gives the tree's age in years. Because growth ring widths normally decrease as the tree ages, widths are usually expressed in terms of a *tree-ring index,* which is the ratio of the actual tree-ring width to the width expected on the basis of the tree's age. The index is relatively low in stressful growing seasons and high in favorable growing seasons.

Trees growing under harsh conditions at climatically marginal sites are the most sensitive to climatic fluctuations, and their growth rings are the most reliable sensors of climate. A simple, hollow drill is used to extract cores from living trees or cut wood. Usually, cores are taken from many trees at one site, and tree-ring indexes are averaged. In western and southwestern North America, the primary locales of dendroclimatic research, scientists sample ponderosa pine, Douglas fir, or the exceptionally long-lived bristlecone pine. By assiduous matching of tree growth ring records from living trees with those from timbers in prehistoric dwellings, detailed tree-ring chronologies are extended back thousands of years. This matching technique is known as *cross-dating.*

Deep-sea sediment cores are also sources of past climatic data. Tiny clay particles and the shell and skeletal remains of marine organisms gradually settle out of ocean water and accumulate as sediment on the sea floor. In the open ocean, far from shoreline turbulence that might disturb sedimentation, an orderly record of environmental conditions accumulates. Scientists aboard specially outfitted ships extract cores from ocean bottom sediments and analyze layers of sediment for their climatic implications.

Much of what we know today about the large-scale climatic shifts that took place over the past 500,000 years came out of the *Climate/Long Range Investigation, Mapping and Prediction (CLIMAP)* project. Sponsored by the National Science Foundation and initiated in 1971, this ambitious research effort was carried out by a cadre of scientists and support personnel from a number of universities who represented a variety of disciplines. CLIMAP's primary objective was to reconstruct global climate for various periods over the past two million years, relying mostly on analyses of shell remains of marine organisms contained in deep-sea sediment cores. One important finding of CLIMAP was the quasi-regularity of shifts between glacial climates and interglacial climates. Glacial-interglacial climatic shifts were reconstructed by measuring the ratio of two isotopes* of oxygen ($^{16}O$ and $^{18}O$) contained in the shells of plankton, tiny free-floating marine organisms.

Water molecules contain both $^{16}O$ and $^{18}O$. When water evaporates from the ocean, however, much of the $^{18}O$ is left behind. This means that snow is richer in $^{16}O$ relative to $^{18}O$. Hence, as ice sheets thicken and spread across the land during major glacial climatic episodes, the lighter oxygen isotope becomes locked up in glacial ice preferentially over the heavier oxygen isotope. Conversely, ocean waters become enriched in $^{18}O$ relative to $^{16}O$. Plankton living during these times incorporate into their shells more $^{18}O$ than do plankton living during interglacial episodes. By analyzing oxygen isotope ratios in plankton shells found at various horizons within deep-sea sediment cores, CLIMAP investigators were able to reconstruct the major glacial-interglacial climatic shifts of the past 500,000 years.

Glacial ice cores are also valuable in climate reconstruction. Where the climate favors more annual snowfall than snowmelt, a glacier eventually forms. The mounting weight of accumulating snow compacts annual snow layers, gradually transforming them into solid ice. If the ice mass flows under the influence of gravity, it is a *glacier.* Scientists use a hollow drill bit to extract a vertical ice core from a glacier. The ice core is a continuous record of past seasonal snowfalls preserved as layers of ice delineated by summer dust. From analysis of those ice layers, scientists can reconstruct changes in temperature and atmospheric chemistry.

Scientists can decipher the record of past air temperature by studying oxygen isotope ratios of ice

---

*Isotopes of a given element differ from one another on the basis of atomic mass.

samples taken from various layers within an ice core. During relatively cold episodes, snowfall is enriched in $^{16}O$ relative to $^{18}O$ (whereas ocean water is enriched in $^{18}O$ relative to $^{16}O$). Hence, the lower the concentration of heavy oxygen in ice core samples, the lower is the indicated temperature.

Furthermore, layers of ice in glacial ice cores also contain bubbles of trapped air that provide clues to changes in atmospheric chemistry (Figure 1). In 1988, Soviet and French scientists reported on their analysis of air bubbles trapped in a 2200-m (7200-ft) ice core extracted from the Antarctic ice sheet at Vostok station. The ice core spanned 160,000 years. Chemical analysis indicated that atmospheric $CO_2$ concentration varied between 200 ppm during glacial climatic episodes and 270 ppm during interglacials.

During the summers of 1991–1993, two independent scientific teams, one American and the other European, drilled the deepest glacial ice cores on record. The two drill sites were located within 30 km (19 mi) of each other on the thickest portion of the Greenland ice sheet, about 650 km (403 mi) north of the Arctic Circle. The two cores were about 3000 m (9850 ft) in length and spanned a period of roughly 200,000 years.

Among findings from analysis of the Greenland ice cores are indications that major climatic shifts took place abruptly rather than gradually. The European team (which was the first to complete drilling, in July 1992) reported about a dozen major oscillations in temperature from cold to warm and back again over the last 30,000 years. Relatively mild episodes developed abruptly with temperature rises of up to 7 Celsius degrees (12.6 Fahrenheit degrees) in only a few decades. These mild episodes occurred at irregular intervals and typically lasted 500 to 2000 years.

In summary, studies of historical documents, pollen, tree growth rings, deep-sea sediments, and glacial ice cores have yielded much information on the climatic

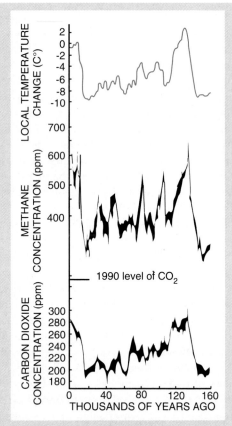

**FIGURE 1**
Analysis of air bubbles trapped within ice cores extracted from the Antarctic ice sheet reveals a strong correlation between reconstructed air temperature and concentrations of the greenhouse gases methane and carbon dioxide. [From J. T. Houghton, G. J. Jenkins, and J. J. Ephraums (eds.), *Climate Change, The IPCC Scientific Assessment.* Cambridge, U.K.: Cambridge University Press, 1990, p. xv]

past. Many other proxy climatic data sources exist, but space does not permit a description of them; readers are urged to consult the references at the close of this chapter.

discoveries reflect climatic conditions millions of years ago when the continents were situated at different latitudes. Drifting continents also impacted climate by altering the flow of heat-transporting ocean currents.

## THE PAST TWO MILLION YEARS

As we focus on the climatic record of the past two million years or so, we do not have to be concerned

**TABLE 19.1**
**Summary of the Earth's Climatic History over the Past 350 Million Years**

| Years before present | Climatic event |
| --- | --- |
| 350–250 million | Pangaean Ice Age |
| 250–53 million | Warming trend |
| 53 million | Onset of cooling |
| 33 million | Glaciation of Antarctica |
| 2–1.5 million | Beginning of Ice Age |
| 1 million | Major interglacial episode |
| 120,000 | Last major interglacial |
| 25,000 | Last major glacial advance |
| 18,000 | Last glacial maximum |
| 12,900–11,600 | Younger Dryas cold episode |
| 10,000 | Glacier retreats from Great Lakes |
| 7000–5000 | Climatic Optimum |
| 1043–743 | Medieval warm period |
| 593–143 | Little Ice Age |

solve the complex oscillations of climate into somewhat simpler fluctuations.

When compared with climatic conditions that prevailed through most of geologic time, the climate of the last two million years was anomalous in favoring the development of huge glacial ice sheets. During most of Earth's history the average global temperature may have been 10 Celsius degrees (18 Fahrenheit degrees) warmer than it was over the past two million years. The trend toward colder conditions began about 53 million years ago and culminated in the Ice Age.

Reconstructions based on deep-sea sediment cores indicate that during the Ice Age the climate shifted numerous times between conditions conducive to expansion of glaciers, a **glacial climate,** and conditions conducive to decay of glaciers, an **interglacial climate.** Because climatic trends are geographically nonuniform, temperature changes were greater in some latitude belts and less in others. A variety of geologic evidence indicates that temperature fluctuations between major glacial and interglacial climatic episodes typically amounted to about 2 Celsius degrees (3.6 Fahrenheit degrees) in the tropics, 6 to 8 Celsius degrees (11 to 14.4 Fahrenheit degrees) in midlatitudes, and 10 Celsius degrees (18 Fahrenheit degrees) or more at high latitudes. Such an increase in the magnitude of a cli-

with the effects of continental drift and mountain building. For all practical purposes, mountain ranges and continents were essentially as they are today. Because climate varies over a wide range of time scales, it is useful to view the climatic record of the past two million years in the perspective provided by progressively narrower time frames. Such an approach helps to re-

**FIGURE 19.1**
Originally, today's continents formed a single supercontinent called Pangaea. (A) Perhaps 200 million years ago the supercontinent began to break up and the fragments drifted apart. (B) The probable configuration of the drifting continents approximately 65 million years ago. (C) Today the continents have this arrangement, but they continue to drift slowly. [From P. J. Wyllie, *The Way the Earth Works.* Copyright © 1976 John Wiley & Sons. Reprinted by permission of John Wiley & Sons, Inc.]

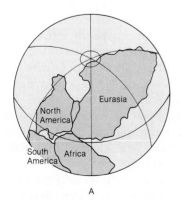
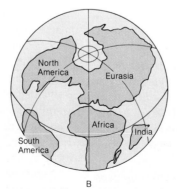

A

B

C

**FIGURE 19.2**
This bedrock exposed in northeastern Wisconsin contains fossil coral that dates from nearly 400 million years ago. Based on what we understand regarding the environmental requirements of modern coral, we conclude that about 400 million years ago, Wisconsin had a tropical marine environment. Continental drift can explain such a drastic change between ancient and modern conditions. [Photograph by J. M. Moran]

matic change with increasing latitude is known as **polar amplification.**

During major glacial climatic episodes, the Laurentide ice sheet developed over central Canada and spread westward to the Rocky Mountains, eastward to the ocean, and southward over the northern tier states of the United States (Figure 19.3). At the same time, a much smaller ice sheet formed over northwestern Europe including the British Isles. Because a vast quantity of water was locked up in these ice sheets, sea level fell by perhaps 130 m (425 ft), exposing portions of the continental shelf, including a land bridge linking Siberia and North America. The Laurentide and European ice sheets thinned and retreated, and may even have disappeared entirely, during relatively mild interglacial episodes, which typically lasted about 10,000 years. Throughout the interglacials, however, glacial ice cover persisted over most of Greenland and Antarctica, as it still does today.

Resolution of the past climatic record improves somewhat when we shift focus to the last 150,000 years. The temperature curve in Figure 19.4 was derived from a combination of midlatitude sea-surface temperature indicators, fossil pollen data, and reconstructed sea-level fluctuations. Many shifts between glacial and interglacial climatic episodes are evident.

In the perspective of the past 22,000 years, even more climatic detail appears. The midlatitude temperature curve in Figure 19.5 is based primarily on reconstructed glacial fluctuations and pollen studies. The last major glacial climatic episode began about 25,000 years ago and reached its peak about 18,000 years ago when global temperatures averaged 4 to 6 Celsius degrees (7.2 to 10.8 Fahrenheit degrees) lower than at present. The last glacial maximum was followed by a warming trend that triggered oscillatory glacial retreat. The Laurentide ice sheet withdrew from the Great Lakes region about 10,000 years ago and finally melted away by about 5500 years ago.

Numerous shifts back to relatively brief glacial climatic episodes punctuated the interglacial climate that

**FIGURE 19.3**
The extent of glaciation over North America about 18,000 years ago, the time of the last glacial maximum.

**FIGURE 19.4**
Generalized curve of midlatitude air temperature over the past 150,000 years based on pollen records, sea-level fluctuations, and reconstructed sea-surface temperatures. [Reprinted from W. L. Gates and Y. Mintz, *Understanding Climatic Change: A Program for Action*, 1975, with permission from the National Academy of Sciences, National Academy Press, Washington, DC.]

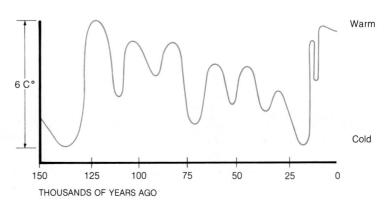

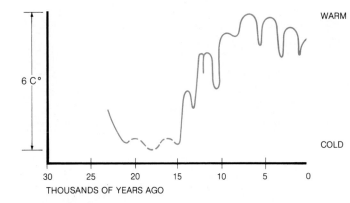

WARM

COLD

**FIGURE 19.5**
Generalized curve of midlatitude air temperature over the past 23,000 years based on pollen records, tree-line changes, and glacial ice fluctuations. [Reprinted from W. L. Gates and Y. Mintz, *Understanding Climatic Change: A Program for Action*, 1975, with permission from the National Academy of Sciences, National Academy Press, Washington, DC.]

has prevailed since 18,000 years ago. A notable example is the cold period from about 12,900 to 11,600 years ago known as the **Younger Dryas.*** Return of glacial climatic conditions triggered short-lived readvances of remnant ice sheets in North America, Scotland, and Scandinavia.

The postglacial warming trend culminated in the so-called **Climatic Optimum** about 7000 to 5000 years ago, a time when the mean global temperature was somewhat higher (perhaps 1 Celsius degree) than at present (Figure 19.6A). A pollen-based climate reconstruction indicates that 6000 years ago, over most of Europe, July temperatures were about 2 Celsius degrees (3.6 Fahrenheit degrees) higher than now.

*Named for the reappearance of *Dryas octopetala,* a polar wildflower, in portions of Europe.

A generalized temperature curve, derived mostly from historical documents, for the past 1000 years is shown in Figure 19.6B. The most notable features of this record are (1) the *medieval warm period* from about 950 to 1250 and (2) the cooling that followed from about 1400 to 1850—an era that has come to be known as the Little Ice Age. During the medieval warm period, the mean global temperature was about 0.5 Celsius degree (0.9 Fahrenheit degree) higher than in 1900. The first European settlements appeared along the southern coast of Greenland, and vineyards thrived in the British Isles. Independent lines of evidence have confirmed that the **Little Ice Age** was indeed a relatively cool period in many parts of the world, with the mean annual global temperatures perhaps 0.5 Celsius degree (0.9 Fahrenheit degree)

**FIGURE 19.6**
Generalized curve of mean global temperature variations on two time scales: (A) the past 10,000 years, (B) the past 1000 years. The dashed horizontal line represents conditions at about the year 1900. [From J. T. Houghton, G. J. Jenkins, and J. J. Ephraums (eds.), *Climate Change, The IPCC Scientific Assessment*. Cambridge, U. K.: Cambridge University Press, 1990, p. 202]

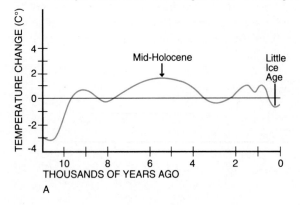

A

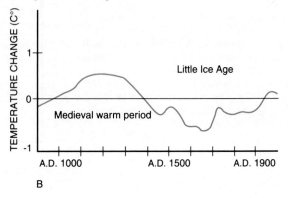

B

lower than it was in 1900. Sea ice expanded equatorward, Alpine glaciers advanced, growing seasons shortened, and erratic harvests caused much hardship for many people.

## THE INSTRUMENT-BASED RECORD

The first European explorers to set foot on North American soil were very interested in the weather and climate of the New World. French colonists apparently were the first to spend an entire winter. In 1604–1605, they established a settlement on the Atlantic Coast near the present border between Maine and New Brunswick. Samuel de Champlain, geographer for that expedition, commented on the sharp contrast in weather with that of his native France.

> Snow fell on the sixth of October. On the third of December we saw ice passing, which came from some frozen river. The cold was severe and more extreme than in France, and lasted much longer. I believe this is caused by the north-west winds, which pass over mountains continually covered with snow. This we had to a depth of three or four feet up to the end of the month of April; and I believe also that it lasts much longer than it would if the land were under cultivation.*

Almost half of the 79 French colonists did not survive that first very severe winter.

Some early weather observers used primitive weather instruments, whereas others made qualitative assessments of the weather, jotting down observations in journals or diaries. The first systematic weather observations in North America were made in 1644–45 at Old Swedes Fort (now Wilmington, Delaware). The observer was John Campanius Holm, chaplain of the Swedish military expedition. Other temperature records were begun in Philadelphia in 1731; in Charleston, South Carolina, in 1738; and in Cambridge, Massachusetts, in 1753. The New Haven, Connecticut, temperature record began in 1781 and continued uninterrupted through the present.

Thomas Jefferson and the Reverend James Madison

*From *The Works of Samuel de Champlain.* H. P. Biggar (ed.). Toronto: The Champlain Society, 1922, pp. 302–303.

(president of the College of William and Mary) are credited with making the first simultaneous weather observations in the United States, in 1778. Over a six-week period, they monitored temperature, air pressure, and wind at Monticello and Williamsburg, Virginia.

On 2 May 1814, James Tilton, M.D., Surgeon General of the United States Army, issued an order that in retrospect marked the first step in the eventual establishment of a national network of weather-observing stations. Tilton directed the army medical corps to maintain a diary of weather conditions at army posts, with responsibility for weather observations falling to the post's chief medical officer or surgeon. Tilton's objective was to learn more about the climate encountered by troops in the then sparsely populated interior of the continent. He also wanted to assess the relationship between weather and health, for it was a popular notion at the time that weather and climate were important factors in the onset of disease.

It took time for Tilton's order to be implemented. The War of 1812 was still raging, and weather instruments had to be distributed along with directions for proper use. Benjamin Waterhouse, M.D., surgeon at Cambridge, Massachusetts, was the first to submit weather data (for March 1816). By 1818, reports of weather observations at several army posts began trickling in to the Surgeon General's office, and under the direction of Tilton's successor, Joseph Lovell, M.D., the data were compiled, summarized, and eventually published. For this reason, Lovell, rather than Tilton, is sometimes credited with being the founder of the government's system of weather observation.

By 1838, 16 army posts had compiled at least 10 complete, though not aways successive, years of weather data. In ensuing years, the number of military weather stations climbed steadily, reaching 60 by 1843, and by the end of the Civil War, weather records had been assembled for varying periods at 143 locations.

Invention of the telegraph in the mid-1800s spurred Professor Joseph Henry, then Secretary of the new Smithsonian Institution in Washington, D.C., to organize a network of 150 volunteer weather observers. Data were wired to the Smithsonian and plotted on maps.

In the 1860s, surprise storms sweeping the Great Lakes caused numerous shipwrecks and an appalling loss of life. In response, the U.S. Congress called on the Army to monitor weather conditions and issue appropriate storm warnings. President Grant signed this resolution into law in early 1870. A new weather-observing network was formed within the Army Signal Services (Corps) that encompassed stations operated by the Surgeon General, the Smithsonian Institution, and the U.S. Army Corps of Engineers. By 1878, 284 stations were operating nationwide.

On 1 October 1890, the military weather network was placed in civilian hands in a new Weather Bureau within the Department of Agriculture, with a special mandate to provide weather and climate guidance to farmers. Later, aviation's need for weather information became increasingly important, and in 1940 the Weather Bureau came under the jurisdiction of the Commerce Department. In 1965, the Weather Bureau was reorganized as the National Weather Service (NWS) and placed under the supervision of the Environmental Science Services Administration (ESSA), which became the National Oceanic and Atmospheric Administration (NOAA) in 1971.

In Canada, the earliest weather records date to sporadic observations by British Army officers in the late 1760s. Reverend Alexander Spark, a Presbyterian minister at Quebec City, began the first long-term weather record in December 1798; it ended with his death in March 1819. At Montreal, Thomas McCord and his son, John S. McCord, kept a weather journal from January 1813 to 1842. In addition, the Hudson's Bay Company maintained early nineteenth-century temperature records at its Whale River (1814 to 1816) and Eastmain (1814 to 1821) posts along the east shore of Hudson Bay.

In 1839, the Meteorological Service of Canada was founded and initially placed under military jurisdiction. In 1853, the Service came under civilian control. The first national weather observation network was established in 1843–44, and the first forecast service began operating in 1876. Cooperation with the U.S. Weather Bureau dates back to that year, and cooperation increased after World War II with establishment of joint Arctic weather stations. Today, Canadian weather services are provided by the Atmospheric Environment Service (AES) in the Department of Fisheries and the Environment.

## HISTORICAL TEMPERATURE TRENDS

With invention of weather instruments and establishment of weather observational networks throughout the world, the climatic record becomes much more detailed and reliable. The most reliable temperature records date from the birth of the World Meteorological Organization (and the National Weather Service) in the late 1800s because by then weather observations were made under standardized conditions. Examination of temperature trends over the past 100 years is instructive as to the basic variability of climate.

Figure 19.7 is a plot of the mean annual temperature of the United States for 1895–1991. A smooth trend line has been fitted to the data. Note that the temperature trend is upward from 1895 through the mid-1930s, then generally downward into the 1970s, and upward again through the 1980s. The total temperature fluctuation amounts to only about ±1.2

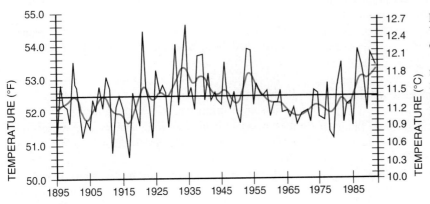

**FIGURE 19.7**
Variation in annual mean temperature of the United States in °F and °C for 1895–1991. Horizontal line is the mean temperature for the entire period of record. Red line shows trend. [National Climate Data Center, NOAA]

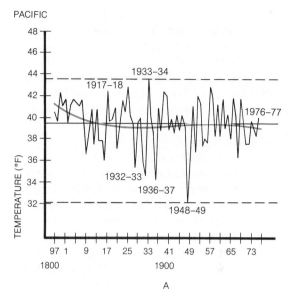

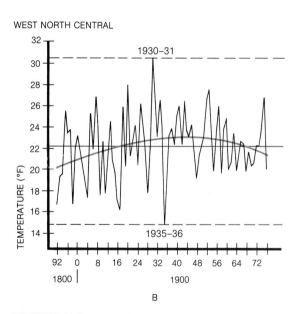

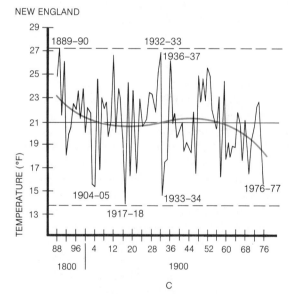

**FIGURE 19.8**

The geographic nonuniformity in climatic trends is illustrated by differences in trends of mean winter (December through February) temperatures in three regions of the United States: (A) Pacific, (B) West North Central, and (C) New England. Data are from the late 1800s to 1977. Red line shows trend. [From H. F. Diaz and R. G. Quayle, "The 1976–77 Winter in the Contiguous United States in Comparison with Past Records," *Monthly Weather Review* 106 (1978):1402–1405. *Monthly Weather Review* is a publication of the American Meteorological Society.]

Celsius degrees (±2.3 Fahrenheit degrees) about the 96-year average. Remember, however, that we are looking at a temperature trend for the United States as a whole, so the trend was amplified or reversed or both in specific regions of the country. For example, Figure 19.8 shows that trends in winter (December through February) mean temperature for much of the same period varied considerably in different regions of the United States.

We gain a somewhat different perspective on temperature variations since the late 1800s when we consider the global scale. Scientists at NASA's Goddard Institute for Space Studies have assembled a global temperature record based on land station data. A gradual warming trend since the 1880s was interrupted by cooling in the middle and high latitudes of the Northern Hemisphere during 1940–1970. Mean global temperature climbed about 0.7 Celsius degree (1.3 Fahrenheit degrees) between the 1880s and 1991, with the greatest warming in midlatitudes of both hemispheres. The eight warmest years on record occurred since 1979: 1980, 1981, 1983, 1987, 1988, 1989, 1990, and 1991.

Similar (although not identical) trends in global mean temperature appear in another global temperature record derived by a British team from the United Kingdom Meteorological Office and the University of East Anglia. The British record is three decades longer than NASA's record and is based on both land station and sea-surface temperatures. In both records, 1990 was the warmest year on record, with 1991 a close second.

The global warming trend was at least temporarily interrupted in 1992 when the global mean temperature dropped to its lowest level since the 1970s. Many climatologists attribute the cooling to the June 1991 eruption of Mt. Pinatubo in the Philippines. We have much more to say about the volcano-climate connection in Chapter 20.

Critics sometimes question the integrity of large-scale (hemispheric or global) temperature trends. They cite as potential sources of error: (1) improved sophistication and reliability of weather instruments through the period of record, (2) changes in location and exposure of instruments at most long-term weather stations, (3) huge gaps in monitoring networks, especially over the oceans, and (4) the moderating influ-

ence of urbanization and expansion of associated heat islands (Chapter 12). By careful statistical treatment of available data, however, large-scale temperature trends have been confirmed with a reasonable degree of certainty.*

## Lessons of the Climatic Record

What does the climatic record tell us about climatic behavior? After all, that is a major reason for journeying into the climatic past. The following are some of the principal lessons of the climatic past.

1. Climate is inherently variable over a broad spectrum of time scales ranging from years to millennia.
2. Variations in climate are geographically nonuniform in both sign (direction) and magnitude.
3. A change in climate may involve an increased or decreased frequency of extreme events (drought, excessive cold, for example) as well as some variation in mean temperature or precipitation.
4. Changes in climate impact society.

Let us explore further this fourth lesson of the climatic record.

### CLIMATE AND SOCIETY

There is little doubt that climate and climatic variations have played a role in the course of history since humans emerged during the Ice Age. Sometimes climate was the key factor.

In their book *Climates of Hunger* (1977), R. A. Bryson and T. J. Murray describe several cases in the distant past when climatic change severely impacted societies. They argue convincingly, for example, that prolonged drought contributed to the decline and fall of (1) the Harappan civilization of the Indus Valley region of northwest India about 1700 B.C., (2) the Mycenaean

---

*For more on this refer to the article by Jones and Wigley cited at the end of this chapter.

civilization of Greece in 1200 B.C., and (3) the Mill Creek culture of northwest Iowa about 1200 A.D. Other scientists propose that a succession of severe droughts forced the Anasazi Indians of Mesa Verde in southwestern Colorado to abandon their homes around 1300 A.D. (Figure 19.9).

In these and other studies of the impact of climatic change on society, one message is very clear. The people most vulnerable to climatic change are those living in areas where the climate is marginal for survival. These are typically climatic zones where barely enough rain falls or where the growing season is just long enough for crops to mature. Even a small change in these critical parameters in the wrong direction can spell disaster. This is apparently what happened to the

early Greenland settlements, as described in the Special Topic "Climatic Change and the Norse Greenland Tragedy."

Interestingly, the fluctuations of climate that seriously impacted past civilizations likely fall within the range of climatic variability that appears in the modern climatic record. Even the very cold summer of 1816, described in a Special Topic in Chapter 18, was within the range of expected climatic variability, albeit at an extreme end. Theoretically, then, the atmospheric system could shift to any anomalous weather pattern at any time. In effect, whatever happened in the climatic past can happen again. This conclusion also underscores the notion that past climatic episodes differ from present climatic episodes in frequency rather than

**FIGURE 19.9**
Drought apparently forced the Anasazi Indians of Mesa Verde, in southwestern Colorado, to abandon their cliff dwellings around 1300 A.D. [Photograph by J. M. Moran]

**SPECIAL TOPIC**

# Climatic Change and the Norse Greenland Tragedy

The late ninth century saw the beginning of a lengthy episode of unusually mild conditions throughout much of the Northern Hemisphere (the medieval warm period). A relatively mild climate enabled Viking explorers to probe the northerly reaches of the Atlantic Ocean. Previously, severe cold and extensive drift ice had proved insurmountable obstacles to European navigators. By 930 A.D., the Vikings established the first permanent settlement in Iceland, some 970 km (600 mi) west of Norway and just south of the Arctic Circle.

Among Iceland's early inhabitants was Eric the Red, a troublesome individual whose exploits eventually caused his banishment from Iceland in 982 A.D. He sailed west and discovered a new land, which he named Greenland. Then as now, much of Greenland was buried under a massive glacial ice sheet. The only habitable lands were small patches scattered along the coast, hemmed in by the sea and ice sheet and separated by treacherous mountain ridges and deep fjords. Some historians speculate that Eric called the new land Greenland to entice others to follow him. At the time, however, the climate was so mild that some sheltered valleys were probably greener than they are today.

In such a place, on Greenland's southwest shore, Eric founded the first of two Norse settlements (Figure 1). Although never prosperous, Norse colonization of Greenland persevered for nearly five centuries. The

**FIGURE 1**
Locations of ancient Norse settlements on the southwest coast of Greenland. [From R. A. Bryson and T. J. Murray, *Climates of Hunger*. Madison, WI: The University of Wisconsin Press, 1977, p. 48]

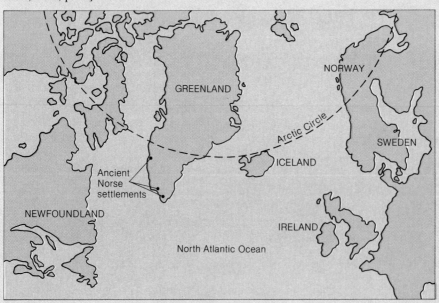

Norse subsistence economy was primarily agrarian and dependent on raising cattle, sheep, and goats. In addition, the Norse hunted migratory harp seals in spring and caribou in autumn. They even embarked on long, dangerous hunting expeditions northward along Greenland's west coast in search of walrus and polar bear, whose valuable tusks and hides they traded to Europeans for durable goods.

Initially, the colonists fared relatively well; the population of the Eastern Settlement climbed to an estimated 4000 to 5000, and that of the Western Settlement peaked at perhaps 1000 to 1500. By about 1350 A.D., however, the Western Settlement was vacant, apparently the victim of some sudden calamity. The larger Eastern Settlement also succumbed, but more gradually, so that by 1500 A.D., Norse society in Greenland had been erased. What happened to the Norse settlements in Greenland can only be inferred from their graves, the ruins of their homes and barns, and a few chronicles. There were no survivors. In 1921, an expedition from Denmark examined the Norse remains and found evidence that the settlers had suffered a painful annihilation. Grazing land was buried under advancing lobes of glacial ice, and most farmland was made useless by permafrost. Near the end, the descendants of a robust and hardy people were ravaged by famine; they were crippled, dwarf-like, and diseased.

Many explanations have been proposed for the extinction of Norse society in Greenland, and there were probably many contributing factors. Climate change and the inability (or unwillingness) of the Norse to adapt to an increasingly stressful climate and unreliable food sources, however, appear to have been the major causes of disaster. By the 1300s, the climate was cooling rapidly, heralding the Little Ice Age. Drift ice expanded over the North Atlantic, hampering and eventually halting navigation between Greenland and Iceland. All contact between the Norse settlements and the outside world ended shortly after 1400 A.D.

Deteriorating climate in Greenland meant wetter summers with poor haying conditions that caused major livestock losses. Snowier and longer winters probably claimed an even greater toll of livestock—especially among newborn animals. Climatic change also disrupted the harp seal migration and decimated caribou herds, further reducing food sources. About the same time, the Norse had to compete for shrinking food sources with the Inuit people who had migrated southward along Greenland's west coast. Competition for food evidently led to hostilities between the two groups.

Unable to survive the climatic stress, the Norse succumbed to famine; yet the neighboring Inuit people survived. The Inuits probably succeeded because their hunting skills and techniques were more suited to the hostile climatic conditions. On the other hand, the Norse refused to adopt the hunting practices of the Inuits and clung tenaciously to their traditional subsistence methods. Unfortunately, those methods were no longer suited to the new climatic regime.

The Norse tragedy in Greenland may be one of only a very few historical examples of the extinction of a European society in North America. What lesson does it teach contemporary society? The lesson is twofold: Climate changes and it can change rapidly—sometimes with serious, even disastrous, consequences. Nowhere are people more vulnerable to climatic shifts than in regions where the climate is marginal for their survival. These are regions where barely enough rain falls to sustain crops and livestock or where mean temperatures are so low and the growing season so short that only a few hardy crops can be cultivated successfully. In such regions, even a small change in climate in the wrong direction can make agriculture impossible. If inhabitants cannot locate and utilize new food sources, or if they cannot or will not migrate to more hospitable lands, their fate may be similar to that of the Greenland Norse.

in type of episode. For example, an exceptionally cold winter is the consequence of a usual westerly wave pattern that occurs with unusually high frequency or persistence (Chapter 10). A climatic shift to drier conditions results from more frequent dry weather circulation patterns. The individual circulation patterns are not themselves unusual, but their greater frequency or persistence may be anomalous.

How rapidly do climatic variations take place? Are climatic shifts abrupt or gradual? A climatic shift that takes 100 years is abrupt in the context of the long-term climatic record, which stretches back a million years, but the same shift is gradual when viewed in the limited human perspective of historical climate. In terms of this rate criterion, the climatic record indicates that significant climatic variations tend to occur abruptly rather than gradually.* Now, as in the past, a significant variation in climate can occur so rapidly that humankind may be unable to adjust to the shift.

We must also be careful to distinguish between a short-lived meandering from the climatic norm and a semipermanent climatic change. The Great Plains drought in the 1930s, the three severe Midwestern winters in the late 1970s, and the tragic Sahelian droughts were all temporary climatic fluctuations. In contrast, the climatic cooling of the late fourteenth century that heralded the Little Ice Age constituted a longer term climatic change.

## RHYTHMS IN THE RECORD

The recorded and reconstructed climatic record has long been the target of painstaking analyses for possible trends or regular rhythms. One formidable challenge in this search is to separate signal from noise. Some climatic elements are so variable through time (the noise) that detection of any cycles or trends (the signal) requires close scrutiny of the climatic record. In recent years, use of high-speed electronic computers programmed with sophisticated statistical routines has greatly facilitated the search for climatic rhythms and trends. The motivation behind this effort is obvious: Identification of any statistically real periodicities

*Climatic reconstructions based on glacial ice cores and deep-sea sediment cores confirm the abrupt change of climate.

or trends in the climatic record would be a powerful tool in climate forecasting. What are the results of such investigations?

So far, few cycles have been identified in the climatic record that are significant in a rigorous statistical sense, and none of the cycles has much practical value for climate forecasting. Cycles established as statistically real are (1) the familiar annual and diurnal radiation/temperature cycles and (2) a less familiar quasibiennial (almost every 2 years) cycle in various climatic elements. The first merely means that winters are cooler than summers and nights are cooler than days. Examples of the quasibiennial cycle include an approximate two-year fluctuation in Midwestern rainfall, a 25.5-month oscillation in a lengthy temperature record (1659 to present) from central England, and an approximately two-year cycle in the strength of the trade winds over the western Pacific and eastern Indian oceans.

A 20- to 22-year recurrence of drought on the western High Plains and the major glacial-interglacial climatic fluctuations are actually only quasiperiodic (not quite regular) and need to be more thoroughly understood before they can be used to reliably forecast future climate. Trends may be visible in the climatic record, but unless a trend is demonstrated to be part of a statistically significant cycle, there is no guarantee that the trend will not end abruptly or reverse direction at any time.

## THE SEARCH FOR ANALOGUES

With the contemporary concern over the potential for global warming due to a buildup of greenhouse gases in the atmosphere, it is not surprising that climatologists have searched the climatic record for possible analogues of a warmer globe. Identification of an appropriate analogue would be valuable in providing some indication of how the climate in specific regions might respond to global-scale warming.

Thus far, the search has not been very fruitful. Proposed analogues include the Climatic Optimum and the last major interglacial, about 120,000 years ago. Upon inspection, these climatic episodes prove to be inappropriate analogues. Warming during the Climatic Optimum and the last interglacial (1) was not globally synchronous and (2) affected seasonal temperatures primarily with only a slight rise in global mean temperature. Furthermore, R. A. Bryson of the University of Wisconsin–Madison points out that during the Climatic Optimum, global ice cover was more extensive, sea level was lower, and dates of perihelion and aphelion were different. Most pre–Ice Age warm episodes are also inappropriate analogues because of differences in topography and land/sea distribution plus the absence of ice sheets.

# Conclusions

In most places, the reliable instrument-based climatic record extends back only 100 years or so. For information on climate prior to this, we must rely on instrumental data gathered under nonstandardized conditions, documentary information, and a variety of geologic and ecological climatic indicators. Although the climatic record loses detail, continuity, and reliability with increasing time before present, it is evident that climate is inherently variable and varies over a wide range of time scales. The potential causes of temporal variations in climate plus prospects for the climatic future are subjects of the next and final chapter.

## Key Terms

| | |
|---|---|
| geologic time | polar amplification |
| continental drift | Younger Dryas |
| glacial climate | Climatic Optimum |
| interglacial climate | Little Ice Age |

## Summary Statements

☐ Because of the rapid deterioration of climatic detail as we go back in time, the most reasonable view of ancient climates is in terms of average climates prevailing over long periods of time.

☐ Continental drift and mountain building complicate our investigation of the Earth's climatic record over the hundreds of millions of years that constitute geologic time.

Both processes can account for large-scale changes in climate.

☐ The climatic record becomes more detailed and reliable as we approach the present, especially since the beginning of instrument-based records and standardized weather observation practices.

☐ When compared with the climate that prevailed through most of geologic time, the climate of the past two million years was unusual in favoring the development of huge glacial ice sheets.

☐ The last glacial maximum occurred about 18,000 years ago. Subsequently, the Laurentide ice sheet began an oscillatory retreat and finally melted away about 5500 years ago.

☐ The post-glacial warming trend reached its peak about 7000 to 5000 years ago (the Climatic Optimum) when global temperatures were somewhat higher than at present.

☐ Following the medieval warm period (950–1250) a drop in global temperatures ushered in the Little Ice Age from about 1400 to 1850.

☐ The first government-sponsored network of weather stations was established among Army posts during the early to mid-1800s. The reliable instrument-based climate record dates back to the late 1800s. Prior to this, weather observations were taken under nonstandardized conditions.

☐ Global temperature records indicate a warming trend from the late 1800s into the early 1990s. In middle and high latitudes of the Northern Hemisphere, the warming was interrupted by a cooling trend from 1940 to about 1970.

☐ Critics have questioned the integrity of large-scale temperature trends because of potential impacts of changes in weather instruments, station relocations, gaps in monitoring networks, and urbanization.

☐ The people most vulnerable to climatic fluctuations are those who live in areas where the climate is marginal for survival. In such areas even small changes in climate can be disastrous.

☐ How we view the rate of climatic change depends on our time frame of reference, but climate can change so rapidly that humankind may be unable to adjust.

☐ Few statistically significant cycles have been identified in the climatic record, and none of these has much practical value for weather or climate forecasting.

## Review Questions

1. How do we know that climate is inherently variable?
2. What happens to the reliability of the climatic record as we go back in time? Why?
3. How does continental drift explain the discovery of fossils of tropical plants in Greenland?
4. How might continental drift and mountain building influence the climate of some locality over millions of years?
5. What was unique about global climate over the past million years compared to the climate of the prior hundreds of millions of years?
6. What is the difference between a glacial climate and an interglacial climate?
7. Describe the geographical extent of the Laurentide ice sheet 18,000 years ago.
8. Describe general global conditions during the Climatic Optimum.
9. What were the prevailing climatic conditions during the Little Ice Age?
10. What motivated the Surgeon General to establish the first network of weather-observing stations at army posts?
11. Describe the trend in mean annual temperature of the United States over the period 1895–1991.
12. What considerations have led some scientists to question the validity of global and hemispheric temperature trends compiled from measurements made over the past 100-plus years?
13. How might urbanization affect the long-term temperature record?
14. In what sense are variations in climate geographically nonuniform?
15. List several lessons of the climatic past.
16. Provide some examples of the societal impact of climatic variability.
17. The people most vulnerable to climatic change are those living in areas where the climate is marginal for survival. What is meant by this statement? Give some illustrations.
18. Is climate change gradual or abrupt? Justify your response.
19. Are any cycles in the climatic record useful for forecasting the climatic future? Explain.
20. What is the potential value in searching the climatic record for possible analogues of greenhouse warming?

## Questions for Critical Thinking

1. In general terms, describe how sensitivity to climatic change varies geographically. Speculate on why polar regions appear to be more sensitive than other areas of the globe to large-scale climatic change.
2. A major climatic change about 18,000 years ago triggered the ultimate demise of the Laurentide ice sheet. Although the climate over most of North America was dominantly interglacial after that, it took more than 12,000 years for the glacier to melt away. Why so long?
3. Some scientists argue that the southern and eastern edge of the Laurentide ice sheet was a zone of stormy weather. From what you know about conditions favorable for cyclone development, comment on this hypothesis.
4. Discovery and exploration of the New World by Europeans took place during an episode of deteriorating climatic conditions. Speculate on what role this climate change might have played.
5. Explain the notion that past climatic episodes differ from present climatic episodes in frequency rather than type of episode. Provide some examples.

## Selected Readings

BROOKS, C. E. P. *Climate Through the Ages.* New York: Dover Publications, 1970, 395 pp. Represents a classic treatment of Earth's climatic history; first published in 1926 and revised in 1949.

BRYSON, R. A., AND T. J. MURRAY. *Climates of Hunger.* Madison, WI.: University of Wisconsin Press, 1977, 171 pp. Describes how past variations in climate have influenced people living in agriculturally marginal regions of the world.

COHMAP MEMBERS. "Climate Changes of the Last 18,000 Years: Observations and Model Simulations," *Science* 241 (1988):1043–1052. Reports on the application of global atmospheric circulation models in an extensive investigation of past climates by a multi-institutional consortium of scientists.

JONES, P. D., AND T. M. L. WIGLEY. "Global Warming Trends," *Scientific American* 263, No. 2 (1990):84–91. Reviews problems involved in computing global temperature trends.

KERR, R. A. "When Climate Twitches, Evolution Takes Great Leaps," *Science* 257 (1992):1622–1624. Examines the association between relatively short-term pulses in climate (during a long-term gradual trend) and evolutionary changes.

LE ROY LADURIE, E. *Times of Feast, Times of Famine: A History of Climate Since the Year 1000.* New York: Doubleday, 1971, 526 pp. Includes a fascinating reconstruction of the climatic past based primarily on documentary sources.

LITTIN, B. "Citizen Weather Observers," *Weatherwise* 43, No. 5 (1990):254–259. Celebrates the contributions of volunteer weather observers to the U.S. climate record.

McGOVERN, T. H. "The Economics of Extinction in Norse Greenland," in T. M. L. Wigley et al. *Climate and History.* New York: Cambridge University Press, 1981, pp. 404–443. Describes the various factors that probably contributed to the decline of the Norse settlements in Greenland.

MONASTERSKY, R. "Tales from Ice Time," *Science News* 140 (1991):168–172. Describes the goals, significance, and difficulties of obtaining ice cores from the Greenland ice sheet.

QUAYLE, R. G., ET AL. "Effects of Recent Thermometer Changes in the Cooperative Station Network." *Bulletin of the American Meteorological Society* 72 (1991):1718–1723. Discusses the effects of the recent shift from liquid-in-glass thermometers to thermistor-based maximum/minimum temperature systems.

ROTBERG, R. I., AND T. K. RABB. *Climate and History.* Princeton, NJ: Princeton University Press, 1981, 280 pp. Contains a collection of informative articles primarily on the methodology of climatic reconstruction.

STAUFFER, B. "The Greenland Ice Core Project." *Science* 260 (1993):1766–1767. Describes recent efforts to drill through the Greenland ice sheet.

# 20 Causes of Climatic Variability

*I had a dream, which was not all
  a dream.
The bright sun was extinguished,
  and the stars
Did wander darkling in the eternal
  space,
Rayless, and pathless, and the icy
  Earth
Swung blind and blackening in the
  moonless air;
Morn came and went—and came,
  and brought no day,
And men forgot their passions in
  the dread
Of this their desolation; and all
  hearts
Were chilled into a selfish prayer
  for light;*

LORD BYRON, 1816
*Darkness*

Recent warming has caused retreat of mountain glaciers such as this one in the Southern Alps of New Zealand. [Photograph by J. M. Moran]

450

A s we saw in the previous chapter, climate is inherently variable over a wide range of time scales. In this closing chapter, we describe what is understood about possible causes of climate variability, and prospects for the climatic future.

# Explaining Climatic Variability

There is no simple explanation for why climate varies. The complex spectrum of climatic variability is a response to the interactions of many processes both internal and external to the Earth-atmosphere system. These processes are displayed schematically in Figure 20.1. Open arrows indicate internal processes, and solid arrows represent external processes.

There are perhaps as many hypotheses about the causes of climatic variations as there are scientists who have seriously investigated the question. Many of these ideas evolved from efforts to explain climates of the Ice Age, but some stemmed from attempts to explain short-term fluctuations of climate. One way to organize the many hypotheses on the cause of climatic variability is to match a possible cause (or forcing) with a specific climatic oscillation based on the period of that oscillation. This approach is shown schematically in Figure 20.2. For example, mountain building and continental drift might explain climatic changes over periods of hundreds of millions of years. Systematic changes in the Earth's orbit about the sun may account for climatic shifts of the order of 10,000 to 100,000 years. Changes in sunspot number and variations in the sun's energy output may be associated with climatic fluctuations of decades to centuries, and volcanic eruptions may induce climatic fluctuations lasting several months or years. Note, however, that matching a possible cause and effect based on similar periods of oscillation is no guarantee of a real physical relationship; we could well be dealing with mere coincidence.

Hypotheses concerning climatic variability can also be organized in the context of the global radiation balance. Energy that enters the Earth-atmosphere system must ultimately equal energy that leaves the system (Chapter 2). That is, the net input of solar radiation must balance the output of infrared radiation. Any change in either energy input or energy output will shift the Earth-atmosphere system to a new equilibrium state and thereby change the planet's climate. The global radiation balance could be altered by fluctuations in (1) the solar constant, (2) the planetary albedo, or (3) the gas and aerosol composition of the atmosphere. We examine this approach to understanding climatic change quantitatively in the Mathematical Note at the end of this chapter.

# Climate and Solar Variability

Fluctuations in the sun's energy output or variations in the Earth's orbit about the sun could alter the solar constant or the distribution of solar radiation over the Earth's surface. Any of these changes could, in turn, alter global climate.

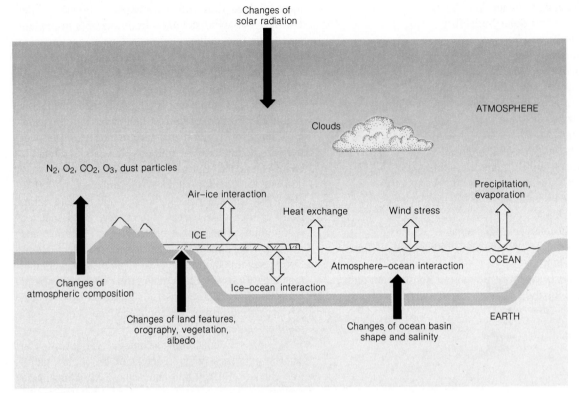

**FIGURE 20.1**
Climatic variability is influenced by many processes, both internal (white arrows) and external (black arrows) to the Earth-atmosphere system. [Reprinted from W. L. Gates and Y. Mintz, *Understanding Climatic Change: A Program for Action*, 1975, with permission from the National Academy of Sciences, National Academy Press, Washington, DC.]

## THE SOLAR CONSTANT

Is the **solar constant*** really constant? As demonstrated in this chapter's Mathematical Note, even a 1% change in the solar constant could significantly alter the radiative equilibrium temperature of the Earth-atmosphere system. In fact, only a 0.5% fluctuation in the solar constant could account for the total variation of the mean annual hemispheric temperature over the past century. Until about 15 years ago, scientists had been unable to monitor small-scale changes in the solar constant. Older ground-based instruments lacked the necessary sensitivity, and high-resolution instruments aboard satellites had not been operating long enough

*Recall from Chapter 2 that the Earth's *solar constant* is defined as the flux of solar radiation on a surface oriented perpendicular to the solar beam at the top of the atmosphere when the planet is at its mean distance from the sun.

to provide reliable records. This situation began to change on 14 February 1980 with the launch of NASA's Solar Maximum Mission (SMM) satellite (Figure 20.3) into orbit 550 km (341 mi) above the Earth's surface.

Extremely sensitive radiation sensors on the SMM satellite could detect changes as small as 0.001% in the sun's total radiative output, called **solar irradiance.** Analysis of measurements by the SMM and other satellites indicates a decline in solar irradiance between 1981 and 1986 amounting to about 0.018% per year. This was a total reduction of about 0.10% in almost six years. Had this trend continued, the dimming of the sun might have been climatically significant by 1990. However, solar irradiance began to increase after the Fall of 1986, about the time of the last sunspot minimum, and continued to increase into the early 1990s. This correspondence suggests that

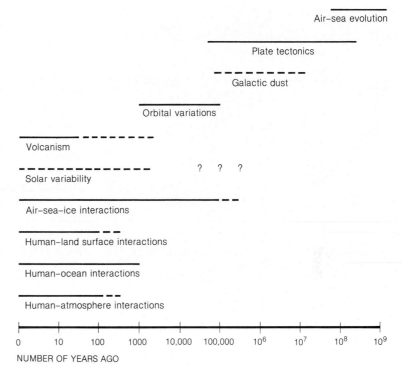

**FIGURE 20.2**
One way to speculate on the possible cause of a climatic oscillation is to match the period of the oscillation with the period of an appropriate forcing phenomenon. Lines are dashed where there is considerable uncertainty. [Reprinted from *Geological Perspectives on Climatic Change,* 1978, with permission from the National Academy of Sciences, National Academy Press, Washington, DC.]

solar irradiance may vary with the 11-year sunspot cycle (discussed later).

Skepticism regarding the reliability of SMM irradiance data appears to be ill-founded. The satellite experienced technical difficulties in November 1980 and was repaired by NASA's space shuttle in April 1984. Nonetheless, SMM findings compare favorably with data from other solar-irradiance-monitoring experiments. Radiation sensors aboard the Nimbus-7 satellite, launched in 1978, detected a 0.015% per year decline in solar irradiance. In addition, probes by rockets and high-altitude balloons found a downward trend of about 0.016% per year. Unfortunately, the SMM satellite finally went out of service and plunged to Earth in 1989. Detailed monitoring of solar irradiance resumed with the 1991 launch of the Upper Atmospheric Research Satellite (UARS).

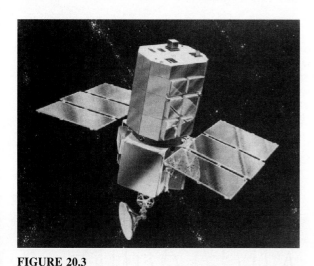

**FIGURE 20.3**
An artist's conception of NASA's Solar Maximum Mission Spacecraft. Instruments aboard this satellite, in a 550-km (340-mi) high orbit around the Earth, provided detailed measurements of solar energy output. [Courtesy of NASA]

## *SUNSPOTS*

Both popular and technical literature contain much speculation on a possible link between the Earth's weather and climate and sunspot activity. A **sunspot** is a relatively large (typically thousands of kilometers in diameter) dark blotch that appears on the face of the sun. As shown in Figure 20.4, a sunspot consists of a dark central area, called an **umbra,** ringed by an outer,

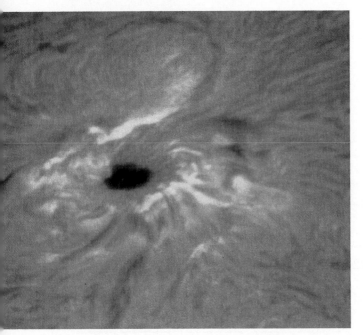

**FIGURE 20.4**

A sunspot. Climate changes on Earth may be linked to variations in number of sunspots on the sun. Although sunspots are relatively cool, dark areas in the sun's photosphere, there is an interesting correlation between periods of reduced sunspot activity and cool episodes on Earth. [Courtesy of Dr. Donat G. Wentzel, University of Maryland at College Park, and the National Optical Astronomical Observatories, Tucson, Arizona]

lighter area, termed a **penumbra.** The umbra radiates at about 4000 K, the penumbra at 5400 K, and the surrounding surface of the sun, the **photosphere,** at 5800 K.

As early as 28 B.C., Chinese astronomers observed sunspots with the unaided eye by viewing the sun's reflection on the surface of a quiet pond. Galileo is credited with being the first to study sunspots telescopically in 1610, and thereafter sunspots became objects of considerable scientific curiosity. Speculation on a possible sunspot-climate link was spurred by Heinrich Schwabe's discovery in 1843 of the regularity of sunspot activity. As shown in Figure 20.5, the number of sunspots varies systematically with an average 11-year period between sunspot maxima. Also an approximate 22-year oscillation *(double sunspot cycle)* characterizes the strong magnetic field that is associated with sunspots. Note, however, that the sunspot record is not precisely periodic and that variations of 10 to 12 years or more occur between successive maxima or minima.

As noted above, recent satellite measurements of solar irradiance indicate that the sun's energy output varies directly with sunspot number; that is, more sunspots mean a brighter sun, perhaps because of a concurrent increase in faculae* activity. Thus more sunspots may translate to a warmer Earth, assuming that all other climate controls are constant. Past records of sunspot ac-

*Faculae* are small, bright, relatively hot spots on the sun.

**FIGURE 20.5**

On this graph of mean annual sunspot number, note the Maunder minimum from 1645 to 1715. Note also that the time between sunspot maxima is not quite regular, but varies between about 10 and 12 years. [Updated from J. A. Eddy, "The Case of the Missing Sunspots," *Scientific American* 236, No. 5 (1977):82–83. Copyright © 1977 by Scientific American, Inc. All rights reserved.]

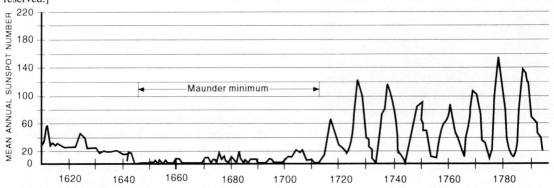

tivity and climate reveal some intriguing correspondences.

In 1893, while searching records at the Old Royal Observatory at Greenwich, England, E. Walter Maunder discovered that sunspot activity was greatly reduced in the 70 years between 1645 and 1715. The total number of sunspots recorded during that period, now called the **Maunder minimum,** was fewer than is typical in a single year today. Strangely, the scientific community largely ignored Maunder's finding until the 1970s, when his work was reinvestigated and confirmed by John A. Eddy, now of the Consortium for International Earth Science Information Network. Eddy pointed out that the Maunder minimum and an earlier episode of reduced sunspot activity, called the **Spörer minimum** (1450–1550), happened to coincide with two relatively cold pulses of the Little Ice Age in Western Europe. However, skeptics dismiss the climatic significance of this correspondence, arguing that anomalous cold did not prevail throughout the sunspot minima and was not global in extent. The late climatologist Helmut Landsberg of the University of Maryland noted that globally the coldest episode of the Little Ice Age occurred 100 years *after* the Maunder minimum. On the other hand, relatively mild weather (the medieval warm period) coincided with a period of considerable sunspot activity between about 1100 and 1250 A.D.

There is also a proposed match between the 22-year *double* sunspot cycle and the frequency of drought on the western High Plains. By analyzing tree growth-ring records from 40 sites in the western United States, Charles Stockton of the University of Arizona's Laboratory of Tree-Ring Research was able to extend the region's drought chronology back to the seventeenth century. He discovered that drought recurred every 20 to 22 years.

Support for a link between solar activity and climate also comes from a 2000-year tree growth-ring record extracted from foxtail pines in California's Sierra Nevada. L. A. Scuderi, a researcher at Boston University, based his reconstruction of solar activity on the concentration of carbon-14 ($^{14}C$) in the tree growth rings. $^{14}C$ is an isotope* of carbon that is generated by an atmospheric chemical reaction powered by solar energy. The more active the sun, the greater is the concentration of $^{14}C$ incorporated in trees via photosynthesis. The relative thickness of tree growth rings served as an index of temperature. Scuderi's year-by-year chronology revealed a close correlation between solar activity and temperature with a period of 125 years. Mild episodes of the medieval warm period and cold episodes during the Little Ice Age are evident in the record.

Are these correlations actual cause-effect relationships or are they coincidental? So far, investigators have not come up with an adequate explanation of the mechanism by which sunspot activity might influence climate. At present, there is no scientific consensus on a definitive solar-climate link.

### ASTRONOMICAL CHANGES

Analysis of deep-sea sediment cores has revealed regular shifts between glacial and interglacial climatic episodes at least over the past 500,000 years. These

*Isotopes* of a given element such as carbon differ from one another on the basis of atomic mass.

**FIGURE 20.5 (continued)**

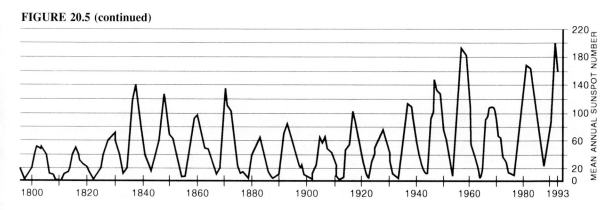

regular oscillations closely match systematic changes in the geometry of the Earth's orbit about the sun. These astronomical changes, first identified in the middle of the nineteenth century, were extensively investigated in the 1920s and 1930s by the Serbian mathematician Milutin Milankovitch.

Milankovitch studied cyclical variations in three elements of Earth-sun geometry: (1) the precession (wobble) of the Earth's axis of rotation such that the spin axis describes a circle, (2) the tilt of the Earth's axis (varying between 22.1 and 24.5 degrees), and (3) the eccentricity (departure from a circle) of the Earth's orbit. The precession cycle alters the dates of perihelion and aphelion and thereby increases the seasonal contrast in one hemisphere and decreases the seasonal contrast in the other hemisphere. As the axial tilt increases, the seasonal contrast increases so that winters are colder and summers are warmer in both hemispheres. Changes in orbital eccentricity affect the Earth-sun distance and, hence, seasonal variations of the solar constant. These so-called **Milankovitch cycles** are illustrated schematically in Figure 20.6.

Although Milankovitch cycles do not alter the solar constant appreciably, they do change significantly the latitudinal and seasonal distribution of solar radiation received by the planet. Milankovitch proposed that glacial climatic episodes were initiated at times when the Earth-sun geometry favored minimum summer insolation at high northern latitudes. During these times, some of the winter snows falling in Canada would presumably survive the summer, and a repetition of many such years would initiate glaciation on a continental scale.

Milankovitch hand-calculated latitudinal variations of insolation for 600,000 years prior to the year 1800. More recently, with some refinement in methodology and use of electronic computers, these calculations were repeated and extended over a longer time period. This research demonstrated that systematic changes in precession, axial tilt, and eccentricity would induce climatic cycles having periods of about 23,000 years, 41,000 years, and 100,000 years, respectively. Discovery of climatic cycles of similar periodicities in deep-sea sediment cores strongly argues for variations in Earth-sun geometry as the principal cause of the regular large-scale fluctuations of the planet's glacial ice cover.

Although the close correspondence between Milankovitch cycles and reconstructed climatic fluctuations of the Ice Age is impressive, the precise physical linkage between glacial-interglacial climates and periodic changes in Earth-sun geometry is not yet fully understood. Furthermore, the Milankovitch cycles do not account for the lengthy periods of climatic quiescence during the millions of years prior to the Ice Age.

Discovery of a possible relationship between systematic changes in Earth-sun geometry and major glacial-interglacial climate shifts has prompted climatologists to search for influences of Milankovitch cycles on smaller-scale climatic fluctuations. For example, J. E. Kutzbach of the University of Wisconsin–Madison has linked the Climatic Optimum to the Earth's precession cycle. Between 8000 and 5000 years ago, Earth was closest to the sun (perihelion) in June rather than in January. In the Northern Hemisphere, this would have meant about 5% more solar radiation during summer and about 5% less solar radiation during winter. The consequent increase in seasonal air temperature contrast probably altered the planetary circulation. For one, evidence suggests that the monsoon circulation of subtropical latitudes was more vigorous during the Climatic Optimum than it is today.

# Volcanoes and Climatic Variability

The spectacular eruption of Mount St. Helens, Washington, on 18 May 1980 (Figure 20.7) spurred considerable interest in the potential climatic impact of volcanic eruptions. Actually, the notion that volcanic eruptions somehow influence weather and climate has been around for more than two centuries. Benjamin Franklin proposed that eruption of Iceland's Laki volcano in the summer of 1783 was responsible for the severe winter of 1783–84. The unusually cool summer of 1816 followed the violent eruption of Tambora, an Indonesian volcano, in the spring of 1815. Several relatively cold years also occurred after Krakatau, also in Indonesia, blew its top in 1883. Is the relationship between volcanic eruptions and atmospheric cooling real or coincidental?

Originally, climatologists were primarily concerned with the possible climatic impact of the relatively fine dust and ash particles thrown high into the stratosphere

Milankovitch Cycles

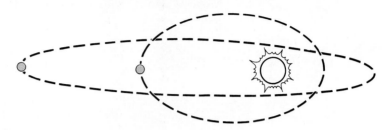

A. Changes in eccentricity: ~100,000 years

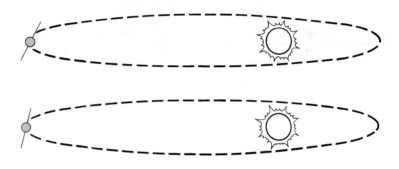

B. Changes in axial tilt: ~41,100 years

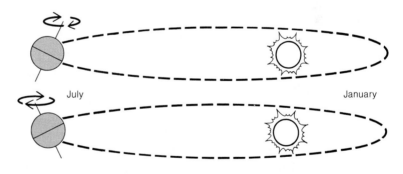

C. Changes in axial precession: ~23,000 years

**FIGURE 20.6**
The Milankovitch cycles in Earth-sun geometry that may determine the timing of major glacial-interglacial climatic shifts. Note that the diagrams greatly exaggerate changes in geometry.

during violent volcanic eruptions. Theoretically, volcanic ash should raise the planetary albedo, and a higher albedo means less solar radiation reaching the Earth's surface and lower air temperatures. Today, however, most climatologists agree that ash particles, being relatively large, quickly settle out of the stratosphere to the ground with little or no long-term effect on the radiation balance. The large-scale climatic impact of violent volcanic eruptions depends more on the volume of sulfur oxide gases ejected.

Once in the stratosphere, sulfur oxide combines with moisture to produce tiny droplets of sulfuric acid ($H_2SO_4$) and sulfate particles. Collectively, these are referred to as *sulfurous aerosols*. The extremely small size of the aerosols (about 0.1 micrometer in diameter), coupled with the stratosphere's stability and absence of

**FIGURE 20.7**
The spectacular eruption of Mount St. Helens on 18 May 1980 apparently had a negligible impact on hemispheric temperatures. [Photograph courtesy of the U.S. Geological Survey, EROS Data Center]

precipitation, means that sulfurous aerosols remain suspended in the stratosphere for many months to perhaps several years before finally settling to the Earth's surface. While in the stratosphere, the aerosol veil interacts with solar radiation, such that a portion of the radiation is absorbed, causing stratospheric warming and reducing the amount of radiation available to the troposphere. Furthermore, the aerosols scatter some solar radiation back to space, also contributing to cooling at the Earth's surface. Over the years, a succession of volcanic eruptions has produced a permanent sulfurous aerosol layer in the stratosphere (at altitudes of 15 to 25 km), and sulfur oxide emissions from an occasional violent volcanic eruption temporarily thicken that layer.*

In addition to sulfur oxide emissions, the latitude and season of a volcanic eruption influence the climatic impact of volcanoes. As a general rule, the prevailing planetary-scale circulation causes plumes from low latitude eruptions to disperse over much greater areas than plumes from high latitude eruptions. Seasonal changes in the planetary circulation mean that in the tropics, for

*Researcher Christian Junge is credited with discovering the stratospheric sulfurous layer in the early 1960s.

example, the plume from a spring eruption does not spread as far as a plume from a summer eruption.

Apparently, only those volcanic eruptions rich in sulfur oxide and sufficiently explosive to send ejecta well into the stratosphere can disturb hemispheric or global climate. Even then, surface hemispheric mean temperatures are not likely to be lowered by more than 1 Celsius degree (1.8 Fahrenheit degrees). A volcanic eruption in this category was the 1963 eruption of Agung in Bali, which, according to one estimate, lowered the mean air temperature of the Northern Hemisphere by about 0.3 Celsius degree (0.5 Fahrenheit degree) for a year or two. Although the 1980 eruption of Mount St. Helens produced about as much ash as the Agung explosion, its ejecta were low in sulfur oxides and had no detectable influence on large-scale climate.

Although the Mount St. Helens eruption did not influence hemispheric or global climate, some short-term localized effects on surface temperatures occurred immediately downwind of the volcano over eastern Washington, Idaho, and western Montana. Over these areas, the ash plume, clearly visible in the satellite photograph in Figure 20.8, was sufficiently thick to alter the local radiation balance for 12 to 24 hours following the eruption. C. Mass and A. Robock of the University of Maryland report that on the day of the eruption, blockage of the sun by volcanic ash lowered surface air temperatures over eastern Washington by up to 8 Celsius degrees (14.4 Fahrenheit degrees). That night, over Idaho and western Montana, low-level ash impeded infrared cooling, thereby elevating temperatures by as much as 8 Celsius degrees.

The violent eruption of the Mexican volcano El Chichón on 4 April 1982 apparently influenced large-scale climate. Although many volcanic eruptions have been more violent and have spewed out much more ejecta, the El Chichón eruption was unusually rich in sulfur oxides. Conversion of these gases to sulfurous aerosols and the gradual spreading of the aerosol veil through the stratosphere of the Northern Hemisphere probably caused cooling on a large scale. Scientists employed computerized global circulation models to assess the influence of El Chichón's aerosol veil on surface temperatures; these studies generally agree on a hemispheric cooling of about 0.2 Celsius degree (0.4 Fahrenheit degree).

The June 1991 eruption of Mount Pinatubo in the Philippines was the most massive since Krakatau in

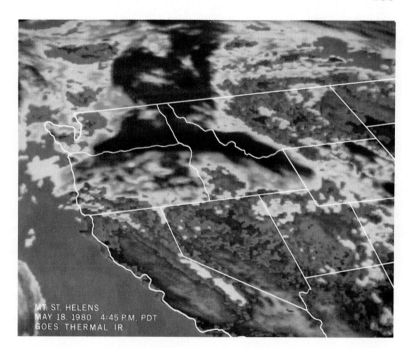

**FIGURE 20.8**
Computer-enhanced GOES infrared image of the ash cloud ejected by Mount Saint Helens. The image was taken at 4:45 P.M. Pacific Daylight Time on the day of the eruption, 18 May 1980. The ash cloud shows up as a dark blotch covering southern Washington, northern Oregon, northern and eastern Idaho, western Montana, and northwestern Wyoming. The dark area to the north and northwest is high clouds. [Courtesy of Michael Matson, National Environmental Satellite Service, NOAA]

1883. In fact, it may turn out to be the single greatest perturber of climate of this century. Mount Pinatubo blasted an estimated 20 to 30 megatons* of sulfurous aerosols—more than twice that of El Chichón—into the stratosphere to altitudes as great as 19 km (11.8 mi). As shown in Figure 20.9, planetary winds dispersed the volcanic veil into both hemispheres.

The Mount Pinatubo eruption interrupted the recent global warming trend (Chapter 19). Scientists at NASA's Langley Research Center reported that in the months after the eruption, sensors aboard the Earth Radiation Budget Satellite detected a 3.8% increase in solar radiation reflected to space by the Earth-atmosphere system. The following year, the global mean temperature was 0.6 Celsius degree (1.1 Fahrenheit degrees) lower than it was in the year preceding the eruption.

Not surprisingly, the climatic fluctuation associated with the Mount Pinatubo eruption was geographically nonuniform (Figure 20.10). Not all locations experienced cooling, and where there was cooling, it varied in magnitude. The summer of 1992 was particularly cool in the upper Midwest where some localities reported the coolest growing season on record (Figure 20.11). In terms of persistence (if not in magni-

*One megaton (Mt) = $10^{12}$ grams.

tude), the summer of 1992 in the upper Midwest was reminiscent of the summer of 1816 in New England.

## Greenhouse Gases and Climatic Variability

Among the many contemporary environmental issues is the possibility that certain human activities may disrupt the Earth-atmosphere radiation balance and thereby contribute to variations in climate. From our study of the fundamentals of weather, we conclude that human activity may influence climate in at least three ways: (1) by altering the radiational properties of the Earth's surface (changing the albedo, for example), (2) by venting waste heat into the atmosphere, and (3) by changing the concentrations of certain key gaseous or aerosol components of the atmosphere. Although many questions remain unanswered, the first two activities appear to be important only on a local scale, that is, in and downwind of large cities. We examined these influences in Chapter 17. In this section, we focus on the large-scale climatic implications of elevated levels of

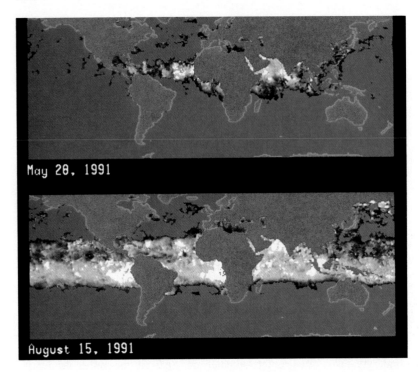

**FIGURE 20.9**
The spread of aerosols from the June 1991 eruption of Mount Pinatubo in the Philippines as detected by instruments aboard NOAA-11 satellite. Top image was obtained 28 May 1991 and the bottom image 15 August 1991. [From J. J. Hidore and J. E. Oliver, *Climatology, An Atmospheric Science.* New York: Macmillan Publishing Company, 1993, plate 6]

greenhouse gases. In the next section, we consider the potential impacts of anthropogenic aerosols.

### TRENDS IN ATMOSPHERIC CARBON DIOXIDE

In recent years, many atmospheric scientists have expressed concern about the possible climatic ramifications of the steadily rising concentrations of atmospheric carbon dioxide ($CO_2$) and other greenhouse gases. Higher levels of these gases are likely to enhance the **greenhouse effect,** that is, increase absorption and reradiation of infrared radiation and consequently warm the lower atmosphere. Actually, the possibility of $CO_2$-induced global warming was first proposed in 1827 by the French mathematician Baron Joseph Fourier.

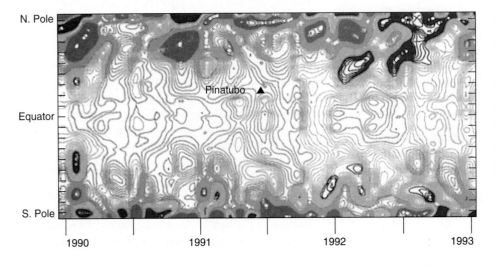

**FIGURE 20.10**
Satellite measurements document the impact of the Mount Pinatubo eruption on surface temperatures. Blue indicates cooling, and red indicates warming. Cooling is likely a response to Mount Pinatubo aerosols, and warming is probably the result of ENSO. [NOAA]

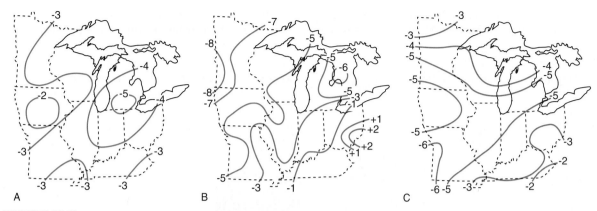

**FIGURE 20.11**
Temperature anomalies in degrees Fahrenheit in the Midwest during (A) June 1992, (B) July 1992, and (C) August 1992. [From W. M. Wendland and J. Dennison, *Weather and Climate Impacts in the Midwest,* Midwestern Climate Center, Illinois State Water Survey, Champaign, IL]

Although G. S. Callendar, a British engineer, first reported an upward trend in atmospheric carbon dioxide in 1939, the increase probably began during the Industrial Revolution when reliance on fossil fuels began to increase (Figure 20.12). Fossil fuel combustion accounts for about 80% of the increase in carbon dioxide concentration, and deforestation is likely responsible for the balance. The burning of coal, oil, and natural gas produces carbon dioxide as a byproduct. Clearing of tropical forests contributes $CO_2$ via burning, decay of wood residue, and reduced photosynthetic removal of carbon dioxide from the atmosphere.

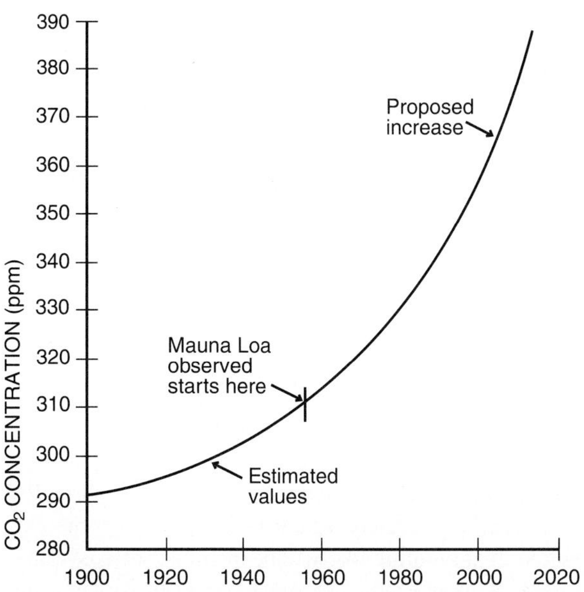

Systematic measurements of atmospheric carbon dioxide levels began in 1957 at NOAA's Mauna Loa Observatory in Hawaii under the direction of Charles D. Keeling of Scripps Institution of Oceanography. At 3400 m (11,155 ft) above sea level in the middle of the Pacific Ocean, Mauna Loa Observatory is far enough away from major industrial sources of air pollution that carbon dioxide levels monitored there are considered representative of at least the Northern Hemisphere. Also since 1957, atmospheric $CO_2$ has been monitored at the South Pole station of the U.S. Antarctic Program. Interestingly, the South Pole carbon dioxide trend closely parallels that at Mauna Loa.

The Mauna Loa and South Pole records show an annual carbon dioxide cycle due to seasonal changes in Northern Hemisphere vegetation; carbon dioxide levels fall during the growing season (when photosynthetic removal exceeds cellular respiration), reaching a minimum in October, and recover in winter (when cellular respiration exceeds photosynthesis), reaching a maximum in May. This annual cycle is superimposed on a nearly exponential rate of growth: The annual rate of increase of carbon dioxide rose from about 0.7 ppm (parts per million) in 1958 to 1.0 to 1.5 ppm during the 1980s. The average carbon dioxide level increased from 315 ppm in 1958 to 353 ppm in 1990. Sketchy

**FIGURE 20.12**
The trend in atmospheric $CO_2$ concentration since 1900 and projected to 2020. [From J. J. Hidore and J. E. Oliver, *Climatology, An Atmospheric Science.* New York: Macmillan Publishing Company, 1993, p. 69]

data suggest that atmospheric $CO_2$ concentration may have climbed by 24% since its estimated preindustrial level of 280 ppm.

What might this rapid rise in atmospheric carbon dioxide mean for the climatic future? Researchers have tried to answer this question by using numerical global climate models and electronic computers. In such experiments, the typical procedure is to test the sensitivity of the model by changing one of the variables—in this case, carbon dioxide concentration. Initially, the model is run to a state of equilibrium using current atmospheric conditions. The model's carbon dioxide concentration is then raised (usually doubled), and the numerical model is run to a new equilibrium state. The difference between the initial and final equilibrium states is assumed to be the consequence of elevated carbon dioxide concentration.

Depending on the specific global climate model used, these experiments predict that the Earth's average surface temperature could rise 1.5 to 4.5 Celsius degrees (2.7 to 8.1 Fahrenheit degrees) with a doubling* of atmospheric $CO_2$—possible by the middle of the next century. Global warming could match the total post–Ice Age temperature rise, albeit at a rate 10 to 100 times faster. Predictive models provide little detail on regional or local climatic change, but the warming is likely to be geographically nonuniform in magnitude. Models predict that polar warming will be up to three times the global average change (*polar amplification*). In addition, models generally agree that annual precipitation will increase poleward of about 30 degrees latitude and within about 5 degrees of the equator, and decrease elsewhere. However, significant summer drying is predicted for the continental interiors of midlatitudes, where much of the world's food is grown.

### TRENDS IN OTHER GREENHOUSE GASES

Warming induced by carbon dioxide may be enhanced by rising levels of other greenhouse gases whose atmospheric concentrations are so small that they are typically expressed in parts per billion (ppb). These gases include methane ($CH_4$) at 1738 ppb, nitrous oxide ($N_2O$) at 310 ppb, and chlorofluorocarbons (CFCs) at less than 0.5 ppb. The climatic significance

*From pre-industrial levels of $CO_2$.

of these trace gases lies in their strong absorption of infrared radiation in the *atmospheric windows,* especially at wavelengths between 7 and 13 micrometers. (Recall from Chapter 2 that within atmospheric windows, infrared absorption by water vapor, carbon dioxide, and ozone is minimal.) Trace-gas absorption is directly proportional to concentration, so that doubling the concentration of a gas doubles the absorption. The combined climatic impact of rising levels of these greenhouse gases could match that of increasing carbon dioxide.

Studies of air bubbles trapped in glacial ice cores suggest that the concentration of methane has more than doubled since the preindustrial era. However, for unknown reasons,** the annual rate of increase in atmospheric methane has declined from about 13.3 ppb in 1983 to 9.5 ppb in 1990. Nitrous oxide concentration is building at about 0.8 ppb per year and is now at 310 ppb, that is, about 8% higher than its preindustrial level. CFCs are currently the fastest growing of the greenhouse trace gases. The two most common CFCs (CFC-11 and CFC-12) have been increasing at about 4% per year but are expected to level off soon (Chapter 17).

The upward trend in methane is probably due mostly to biological decay in wetlands and bacteria in the digestive systems of ruminants (cud-chewing animals) such as cattle and sheep. Termites, rice paddies, and sanitary landfills are additional sources of methane. The increase in nitrous oxide is probably linked to industrial and agricultural activity. Recall from our discussion in Chapter 17 that there are several sources of CFCs.

### POTENTIAL SOCIETAL IMPACTS

What are the potential societal impacts of an enhanced greenhouse effect? The consequent global climatic change could be greater than any experienced thus far in the 10,000 years of civilization. Virtually every sector of society would be affected to some extent, with agriculture likely to be the most seriously disrupted. Heat and moisture stress would likely cut crop yields, and traditional farming practices would have to

**Proposed explanations include reduced methane loss during oil extraction, a leveling off in cattle populations, and a slowing in rice cultivation.

change. For the North American grain belt, for example, higher temperatures and more frequent drought might necessitate a switch from corn to wheat and require more irrigation.

Predicted amplification of a global warming trend in polar latitudes has prompted much speculation on the fate of the Greenland and Antarctic ice sheets. Some studies contend that the warming would melt enough ice that sea level would rise and inundate densely populated coastal areas. In addition, warming causes ocean water to expand, and this would also contribute to a sea-level rise. The combination of melting glaciers plus thermal expansion of ocean water would raise sea level an estimated 0.2 to 1.5 m (0.7 to 4.9 ft) in 60 years.

Some researchers hasten to point out that an enhanced greenhouse effect may actually yield some benefits. For one, warmer winters would reduce fuel demand for space heating in middle and high latitudes. In their 1988 summary of potential socioeconomic impacts in the Great Lakes region (due to a doubling of $CO_2$), S. J. Cohen and T. R. Allsopp of the Atmospheric Environment Service of Canada reported that milder winters in Ontario would reduce energy demand for space heating by 45% (more than offsetting an estimated 7% increase in energy demand for summer air conditioning). Warmer winters would also lengthen the navigation season on lakes, rivers, and harbors where ice cover is a problem.

Has the warming begun? At many long-term weather stations in the Midwest and Great Plains, the summer of 1988 was one of the driest and hottest on record (Chapter 10). The extreme heat triggered much speculation that a major climatic change was under way and that the much heralded $CO_2$-induced global warming had at last begun. Although acknowledging that one long hot summer does not constitute a climatic change, some atmospheric scientists are quick to point to recent trends in climate that are similar to changes that would attend a greenhouse warming as predicted by global climate models. These trends include

1. On a global scale, the eight warmest years since the 1880s have occurred since 1979.

2. In Alaska, soil temperature at the top of the permafrost layer (perennially frozen ground) has risen 2 to 4 Celsius degrees (3.6 to 7.2 Fahrenheit degrees) since 1900.

3. Many mountain glaciers in tropical and subtropical latitudes are retreating at accelerated rates.

4. Since 1940, in the Northern Hemisphere, average annual precipitation has increased at high latitudes (35 to 70 degrees N) and decreased at low latitudes (5 to 35 degrees N).

But before residents of Boston and New Orleans quit the city and head for high ground to avoid rising ocean water, a note of caution is warranted. Even though some recent climatic trends are consistent with the changes that are predicted to accompany $CO_2$-induced warming, there is as yet no evidence of a direct cause-effect relationship. Climate controls other than carbon dioxide may well be responsible for these observed trends. Furthermore, the atmosphere is a complex and highly interactive system, so that other processes may compensate for an enhanced greenhouse effect. For instance, the ocean has great thermal stability, which for a time may slow the warming. In addition, warmer conditions in high latitudes may mean more snowfall and eventually more, not less, glacial ice!

As pointed out in Chapter 19, serious questions surround the integrity of the global mean temperature record that shows a gradual warming trend between the 1880s and the 1990s. Criticism is also directed at the computerized numerical models that have been used to predict the climatic impact of rising levels of atmospheric carbon dioxide. For one, models vary in sophistication and in their ability to simulate the climatic influence of ocean currents, cloud cover, and atmospheric water vapor. Furthermore, the current generation of climate models does not do well in forecasting the regional response to global or hemispheric climatic change.

In spite of current scientific uncertainties, some experts argue that so much is at stake that action must be taken now to head off enhanced greenhouse warming. They call for (1) at least a 50% reduction in global fossil-fuel consumption, (2) greater reliance on nonfossil fuels (solar power, for example), (3) higher energy efficiencies (more vehicle miles per gallon, for example), (4) massive reforestation and a halt to deforestation, and (5) phasing out of CFCs.* They hasten to

---

*As noted in Chapter 17, CFC-producing nations have already agreed to a complete phaseout of CFCs because of the threat to the stratospheric ozone layer.

point out that these actions are advisable even if greenhouse warming fails to materialize because they will help alleviate other serious environmental problems. For example, phasing down our reliance upon fossil fuels will also reduce acid deposition and other forms of air pollution.

## Aerosols and Climatic Variability

Some atmospheric scientists argue that an increase in the aerosol content of the atmosphere, called **turbidity,** has the opposite effect of elevated levels of greenhouse gases. Instead of warming the Earth, an increase in turbidity adds to the reflectivity of the Earth-atmosphere system, reducing insolation and thereby cooling the lower atmosphere.

In recent years, attention has focused on the potential climatic impact of sulfurous aerosols of anthropogenic origin. Perhaps 90% of these aerosols are byproducts of fossil fuel burning in the Northern Hemisphere. Sulfur oxides emitted by power plant smokestacks and boiler vent pipes combine with atmospheric water vapor to produce sulfurous aerosols (sulfate particles and droplets of sulfuric acid). Some scientists propose that these aerosols raise the albedo directly by reflecting sunlight to space and indirectly by acting as cloud condensation nuclei and increasing cloud cover. If true, this might explain the difference in trends of annual mean maximum and minimum temperatures in the United States (Figure 20.13). While the annual mean minimum temperature trended upward from 1950 to 1989, the annual mean maximum temperature showed little change. This contrast

in temperature behavior is consistent with greenhouse warming that is partially offset during the day by the cooling effect of sulfurous aerosols.

Other atmospheric scientists contend that increased turbidity resulting from human activity, such as industry and agriculture, promotes warming at the Earth's surface. They argue that these aerosols tend to scatter solar radiation back toward the surface, where it is absorbed, and that the larger dust particles absorb and reemit infrared radiation to enhance the greenhouse effect.

Resolution of this disagreement hinges on a better understanding of the net effect of aerosols on the global radiation balance. It is known that the percentage of insolation that is scattered downward toward the Earth's surface versus the percentage that is scattered back into space varies with the optical properties of the aerosol. In addition, the albedo of the Earth's surface underlying an aerosol layer may help determine whether the net effect is heating or cooling. For example, viewed from an airplane, a plume of smoke over a lake (low albedo) appears lighter than the lake and therefore has a cooling effect. On the other hand, the same plume of smoke over a snow-covered surface (high albedo) appears darker than the snow and hence causes warming.

The impact of aerosols on climate is thus a complex and unresolved problem. Although sulfurous aerosols of volcanic origin likely trigger cooling at the Earth's surface, the impact of aerosols produced by human activity is uncertain. Most anthropogenic aerosols never reach the stratosphere, and even the smallest ones remain in the troposphere only a few days before being removed by precipitation, impaction, or gravitational settling (Chapter 17).

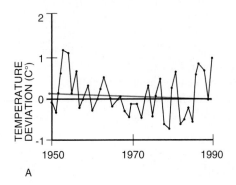

A

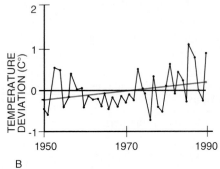

B

**FIGURE 20.13**
Trends in daily maximum (A) and minimum (B) air temperature between 1950 and 1990 for the United States. Horizontal line is the period average; temperatures are plotted as the number of degrees Celsius above or below the period average. Trend lines are shown in red. [NOAA]

## Earth's Surface and Climatic Variability

In Chapter 2, we saw that the Earth's surface, which is mostly water, is the prime absorber of solar radiation. Any change in the characteristics of the land and water surface or in the relative distribution of land and sea may affect the radiation balance and hence the climate.

On land, variations in regional snow cover may trigger important climatic fluctuations. This is because an extensive snow cover has a refrigerating effect on the atmosphere (Chapter 4). Fresh-fallen snow typically reflects 80% or more of incident solar radiation, thereby substantially reducing the amount of solar heating and lowering the daily maximum temperature. Snow is also an excellent emitter of infrared radiation, so heat is quickly radiated off to space at night, especially on nights when skies are clear. Because of this radiative feedback, a snow cover tends to be self-sustaining. This effect may be further enhanced by storm tracks that often follow the periphery of a regional snow cover, where horizontal air temperature gradients are relatively steep. This places the snow-covered area on the cold, snowy side of migrating cyclones, thereby enforcing the chill. Hence, an unusually extensive snow cover favors persistence of an anomalously cold episode.

Although large-scale changes in surface characteristics of the continents may affect climate, changes in ocean characteristics and circulation may be much more important. This follows from our discussion of radiative transfer within the Earth-atmosphere system (Chapter 2). Because ocean waters cover nearly three-quarters of the planet's surface and exhibit a very low albedo for solar radiation, the ocean is the principal absorber of insolation. Anything that alters this strong absorption, such as fluctuations in sea-ice cover, is likely to affect radiative equilibrium and climate.

As noted in Chapter 16, there are linkages between anomalies in sea-surface temperatures and atmospheric long-wave circulation patterns. Shifts in sea-surface temperature anomaly patterns may therefore alter the prevailing circulation and climate. The atmosphere-ocean system involves a two-way interaction, however, and determining which dominates is difficult: the ocean's influence on the atmosphere or the atmosphere's influence on the ocean. Atmospheric impacts on the ocean are relatively short-term, whereas the ocean's impacts on the atmosphere are relatively long-term because of the great thermal stability of the ocean.

## Factor Interaction

We have now identified and elaborated on many of the potential causes of climatic variability. Note, however, that this is not an exhaustive list of all possible controls of climatic variability. For example, subtle astronomical shifts may contribute to short-term climatic fluctuations. We have also treated the various climatic controls as if each acted independently of the others. This is clearly not the case. Within the Earth-atmosphere system, many factors are linked in complex cause-effect chains, as illustrated in Figure 20.14. Internal and external forces work together to bring about shifts in climate.

Factor interactions involve feedback loops that may at one extreme amplify (positive feedback) and at the other extreme weaken (negative feedback) climatic fluctuations (Figure 20.15). Consider an illustration. If rising levels of atmospheric carbon dioxide enhance the greenhouse effect, temperatures in the lower troposphere will rise, and the pack ice cover in polar regions will shrink. Less pack ice, in turn, lowers the albedo of polar seas, further warming the troposphere (positive feedback). On the other hand, higher temperatures at the Earth's surface mean more evaporation, which may result in thicker and more persistent cloud cover. Recall from Chapter 2 that clouds have both cooling and warming effects on the lower atmosphere. The relatively high albedo of cloud tops causes cooling, and cloud absorption and reradiation of terrestrial IR contributes to the greenhouse effect. Recent satellite measurements indicate that the cooling effect dominates. Thus, a thicker and more persistent cloud cover would cool the Earth's surface (negative feedback).

In part because of factor interaction, it is difficult to isolate simple cause-effect relationships. Even when a near-perfect match in periodicities occurs between some internal or external forcing (such as a volcanic eruption) and a climatic response (such as global cool-

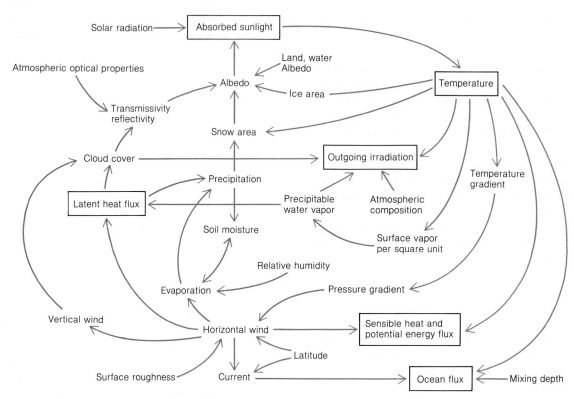

**FIGURE 20.14**
Climatic variability is influenced by the complex interplay of many processes operating within the Earth-atmosphere system. [From W. W. Kellogg and S. H. Schneider, "Climate Stabilization: For Better or for Worse?" *Science* 186 (1974):1164. Copyright 1974 by AAAS.]

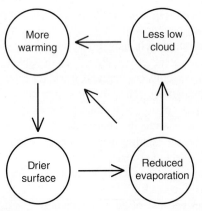

**FIGURE 20.15**
Schematic representation of some feedback loops operating within the Earth-atmosphere system. [From J. T. Houghton, G. J. Jenkins, and J. J. Ephraums (eds.), *Climate Change, The IPCC Scientific Assessment.* Cambridge, U. K.: Cambridge University Press, 1990, p. 143]

ing), there is no guarantee that the relationship is actually one of cause and effect. In the absence of a demonstrated physical linkage, the relationship may well be coincidental.

## The Climatic Future

What does the climatic future hold for us? Theoretically, we could take two approaches in attempting to answer this question. The first is numerical modeling of the atmosphere. Numerical models have been used to forecast both the climatic future and the climatic past. Typically, these predictions identify broad regions of expected positive and negative climatic anomalies. Earlier, we discussed the use of numerical models to predict the impact of rising levels of atmospheric carbon dioxide. Numerical models have also been used to *predict* the global climate of 18,000 years ago, the last glacial maximum.

Most modelers agree that global climate models are in need of considerable refinement. For one, today's models may not adequately simulate the role of oceans in climate variability. Second, they may not accurately portray local and regional conditions and thus may miss important feedback processes such as produced by clouds. The latter problem stems from the fact that climate models use data grids that partition the atmosphere into cells, each having an area about the size of Colorado. A mean state of the atmosphere (temperature, humidity, cloud cover, for example) is computed for each cell. This problem of low resolution can only be solved by faster and larger supercomputers. Furthermore, as discussed in this chapter's Special Topic, the accuracy of numerical models is limited by the chaotic nature of the Earth-atmosphere system.

A second approach to climate forecasting is to identify the various factors that may have contributed to past climatic fluctuations and to extrapolate their influence into the future. This is an empirical method. Atmospheric scientists have long probed climatic records in search of (1) cycles that might be extended into the future and (2) analogues that might provide clues to regional responses to large-scale climate change. As noted in Chapter 19, however, none of the statistically significant oscillations that appear in the climatic record has much practical value for climate (or weather) forecasting. Furthermore, the search for appropriate analogues for global warming has been fruitless thus far.

If all other controls of climate remain the same, then increasing concentrations of atmospheric carbon dioxide and other greenhouse gases are likely to cause global warming. To what extent all other controls remain the same, however, is open to question. The climatic future will evidently remain a mystery until we have more accurate numerical models of the atmosphere, denser weather observation networks, and a better understanding of how climatic controls interact. It is likely that the climatic future, like the climatic past, will be shaped by many interacting factors both external and internal to the Earth-atmosphere system. At present, we are confident only that our climate will vary (as it has in the past), but precisely how it will vary is simply not known.

## Conclusions

The interaction of many factors causes climate to vary on many time scales. Although we can isolate specific climatic controls that are internal or external to the Earth-atmosphere system, our understanding of how these controls interact is far from complete. This state of the art limits the ability of atmospheric scientists to forecast the climatic future. We should therefore be wary of simplistic scenarios of the climatic future, for not enough is known about the causes of climatic variations.

Continued research on climate is needed. It is reasonable to assume that climatic change is controlled by physical laws; that is, variations in climate are not arbitrary, random events. As scientists more fully comprehend the laws governing climatic variability, their ability to predict the climatic future will improve. Meanwhile, trends in climate must be monitored closely, especially in view of the strong dependence of our food and water supplies and energy demand on climate. In spite of the scientific uncertainty, some experts argue that we should plan now for the worst possible future climate scenarios.

# Chaos and Predicting the Climatic Future

Can the fluttering of a butterfly's wings in Brazil eventually spawn a tornado in Texas? This question, originally attributed to Edward N. Lorenz, an MIT meteorologist, provides the key to understanding the limits to weather and climate forecasting and why modern science can fly a person to the Moon but cannot accurately predict the weather more than 10 days from now. The so-called *butterfly effect* refers to the fact that future states of certain systems (e.g., the Earth-atmosphere system) are highly sensitive to initial conditions. Even the slightest error in initial conditions can rapidly magnify with time making long-range forecasting untenable.

A system consists of a number of components that function and interact in an orderly manner. The human respiratory and digestive systems are familiar examples. A *dynamical system* is one whose evolution from an initial state can be described by one or more mathematical equations. The Earth-atmosphere is a dynamical system and, as noted elsewhere in this book, meteorologists have built computerized numerical models that attempt to predict its future behavior.

As the forecast period lengthens beyond a few days or so, however, weather forecasts derived from numerical models quickly go awry. The problem is not so much the accuracy of the component equations as it is the imprecise description of the initial state of the system. That description is based on weather observations from a network of widely spaced stations utilizing instruments whose measurements are not always reliable or representative. The system is so sensitive that even slight changes in initial conditions can give rise to totally different forecasts. Hence, meteorologists cannot accurately forecast the weather more than 10 days from now. On the other hand, scientists know much more precisely the initial conditions and governing physical laws that enabled them to send a manned rocket to the Moon.

Researchers with the National Weather Service have experimented with two techniques to detect the butterfly effect: (1) ensemble forecasting and (2) model comparison. In the first, a numerical model generates several forecasts, each based on a slightly different set of initial conditions. If the forecasts are consistent, then the forecast is considered reliable. In the second technique, forecasts generated by several different models are compared. If they all agree, then the forecast is issued with relatively high confidence. If, however, through either technique, forecasts are inconsistent, then any one forecast is considered to be unreliable.

Application of chaos theory to the Earth-atmosphere system provides some insight as to the behavior of that system. Professor Lorenz pioneered this application in 1963, but it was not widely recognized or utilized until the late 1970s and early 1980s.

According to *chaos theory,* internal instabilities cause complex behavior in a system. Because its initial conditions are imperfectly known, the Earth-atmosphere is a chaotic dynamical system. The term chaotic should not be construed as implying random behavior, however. Certain physical laws govern the system so that there are boundaries within which chaos takes place, that is, the chaos has a certain underlying order or structure. For example, in the Northern Hemisphere, winds about a cyclone will always blow in a counterclockwise direction.

A dynamical system whose initial conditions are known with precision may be drawn toward some stable state known as an *attractor.* Hence, an attractor imposes predictability. In a chaotic dynamical system, on the other hand, there is some range of possible future states that is encompassed by so-called *strange attractors*. Strange attractors contain the system's preferred behaviors and hence, if identified, can be used to predict the future state of the system.

Not surprisingly, some scientists are searching for strange attractors, that is, the structure in complex climate data. Others are searching for multiple attractors. The latter effort is inspired by the fact that the atmosphere shifts between a finite number of circulation patterns. This may imply that the system follows one attractor for a period of time and then shifts to another.

The dimensions of a strange attractor indicate the complexity of a system. In effect, the number of dimensions provides modelers with an estimate of the number of variables needed to describe the behavior of the system. From studies to date, the Earth-atmosphere system appears to have a relatively high dimensional strange attractor. Chaos theory thus raises the question as to whether there are internal constraints on the ultimate predictability of weather and climate. It also prompts the question as to whether the recent global warming trend might be merely an expression of the natural variability of the Earth-atmosphere system.

# MATHEMATICAL NOTE

## Radiative Equilibrium and Climatic Change

We can express the condition of radiative equilibrium within the Earth-atmosphere system in a mathematical expression that equates energy input to energy output. Then by perturbing the variables in the equation, we can simulate climatic change.

Radiation incident on the Earth-atmosphere system is given by the solar constant, $S$. A portion of this energy is reflected or scattered back into space, the amount depending on the planetary albedo, $\alpha$. The energy actually available to drive the atmosphere is

$$S(1 - \alpha)$$

Planet Earth intercepts this energy as a disk having an area of $\pi R^2$, where $R$ is the radius of the Earth. The net energy input is then given by

$$(\pi R^2)S(1 - \alpha)$$

Energy is emitted by the Earth-atmosphere system in the form of infrared radiation. If we assume the Earth-atmosphere system to be a perfect radiator, then we can describe the energy output by the *Stefan–Boltzmann law* (see the Mathematical Note at the end of Chapter 2). This law states that the total radiational energy output (at all wavelengths) of a blackbody is proportional to the radiating temperature, $T$ (in kelvins), raised to the fourth power, that is, $T^4$. In this relationship, the constant of porportionality is $\sigma$, the Stefan–Boltzmann constant. However, the Earth-atmosphere system is not quite a blackbody, so we must introduce a correction factor to account for its actual radiating properties. The correction factor, $\epsilon$, is called the *effective emissivity*. The energy output of the Earth-atmosphere system is therefore

$$\epsilon \sigma T^4$$

This energy is emitted from the entire surface area $(4\pi R^2)$ of the spherical Earth. The total energy output is thus given by

$$(4\pi R^2)\epsilon \sigma T^4$$

At radiative equilibrium, what comes in (energy input) must equal what goes out (energy output), so

$$(\pi R^2)S(1 - \alpha) = (4\pi R^2)\epsilon \sigma T^4$$

This simplifies to

$$S(1 - \alpha) = 4\epsilon \sigma T^4$$

Solving for $T$, the temperature of the Earth-atmosphere system at radiative equilibrium, we have

$$T = [S(1 - \alpha)/4\epsilon \sigma]^{1/4}$$

The radiative equilibrium temperature thus depends on (1) the solar constant, (2) the planetary albedo, and (3) the effective emissivity. A change in any one or any combination of these variables will change the value or $T$ and hence the climate. Consider some illustrations.

First, keep $\alpha$ and $\epsilon$ constant. By mathematical manipulation, we can show that a 1% change in the solar constant translates into a 0.6 Celsius degree (1.1 Fahrenheit degrees) change in radiative equilibrium temperature. This does not appear to be a major temperature change, but recall that climatic trends are geographically nonuniform in magnitude (and direction). In some localities, therefore, this temperature change will be amplified considerably. If we keep $S$ and $\epsilon$ constant and increase the planetary albedo from its present value of 31% up to 36% (perhaps by increasing the ocean's ice cover), the radiative equilibrium temperature will drop by about 4.5 Celsius degrees (8.1 Fahrenheit degrees).

As a third illustration, we could hold $S$ and $\alpha$ constant and vary the effective emissivity. As noted earlier, the effective emissivity depends on the radiative properties of the Earth-atmosphere system, so any change in the chemistry of the Earth-atmosphere system could alter $\epsilon$. Such chemical changes might include, for example, alterations in levels of greenhouse gases.

## Key Terms

| | |
|---|---|
| solar constant | **Maunder minimum** |
| solar irradiance | **Spörer minimum** |
| sunspot | **Milankovitch cycles** |
| umbra | **greenhouse effect** |
| penumbra | **turbidity** |
| photosphere | |

## Summary Statements

☐ Climatic variability may be explained by forces both internal and external to the Earth-atmosphere system that affect the global radiation balance.

☐ One approach to explaining climatic variability is to match an appropriate cause with a specific climatic oscillation on the basis of the period of that oscillation. Such a match is, however, no guarantee of a cause-effect relationship.

☐ The global radiation balance could shift to a new equilibrium as a consequence of changes in the solar constant, planetary albedo, or gas and aerosol content of the atmosphere.

☐ Measurements by instruments aboard satellites indicate that the solar constant is not constant. Changes in solar irradiance may be linked to sunspot activity.

☐ The precise linkage between sunspots and climate variability is not known, although there are correspondences between prolonged sunspot minima and relatively cool periods and between the double sunspot cycle and drought on the western High Plains.

☐ Regular long-term changes in Earth-sun geometry (precession, axial tilt, eccentricity) may explain large-scale glacial-interglacial climatic shifts as recorded in deep-sea sediment cores.

☐ The potential impact of violent volcanic eruptions on large-scale climate is negligible unless the eruptive cloud is relatively rich in sulfur oxide gases. In the stratosphere, these gases convert to sulfate particles and sulfuric acid droplets that increase the planetary albedo.

☐ Fossil fuel combustion and, to a much lesser extent, deforestation are raising atmospheric carbon dioxide levels. Uncompensated, this trend will likely intensify the greenhouse effect. $CO_2$-induced warming may be enhanced by rising levels of other infrared-absorbing trace gases that are active within atmospheric windows.

☐ The potential global climatic impact of elevated aerosol levels in the troposphere is not known.

☐ Changes in the radiative characteristics of the Earth's land and water surfaces or in the relative distribution of land and sea may influence the global radiation balance and, hence, the climate.

☐ Factor interactions involve feedback loops that might at one extreme amplify (positive feedback) and at the other extreme weaken (negative feedback) climatic oscillations.

☐ With regard to the climatic future, atmospheric scientists are confident only that the climate will vary (as it has in the past), but precisely how it will vary is simply not known.

## Review Questions

1. What is meant by the globe's radiative equilibrium?
2. Is the *solar constant* really constant? How do we know?
3. What is a sunspot? Describe its structure.
4. Is the sunspot record really periodic in a rigorous statistical sense? Explain why or why not.
5. What is the Maunder minimum, and how might it have influenced climate?
6. What are the Milankovitch cycles?
7. How does the Milankovitch theory seek to explain the major climatic shifts of the Ice Age?
8. Identify some weaknesses of the Milankovitch theory.
9. How might violent volcanic eruptions influence global climate?
10. How did the 1980 eruption of Mount St. Helens affect the weather downwind?
11. Identify those climatic forcing phenomena that might contribute to short-term variations in climate, that is, variations over years and decades.
12. Evaluate humankind's contributions thus far to climatic variability. How might this change in the future?
13. What is the greenhouse effect? Identify the principal greenhouse gas.
14. What human activities are contributing to the upward trend in atmospheric carbon dioxide concentration?
15. How does deforestation contribute to the buildup of atmospheric $CO_2$?
16. How are numerical models of the atmosphere used to predict the climatic consequences of increasing atmospheric $CO_2$? What are the results of such experiments?
17. Methane, nitrous oxide, and CFCs occur in very low concentrations in the atmosphere and yet their contributions to the greenhouse effect are quite significant. Explain why.

**18.** How is the ocean involved in the controversy now surrounding the potential climatic impacts of rising carbon dioxide levels?

**19.** Explain why aerosols injected into the stratosphere may have a more prolonged impact on climate than aerosols that are confined to the troposphere.

**20.** Provide examples of positive and negative feedback in the climatic system.

## Questions for Critical Thinking

**1.** A high statistical correlation between two factors does not necessarily indicate a cause-effect relationship. Give an example of how this statement applies to the search for the causes of climatic variability.

**2.** If the Milankovitch theory is correct, then the next ice age will peak about 23,000 years from now. Does this imply that each successive year from now on will get progressively colder? Explain your response.

**3.** How does the geographic nonuniformity of climatic trends complicate the search for causes of climatic variability?

**4.** In attempting to explain climatic variability, we cannot focus simply on changes that occur in the atmosphere alone. We must also consider changes in the oceans and in the Earth's land-surface characteristics. Explain.

**5.** From what you have learned about conditions required for hurricane development (Chapter 15), speculate on the possible effects of $CO_2$-induced global warming on the intensity and frequency of hurricanes.

## Selected Readings

AMERICAN GEOPHYSICAL UNION. *Volcanism and Climate Change*. AGU Special Report. Washington, DC: American Geophysical Union, 1992, 27 pp. Summarizes the impact of volcanic eruptions on hemispheric and global-scale climate.

AMERICAN METEOROLOGICAL SOCIETY. "Policy Statement of the American Meteorological Society on Global Climate Change," *Bulletin of the American Meteorological Society* 72 (1991):57–59. Calls for more research on the nature of climate variability.

CHAPMAN, W. L., AND J. E. WALSH. "Recent Variations of Sea Ice and Air Temperature in High Latitudes," *Bulletin of the American Meteorological Society* 74 (1993):33–47. Concludes that warming has reduced sea ice cover in both the Arctic and Antarctic.

COHEN, S. J., AND T. R. ALLSOPP. "The Potential Impacts of a Scenario of $CO_2$-Induced Climatic Change on Ontario, Canada," *Journal of Climate* 1 (1988):669–681. Provides a fascinating summary of the potential economic impacts of climatic change that may accompany a doubling of atmospheric carbon dioxide.

COVEY, C. "The Earth's Orbit and the Ice Ages," *Scientific American* 250, No. 2 (1984):58–66. Describes the Milankovitch theory of the link between Earth-sun geometry and major fluctuations in continental-scale glaciers.

HOUGHTON, J. T., ET AL. (EDS.). *Climate Change, The IPCC Scientific Assessment*. New York: Cambridge University Press, 1990, 365 pp. Summarizes in some detail what is understood about the nature of climatic variability and prospects for the climatic future.

KERR, R. A. "Pollutant Haze Cools the Greenhouse," *Science* 255 (1992):682–683. Considers the possible countering of greenhouse warming by aerosols of anthropogenic origin.

RAMANATHAN, V. "The Greenhouse Theory of Climate Change: A Test by an Inadvertent Global Experiment," *Science* 240 (1988):293–299. Thoroughly reviews the possible climatic impacts of rising levels of atmospheric $CO_2$ and other greenhouse gases.

SCHNEIDER, S. H. "Climate Modeling," *Scientific American* 256, No. 5 (1987):72–80. Presents an interesting critique of climate modeling as applied to projected trends in atmospheric carbon dioxide level.

SCHNEIDER, S. H. *Global Warming*. San Francisco, CA: Sierra Club Books, 1989, 317 pp. Discusses the causes and possible consequences of an enhanced greenhouse effect.

SCUDERI, L. A. "A 2,000-Year Tree Ring Record of Annual Temperature in the Sierra Nevada Mountains," *Science* 259 (1993):1433–1439. Describes a close correlation between a reconstructed solar activity index and temperature.

TURCO, R. P., ET AL. "Climate and Smoke: An Appraisal of Nuclear Winter," *Science* 247 (1990):166–175. Updates what is understood about the potential global climatic impact of nuclear war.

WHITE, R. M. "Greenhouse Policy and Climate Uncertainty," *Bulletin of the American Meteorological Society* 70 (1989):1123–1127. Describes the problems involved when public policy-making decisions are based on uncertain climate forecasts.

# Milestones in the History of Atmospheric Science

| | | |
|---|---|---|
| ca. | 500 B.C. | Parmenides classified world climates by latitude as torrid, temperate, or frigid. |
| ca. | 400 B.C. | Rainfall measured in India. |
| | 334 B.C. | Aristotle produced his *Meteorologica,* the first work on the atmospheric sciences. |
| | 61 A.D. | Seneca complained of the air pollution of Rome. |
| | 1442 | Rain gauges used in Korea. |
| ca. | 1500 | Leonardo Da Vinci conceived of a hygrometer. |
| | 1593 | Galileo invented the thermometer. |
| | 1643 | Torricelli invented the barometer. |
| | 1644 | Rev. John Campanius Holm took the first weather records in America near present site of Wilmington, DE. |
| | 1658 | Florin Périer ascended the Puy-de-Dome, France, and demonstrated that air pressure drops with increasing altitude. |
| | 1660 | Von Guericke noted that a severe storm follows a sudden drop in barometric pressure. |
| ca. | 1670 | Mercury used in thermometers for the first time. |
| | 1683 | Edmund Halley published first comprehensive map of winds. |
| | 1686 | Edmund Halley proposed explanation for monsoons. |
| | 1687 | Isaac Newton developed his three laws of motion. |
| | 1714 | Gabriel Fahrenheit introduced the Fahrenheit temperature scale. |
| | 1735 | George Hadley proposed the Hadley cell circulation. |
| | 1736 | Anders Celsius introduced the Celsius temperature scale. |
| | 1743 | Benjamin Franklin deduced the progressive movement of a hurricane. |

| | |
|---|---|
| 1749 | In Glasgow, Scotland, Alexander Wilson used thermometers attached to a kite to obtain the first temperature profile of the lower atmosphere. |
| 1752 | Benjamin Franklin demonstrated that lightning is an electrical phenomenon. |
| 1760 | Joseph Black formulated the concept of specific heat. |
| 1768 | Johann Heinrich Lambert, a German physicist, invented the hygrometer. |
| 1781 | Systematic weather records began at New Haven, CT. |
| 1800 | Sir William Herschel discovered energy transfer in the infrared. |
| 1802 | Luke Howard developed his classification of cloud types. |
| 1804 | J. L. Gay-Lussac conducted the first manned balloon exploration of the atmosphere. |
| 1806 | Sir Francis Beaufort developed a wind scale. |
| 1825 | E. F. August developed the psychrometer. |
| 1835 | Gustav Gaspard de Coriolis demonstrated quantitatively the effects of the Earth's rotational forces. |
| 1842 | Johann Christian Doppler first explained Doppler effect. |
| 1843 | Sunspot cycle discovered. |
| 1844 | Aneroid barometer invented. |
| 1856 | William Ferrel proposed a scheme of the general circulation of the atmosphere using three cells. |
| 1857 | Lorin Blodget published *Climatology of the United States.* |
| 1862 | James Glaisher and Henry Coxwell set manned balloon altitude record of 9000 m (29,000 ft) in England. |

| | |
|---|---|
| 1869 | Cleveland Abbe in Cincinnati prepared first regular weather maps for part of the United States. |
| 1878 | The International Meteorological Organization (IMO) founded. |
| 1884 | Ludwig Boltzmann derived the Stefan–Boltzmann law. |
| 1894 | Wilhelm Wien developed the displacement law of radiation. |
| 1894 | First sounding of the atmosphere using a self-recording thermometer attached to a kite, made at Blue Hill Observatory, Milton, MA. |
| 1916 | Pyranometer for measuring global radiation developed. |
| 1917 | Vilhelm Bjerknes formulated polar front theory. |
| 1918 | Wladimir Köppen introduced his climate classification scheme. |
| 1924 | Sir Gilbert Walker discovered the southern oscillation. |
| ca. 1928 | First radiosonde developed. |
| 1929 | Robert H. Goddard conducted the first rocket probe of the atmosphere. |
| 1933 | Tor Bergeron published paper on "Physics of Cloud and Precipitation." |
| 1937 | Carl-Gustaf Rossby introduced techniques for forecasting upper westerly waves. |
| 1941 | Weather radar developed. |

| | |
|---|---|
| 1944 | Hurd Curtis Willett produced atmospheric cross sections showing the jet stream. |
| 1946 | Vincent J. Schaefer and Irving Langmuir performed the first cloud-seeding experiments. |
| 1946 | John von Neumann began mathematical modeling of weather. |
| 1947 | World Meteorological Organization (WMO) founded. |
| 1947 | First photographs of Earth's cloud cover obtained by a V2 rocket at altitudes of 110 to 165 km (68 to 102 mi). |
| 1948 | Rocket probes of the upper atmosphere begun. |
| 1957 | $CO_2$ measurements at Mauna Loa and Antarctica begun. |
| 1960 | First weather satellite, TIROS-I, orbited by United States. |
| 1963 | Edward N. Lorenz applied chaos theory to meteorology. |
| 1966 | Jacob Bjerknes demonstrated relationship between El Niño and southern oscillation. |
| 1967 | Verner E. Suomi, the University of Wisconsin–Madison, processed first geostationary satellite image. |
| 1985 | Antarctic ozone hole reported. |

Modified after J. E. Griffiths, "A Chronology of Items of Meteorological Interest," *Bulletin of the American Meteorological Society* 58 (1977):1058–1067.

# The Standard Atmosphere

| Altitude (km) | Temperature (°C) | Pressure (mb) | $P/P_0$* | Density $(kg/m^3)$ | $D/D_0$* |
|---|---|---|---|---|---|
| 30.00 | −46.60 | 11.97 | 0.01 | 0.02 | 0.02 |
| 25.00 | −51.60 | 25.49 | 0.03 | 0.04 | 0.03 |
| 20.00 | −56.50 | 55.29 | 0.05 | 0.09 | 0.07 |
| 19.00 | −56.50 | 64.67 | 0.06 | 0.10 | 0.08 |
| 18.00 | −56.50 | 75.65 | 0.07 | 0.12 | 0.09 |
| 17.00 | −56.50 | 88.49 | 0.09 | 0.14 | 0.12 |
| 16.00 | −56.50 | 103.52 | 0.10 | 0.17 | 0.14 |
| 15.00 | −56.50 | 121.11 | 0.12 | 0.20 | 0.16 |
| 14.00 | −56.50 | 141.70 | 0.14 | 0.23 | 0.19 |
| 13.00 | −56.50 | 165.79 | 0.16 | 0.27 | 0.22 |
| 12.00 | −56.50 | 193.99 | 0.19 | 0.31 | 0.25 |
| 11.00 | −56.40 | 226.99 | 0.22 | 0.37 | 0.30 |
| 10.00 | −49.90 | 264.99 | 0.26 | 0.41 | 0.34 |
| 9.50 | −46.70 | 285.84 | 0.28 | 0.44 | 0.36 |
| 9.00 | −43.40 | 308.00 | 0.30 | 0.47 | 0.38 |
| 8.50 | −40.20 | 331.54 | 0.33 | 0.50 | 0.40 |
| 8.00 | −36.90 | 356.51 | 0.35 | 0.53 | 0.43 |
| 7.50 | −33.70 | 382.99 | 0.38 | 0.56 | 0.45 |
| 7.00 | −30.50 | 411.05 | 0.41 | 0.59 | 0.48 |
| 6.50 | −27.20 | 440.75 | 0.43 | 0.62 | 0.50 |
| 6.00 | −23.90 | 472.17 | 0.47 | 0.66 | 0.54 |
| 5.50 | −20.70 | 505.39 | 0.50 | 0.70 | 0.57 |
| 5.00 | −17.50 | 540.48 | 0.53 | 0.74 | 0.60 |
| 4.50 | −14.20 | 577.52 | 0.57 | 0.78 | 0.63 |
| 4.00 | −11.00 | 616.60 | 0.61 | 0.82 | 0.67 |
| 3.50 | −7.70 | 657.80 | 0.65 | 0.86 | 0.70 |
| 3.00 | −4.50 | 701.21 | 0.69 | 0.91 | 0.74 |
| 2.50 | −1.20 | 746.91 | 0.74 | 0.96 | 0.78 |
| 2.00 | 2.00 | 795.01 | 0.78 | 1.01 | 0.82 |
| 1.50 | 5.30 | 845.59 | 0.83 | 1.06 | 0.86 |
| 1.00 | 8.50 | 898.76 | 0.89 | 1.11 | 0.91 |
| 0.50 | 11.80 | 954.61 | 0.94 | 1.17 | 0.95 |
| 0.00 | 15.00 | 1013.25 | 1.00 | 1.23 | 1.00 |

*$P/P_0$ = ratio of air pressure to sea-level value; $D/D_0$ = ratio of air density to sea-level value.

# Weather Map Symbols

By international agreement, a standard set of symbols is plotted on weather maps to represent the state of the atmosphere at a specified time. This standardization facilitates the international exchange of weather information. Presented here is an abridged listing of weather symbols. We also include in this appendix a hurricane tracking chart.

## EXPLANATION OF CODES

$$
\begin{array}{c}
\text{ff} \\
\text{TT dd } C_M^{C_H} \text{PPP} \\
\text{VV ww } \textcircled{N} \pm \text{pp a} \\
T_d T_d \, C_L \, N_h \, W \, R_t \\
\text{h} \quad \text{RR}
\end{array}
$$

Symbolic station model

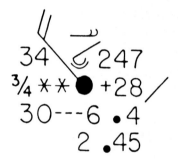

Sample report

| | |
|---|---|
| N | Total cloud cover—Table E |
| dd | Wind direction |
| ff | Wind speed in knots or mi/hr—Table F |
| VV | Visibility in miles |
| ww | Present weather—Table H |
| W | Past weather—Table H |
| PPP | Barometric pressure reduced to sea level (add an initial 9 or 10 and place a decimal point to the left of last number) |
| TT | Current air temperature in °F |
| $N_h$ | Fraction of sky covered by low or middle clouds—Table E (ranges from 0 for no clouds to 9 for sky obscured) |
| $C_L$ | Low clouds or clouds with vertical development—Table C |

| | |
|---|---|
| h | Height in feet of the base of the lowest clouds—Table D |
| $C_M$ | Middle clouds—Table C |
| $C_H$ | High clouds—Table C |
| $T_d T_d$ | Dew point temperature in °F |
| a | Pressure tendency—Table A |
| pp | Pressure change in mb in preceding 3 hr (+28 = +2.8) |
| RR | Amount of precipitation in last 6 hr |
| $R_t$ | Time precipitation began or ended (0 = none; 1 = <1 hr ago; 2 = 1–2 hr ago; 3 = 2–3 hr ago; 4 = 3–4 hr ago; 5 = 4–5 hr ago; 6 = 5–6 hr ago; 7 = 6–12 hr ago; 8 = >12 hr ago; 9 = unknown) |

477

**TABLE A**
**Air Pressure Tendency**

| | | | |
|---|---|---|---|
| ⟋ | Rising, then falling; same as or higher than 3 hours ago | ⟍ | Falling, then rising; same as or lower than 3 hours ago |
| ⌐ | Rising, then steady; or rising, then rising more slowly | ⌐\ | Falling, then steady; or falling, then falling more slowly |
| ⟋ | Rising steadily, or unsteadily | ⟍ | Falling steadily, or unsteadily |
| ⩗ | Falling or steady, then rising; or rising, then rising more rapidly | ⋀ | Steady or rising, then falling; or falling, then falling more rapidly |
| — | Steady; same as 3 hours ago | | |

**TABLE B**
**Cloud Abbreviations**

| | |
|---|---|
| St | stratus |
| Fra | fractus |
| Sc | stratocumulus |
| Ns | nimbostratus |
| As | altostratus |
| Ac | altocumulus |
| Ci | cirrus |
| Cs | cirrostratus |
| Cc | cirrocumulus |
| Cu | cumulus |
| Cb | cumulonimbus |

**TABLE C**
**Cloud Types**

Cu of fair weather, little vertical development and seemingly flattened

Cu of considerable development, generally towering, with or without other Cu or Sc bases all at same level

Cb with tops lacking clear-cut outlines, but distinctly not cirriform or anvil shaped; with or without Cu, Sc, or St

Sc formed by spreading out of Cu; Cu often present also

Sc not formed by spreading out of Cu

St or StFra, but no StFra of bad weather

StFra and/or CuFra of bad weather (scud)

Cu and Sc (not formed by spreading out of Cu) with bases at different levels

Cb having a clearly fibrous (cirriform) top, often anvil shaped, with or without Cu, Sc, St, or scud

Thin As (most of cloud layer semitransparent)

Thick As, greater part sufficiently dense to hide sun (or moon), or Ns

Thin Ac, mostly semitransparent; cloud elements not changing much and at a single level

Thin Ac in patches; cloud elements continually changing and/or occurring at more than one level

Thin Ac in bands or in a layer gradually spreading over sky and usually thickening as a whole

Ac formed by the spreading out of Cu or Cb

Double-layered Ac, or a thick layer of Ac, not increasing; or Ac with As and/or Ns

Ac in the form of Cu-shaped tufts or Ac with turrets

Ac of a chaotic sky, usually at different levels; patches of dense Ci usually present also

Filaments of Ci, or "mares tails," scattered and not increasing

Dense Ci in patches or twisted sheaves, usually not increasing, sometimes like remains of Cb; or towers or tufts

Dense Ci, often anvil shaped, derived from or associated with Cb

Ci, often hook shaped, gradually spreading over the sky and usually thickening as a whole

Ci and Cs, often in converging bands, or Cs alone; generally overspreading and growing denser; the continuous layer not reaching 45° altitude

Ci and Cs, often in converging bands, or Cs alone; generally overspreading and growing denser; the continuous layer exceeding 45° altitude

Veil of Cs covering the entire sky

Cs not increasing and not covering entire sky

Cc alone or Cc with some ci or Cs, but the Cc being the main cirriform cloud

**TABLE D**
**Height of Base of Lowest Cloud**

| Code | Feet | Meters |
|---|---|---|
| 0 | 0–149 | 0–49 |
| 1 | 150–299 | 50–99 |
| 2 | 300–599 | 100–199 |
| 3 | 600–999 | 200–299 |
| 4 | 1000–1999 | 300–599 |
| 5 | 2000–3499 | 600–999 |
| 6 | 3500–4999 | 1000–1499 |
| 7 | 5000–6499 | 1500–1999 |
| 8 | 6500–7999 | 2000–2499 |
| 9 | 8000 or above or no clouds | 2500 or above or no clouds |

**TABLE E**
**Cloud Cover**

| Symbol | Description |
|---|---|
| 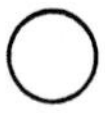 ○ | No clouds |
|  ◔ (one tenth) | One tenth or less |
|  | Two tenths or three tenths |
|  | Four tenths |
|  ◑ | Five tenths |
|  | Six tenths |
|  | Seven tenths or eight tenths |
|  | Nine tenths or overcast with openings |
|  ● | Completely overcast (ten tenths) |
|  ⊗ | Sky obscured |

**TABLE F**
**Wind Speed**

| Symbol | Knots | Miles per hour | Kilometers per hour |
|---|---|---|---|
|  | Calm | Calm | Calm |
| | 1–2 | 1–2 | 1–3 |
| | 3–7 | 3–8 | 4–13 |
| | 8–12 | 9–14 | 14–19 |
|  | 13–17 | 15–20 | 20–32 |
| | 18–22 | 21–25 | 33–40 |
|  | 23–27 | 26–31 | 41–50 |
|  | 28–32 | 32–37 | 51–60 |
|  | 33–37 | 38–43 | 61–69 |
|  | 38–42 | 44–49 | 70–79 |
|  | 43–47 | 50–54 | 80–87 |
| | 48–52 | 55–60 | 88–96 |
|  | 53–57 | 61–66 | 97–106 |
|  | 58–62 | 67–71 | 107–114 |
|  | 63–67 | 72–77 | 115–124 |
|  | 68–72 | 78–83 | 125–134 |
|  | 73–77 | 84–89 | 135–143 |
|  | 103–107 | 119–123 | 192–198 |

**TABLE G**
**Fronts**

Fronts are shown on surface weather maps by the symbols below. (Arrows—not shown on maps—indicate direction of motion of front.)

Cold front (surface)

Warm front (surface)

Occluded front (surface)

Stationary front (surface)

Warm front (aloft)

Cold front (aloft)

**TABLE H**
**Weather Conditions**

| | | | | |
|---|---|---|---|---|
| Cloud development NOT observed or NOT observable during past hour | Clouds generally dissolving or becoming less developed during past hour | State of sky on the whole unchanged during past hour | Clouds generally forming or developing during past hour | Visibility reduced by smoke |
| Light fog (mist) | Patches of shallow fog at station, NOT deeper than 6 feet on land | More or less continuous shallow fog at station, NOT deeper than 6 feet on land | Lightning visible, no thunder heard | Precipitation within sight, but NOT reaching the ground |
| Drizzle (NOT freezing) or snow grains (NOT falling as showers) during past hour, but NOT at time of observation | Rain (NOT freezing and NOT falling as showers) during past hour, but NOT at time of observation | Snow (NOT falling as showers) during past hour, but NOT at time of observation | Rain and snow or ice pellets (NOT falling as showers) during past hour, but NOT at time of observation | Freezing drizzle or freezing rain (NOT falling as showers) during past hour, but NOT at time of observation |
| Slight or moderate dust storm or sandstorm, has decreased during past hour | Slight or moderate dust storm or sandstorm, no appreciable change during past hour | Slight or moderate dust storm or sandstorm has begun or increased during past hour | Severe dust storm or sandstorm, has decreased during past hour | Severe dust storm or sandstorm, no appreciable change during past hour |
| Fog or ice fog at distance at time of observation, but NOT at station during past hour | Fog or ice fog in patches | Fog or ice fog, sky discernible, has become thinner during past hour | Fog or ice fog, sky NOT discernible, has become thinner during past hour | Fog or ice fog, sky discernible, no appreciable change during past hour |
| Intermittent drizzle (NOT freezing), slight at time of observation | Continuous drizzle (NOT freezing), slight at time of observation | Intermittent drizzle (NOT freezing), moderate at time of observation | Continuous drizzle (NOT freezing), moderate at time of observation | Intermittent drizzle (NOT freezing), heavy at time of observation |
| Intermittent rain (NOT freezing), slight at time of observation | Continuous rain (NOT freezing), slight at time of observation | Intermittent rain (NOT freezing), moderate at time of observation | Continuous rain (NOT freezing), moderate at time of observation | Intermittent rain (NOT freezing), heavy at time of observation |
| Intermittent fall of snowflakes, slight at time of observation | Continuous fall of snowflakes, slight at time of observation | Intermittent fall of snowflakes, moderate at time of observation | Continuous fall of snowflakes, moderate at time of observation | Intermittent fall of snowflakes, heavy at time of observation |
| Slight rain shower(s) | Moderate or heavy rain shower(s) | Violent rain shower(s) | Slight shower(s) of rain and snow mixed | Moderate or heavy shower(s) of rain and snow mixed |
| Moderate or heavy shower(s) of hail, with or without rain, or rain and snow mixed, not associated with thunder | Slight rain at time of observation; thunderstorm during past hour, but NOT at time of observation | Moderate or heavy rain at time of observation; thunderstorm during past hour, but NOT at time of observation | Slight snow, or rain and snow mixed, or hail at time of observation; thunderstorm during past hour, but NOT at time of observation | Moderate or heavy snow, or rain and snow mixed, or hail at time of observation; thunderstorm during past hour, but NOT at time of observation |

**TABLE H**
**Weather Conditions**

| Symbol | Description | Symbol | Description | Symbol | Description | Symbol | Description | Symbol | Description |
|---|---|---|---|---|---|---|---|---|---|
| ∞ | Haze | S | Widespread dust in suspension in the air, NOT raised by wind, at time of observation | $ | Dust or sand raised by wind at time of observation | (8) | Well-developed dust whirl(s) within past hour | (S→) | Dust storm or sandstorm within sight of or at station during past hour |
| )•( | Precipitation within sight, reaching the ground but distant from station | (•) | Precipitation within sight, reaching the ground, near to but NOT at station | R | Thunderstorm, but no precipitation at the station | V | Squall(s) within sight during past hour or at time of observation | ][ | Funnel cloud(s) within sight of station at time of observation |
| ∇] | Showers of rain during past hour, but NOT at time of observation | ※∇ | Showers of snow, or of rain and snow, during past hour, but NOT at time of observation | △∇ | Showers of hail, or of hail and rain, during past hour, but NOT at time of observation | ═] | Fog during past hour, but NOT at time of observation | R] | Thunderstorm (with or without precipitation) during past hour, but NOT at time of observation |
| ⇆ | Severe dust storm or sandstorm has begun or increased during past hour | → | Slight or moderate drifting snow, generally low (less than 6 ft) | ⇉ | Heavy drifting snow, generally low | ↑ | Slight or moderate blowing snow, generally high (more than 6 ft) | ⇈ | Heavy blowing snow, generally high |
| ═ | Fog or ice fog, sky NOT discernible, no appreciable change during past hour | ⌐═ | Fog or ice fog, sky discernible, has begun or become thicker during past hour | ⌐═ | Fog or ice fog, sky NOT discernible, has begun or become thicker during past hour | ⩛ | Fog depositing rime, sky discernible | ⩛ | Fog depositing rime, sky NOT discernible |
| ,, | Continuous drizzle (NOT freezing), heavy at time of observation | ⌒ | Slight freezing drizzle | ⌒) | Moderate or heavy freezing drizzle | •, | Drizzle and rain, slight | •, | Drizzle and rain, moderate or heavy |
| •• | Continuous rain (NOT freezing), heavy at time of observation | ⊙⌒ | Slight freezing rain | ⊙⌒• | Moderate or heavy freezing rain | •* | Rain or drizzle and snow, slight | *•* | Rain or drizzle and snow, moderate or heavy |
| ** | Continuous fall of snowflakes, heavy at time of observation | ↔ | Ice prisms (with or without fog) | ▵→ | Snow grains (with or without fog) | ⤬ | Isolated starlike snow crystals (with or without fog) | △ | Ice pellets or snow pellets |
| ∇ | Slight snow shower(s) | ∇ | Moderate or heavy snow shower(s) | △∇ | Slight shower(s) of snow pellets, or ice pellets with or without rain, or rain and snow mixed | △∇ | Moderate or heavy shower(s) of snow pellets, or ice pellets, or ice pellets with or without rain or rain and snow mixed | ▲∇ | Slight shower(s) of hail, with or without rain or rain and snow mixed, not associated with thunder |
| •/*R | Slight or moderate thunderstorm without hail, but with rain and/or snow at time of observation | △R | Slight or moderate thunderstorm, with hail at time of observation | •/*R | Heavy thunderstorm, without hail, but with rain and/or snow at time of observation | S→R | Thunderstorm combined with dust storm or sandstorm at time of observation | △R | Heavy thunderstorm with hail at time of observation |

# KEY TO AVIATION WEATHER OBSERVATIONS

| LOCATION IDENTIFIER TYPE AND TIME OF REPORT * | SKY AND CEILING | VISIBILITY WEATHER AND OBSTRUCTION TO VISION | SEA-LEVEL PRESSURE | TEMPERATURE AND DEW POINT | WIND | ALTIMETER SETTING | REMARKS AND CODED DATA |
|---|---|---|---|---|---|---|---|
| MCI SA Ø758 | 15 SCT M25 OVC | 1R-F | 132 | /58/56 | /1897 | /993/ | RØ1VR2ØV4Ø |

## SKY AND CEILING

Sky cover contractions are for each layer in ascending order. Figures preceding contractions are base heights in hundreds of feet above station elevation. Sky cover contractions used are:

**CLR** = Clear: Less than Ø.1 sky cover.
**SCT** = Scattered: Ø.1 to Ø.5 sky cover.
**BKN** = Broken: Ø.6 to Ø.9 sky cover.
**OVC** = Overcast: More than Ø.9 sky cover.

— = Thin (When prefixed to SCT, BKN, OVC).
—X = Partly obscured: Ø.9 or less of sky hidden by precipitation or obstruction to vision (bases at surface).
X = Obscured: 1.Ø sky hidden by precipitation or obstruction to vision (bases at surface).

A letter preceding the height of a base identifies a ceiling layer and indicates how ceiling height was determined. Thus:

E = Estimated
M = Measured
W = Vertical visibility into obscured sky
V = Immediately following the height of a base indicates a variable ceiling.

## VISIBILITY

Reported in statute miles and fractions
(V = Variable)

## WEATHER AND OBSTRUCTION TO VISION SYMBOLS

| | | | |
|---|---|---|---|
| A | Hail | IC | Ice crystals |
| BD | Blowing dust | IF | Ice-fog |
| BN | Blowing sand | IP | Ice pellets |
| BS | Blowing snow | IPW | Ice pellet showers |
| D | Dust | K | Smoke |
| F | Fog | L | Drizzle |
| GF | Ground fog | R | Rain |
| H | Haze | RW | Rain showers |

| | | | |
|---|---|---|---|
| S | Snow |
| SG | Snow grams |
| SP | Snow pellets |
| SW | Snow showers |
| T | Thunderstorms |
| T+ | Severe thunderstorm |
| ZL | Freezing drizzle |
| ZR | Freezing rain |

Precipitation intensities are indicated thus: - Light: (no sign) Moderate: + Heavy

**WIND** Direction in tens of degrees from true north, speed in knots; ØØØØ indicates calm. G indicates gusty. Q indicates Squalls. Peak wind speed in the past 1Ø minutes follows G or Q when gusts or squalls are reported. The contraction WSHFT, followed by GMT time group in remarks, indicates windshift and its time of occurrence. (Knots x 1.15 = statute mi/hr).

**EXAMPLES:** 3627 = wind from 36Ø Degrees at 27 knots;
3627G4Ø = wind from 36Ø Degrees at 27 knots. peak speed in gusts 4Ø knots

## ALTIMETER SETTING

The first figure of the actual altimeter setting is always omitted from the report.

## RUNWAY VISUAL RANGE (RVR)

RVR is reported from some stations. For planning purposes. the value range during 1Ø minutes prior to observations and based on runway light setting 5 are reported in hundreds of feet. Runway identification precedes RVR report.

## PILOT REPORTS (PIREPs)

When available. PIREPs in fixed-format may be appended to weather observations. PIREPs are designated by UA or UUA for urgent PIREPs.

## DECODED REPORT

Kansas City International: Record observation completed at Ø758 GMT 15ØØ feet scattered clouds. measured ceiling 25ØØ feet overcast. visibility 1 mile. light rain, fog. sea-level pressure 1Ø13.2 millibars. temperature 58ºF. dewpoint 56ºF. wind from 18ØØ. at 7 knots, altimeter setting 29.93 inches. Runway Ø1, visual range 2ØØØ feet lowest 4ØØØ feet highest in the past 1Ø minutes.

## * TYPE OF REPORT

SA = a scheduled record observation
SP = an unscheduled special observation indicating a significant change in one or more elements
RS = a scheduled record observation that also qualifies as a special observation

The designator for all three types of observations (SA. SP. RS) is followed by a 24 hour-clock-time-group in Greenwich Mean Time (GMT or Z).

**U.S. DEPARTMENT OF COMMERCE**—NATIONAL OCEANIC AND ATMOSPHERIC ADMINISTRATION—NATIONAL WEATHER SERVICE

# KEY TO AVIATION WEATHER FORECASTS

**TERMINAL FORECASTS** contain information for specific airports on expected ceiling, cloud heights, cloud amounts, visibility, weather and obstructions to vision, and surface wind. They are issued 3 times/day, amended as needed, and are valid for up to 24 hours. The last six hours of each forecast period are covered by a categorical statement indicating whether VFR, MVFR, IFR or LIFR conditions are expected (L in LIFR and M in MVFR indicate "low" and "marginal"). Terminal forecasts are written in the following form:

CEILING: Identified by the letter "C" (for lowest layer with cumulative sky cover greater than 5/10)
CLOUD HEIGHTS: In hundreds of feet above the station (ground)
SKY COVER AMOUNT (including any obscuration)
CLOUD LAYERS: Stated in ascending order of height
VISIBILITY: In statute miles (omitted if over 6 miles)
WEATHER AND OBSTRUCTION TO VISION: Standard weather and obstruction to vision symbols are used
SURFACE WIND: In tens of degrees and knots (omitted when less than 6 knots)

### EXAMPLE OF TERMINAL FORECAST

**DCA 221010:** DCA Forecast 22nd day of month - valid time 10Z-10Z.
**10SCT C18 BKN 5SW—3415G25 OCNL C8 X 1/2SW:** Scattered clouds at 1000 feet, ceiling 1800 feet broken, visibility 5 miles, light snow showers, surface wind from 340 degrees at 15 knots. Gusts to 25 knots, occasional ceiling 8 hundred feet sky totally obscured, visibility 1/2 mile in moderate snow showers.
**12Z C50 BKN 3312G22:** By 12Z becoming ceiling 5000 feet broken, surface wind 330 degrees at 12 knots. Gusts to 22.
**04Z MVFR CIG:** Last 6 hours of FT after 04Z marginal VFR due to ceiling.

**AREA FORECASTS** are 12-hour aviation forecasts plus a 6-hour categorical outlook prepared 3 times/day, with each section amended as needed, giving general descriptions of potential hazards, airmass and frontal conditions, icing and freezing level turbulence and low-level windshear and significant clouds and weather for an area the size of several states. Heights of cloud bases and tops, turbulence and icing are referenced ABOVE MEAN SEA LEVEL (MSL); unless indicated by Ceiling (CIG) or ABOVE GROUND LEVEL (AGL). Each SIGMET OR AIRMET affecting an FA area will also serve to amend the Area Forecast.

**SIGMET, AIRMET and CWA** messages (In-flight advisories) broadcast by FAA on NAVAID voice channels warn pilots of potentially hazardous weather. SIGMET's concern severe and extreme conditions of importance to all aircraft (i.e., icing, turbulence and dust storms/sandstorms or volcanic ash). Convective SIGMET's are issued for thunderstorms if they are sufficiently strong, wide spread or embedded. AIRMET's concern less severe conditions which may be hazardous to aircraft particularly smaller aircraft and less experienced or VFR only pilots. CWA's (Center Weather Advisories) concern both SIGMET and AIRMET type conditions described in greater detail and relating to a specific ARTCC area.

**WINDS AND TEMPERATURES ALOFT (FD) FORECASTS** are 6, 12, and 24-hour forecasts of wind direction (nearest 10° true N) and speed (knots) for selected flight levels. Forecast Temperatures Aloft (°C) are included for all but the 3000-foot level.

EXAMPLES OF WINDS AND TEMPERATURES ALOFT (FD) FORECASTS:
FD WBC 121645
BASED ON 121200Z DATA
VALID 130000Z FOR USE 2100-0600Z. TEMPS NEG ABV 24000 FT

|      | 3000 | 6000 | 9000 | 12000 | 18000 | 24000 | 30000 | 34000 | 39000 |
|------|------|------|------|-------|-------|-------|-------|-------|-------|
| BOS  | 3127 | 3425-17 | 3429-11 | 3421-16 | 3516-27 | 3512-38 | 311649 | 292451 | 283451 |
| JFK  | 3026 | 3327-08 | 3324-12 | 3322-16 | 3312-27 | 2923-38 | 284248 | 285150 | 285749 |

At 6000 feet MSL over JFK wind from 330° at 27 knots and temperature minus 8°C.

**TWEB (CONTINUOUS TRANSCRIBED WEATHER BROADCAST)** - Individual route forecasts covering a 25-nautical-mile zone either side of the route. By requesting a specific route number, detailed en route weather for a 12- or 18 hour period (depending on forecast issuance) plus a synopsis can be obtained.

**PILOTS.... report in-flight weather to nearest FSS. The latest surface weather reports are available by phone at the nearest pilot weather briefing office by calling at H+10.**

**U.S. DEPARTMENT OF COMMERCE**—NATIONAL OCEANIC AND ATMOSPHERIC ADMINISTRATION—NATIONAL WEATHER SERVICE—REVISED JANUARY 1984

NOAA PA 73029

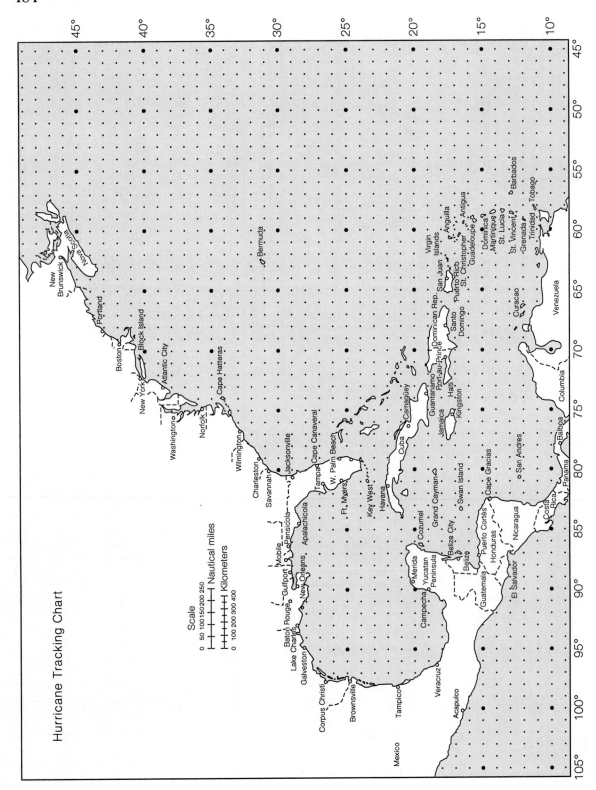

Hurricane Tracking Chart

# Weather Extremes

Temperature
  Maximum
    United States     57° C (134° F), Death Valley, CA, 10 July 1913
    Canada            45° C (113° F), Midale and Yellow Grass, SK, 5 July 1937
    World             58° C (136° F), El Azizia, Libya, 13 September 1922
  Minimum
    United States     −62.1° C (−79.8° F), Prospect Creek, AK, 23 January
                        1971
                      −56.5° C (−70° F), Rogers Pass, MT, 20 January 1954
    Canada            −63° C (−81.4° F), Snag, YT, 3 February 1947
    World             −89° C (−129° F), Vostok, Antarctica, 21 July 1983
Precipitation
  24-hour maximum
    United States     1092 mm (43 in.), Alvin, TX, 25–26 July 1979
    Canada            489 mm (19.3 in.), Ucluelet, BC, 6 October 1967
    World             1880 mm (73.62 in.), Cilaos, La Reunion, Indian Ocean,
                        15–16 March 1952

  One-month maximum
    United States     1817 mm (71.54 in.), Helin Mine, CA, January 1909
                      2718 mm (107 in.), Kuki, Maui, HI, March 1942
    Canada            2235.5 mm (88.01 in.), Swanson Bay, BC, November 1917
    World             9300 mm (366.14 in.), Cherrapunji, Assam, India, July
                        1861

  One-year maximum
    United States     1878 cm (739 in.), Kuki, Maui, HI, December 1981–
                        December 1982
    Canada            8122 mm (319.8 in.), Henderson Lake, BC, 1931
    World             26461 mm (1041.78 in.), Cherrapunji, Assam, India, August
                        1860–August 1861

  One-year minimum
    United States     0.0 mm (0.0 in.), Death Valley, CA, 1919
                      0.0 mm (0.0 in.,), Bagdad, CA, 3 October 1912–8 November
                        1914
    Canada            81.2 mm (1.23 in.), Eureka, NT, 1956
    World             0.0 mm (0.0 in.), Arica, Chile, October 1903–December
                        1917

Snowfall

  24-hour maximum

    United States    192.5 cm (75.8 in.), Silver Lake, Boulder County, CO, 14–15 April 1921

    Canada    118 cm (46 in.), Lakelse Lake, BC, 17 January 1974

  Single storm maximum

    United States    480 cm (189 in.), Mt. Shasta Ski Bowl, CA, 13–19 February 1959

  One-month maximum

    United States    991 cm (390 in.), Tamarack, CA, January 1911

  Season maximum

    United States    2850 cm (1122 in.), Paradise Ranger Station, Mt. Rainier, WA, 1971–72

    Canada    2446.5 cm (964 in.), Revelstoke, Mt. Copeland, BC, 1971–72

Atmospheric pressure

  Maximum

    United States    1068 mb (31.43 in.), Barrow, AK, 3 January 1970

        1063 mb (31.40 in.), Helena, MT, 9 January 1962

    Canada    1068 mb (31.43 in.), Mayo, YT, 1 January 1974

    World    1083.8 mb (32.005 in.), Agata, Siberia, USSR, 31 December 1968

  Minimum

    United States    892 mb (26.35 in.), Matecumbe Key, FL, 3 September 1935

    Canada    940.2 mb (27.77 in.), St. Anthony, NF, 20 January 1977

    World    870 mb (25.69 in.), Typhoon Tip, Pacific Ocean, 12 October 1979

# Selected Climatic Data for United States and Canada

T = average temperature (degrees Celsius); P = total precipitation (millimeters).

| | | Jan | Feb | Mar | Apr | May | Jun | Jul | Aug | Sep | Oct | Nov | Dec | Annual |
|---|---|---|---|---|---|---|---|---|---|---|---|---|---|---|
| Evergreen, AL | T | 8.5 | 10.1 | 14.0 | 18.5 | 22.4 | 25.7 | 27.1 | 26.8 | 24.6 | 18.4 | 12.8 | 9.6 | 18.2 |
| | P | 133.1 | 135.6 | 171.2 | 138.2 | 120.9 | 137.2 | 173.2 | 104.1 | 131.6 | 59.4 | 93.5 | 147.6 | 1549.4 |
| Fayette, AL | T | 5.5 | 7.4 | 11.7 | 16.7 | 20.8 | 24.6 | 26.4 | 26.0 | 23.0 | 16.6 | 10.8 | 6.9 | 16.4 |
| | P | 142.7 | 130.3 | 181.1 | 153.4 | 114.6 | 96.0 | 126.7 | 79.5 | 100.6 | 75.4 | 98.8 | 143.8 | 1447.8 |
| Fairbanks, AK | T | −24.9 | −20.0 | −13.1 | −1.0 | 9.0 | 15.2 | 16.4 | 13.7 | 7.2 | −3.9 | −15.6 | −23.4 | −3.4 |
| | P | 13.5 | 10.7 | 10.2 | 6.9 | 14.5 | 33.5 | 45.0 | 47.2 | 27.7 | 18.8 | 17.0 | 18.5 | 254.0 |
| Juneau, AK | T | −5.7 | −2.3 | −0.4 | 3.9 | 8.1 | 11.5 | 13.2 | 12.6 | 9.6 | 5.4 | 0.4 | −2.9 | 4.4 |
| | P | 93.7 | 95.0 | 84.8 | 74.2 | 86.6 | 75.7 | 104.9 | 127.5 | 162.6 | 195.8 | 130.8 | 117.9 | 1350.0 |
| Prescott, AZ | T | 2.3 | 4.0 | 6.0 | 9.7 | 14.0 | 19.3 | 22.8 | 21.3 | 18.4 | 12.8 | 6.7 | 2.9 | 2.0 |
| | P | 43.7 | 38.4 | 38.9 | 19.3 | 12.7 | 13.5 | 80.0 | 87.6 | 37.8 | 31.0 | 33.8 | 41.9 | 477.5 |
| Tombstone, AZ | T | 8.9 | 10.1 | 12.3 | 16.2 | 20.6 | 25.8 | 26.5 | 25.2 | 23.7 | 18.9 | 12.9 | 9.3 | 17.6 |
| | P | 21.3 | 14.0 | 17.8 | 6.4 | 4.3 | 11.7 | 89.2 | 82.0 | 27.4 | 21.6 | 12.2 | 17.5 | 325.1 |
| Benton, AR | T | 4.7 | 7.0 | 11.4 | 16.7 | 20.7 | 24.7 | 26.8 | 26.3 | 22.8 | 16.7 | 10.4 | 6.3 | 16.2 |
| | P | 96.8 | 97.8 | 129.5 | 141.7 | 140.5 | 99.8 | 103.1 | 81.3 | 120.1 | 83.6 | 120.9 | 112.4 | 1328.4 |
| Fayetteville, AR | T | 1.6 | 4.0 | 8.2 | 14.4 | 18.6 | 23.1 | 25.7 | 24.8 | 20.9 | 14.8 | 8.1 | 3.8 | 14.0 |
| | P | 49.8 | 69.1 | 97.5 | 116.3 | 138.7 | 114.8 | 94.2 | 89.2 | 106.4 | 84.3 | 86.4 | 68.6 | 1115.1 |
| Pasadena, CA | T | 12.8 | 13.8 | 14.3 | 15.9 | 17.8 | 20.4 | 23.7 | 23.9 | 23.1 | 19.9 | 15.9 | 13.3 | 17.9 |
| | P | 119.4 | 101.6 | 78.7 | 40.6 | 10.2 | 2.5 | 0.3 | 2.5 | 7.6 | 10.2 | 58.4 | 61.0 | 490.2 |
| Redding, CA | T | 8.1 | 10.7 | 12.4 | 15.7 | 20.4 | 24.9 | 28.6 | 27.3 | 24.6 | 19.0 | 12.1 | 8.5 | 17.7 |
| | P | 216.2 | 157.2 | 126.0 | 71.6 | 32.5 | 21.1 | 4.6 | 13.0 | 26.7 | 51.6 | 141.2 | 178.6 | 1040.1 |
| San Jose, CA | T | 9.7 | 11.6 | 12.6 | 14.3 | 16.7 | 19.0 | 20.4 | 20.3 | 19.9 | 17.1 | 12.9 | 9.8 | 15.4 |
| | P | 76.2 | 55.9 | 50.8 | 30.5 | 7.6 | 2.5 | 2.5 | 2.5 | 5.1 | 17.8 | 43.2 | 58.4 | 353.1 |
| Aspen, CO | T | −6.3 | −4.7 | −1.7 | 3.6 | 8.9 | 13.6 | 16.9 | 15.8 | 11.9 | 6.7 | −0.7 | −5.7 | 4.9 |
| | P | 49.8 | 40.9 | 48.3 | 42.9 | 38.6 | 30.7 | 37.1 | 47.0 | 39.6 | 39.1 | 39.4 | 46.2 | 499.6 |
| Boulder, CO | T | 0.3 | 2.4 | 4.4 | 9.4 | 14.8 | 20.1 | 23.3 | 22.2 | 17.7 | 12.2 | 5.1 | 2.1 | 11.2 |
| | P | 15.2 | 20.3 | 38.1 | 58.4 | 83.8 | 50.8 | 45.7 | 38.1 | 40.6 | 30.5 | 25.4 | 17.8 | 459.7 |
| Danbury, CT | T | −3.0 | −1.9 | 2.7 | 9.0 | 14.6 | 19.4 | 22.1 | 21.2 | 17.1 | 11.3 | 5.6 | −0.7 | 9.8 |
| | P | 91.4 | 81.3 | 106.7 | 104.1 | 101.6 | 91.4 | 96.5 | 116.8 | 111.8 | 101.6 | 114.3 | 106.7 | 1224.3 |
| Middleton, CT | T | −2.6 | −1.6 | 2.9 | 9.1 | 14.7 | 19.7 | 22.5 | 21.5 | 17.2 | 11.7 | 6.2 | 3.4 | 10.1 |
| | P | 101.6 | 86.4 | 116.8 | 111.8 | 101.6 | 91.4 | 99.1 | 109.2 | 119.4 | 106.7 | 114.3 | 116.8 | 1272.5 |
| Dover, DE | T | 1.3 | 2.4 | 6.8 | 12.6 | 17.9 | 22.6 | 25.0 | 24.3 | 20.9 | 14.8 | 9.0 | 3.6 | 13.4 |
| | P | 81.3 | 73.7 | 99.1 | 81.3 | 88.9 | 91.4 | 109.2 | 129.5 | 104.1 | 91.4 | 88.9 | 88.9 | 1127.8 |
| Fort Lauderdale, FL | T | 19.0 | 19.2 | 21.5 | 23.6 | 25.4 | 27.1 | 27.9 | 28.1 | 27.5 | 25.3 | 22.4 | 19.8 | 23.9 |
| | P | 55.9 | 63.5 | 55.9 | 86.4 | 162.6 | 231.1 | 152.4 | 175.3 | 205.7 | 195.6 | 96.5 | 71.1 | 1544.3 |
| Gainesville, FL | T | 13.6 | 14.3 | 17.7 | 20.9 | 24.2 | 26.6 | 27.4 | 27.4 | 26.3 | 21.9 | 17.4 | 14.3 | 21.0 |
| | P | 83.8 | 99.1 | 94.0 | 73.7 | 106.7 | 167.6 | 180.3 | 203.2 | 142.2 | 58.4 | 50.8 | 81.3 | 1341.1 |
| Dalton, GA | T | 3.9 | 5.7 | 9.9 | 15.6 | 19.9 | 23.7 | 25.6 | 25.2 | 22.2 | 15.7 | 9.9 | 5.5 | 15.2 |
| | P | 137.2 | 127.0 | 167.6 | 124.5 | 116.8 | 101.6 | 132.1 | 96.5 | 116.8 | 81.3 | 109.2 | 127.0 | 1435.1 |

| | | Jan | Feb | Mar | Apr | May | Jun | Jul | Aug | Sep | Oct | Nov | Dec | Annual |
|---|---|---|---|---|---|---|---|---|---|---|---|---|---|---|
| Waycross, GA | T | 10.1 | 11.1 | 14.9 | 19.0 | 22.8 | 25.9 | 27.3 | 27.2 | 25.1 | 19.5 | 14.4 | 10.7 | 19.0 |
| | P | 99.1 | 91.4 | 111.8 | 88.9 | 116.8 | 124.5 | 165.1 | 144.8 | 121.9 | 61.0 | 55.9 | 88.9 | 1264.9 |
| Hilo, HI | T | 21.9 | 21.8 | 21.9 | 22.4 | 23.1 | 23.8 | 24.1 | 24.4 | 24.3 | 24.1 | 23.1 | 22.2 | 23.1 |
| | P | 238.8 | 342.9 | 345.4 | 332.7 | 238.8 | 154.9 | 221.0 | 254.0 | 167.6 | 254.0 | 378.0 | 326.6 | 3255.0 |
| Honolulu, HI | T | 22.6 | 22.7 | 23.6 | 24.3 | 25.9 | 26.2 | 26.7 | 27.2 | 27.0 | 26.4 | 24.8 | 23.3 | 25.0 |
| | P | 96.5 | 68.6 | 88.9 | 38.1 | 30.5 | 12.7 | 12.7 | 15.2 | 15.2 | 48.3 | 81.3 | 86.4 | 596.9 |
| Caldwell, ID | T | −1.0 | 2.7 | 6.3 | 10.7 | 15.5 | 19.8 | 23.8 | 22.3 | 17.1 | 10.7 | 4.2 | 0.1 | 11.0 |
| | P | 38.1 | 27.9 | 22.9 | 25.4 | 25.4 | 22.9 | 5.1 | 10.2 | 15.2 | 20.3 | 30.5 | 33.0 | 276.9 |
| Coeur D' Alene, ID | T | −2.1 | 1.0 | 3.2 | 7.8 | 12.8 | 16.7 | 20.7 | 20.3 | 15.7 | 9.7 | 3.1 | −0.1 | 9.1 |
| | P | 99.1 | 61.0 | 53.3 | 43.2 | 53.3 | 48.3 | 17.8 | 30.5 | 27.9 | 48.3 | 78.7 | 96.5 | 655.3 |
| Mount Vernon, IL | T | −1.1 | 1.3 | 6.7 | 13.6 | 18.6 | 23.4 | 25.5 | 24.6 | 20.8 | 14.3 | 7.3 | 1.8 | 13.1 |
| | P | 58.4 | 58.4 | 99.1 | 106.7 | 104.1 | 91.4 | 104.1 | 83.8 | 78.7 | 58.4 | 88.9 | 81.3 | 1013.5 |
| Park Forest, IL | T | −5.9 | −3.4 | 2.1 | 9.3 | 15.2 | 20.8 | 23.0 | 22.1 | 18.3 | 12.0 | 4.2 | −2.5 | 9.6 |
| | P | 38.1 | 35.6 | 66.0 | 104.1 | 94.0 | 109.2 | 104.1 | 91.4 | 76.2 | 68.6 | 55.9 | 50.8 | 894.1 |
| Evansville, IN | T | 0.6 | 2.9 | 8.1 | 14.7 | 19.6 | 24.3 | 26.2 | 25.3 | 21.8 | 15.1 | 8.4 | 3.1 | 14.2 |
| | P | 83.8 | 81.3 | 121.9 | 106.7 | 109.2 | 96.5 | 111.8 | 83.3 | 76.2 | 63.5 | 96.5 | 91.4 | 1120.1 |
| Huntington, IN | T | −4.1 | −2.2 | 3.4 | 10.4 | 16.0 | 21.1 | 23.1 | 22.2 | 18.6 | 12.2 | 5.1 | −1.1 | 10.4 |
| | P | 55.9 | 50.8 | 81.3 | 94.0 | 96.5 | 106.7 | 91.4 | 83.8 | 73.7 | 68.6 | 68.6 | 68.6 | 939.8 |
| Cedar Rapids, IA | T | −7.5 | −4.2 | 1.6 | 9.9 | 16.3 | 21.3 | 23.5 | 22.3 | 17.8 | 11.8 | 3.4 | −3.7 | 9.4 |
| | P | 33.0 | 27.9 | 63.5 | 96.5 | 111.8 | 121.9 | 111.8 | 101.6 | 99.1 | 66.0 | 48.3 | 38.1 | 914.4 |
| Fort Dodge, IA | T | −9.0 | −5.5 | 0.3 | 9.1 | 15.8 | 21.0 | 23.3 | 21.9 | 17.1 | 11.1 | 2.4 | −4.9 | 8.6 |
| | P | 22.9 | 25.4 | 55.9 | 76.2 | 94.0 | 129.5 | 109.2 | 111.8 | 83.8 | 53.3 | 35.6 | 25.4 | 820.4 |
| Great Bend, KS | T | −1.4 | 1.9 | 6.3 | 13.3 | 18.8 | 24.5 | 27.4 | 26.4 | 21.3 | 15.2 | 6.6 | 1.3 | 13.3 |
| | P | 15.2 | 22.9 | 45.7 | 50.8 | 83.8 | 101.6 | 88.9 | 73.7 | 66.0 | 48.3 | 27.9 | 20.3 | 645.2 |
| Lawrence, KS | T | −2.0 | 1.3 | 6.5 | 13.8 | 19.1 | 23.9 | 26.7 | 25.8 | 21.3 | 15.3 | 7.2 | 1.2 | 13.3 |
| | P | 27.9 | 33.0 | 63.5 | 83.8 | 106.7 | 139.7 | 114.3 | 99.1 | 99.1 | 81.3 | 50.8 | 35.6 | 934.7 |
| Bowling Green, KY | T | 1.1 | 3.0 | 7.9 | 14.1 | 19.1 | 23.6 | 25.7 | 25.0 | 21.3 | 14.4 | 7.9 | 3.4 | 13.9 |
| | P | 116.8 | 101.6 | 139.7 | 106.7 | 106.7 | 114.3 | 109.2 | 83.8 | 81.3 | 71.1 | 99.1 | 114.3 | 1244.6 |
| Frankfort, KY | T | −0.6 | 0.8 | 5.8 | 11.9 | 17.2 | 21.9 | 24.2 | 23.6 | 20.2 | 13.3 | 7.0 | 2.0 | 12.3 |
| | P | 83.8 | 76.2 | 114.3 | 99.1 | 106.7 | 96.5 | 111.8 | 83.8 | 66.0 | 61.0 | 83.8 | 86.4 | 1089.7 |
| Lafayette, LA | T | 10.9 | 12.3 | 16.1 | 20.4 | 23.9 | 27.0 | 27.8 | 27.6 | 25.7 | 20.4 | 15.3 | 12.3 | 20.0 |
| | P | 119.9 | 115.6 | 105.7 | 129.5 | 133.1 | 106.2 | 182.6 | 136.7 | 135.9 | 81.3 | 91.4 | 127.5 | 1465.3 |
| Minden, LA | T | 7.2 | 9.2 | 13.1 | 18.3 | 22.3 | 26.1 | 27.9 | 27.6 | 24.6 | 18.6 | 12.6 | 8.6 | 18.0 |
| | P | 110.7 | 102.4 | 111.3 | 117.1 | 138.2 | 95.5 | 108.0 | 75.9 | 83.8 | 63.5 | 106.7 | 112.8 | 1225.8 |
| Augusta, ME | T | −6.9 | −5.8 | −0.3 | 6.2 | 12.6 | 17.7 | 20.8 | 19.7 | 15.0 | 9.2 | 3.1 | −4.4 | 7.2 |
| | P | 83.8 | 81.3 | 88.9 | 94.0 | 86.4 | 83.8 | 83.8 | 81.3 | 81.3 | 99.1 | 114.3 | 99.1 | 1079.5 |
| Presque Isle, ME | T | −11.0 | −9.7 | −3.7 | 3.2 | 10.6 | 16.3 | 18.9 | 17.4 | 12.8 | 6.9 | 0.1 | −8.4 | 4.4 |
| | P | 61.0 | 55.9 | 58.4 | 58.4 | 76.2 | 86.4 | 106.7 | 96.5 | 91.4 | 81.3 | 78.7 | 71.1 | 916.9 |
| Cumberland, MD | T | −0.6 | 0.7 | 5.6 | 11.9 | 17.1 | 21.3 | 23.6 | 22.9 | 19.2 | 12.8 | 6.7 | 1.3 | 11.9 |
| | P | 63.5 | 61.0 | 91.4 | 86.4 | 91.4 | 96.5 | 81.3 | 83.8 | 73.7 | 68.6 | 61.0 | 66.0 | 927.1 |
| Salisbury, MD | T | 2.6 | 3.6 | 7.8 | 13.2 | 18.2 | 22.6 | 25.0 | 24.4 | 21.1 | 15.1 | 9.7 | 4.6 | 14.0 |
| | P | 88.9 | 86.4 | 106.7 | 83.8 | 83.8 | 91.4 | 96.5 | 144.8 | 91.4 | 91.4 | 78.7 | 91.4 | 1137.9 |
| New Bedford, MA | T | −0.2 | 0.2 | 4.1 | 9.3 | 14.7 | 19.6 | 23.0 | 22.6 | 18.7 | 13.5 | 8.0 | 2.0 | 11.3 |
| | P | 104.1 | 96.5 | 106.7 | 96.5 | 86.4 | 68.6 | 61.0 | 109.2 | 86.4 | 81.3 | 106.7 | 119.4 | 1115.1 |
| Springfield, MA | T | −3.0 | −1.9 | 3.0 | 9.6 | 15.3 | 20.2 | 23.1 | 22.2 | 17.7 | 12.1 | 6.1 | −0.8 | 10.3 |
| | P | 83.8 | 73.7 | 101.6 | 104.1 | 91.4 | 99.1 | 91.4 | 101.6 | 99.1 | 88.9 | 101.6 | 106.7 | 1140.5 |
| Kalamazoo, MI | T | −4.6 | −3.1 | 2.2 | 9.4 | 15.7 | 20.7 | 22.9 | 22.0 | 18.1 | 11.9 | 4.6 | −1.7 | 9.8 |
| | P | 53.3 | 43.2 | 61.0 | 91.4 | 78.7 | 96.5 | 91.4 | 81.3 | 81.3 | 76.2 | 68.6 | 66.0 | 883.9 |
| Ishpeming, MI | T | −10.2 | −8.6 | −3.7 | 4.2 | 11.1 | 16.3 | 19.0 | 17.9 | 13.3 | 7.9 | −0.2 | −7.0 | 5.0 |
| | P | 35.6 | 33.0 | 48.3 | 63.5 | 83.8 | 96.5 | 96.5 | 88.9 | 96.5 | 71.1 | 58.4 | 43.2 | 810.3 |
| Bemidji, MN | T | −16.8 | −12.9 | −6.0 | 3.8 | 11.3 | 16.8 | 19.7 | 18.3 | 12.4 | 6.6 | −3.3 | −12.2 | 3.2 |
| | P | 15.2 | 12.7 | 20.3 | 48.3 | 71.1 | 94.0 | 88.9 | 86.4 | 63.5 | 40.6 | 20.3 | 17.8 | 576.6 |
| Fairmont, MN | T | −11.1 | −7.5 | −1.5 | 7.8 | 15.1 | 20.6 | 22.7 | 21.4 | 16.3 | 10.2 | 0.9 | −7.0 | 7.3 |
| | P | 17.8 | 22.9 | 45.7 | 68.6 | 94.0 | 109.2 | 99.1 | 106.7 | 76.2 | 48.3 | 30.5 | 22.9 | 741.7 |

| | | Jan | Feb | Mar | Apr | May | Jun | Jul | Aug | Sep | Oct | Nov | Dec | Annual |
|---|---|---|---|---|---|---|---|---|---|---|---|---|---|---|
| Columbia, MS | T | 9.7 | 11.4 | 15.2 | 19.6 | 23.1 | 26.4 | 27.7 | 27.3 | 24.9 | 19.1 | 14.0 | 10.9 | 19.1 |
| | P | 137.2 | 144.8 | 152.4 | 144.8 | 129.5 | 106.7 | 134.6 | 129.5 | 104.1 | 73.7 | 114.3 | 160.0 | 1529.1 |
| Tupelo, MS | T | 5.1 | 7.2 | 11.4 | 17.0 | 21.3 | 25.4 | 27.2 | 26.7 | 23.4 | 16.8 | 10.6 | 6.7 | 16.6 |
| | P | 144.8 | 116.8 | 175.3 | 144.8 | 132.1 | 94.0 | 116.8 | 71.1 | 91.4 | 76.2 | 116.8 | 142.2 | 1424.9 |
| Jefferson City, MO | T | −1.4 | 1.3 | 6.4 | 13.5 | 18.2 | 22.8 | 25.6 | 24.7 | 20.5 | 14.2 | 6.9 | 1.3 | 12.8 |
| | P | 35.6 | 40.6 | 71.1 | 86.4 | 119.4 | 114.3 | 81.3 | 76.2 | 101.6 | 88.9 | 53.3 | 48.3 | 916.9 |
| Neosho, MO | T | 1.3 | 4.1 | 8.6 | 15.1 | 19.3 | 23.6 | 26.2 | 25.4 | 21.4 | 15.6 | 8.4 | 3.7 | 14.4 |
| | P | 38.4 | 53.8 | 86.9 | 105.2 | 118.1 | 122.4 | 87.9 | 83.8 | 113.0 | 95.0 | 75.9 | 57.7 | 1038.1 |
| Kalispell, MT | T | −5.4 | −1.7 | 1.2 | 6.7 | 11.8 | 15.6 | 19.2 | 18.3 | 13.1 | 7.1 | 0.4 | −3.2 | 6.9 |
| | P | 35.1 | 25.1 | 20.3 | 25.7 | 45.0 | 50.8 | 25.7 | 41.7 | 27.9 | 26.7 | 30.5 | 35.6 | 390.1 |
| Miles City, MT | T | −9.2 | −4.6 | 0.6 | 8.2 | 14.6 | 19.7 | 23.8 | 22.3 | 15.8 | 9.5 | 0.6 | −5.4 | 8.0 |
| | P | 12.7 | 10.2 | 13.7 | 32.0 | 51.8 | 67.3 | 38.1 | 31.8 | 27.9 | 22.4 | 14.0 | 11.4 | 330.2 |
| Fremont, NB | T | −6.2 | −2.5 | 3.0 | 11.2 | 17.3 | 22.4 | 25.0 | 23.7 | 18.7 | 12.6 | 4.1 | −2.4 | 10.6 |
| | P | 22.9 | 25.4 | 53.3 | 73.7 | 104.1 | 119.4 | 76.2 | 114.3 | 81.3 | 53.3 | 27.9 | 22.1 | 770.9 |
| Kimball, NB | T | −3.6 | −0.9 | 1.5 | 7.2 | 12.7 | 18.4 | 22.1 | 20.8 | 15.5 | 9.4 | 2.0 | −1.7 | 8.6 |
| | P | 12.7 | 10.2 | 30.5 | 40.6 | 78.7 | 73.7 | 68.6 | 43.2 | 30.5 | 22.9 | 15.2 | 12.7 | 434.3 |
| Boulder City, NV | T | 7.9 | 10.9 | 13.6 | 17.8 | 22.8 | 28.2 | 31.6 | 30.4 | 27.0 | 20.7 | 13.0 | 8.6 | 19.4 |
| | P | 15.2 | 12.7 | 17.8 | 10.2 | 7.6 | 2.5 | 12.7 | 20.3 | 12.7 | 10.2 | 12.7 | 10.2 | 144.8 |
| Carson City, NV | T | 0.8 | 3.3 | 5.3 | 8.3 | 12.7 | 17.1 | 20.9 | 19.8 | 16.1 | 10.7 | 4.8 | 1.2 | 10.1 |
| | P | 53.3 | 38.1 | 25.4 | 12.7 | 15.2 | 10.2 | 7.6 | 7.4 | 9.7 | 12.7 | 26.7 | 53.8 | 274.1 |
| Keene, NH | T | −5.8 | −4.4 | 0.8 | 7.6 | 13.7 | 18.7 | 21.2 | 20.2 | 15.8 | 9.8 | 3.7 | −3.3 | 8.2 |
| | P | 81.3 | 68.6 | 86.4 | 83.8 | 88.9 | 91.4 | 86.4 | 91.4 | 86.4 | 78.7 | 91.4 | 91.4 | 1024.4 |
| Woodstock, NH | T | −7.3 | −6.2 | −0.8 | 5.8 | 12.3 | 17.2 | 19.7 | 18.6 | 14.2 | 8.4 | 2.2 | −4.9 | 6.6 |
| | P | 78.7 | 73.7 | 86.4 | 91.4 | 99.1 | 96.5 | 109.2 | 100.3 | 99.1 | 101.6 | 104.1 | 104.1 | 1148.1 |
| Jersey City, NJ | T | −0.8 | 0.1 | 4.4 | 10.3 | 15.8 | 20.8 | 23.7 | 22.9 | 19.1 | 13.2 | 7.5 | 1.4 | 11.5 |
| | P | 83.3 | 72.6 | 107.7 | 96.8 | 93.2 | 82.8 | 96.8 | 109.2 | 97.0 | 85.3 | 93.2 | 93.7 | 1111.8 |
| Sussex, NJ | T | −4.4 | −3.4 | 1.8 | 8.5 | 14.0 | 18.9 | 21.5 | 20.6 | 16.4 | 10.4 | 4.8 | −1.8 | 8.9 |
| | P | 84.6 | 68.1 | 90.2 | 103.4 | 90.9 | 105.2 | 108.2 | 126.2 | 97.3 | 90.2 | 97.8 | 90.9 | 1152.9 |
| Alamogordo, NM | T | 5.8 | 7.9 | 11.3 | 16.1 | 20.8 | 26.1 | 26.8 | 25.7 | 22.6 | 16.7 | 9.9 | 6.1 | 16.3 |
| | P | 15.7 | 13.5 | 13.2 | 6.1 | 10.4 | 19.8 | 55.4 | 54.6 | 41.4 | 28.7 | 10.9 | 14.2 | 284.0 |
| Carlsbad, NM | T | 6.2 | 8.6 | 12.4 | 17.6 | 22.3 | 27.0 | 28.1 | 27.0 | 23.3 | 17.3 | 10.4 | 6.8 | 17.3 |
| | P | 8.6 | 8.9 | 8.4 | 10.2 | 23.6 | 18.0 | 43.2 | 47.8 | 54.9 | 29.5 | 11.2 | 6.6 | 270.8 |
| Lockport, NY | T | −4.7 | −3.9 | 0.7 | 7.6 | 13.6 | 18.9 | 21.5 | 20.6 | 16.7 | 10.9 | 4.6 | −1.8 | 8.7 |
| | P | 67.1 | 60.5 | 70.4 | 79.5 | 73.7 | 72.6 | 70.1 | 99.1 | 85.6 | 71.9 | 79.2 | 77.2 | 906.8 |
| Utica, NY | T | −6.6 | −5.7 | −0.4 | 7.0 | 13.2 | 18.4 | 21.0 | 20.1 | 15.8 | 9.7 | 3.6 | −3.6 | 7.7 |
| | P | 85.3 | 76.5 | 85.6 | 88.6 | 91.4 | 95.3 | 106.2 | 91.2 | 95.3 | 85.6 | 99.6 | 102.9 | 1103.4 |
| Asheville, NC | T | 2.9 | 4.2 | 8.3 | 13.7 | 17.8 | 21.3 | 23.2 | 22.9 | 19.7 | 13.7 | 8.3 | 4.4 | 13.4 |
| | P | 69.3 | 74.4 | 110.2 | 85.3 | 84.1 | 83.1 | 73.9 | 95.5 | 89.7 | 69.1 | 68.8 | 70.9 | 974.3 |
| Durham, NC | T | 4.2 | 5.2 | 9.5 | 15.2 | 19.5 | 23.3 | 25.4 | 24.9 | 21.4 | 15.2 | 9.8 | 5.2 | 14.9 |
| | P | 100.1 | 103.1 | 105.9 | 88.9 | 101.6 | 109.5 | 116.1 | 121.2 | 96.5 | 80.8 | 83.3 | 88.4 | 1195.3 |
| Dickinson, ND | T | −11.7 | −7.9 | −3.0 | 5.1 | 11.8 | 17.0 | 20.8 | 20.1 | 13.7 | 7.8 | −1.5 | −7.4 | 5.4 |
| | P | 10.4 | 12.4 | 17.3 | 43.4 | 59.9 | 84.8 | 52.3 | 40.1 | 38.9 | 19.3 | 11.9 | 9.1 | 400.0 |
| Grand Forks, ND | T | −16.6 | −12.6 | −5.5 | 4.8 | 12.4 | 17.7 | 20.4 | 19.4 | 13.4 | 7.2 | −3.1 | −11.6 | 3.8 |
| | P | 19.3 | 13.0 | 19.6 | 34.0 | 49.5 | 73.4 | 72.6 | 66.5 | 51.8 | 28.2 | 20.3 | 16.3 | 464.6 |
| Hamilton, OH | T | −1.2 | 0.7 | 5.9 | 12.3 | 17.6 | 22.2 | 24.1 | 23.3 | 19.9 | 13.3 | 6.8 | 1.4 | 12.2 |
| | P | 72.4 | 62.5 | 87.6 | 91.2 | 90.4 | 95.8 | 107.2 | 85.3 | 79.5 | 61.2 | 71.1 | 66.0 | 970.3 |
| Wooster, OH | T | −4.0 | −2.7 | 2.6 | 8.9 | 14.4 | 19.4 | 21.4 | 20.7 | 17.0 | 10.7 | 4.6 | −1.3 | 9.3 |
| | P | 63.0 | 49.0 | 78.0 | 82.6 | 94.5 | 85.1 | 106.2 | 92.5 | 81.0 | 52.3 | 62.2 | 60.7 | 906.8 |
| Kingfisher, OK | T | 2.2 | 5.1 | 9.8 | 16.0 | 20.8 | 25.9 | 28.7 | 28.0 | 23.4 | 17.2 | 9.4 | 4.4 | 15.9 |
| | P | 21.1 | 28.7 | 44.7 | 61.5 | 125.5 | 95.5 | 65.3 | 60.7 | 91.4 | 62.0 | 38.9 | 28.7 | 723.9 |
| McAlester, OK | T | 3.4 | 6.2 | 10.7 | 16.6 | 20.8 | 25.4 | 28.2 | 27.6 | 23.4 | 17.3 | 10.4 | 5.6 | 16.3 |
| | P | 41.1 | 57.4 | 97.8 | 115.3 | 142.7 | 93.0 | 86.6 | 82.6 | 126.0 | 99.1 | 78.0 | 60.5 | 1080.0 |
| Baker, OR | T | −3.8 | −0.2 | 2.8 | 6.8 | 11.4 | 15.3 | 19.2 | 18.1 | 13.8 | 7.8 | 1.7 | −2.4 | 7.6 |
| | P | 25.9 | 17.0 | 20.3 | 21.1 | 36.3 | 35.1 | 11.9 | 18.8 | 17.5 | 17.3 | 22.4 | 26.4 | 270.0 |

| | | Jan | Feb | Mar | Apr | May | Jun | Jul | Aug | Sep | Oct | Nov | Dec | Annual |
|---|---|---|---|---|---|---|---|---|---|---|---|---|---|---|
| Oregon City, OR | T | 4.6 | 6.8 | 8.2 | 10.8 | 14.3 | 17.4 | 20.2 | 19.9 | 17.6 | 12.7 | 7.9 | 5.4 | 12.2 |
| | P | 206.0 | 130.6 | 126.7 | 85.6 | 67.1 | 47.5 | 14.2 | 31.5 | 52.8 | 96.3 | 167.4 | 203.7 | 1229.4 |
| Gettysburg, PA | T | −1.4 | −0.4 | 4.6 | 11.1 | 16.4 | 21.3 | 23.9 | 23.1 | 19.1 | 12.6 | 6.6 | 0.8 | 11.4 |
| | P | 78.2 | 70.6 | 99.6 | 93.7 | 90.4 | 101.3 | 80.0 | 95.0 | 96.0 | 79.8 | 82.3 | 80.0 | 1047.0 |
| Meadville, PA | T | −4.7 | −4.3 | 1.1 | 7.6 | 13.2 | 18.2 | 20.4 | 19.7 | 16.2 | 10.2 | 4.3 | −1.8 | 8.3 |
| | P | 74.7 | 62.7 | 87.1 | 93.2 | 96.3 | 109.7 | 111.5 | 111.8 | 86.1 | 88.6 | 90.2 | 85.1 | 1097.0 |
| Kingston, RI | T | −2.1 | −1.5 | 2.7 | 8.1 | 13.2 | 18.2 | 21.2 | 20.7 | 16.8 | 11.3 | 6.0 | 0.2 | 9.6 |
| | P | 107.4 | 93.7 | 118.1 | 104.9 | 104.6 | 74.2 | 75.9 | 113.3 | 104.4 | 100.6 | 118.1 | 116.3 | 1231.6 |
| Anderson, SC | T | 5.7 | 7.1 | 11.2 | 16.5 | 20.9 | 24.8 | 26.6 | 26.2 | 22.9 | 16.7 | 11.2 | 7.0 | 16.4 |
| | P | 111.3 | 101.1 | 144.0 | 100.1 | 99.8 | 88.1 | 101.6 | 94.5 | 100.6 | 69.6 | 78.5 | 99.3 | 1188.5 |
| Georgetown, SC | T | 8.4 | 9.4 | 13.3 | 17.8 | 22.1 | 25.3 | 26.9 | 26.7 | 24.2 | 18.8 | 13.9 | 9.7 | 18.1 |
| | P | 90.7 | 86.4 | 107.4 | 56.9 | 106.4 | 129.8 | 173.2 | 156.0 | 149.9 | 94.2 | 64.0 | 85.1 | 1300.0 |
| Hot Springs, SD | T | −4.9 | −1.9 | 1.3 | 7.5 | 13.3 | 18.8 | 22.7 | 21.7 | 16.1 | 9.9 | 1.9 | −2.4 | 8.7 |
| | P | 8.6 | 11.2 | 18.3 | 38.9 | 65.5 | 73.7 | 64.5 | 32.0 | 32.0 | 20.6 | 9.1 | 8.6 | 383.0 |
| Mitchell, SD | T | −9.8 | −5.9 | −0.2 | 8.9 | 15.6 | 21.0 | 24.2 | 23.1 | 17.3 | 10.8 | 1.6 | −5.9 | 8.4 |
| | P | 9.4 | 16.5 | 30.7 | 59.2 | 80.0 | 88.9 | 62.5 | 68.6 | 52.3 | 35.3 | 19.0 | 13.2 | 535.7 |
| Dyersburg, TN | T | 2.9 | 5.2 | 10.0 | 16.3 | 21.0 | 25.4 | 27.1 | 26.2 | 22.6 | 16.4 | 9.8 | 5.2 | 15.7 |
| | P | 104.9 | 111.8 | 137.2 | 115.3 | 117.1 | 90.7 | 100.3 | 83.1 | 89.2 | 57.7 | 110.0 | 112.3 | 1229.4 |
| Springfield, TN | T | 1.2 | 3.1 | 8.0 | 14.2 | 18.8 | 23.2 | 25.2 | 24.6 | 21.2 | 14.6 | 8.3 | 3.6 | 13.8 |
| | P | 117.6 | 106.7 | 138.9 | 114.6 | 116.6 | 96.3 | 105.2 | 87.1 | 84.6 | 68.1 | 106.4 | 116.3 | 1258.3 |
| Corpus Christi, TX | T | 13.4 | 15.5 | 19.0 | 22.7 | 25.7 | 28.3 | 29.1 | 29.1 | 27.6 | 23.7 | 18.7 | 15.1 | 22.3 |
| | P | 40.1 | 42.7 | 22.9 | 49.0 | 82.0 | 79.5 | 39.1 | 82.3 | 153.9 | 84.6 | 40.9 | 36.8 | 753.9 |
| Graham, TX | T | 5.3 | 7.9 | 12.3 | 18.0 | 22.1 | 26.7 | 29.2 | 28.8 | 24.7 | 18.6 | 11.6 | 7.2 | 17.7 |
| | P | 33.0 | 34.5 | 37.3 | 78.5 | 104.4 | 72.1 | 51.8 | 59.2 | 100.1 | 65.8 | 43.4 | 31.2 | 711.2 |
| Stratford, TX | T | 0.6 | 2.9 | 6.4 | 12.2 | 17.5 | 23.1 | 25.6 | 24.6 | 20.1 | 14.1 | 6.3 | 2.4 | 12.9 |
| | P | 7.6 | 10.2 | 19.8 | 31.0 | 69.1 | 57.2 | 76.5 | 67.8 | 38.9 | 25.9 | 16.3 | 8.1 | 428.2 |
| Blanding, UT | T | −2.6 | 0.6 | 3.8 | 8.4 | 13.8 | 19.4 | 23.1 | 21.6 | 17.3 | 11.0 | 3.6 | −1.4 | 9.9 |
| | P | 34.0 | 24.1 | 20.3 | 17.0 | 15.0 | 9.4 | 26.4 | 35.8 | 22.6 | 37.1 | 22.6 | 32.8 | 297.2 |
| Ogden, UT | T | −1.9 | 0.9 | 4.4 | 9.4 | 15.0 | 20.0 | 25.0 | 23.5 | 18.2 | 11.7 | 4.1 | −0.8 | 10.8 |
| | P | 59.9 | 48.3 | 52.1 | 64.0 | 54.4 | 40.1 | 16.5 | 24.9 | 30.5 | 40.1 | 43.9 | 48.0 | 522.7 |
| Montpelier, VT | T | −9.1 | −8.1 | −2.3 | 4.9 | 11.7 | 16.7 | 19.2 | 17.9 | 13.6 | 7.9 | 1.4 | −6.3 | 5.6 |
| | P | 59.7 | 60.7 | 63.0 | 65.8 | 75.4 | 81.3 | 77.0 | 84.6 | 73.7 | 73.2 | 75.2 | 72.6 | 862.1 |
| Rutland, VT | T | −6.3 | −5.1 | 0.4 | 7.3 | 13.7 | 18.6 | 20.9 | 19.9 | 15.6 | 9.9 | 3.8 | −3.3 | 7.9 |
| | P | 57.2 | 48.3 | 57.9 | 69.1 | 81.5 | 91.7 | 87.1 | 98.8 | 86.9 | 68.6 | 72.6 | 67.1 | 886.7 |
| Danville, VA | T | 3.1 | 4.3 | 8.9 | 14.8 | 19.3 | 23.2 | 25.4 | 25.1 | 21.4 | 14.8 | 9.3 | 4.4 | 14.5 |
| | P | 87.1 | 84.3 | 101.9 | 87.6 | 94.7 | 98.0 | 103.1 | 102.6 | 93.5 | 82.0 | 71.6 | 84.1 | 1090.7 |
| Hopewell, VA | T | 4.3 | 5.5 | 9.9 | 15.7 | 20.1 | 23.9 | 26.1 | 25.7 | 22.3 | 16.2 | 10.9 | 5.9 | 15.6 |
| | P | 90.7 | 82.6 | 93.5 | 80.5 | 97.0 | 87.9 | 128.3 | 117.3 | 103.9 | 92.2 | 78.2 | 86.1 | 1138.2 |
| Coulee Dam, WA | T | −3.3 | 0.4 | 4.3 | 9.4 | 14.3 | 18.5 | 22.4 | 21.7 | 17.1 | 10.3 | 2.9 | −1.2 | 9.8 |
| | P | 29.7 | 22.1 | 18.0 | 19.3 | 27.7 | 18.3 | 9.7 | 14.7 | 13.5 | 16.5 | 32.0 | 38.4 | 259.8 |
| Tacoma, WA | T | 4.6 | 6.6 | 7.3 | 9.9 | 13.2 | 15.9 | 18.3 | 18.2 | 15.9 | 11.9 | 7.6 | 5.6 | 11.3 |
| | P | 145.8 | 103.1 | 85.9 | 63.2 | 37.6 | 33.3 | 19.0 | 31.8 | 49.5 | 83.1 | 138.9 | 152.9 | 944.1 |
| Parkersburg, WV | T | −0.9 | 0.6 | 5.9 | 12.0 | 17.2 | 21.4 | 23.6 | 23.0 | 19.5 | 12.9 | 6.8 | 1.7 | 11.9 |
| | P | 81.0 | 72.4 | 97.5 | 90.4 | 94.5 | 95.5 | 112.8 | 101.3 | 74.7 | 68.8 | 69.9 | 74.2 | 1033.0 |
| Pineville, WV | T | −0.2 | 1.2 | 5.9 | 11.8 | 16.8 | 21.0 | 23.1 | 22.7 | 19.3 | 12.6 | 6.4 | 1.7 | 11.8 |
| | P | 94.7 | 83.1 | 110.1 | 103.1 | 102.1 | 102.9 | 126.5 | 103.6 | 88.6 | 75.2 | 78.2 | 84.3 | 1152.4 |
| Waukesha, WI | T | −7.7 | −5.2 | 0.2 | 7.8 | 14.1 | 19.4 | 22.1 | 21.2 | 16.8 | 10.7 | 2.8 | −4.1 | 8.2 |
| | P | 35.6 | 27.9 | 61.5 | 87.6 | 76.5 | 92.2 | 94.7 | 96.3 | 80.0 | 60.2 | 54.6 | 46.2 | 813.3 |
| Wausau, WI | T | −11.6 | −8.9 | −2.8 | 6.2 | 13.0 | 18.1 | 20.8 | 19.4 | 14.3 | 8.5 | 0.0 | −7.6 | 5.8 |
| | P | 23.6 | 24.1 | 48.3 | 72.6 | 95.8 | 99.8 | 100.8 | 105.2 | 98.6 | 57.4 | 45.2 | 31.8 | 803.1 |
| Newcastle, WY | T | −6.0 | −3.1 | 0.6 | 6.9 | 13.1 | 18.7 | 22.9 | 21.6 | 15.4 | 9.2 | 0.8 | −3.7 | 8.1 |
| | P | 10.7 | 12.4 | 14.7 | 36.6 | 62.5 | 66.8 | 49.8 | 38.4 | 23.6 | 15.7 | 11.9 | 13.5 | 356.6 |
| Rock Springs, WY | T | −5.8 | −3.2 | 0.1 | 5.5 | 11.1 | 16.2 | 20.6 | 19.1 | 13.7 | 7.7 | −0.1 | −4.3 | 6.7 |
| | P | 11.4 | 10.7 | 15.2 | 24.6 | 32.5 | 25.1 | 14.2 | 18.3 | 18.3 | 18.8 | 13.2 | 11.7 | 214.1 |

| | | Jan | Feb | Mar | Apr | May | Jun | Jul | Aug | Sep | Oct | Nov | Dec | Annual |
|---|---|---|---|---|---|---|---|---|---|---|---|---|---|---|
| Calgary, Alb. | T | −11.8 | −7.3 | −4.0 | 3.3 | 9.4 | 13.5 | 16.4 | 15.2 | 10.6 | 5.5 | −2.7 | −7.8 | 3.4 |
| | P | 16.2 | 15.5 | 16.1 | 32.6 | 48.7 | 89.4 | 65.4 | 55.4 | 38.2 | 17.6 | 12.7 | 16.0 | 423.8 |
| Edmonton, AB | T | −15.0 | −9.6 | −5.0 | 4.2 | 11.3 | 15.1 | 17.4 | 16.2 | 11.0 | 5.8 | −3.7 | −10.4 | 3.1 |
| | P | 24.6 | 18.8 | 18.5 | 21.7 | 42.5 | 77.3 | 88.7 | 77.9 | 39.1 | 16.6 | 15.7 | 24.7 | 466.1 |
| Prince George, BC | T | −12.1 | −6.1 | −1.8 | 4.3 | 9.3 | 12.9 | 15.1 | 14.1 | 9.7 | 4.8 | −2.9 | −7.9 | 3.3 |
| | P | 57.4 | 39.2 | 36.8 | 27.4 | 47.3 | 66.9 | 59.7 | 68.2 | 58.7 | 59.2 | 50.5 | 57.0 | 628.3 |
| Vancouver, BC | T | 2.5 | 4.6 | 5.8 | 8.8 | 12.2 | 15.1 | 17.3 | 17.1 | 14.2 | 10.0 | 5.9 | 3.9 | 9.8 |
| | P | 153.8 | 114.7 | 101.0 | 59.6 | 51.6 | 45.2 | 32.0 | 41.1 | 67.1 | 114.0 | 150.1 | 182.4 | 1112.6 |
| Churchill, MB | T | −27.5 | −25.9 | −20.4 | −10.1 | −1.5 | 6.2 | 11.8 | 11.3 | 5.4 | −1.5 | −12.1 | −22.2 | −7.2 |
| | P | 15.3 | 13.1 | 18.1 | 22.9 | 31.9 | 43.5 | 45.6 | 58.3 | 50.9 | 43.0 | 38.8 | 20.9 | 402.3 |
| Winnipeg, MB | T | −19.3 | −15.6 | −8.2 | 3.4 | 11.3 | 16.8 | 19.6 | 18.3 | 12.4 | 6.1 | −4.5 | −14.0 | 2.2 |
| | P | 21.3 | 17.5 | 22.7 | 38.5 | 65.7 | 80.1 | 75.9 | 75.2 | 53.3 | 30.9 | 25.2 | 19.2 | 525.5 |
| Chatham, NB | T | −9.7 | −8.8 | −3.3 | 3.0 | 9.5 | 15.7 | 19.2 | 18.0 | 13.0 | 7.1 | 0.9 | −6.9 | 4.8 |
| | P | 98.8 | 86.8 | 97.1 | 84.5 | 81.9 | 82.0 | 91.0 | 83.5 | 85.2 | 95.6 | 102.4 | 107.9 | 1096.7 |
| Moncton, NB | T | −8.0 | −7.6 | −2.5 | 3.6 | 9.9 | 15.5 | 18.8 | 17.9 | 13.2 | 7.7 | 2.2 | −5.2 | 5.5 |
| | P | 112.9 | 88.7 | 92.2 | 81.5 | 81.9 | 87.8 | 97.5 | 76.8 | 75.7 | 92.8 | 99.3 | 107.5 | 1094.6 |
| Gander, NF | T | −6.2 | −6.8 | −3.5 | 0.9 | 6.2 | 11.8 | 16.5 | 15.6 | 11.4 | 6.0 | 1.8 | −3.8 | 4.2 |
| | P | 109.1 | 99.7 | 110.1 | 93.2 | 70.0 | 80.3 | 69.0 | 97.3 | 81.2 | 104.7 | 107.3 | 108.2 | 1130.1 |
| St. John's, NF | T | −3.9 | −4.5 | −2.3 | 1.2 | 5.4 | 10.9 | 15.5 | 15.3 | 11.6 | 6.9 | 3.4 | −1.5 | 4.8 |
| | P | 155.8 | 140.1 | 131.9 | 115.6 | 101.8 | 85.6 | 75.3 | 121.6 | 116.7 | 145.5 | 162.5 | 161.2 | 1513.6 |
| Fort Smith, NT | T | −26.8 | −21.8 | −14.8 | −2.2 | 7.9 | 13.6 | 16.0 | 14.2 | 7.5 | 0.3 | −11.6 | −21.6 | −3.3 |
| | P | 18.5 | 15.9 | 14.4 | 16.2 | 27.8 | 41.2 | 56.9 | 42.5 | 41.1 | 26.5 | 26.1 | 22.2 | 349.3 |
| Frobisher, NT | T | −25.6 | −25.9 | −22.7 | −14.3 | −3.2 | 3.4 | 7.6 | 6.9 | 2.4 | −5.0 | −13.0 | −21.8 | −9.3 |
| | P | 26.1 | 23.3 | 23.3 | 26.4 | 25.3 | 39.4 | 63.3 | 58.9 | 46.0 | 44.1 | 34.4 | 22.1 | 432.6 |
| Greenwood, NS | T | −5.0 | −5.4 | −0.9 | 4.6 | 10.5 | 15.9 | 19.1 | 18.3 | 13.8 | 8.6 | 3.9 | −2.3 | 6.8 |
| | P | 125.6 | 90.1 | 84.1 | 75.5 | 73.9 | 71.6 | 77.6 | 90.0 | 84.0 | 98.0 | 108.4 | 120.2 | 1099.0 |
| Yarmouth, NS | T | −2.7 | −3.2 | 0.3 | 4.7 | 9.2 | 13.4 | 16.3 | 16.4 | 13.6 | 9.5 | 5.2 | −0.3 | 6.9 |
| | P | 141.0 | 114.2 | 98.5 | 96.3 | 92.4 | 81.3 | 77.8 | 97.3 | 89.4 | 116.5 | 134.7 | 142.2 | 1281.6 |
| Armstrong, ON | T | −21.1 | −18.3 | −11.2 | −0.8 | 6.9 | 12.8 | 16.1 | 14.2 | 8.7 | 3.0 | −6.7 | −16.8 | −1.1 |
| | P | 35.8 | 29.6 | 37.5 | 46.5 | 63.4 | 89.5 | 93.7 | 91.2 | 86.4 | 68.6 | 57.2 | 39.0 | 738.4 |
| Toronto, ON | T | −4.6 | −3.9 | 0.7 | 7.6 | 13.6 | 19.1 | 22.0 | 21.2 | 17.1 | 11.10 | 4.9 | −1.6 | 8.9 |
| | P | 60.9 | 51.8 | 69.7 | 72.9 | 65.8 | 63.9 | 74.0 | 73.1 | 66.2 | 61.0 | 68.4 | 72.8 | 800.5 |
| Charlottetown, PE | T | −7.1 | −7.5 | −3.1 | 2.3 | 8.5 | 14.5 | 18.3 | 17.8 | 13.5 | 8.1 | 2.9 | −3.9 | 5.4 |
| | P | 116.8 | 97.4 | 95.3 | 81.8 | 83.6 | 79.9 | 84.3 | 88.1 | 86.3 | 106.4 | 120.5 | 129.0 | 1169.4 |
| Summerside, PE | T | −7.2 | −7.2 | −2.8 | 2.6 | 9.0 | 14.9 | 18.9 | 18.4 | 14.1 | 8.6 | 3.0 | −4.0 | 5.7 |
| | P | 102.9 | 82.4 | 84.4 | 75.4 | 81.3 | 74.1 | 78.0 | 80.1 | 78.8 | 94.0 | 100.0 | 107.3 | 1038.7 |
| Montreal, PQ | T | −10.2 | −9.0 | −2.5 | 5.7 | 13.0 | 18.3 | 20.9 | 19.6 | 14.8 | 8.7 | 2.0 | −6.9 | 6.2 |
| | P | 72.0 | 65.2 | 73.6 | 74.1 | 65.6 | 82.2 | 90.0 | 91.9 | 88.4 | 75.5 | 81.0 | 86.7 | 946.2 |
| Quebec, PQ | T | −12.1 | −10.8 | −4.5 | 3.3 | 10.8 | 16.4 | 19.1 | 17.5 | 12.6 | 6.6 | −0.2 | −9.0 | 4.1 |
| | P | 89.8 | 78.1 | 82.0 | 72.8 | 86.9 | 109.9 | 116.6 | 117.1 | 119.4 | 90.7 | 98.1 | 113.6 | 1174.0 |
| Regina, SK | T | −17.9 | −13.6 | −7.8 | 3.3 | 11.1 | 15.9 | 18.9 | 17.8 | 11.7 | 5.2 | −5.1 | −12.8 | 2.2 |
| | P | 16.6 | 16.1 | 17.8 | 23.7 | 46.4 | 79.6 | 53.3 | 44.8 | 36.7 | 18.8 | 13.5 | 16.7 | 384.0 |
| Swift Current, SK | T | −14.7 | −10.7 | −5.7 | 3.5 | 10.5 | 15.1 | 18.3 | 17.5 | 11.7 | 5.8 | −3.7 | −9.9 | 3.2 |
| | P | 21.1 | 17.2 | 20.1 | 28.3 | 39.9 | 75.6 | 46.9 | 43.0 | 34.1 | 18.1 | 15.8 | 19.9 | 380.0 |
| Watson Lake, YT | T | −26.7 | −18.7 | −11.3 | −0.6 | 6.9 | 12.7 | 14.9 | 13.1 | 7.6 | −0.1 | −13.8 | −23.5 | −3.3 |
| | P | 33.1 | 25.3 | 23.2 | 15.1 | 29.4 | 51.6 | 58.2 | 42.0 | 43.7 | 35.0 | 31.8 | 36.8 | 425.2 |
| Whitehorse, YT | T | −20.7 | −13.2 | −8.2 | 0.3 | 6.7 | 12.0 | 14.1 | 12.5 | 7.5 | 0.6 | −8.8 | −16.6 | −1.2 |
| | P | 17.7 | 13.3 | 13.5 | 9.5 | 12.9 | 30.7 | 33.9 | 37.9 | 30.3 | 21.5 | 19.8 | 20.2 | 261.2 |

**Freeze-free Season (0 °C, 50 Percentile)**

| City | Freeze-free period | Last freeze spring | First freeze fall | City | Freeze-free period | Last freeze spring | First freeze fall |
|------|------|------|------|------|------|------|------|
| Birmingham, AL | 221 | Mar 29 | Nov 6 | Albany, NY | 144 | May 7 | Sep 29 |
| Fairbanks, AK | 112 | May 17 | Sep 6 | Raleigh, NC | 198 | Apr 11 | Oct 27 |
| Phoenix, AZ | 308 | Feb 5 | Dec 15 | Bismark, ND | 129 | May 14 | Sep 20 |
| Benton, AR | 201 | Apr 7 | Oct 25 | Cincinnati, OH | 195 | Apr 14 | Oct 27 |
| Pasadena, CA | >365 | * | * | Tulsa, OK | 218 | Mar 30 | Nov 4 |
| San Francisco, CA | >365 | * | * | Portland, OR | 217 | Apr 3 | Nov 7 |
| Denver, CO | 157 | May 3 | Oct 8 | Allentown, PA | 179 | Apr 21 | Oct 18 |
| Hartford, CT | 167 | Apr 25 | Oct 10 | Kingston, RI | 144 | May 8 | Sep 30 |
| Dover, DE | 202 | Apr 9 | Oct 28 | Charleston, SC | 239 | Mar 18 | Nov 12 |
| Miama, FL | >365 | * | * | Mitchell, SD | 149 | May 7 | Oct 3 |
| Waycross, GA | 234 | Mar 16 | Nov 6 | Nashville, TN | 207 | Apr 5 | Oct 29 |
| Hilo, HI | >365 | * | * | Galveston, TX | 345 | Jan 28 | Jan 9 |
| Boise, ID | 153 | May 8 | Oct 9 | San Antonio, TX | 265 | Mar 3 | Nov 24 |
| Chicago, IL | 187 | Apr 22 | Oct 26 | Ogden, UT | 161 | May 4 | Oct 13 |
| Indianapolis, IN | 180 | Apr 22 | Oct 20 | Burlington, VT | 142 | May 11 | Oct 1 |
| Dubuque, IA | 185 | Apr 20 | Oct 22 | Richmond, VA | 198 | Apr 10 | Oct 26 |
| Topeka, KS | 175 | Apr 21 | Oct 14 | Seattle, WA | 232 | Mar 24 | Nov 11 |
| Lexington, KY | 190 | Apr 17 | Oct 25 | Parkersburg, WV | 175 | Apr 25 | Oct 18 |
| New Orleans, LA | 288 | Feb 20 | Dec 5 | Green Bay, WI | 143 | May 12 | Oct 2 |
| Presque Isle, ME | 116 | May 24 | Sep 17 | Casper, WY | 123 | May 22 | Sep 22 |
| Cumberland, MD | 167 | Apr 27 | Oct 12 | Edmonton, AB | 140 | May 6 | Sep 24 |
| Springfield, MA | 179 | Apr 19 | Oct 15 | Vancouver, BC | 216 | Mar 31 | Nov 3 |
| Ann Arbor, MI | 175 | Apr 28 | Oct 20 | Winnipeg, MB | 121 | May 23 | Sep 22 |
| Marquette, MI | 159 | May 12 | Oct 19 | Saint John, NB | 139 | May 16 | Oct 3 |
| Duluth, MN | 122 | May 21 | Sep 21 | St. John's, NF | 131 | May 1 | Oct 11 |
| Tupelo, MS | 206 | Apr 1 | Oct 25 | Yellowknife, NT | 111 | May 27 | Sep 16 |
| Jefferson City, MO | 173 | Apr 26 | Oct 16 | Halifax, NS | 155 | May 12 | Oct 15 |
| Helena, MT | 122 | May 18 | Sep 18 | Toronto, ON | 149 | May 8 | Oct 5 |
| North Platte, NE | 136 | May 11 | Sep 24 | Charlottetown, PE | 151 | May 16 | Oct 15 |
| Las Vagas, NV | 259 | Mar 7 | Nov 21 | Montreal, PQ | 157 | May 3 | Oct 8 |
| Concord, NH | 121 | May 23 | Sep 22 | Regina, SK | 109 | May 24 | Sep 11 |
| Trenton, NJ | 214 | Apr 6 | Nov 7 | Whitehorse, YT | 82 | Jun 8 | Aug 30 |
| Carlsbad, NM | 223 | Mar 29 | Nov 7 | | | | |

*Indicates a freeze doesn't occur every year.

# Glossary

**absolute humidity**   The mass of water vapor per unit volume of air containing the water vapor; usually expressed as grams of water vapor per cubic meter of air.

**absolute instability**   Property of an ambient air layer that is unstable for both saturated and unsaturated air parcels.

**absolute stability**   Property of an ambient air layer that is stable for both saturated and unsaturated air parcels.

**absolute zero**   The theoretical temperature at which a body emits no electromagnetic radiation and all molecular activity ceases (actually some atomic-level activity takes place); 0 K.

**absorption**   The process whereby a portion of the radiation incident on an object is converted to heat.

**acid deposition**   The combination of acid rain (or snow) plus dry deposition.

**acid rain**   Rain having a pH lower than 5.6, often due to the presence of sulfuric acid and nitric acid.

**adiabatic process**   Expansional cooling or compressional warming of air parcels in which there is no net heat exchange between the air parcels and the surrounding (ambient) air.

**Advanced Weather Interactive Processing System (AWIPS)**   A computerized workstation that enables National Weather Service meteorologists to integrate a variety of weather data and displays.

**advection**   Horizontal movement of air from one place to another.

**advection fog**   Ground-level clouds generated by advective cooling of a mild, humid air mass as it travels over a relatively cool surface.

**aerosols**   Tiny liquid or solid particles of various composition that occur suspended in the atmosphere.

**agroclimatic compensation**   Poor growing weather in one area is offset to some extent by better growing weather in other areas.

**air density**   Mass per unit volume of air; about 1.275 kg per cubic meter at 0 °C and 1000 millibars.

**air mass**   A huge volume of air covering thousands of square kilometers that is relatively uniform horizontally in temperature and water vapor concentration.

**air mass advection**   Horizontal movement of air or air masses from one place to another.

**air mass climatology**   Description of climate in terms of frequency of occurrence of various types of air masses.

**air mass modification**   Changes in the temperature, humidity, or stability of an air mass as it travels away from its source region.

**air mass thunderstorm**   A thunderstorm that develops almost randomly within a mass of maritime tropical air; usually most common during the warmest time of day.

**air pollutants**   Atmospheric gases and aerosols that occur in concentrations that threaten the well-being of living organisms or that disrupt the orderly functioning of the environment.

**air pollution episode**   Period when atmospheric conditions inhibit dilution and dispersal of air pollutants; pollutants thus pose a hazard to human health.

**air pressure**   The cumulative force exerted on any surface by the molecules composing air; usually expressed as the weight of a column of air.

**air pressure gradient**   Change in air pressure with distance.

**air pressure tendency**   Change in air pressure with time; on a surface weather map, the air pressure change over the prior 3 hours.

**albedo**   The fraction of radiation striking a surface that is reflected by that surface.

**altimeter**   An aneroid barometer calibrated to read altitude or elevation.

**altocumulus**   Middle clouds consisting of roll-like patches or puffs forming a wavy pattern.

**altocumulus lenticularis**   A lens-shaped altocumulus cloud; a mountain-wave cloud generated by the disturbance of horizontal airflow by a prominent mountain range.

**altostratus**   Middle layer clouds that are uniformly gray or white.

**aneroid barometer**   A portable instrument that utilizes a flexible metal chamber and spring to measure air pressure; may be used as an altimeter.

**anomalies**   Departures of temperature, precipitation, or other weather element from long-term average values.

**Antarctic Circle**   Latitude 66 degrees 33 minutes S. Poleward of this latitude, there are 24 hours of sunlight at the summer solstice and 24 hours of darkness at the winter solstice.

**Antarctic ozone hole**   A depletion of ozone in the Antarctic stratosphere that occurs in spring; attributed to the action of chlorine liberated from CFCs.

**anticyclone**   A dome of air that exerts relatively high pressure compared with the surrounding air; same as a *high*.

In the Northern Hemisphere, surface winds in an anticyclone blow clockwise and outward.

**aphelion**   The time of the year when the Earth is farthest from the sun (about 4 July).

**arctic air (A)**   A very cold and dry air mass that forms primarily in winter over the Arctic Basin, Greenland, and the northern interior of North America.

**Arctic Circle**   Latitude 66 degrees 33 minutes N. Poleward of this latitude, there are 24 hours of sunlight at the summer solstice and 24 hours of darkness at the winter solstice.

**arctic high**   An anticyclone that forms in the source regions for cold, dry arctic air.

**Arctic sea smoke**   Fog that develops when extremely cold, dry air flows over a body of relatively warm open water; a type of steam fog.

**atmosphere**   A thin envelope of gases (also containing suspended solid and liquid particles and clouds) that encircles the globe.

**atmospheric stability**   Property of ambient air that either enhances (unstable) or suppresses (stable) vertical motion of air parcels; depends on the vertical temperature profile of the ambient air and whether the air parcels are saturated (cloudy) or unsaturated (clear).

**atmospheric windows**   Infrared wavelength bands within which there is little or no absorption by the major greenhouse gases (e.g., $H_2O$, $CO_2$, $O_3$).

**aurora australis**   Southern Hemisphere equivalent of the aurora borealis.

**aurora borealis**   Lights, visible at night in the Northern Hemisphere, produced by electrical activity in the ionosphere; northern lights.

**auroral ovals**   Geographical areas where the aurora is visible; in the Northern Hemisphere, centered on northwestern Greenland.

**Automated Surface Observing System (ASOS)**   Meteorological sensors that record and transmit atmospheric conditions automatically; a component of the modernization program of the National Weather Service.

**back-door cold front**   A surge of relatively cold air that advances from the east or northeast; usually along the North Atlantic coast.

**banner cloud**   A cloud that forms over the summit of a lofty mountain peak.

**barograph**   A recording instrument that provides a continuous trace of air pressure variation with time.

**barometer**   An instrument used to monitor variations in air pressure. *See also* aneroid barometer; mercurial barometer.

**Beaufort scale**   A scale of wind strength originally based on visual assessment of the effects of the wind on seas.

**bent-back occlusion**   Occurs when an occluded front begins to rotate in a counterclockwise sense (in the Northern Hemisphere) about the center of an extratropical cyclone.

**Bergeron process**   Precipitation formation in cold clouds whereby ice crystals grow at the expense of supercooled water droplets.

**blackbody**   A perfect radiator and absorber; a material that absorbs 100% of the radiation striking it and emits the maximum possible radiation at all wavelengths.

**blizzard warning**   Issued when falling or blowing snow is accompanied by winds in excess of 55 km (34 mi) per hour and reduced visibility.

**blocking system**   A cutoff cyclone or anticyclone that blocks the usual west-to-east progression of weather systems.

**bora**   A cold katabatic wind that originates in Yugoslavia and flows onto the coastal plain of the Adriatic Sea.

**boundary layer (skin)**   A very thin layer of still air in immediate contact with the skin that helps insulate the body from excess heat loss.

**Bowen ratio**   The ratio of heat energy used for sensible heating to heat energy used for latent heating.

**British thermal unit (Btu)**   The quantity of heat needed to raise the temperature of one pound of water one Fahrenheit degree (from 62 to 63 °F).

**calorie (cal)**   The amount of heat needed to raise the temperature of one gram of water one Celsius degree (from 14.5 to 15.5 °C).

**central pressure**   The lowest surface pressure in a cyclone.

**centripetal force**   An inward-directed force that confines an object to a curved path; the resultant of other forces.

**chinook wind**   Air that is adiabatically compressed as it is drawn down the leeward slopes of a mountain range. As a consequence, the air is warm and dry.

**chromosphere**   Portion of the sun above the photosphere; consists of ionized hydrogen and helium at 4000 to 40,000 °C.

**circumpolar vortex**   Belt of planetary-scale westerlies that surrounds the Antarctic continent.

**cirrocumulus**   A high cloud exhibiting a wavelike pattern of small white puffs; composed of ice crystals.

**cirrostratus**   A high, layered cloud composed of ice crystals; forms a thin white veil over the sky.

**cirrus**   High thin clouds occurring as silky strands and composed of ice crystals.

**climate**   Weather of some locality averaged over some time period plus extremes in weather behavior observed during the same period.

**climatic norm (or normal)**   Encompasses the total variation in the climatic record, that is, averages plus extremes.

**Climatic Optimum**   A period about 5000 to 7000 years ago when global temperatures were somewhat higher than at present.

**climatology**   The study of climate, its controls and variability.

**cloud condensation nuclei (CCN)**   Tiny solid and liquid particles on which water vapor condenses.

**cloud seeding**   An attempt to stimulate natural precipitation processes by injecting nucleating agents, such as silver iodide, into clouds.

**cold air advection**   Flow of air from relatively cold localities to relatively warm localities.

**cold clouds**   Clouds at temperatures below 0 °C (32 °F); composed of ice crystals or supercooled water droplets or a mixture of both.

**cold-core anticyclones (or highs)**   Shallow high-pressure systems that coincide with domes of relatively cold, dry air.

**cold-core lows**   Cyclones that occupy relatively cold columns of air; migrating midlatitude low-pressure systems whose circulation intensifies with altitude.

**cold front**   A narrow zone of transition between relatively cold, dense air that is advancing and relatively warm, less dense air that is retreating.

**cold-type occlusion**   A front formed when a cold front overtakes a warm front and the air behind the front is colder than the air ahead of the front.

**collision–coalescence process**   The growth of cloud droplets into raindrops within warm clouds; droplets merge upon impact.

**comma cloud**   The pattern of cloudiness associated with a wave cyclone; the head of the comma stretches from the low center to the northwest and its tail follows along the cold front.

**compressional warming**   A temperature rise that accompanies a pressure increase on a parcel of air, as when air parcels descend within the atmosphere.

**conceptual model**   Describes the general functional relationships among components of a system.

**condensation**   Process by which water changes phase from a vapor to a liquid.

**conditional stability**   Property of ambient air that is stable for unsaturated (clear) air parcels and unstable for saturated (cloudy) air parcels.

**conduction**   Flow of heat in response to a temperature gradient within an object or between objects that are in physical contact.

**confluence zone**   Zone where air streams having different origins (and often different properties) come together.

**continental drift**   The slow movement of continents across the face of the globe; continents are part of gigantic plates.

**continental polar (cP) air**   Relatively dry air mass that develops over the northern interior of North America; very cold in winter and relatively mild in summer.

**continental tropical (cT) air**   Warm, dry air mass that forms over the subtropical deserts of the southwestern United States and Mexico.

**convection**   Vertical air circulation in which warm air rises and cool air sinks.

**convective condensation level (CCL)**   The altitude at which air rising in convection cells reaches saturation; coincides with the base of cumuliform clouds.

**convergence**   A wind pattern whereby there is a net inflow of air toward some central point.

**conveyor-belt model**   A three-dimensional depiction of the circulation in fronts and cyclones in terms of three interacting airstreams, often referred to as conveyor belts.

**cooling degree-days**   A measure of the need for air conditioning when the average daily temperature is above 65 °F (18 °C); computed by subtracting 65 °F from the average daily temperature in °F.

**Coriolis effect**   A deflective force arising from the rotation of the Earth on its axis; affects principally synoptic-scale and planetary-scale winds. Winds are deflected to the right of their initial direction in the Northern Hemisphere and to the left in the Southern Hemisphere.

**corona**   Outermost region of the sun; consists of highly rarefied gases at 1 to 4 million °C.

**corona (optical)**   Colored rings that appear about the moon or sun; due to diffraction of light by spherical cloud droplets.

**cumuliform clouds**   Exhibit significant vertical development; often produced by updrafts in convection cells.

**cumulonimbus**   Thunderstorm cloud that forms as a consequence of deep convection in the atmosphere.

**cumulus**   Cloud that develops as a consequence of the updraft in a convection cell; resembles huge puff of cotton floating in the sky.

**cumulus congestus**   An upward-building convective cloud with vertical development between that of a cumulus cloud and that of a cumulonimbus cloud.

**cumulus stage**   Initial stage in the life cycle of a thunderstorm cell; consists of towering cumulus clouds with updrafts throughout the system.

**cup anemometer**   An instrument used to monitor wind speed. Wind rotates cups and generates an electric current that is calibrated in wind speed.

**cyclogenesis**   Birth and development of a cyclone, a low-pressure system.

**cyclolysis**   Process whereby a cyclone weakens, its central pressure rises, and its winds slacken.

**cyclone**   A weather system characterized by relatively low surface air pressure compared with the surrounding air;

same as a *low.* Surface winds blow counterclockwise and inward in the Northern Hemisphere.

**Dalton's law**　The total pressure exerted by a mixture of gases is equal to the sum of the partial pressures of the constituent gases.

**dart leaders**　Surges of negative electrical charge that follow the conductive path formed by the initial stepped leaders and return stroke of a lightning bolt.

**deepening**　Process whereby the central pressure of a developing cyclone drops.

**degree-day**　*See* cooling degree-days, heating degree-days.

**deposition**　Process by which water changes phase directly from a vapor into a solid. An example is frost formation.

**deposition nuclei**　Tiny particles on which water vapor deposits directly as ice.

**dew**　Water droplets formed by condensation of water vapor on a relatively cold surface.

**dew point** or **dew-point temperature**　Temperature to which air must be cooled at constant pressure to achieve saturation (if above 0 °C or 32 °F).

**diffraction**　Slight bending of a light wave as it moves along the boundary of an object such as a water droplet.

**diffuse insolation**　Solar radiation that is scattered or reflected by atmospheric components (e.g., clouds or aerosols) to the Earth's surface.

**direct insolation**　Solar radiation that is transmitted directly through the atmosphere to the Earth's surface without interacting with atmospheric components.

**dissipating stage (thunderstorm)**　The final phase in the life cycle of a thunderstorm cell; features downdrafts throughout the system and vaporization of clouds.

**divergence**　A wind pattern whereby there is a net outflow of air from some central point.

**doldrums**　An east-west belt of light and variable surface winds where the trade winds of the two hemispheres converge.

**Doppler effect**　A shift in frequency of an electromagnetic or sound wave due to the relative movement of the source or the observer.

**Doppler radar**　Conventional weather radar that has the added capability of determining the detailed motion of target precipitation (or aerosols) based on the frequency shift between the outgoing and returning radar beam.

**downburst**　A strong and potentially destructive thunderstorm downdraft; depending on size, classified as either a microburst or a macroburst.

**drizzle**　A form of liquid precipitation consisting of water droplets less than 0.5 mm (0.02 in.) in diameter; falls from low stratus clouds.

**dropwindsonde**　A small instrument package equipped with a radio transmitter that is dropped from an aircraft and

measures vertical profiles of temperature, pressure, relative humidity, and wind.

**drought**　An extended period of anomalous moisture deficit.

**dry adiabatic lapse rate**　Rising unsaturated air parcels cool at the rate of about 10 Celsius degrees per 1000 m of uplift (or 5.5 Fahrenheit degrees per 1000 ft).

**dry deposition**　Removal of suspended particulates from the air through impaction and gravitational settling; a natural means whereby air is cleansed of pollutants.

**dry line**　Boundary between warm, dry air and warm, humid air in the southeast sector of a mature midlatitude cyclone; likely site for severe thunderstorm development.

**dust devil**　Swirling mass of dust caused by intense solar heating of dry surface areas.

**dust dome**　An accumulation of visibility-restricting aerosols in the air over an urban-industrial area.

**dust plume**　A dust dome that elongates downwind.

**eddy viscosity**　Frictional resistance arising from eddies (irregular whirls) within a fluid such as air or water.

**El Niño**　Anomalous warming of surface ocean waters in the eastern tropical Pacific; accompanied by suppression of upwelling off the coasts of Ecuador and northern Peru.

**El Niño/Southern Oscillation (ENSO)**　An episode of anomalously high sea-surface temperatures in the equatorial and tropical eastern Pacific; associated with large-scale swings in surface air pressure between the western and eastern tropical Pacific.

**electromagnetic radiation**　Energy transfer in the form of waves that have both electrical and magnetic properties; takes place even in a vacuum.

**electromagnetic spectrum**　Range of radiation types arranged by wavelength and frequency.

**ENSO**　*See* El Niño/Southern Oscillation.

**entrainment**　Mixing of saturated (cloudy) air with unsaturated air that surrounds the cloud.

**equinoxes**　The first days of spring and autumn when day and night are of equal length at all latitudes. The noon sun is directly over the equator.

**evaporation**　Process by which water changes phase from a liquid to a vapor at a temperature below the boiling point of water.

**evapotranspiration**　Vaporization of water through direct evaporation from wet surfaces plus the release of water vapor by vegetation.

**expansional cooling**　A temperature drop that accompanies a pressure reduction on an air parcel, as when air parcels ascend within the atmosphere.

**eye (of a hurricane)**　An area of almost cloudless skies, light winds, and gently subsiding air at the center of a hurricane.

**eye wall**   A circle of cumulonimbus clouds surrounding the eye of a mature hurricane.

**F-scale**   Tornado intensity scale, developed by T. T. Fujita, that rates tornadoes from F0 to F5 on the basis of estimated wind speed.

**faculae**   Small, bright, relatively hot spots on the sun.

**fair-weather bias**   The observation that fair-weather days out-number stormy days almost everywhere.

**filling**   Process whereby a cyclone weakens, its central pressure rises, and its winds slacken.

**first law of thermodynamics**   Energy cannot be created or destroyed but can change form; also known as the law of energy conservation.

**flash flood**   A sudden rise in river or stream levels causing flooding.

**foehn**   European term for chinook wind; warm, dry wind that is drawn down the Alpine valleys of Austria and Germany.

**fog**   A cloud in contact with the Earth's surface that reduces visibility to less than 1.0 km (0.62 mi).

**fog dispersal**   Clearing of fog either by increasing the air temperature (thereby lowering the relative humidity) or by cloud seeding.

**forced convection**   Convection aided by topographic uplift.

**free convection**   Convection triggered by intense solar heating of the Earth's surface.

**freezing rain**   Supercooled raindrops that freeze on contact with cold surfaces.

**friction**   The resistance an object encounters as it comes into contact with other objects; in fluids known as *viscosity*.

**friction layer**   Zone of the atmosphere, between the Earth's surface and an altitude of about 1000 m (3280 ft), where most frictional resistance is confined.

**front**   A narrow zone of transition between air masses of contrasting density, that is, air masses of different temperatures or different water vapor concentrations or both.

**frontal fog**   A cloud in contact with the Earth's surface, formed when precipitation falls from relatively warm air into a wedge of relatively cool air near the surface and raises the vapor pressure in the cool air to saturation; occurs either just ahead of a warm front or just behind a cold front.

**frontal thunderstorm**   A thunderstorm associated with lifting of air along the surface of a front, usually a cold front.

**frontogenesis**   Development or strengthening of a front.

**frontolysis**   Dissipation of a front.

**frost**   Ice crystals formed by deposition of water vapor on a relatively cold surface.

**frost point** or **frost-point temperature**   Temperature to which air must be cooled at constant pressure to achieve saturation at or below 0°C (32°F).

**fumigation**   Atmospheric stability conditions that favor trapping of air pollutants within an air layer adjacent to the Earth's surface.

**funnel cloud**   A tornadic circulation extending below cloud base but not reaching the ground.

**gamma radiation**   Electromagnetic radiation having very short wavelength and great penetrating power.

**gas law**   Relationship among the variables of state; pressure is proportional to the product of density and temperature.

**geologic time**   A span of millions or billions of years in the past.

**Geostationary Operational Environmental Satellites (GOES)**   Geosynchronous satellites designed to monitor weather systems.

**geostrophic wind**   Unaccelerated horizontal wind that flows in straight paths above the friction layer; results from a balance between the horizontal pressure gradient force and the Coriolis effect.

**geosynchronous satellite**   A satellite that orbits Earth at the same rate as the Earth's rotation, so it always scans the same region of the planet.

**glacial climate**   Conditions favorable to the initiation and growth of glacial ice.

**glaze**   A transparent layer of ice caused by slow cooling of supercooled water.

**global radiative equilibrium**   The balance between net incoming solar radiation and the infrared radiation emitted to space by the Earth–atmosphere system.

**global water budget**   Balance sheet for the inputs and outputs of water to and from the various global water reservoirs; shows a net flow of water from land to sea.

**glory**   Concentric rings of color about the shadow of an observer's head that appear on the top of a cloud situated below the observer; caused by the same optics as a rainbow plus diffraction.

**gradient wind**   Large-scale, horizontal, and frictionless wind that describes a curved path.

**granules**   A network of huge, irregularly shaped convective cells in the sun's photosphere.

**graphical model**   A compilation or display of data in a form that can be readily useful; a weather map is an example.

**graupel**   Tiny granules of ice or compact snow; a form of precipitation.

**gravitation**   Force of attraction between objects that is directly proportional to the product of the masses of the objects and inversely proportional to the square of the distance between them.

**gravitational settling**   Drifting of aerosols downward toward the Earth's surface under the influence of gravity.

**gravity**   The force that holds all objects on the Earth's surface; the net effect of gravitation and the centripetal force due to the Earth's rotation.

**greenhouse effect** Although nearly transparent to solar radiation, the atmosphere is much less transparent to infrared radiation. Terrestrial infrared radiation is absorbed and reradiated primarily by water vapor and to a lesser extent by carbon dioxide and other gases, thereby slowing the escape of heat to space from the Earth–atmosphere system.

**greenhouse gases** Gases such as water vapor and carbon dioxide that contribute to the greenhouse effect in the Earth–atmosphere system.

**Greenwich Mean Time (GMT)** A worldwide time reference used for synchronizing weather observations; the time at 0 degrees longitude, the prime meridian, which passes through Greenwich, England.

**ground clutter** Reflection of radar signals by fixed objects on the Earth's surface.

**gust front** The leading edge of a mass of relatively cool air that flows out of the base of a thunderstorm cloud (downdraft) and spreads along the ground well in advance of the parent thunderstorm cell; a mesoscale cold front.

**haboob** A dust storm caused by the downdraft of a desert thunderstorm.

**Hadley cell** Air circulation in tropical and subtropical latitudes of both hemispheres resembling a huge convective cell with rising air over the equator and sinking air in the subtropical anticyclones.

**hail** or **hailstone** Precipitation in the form of rounded or jagged chunks of ice; often characterized by internal concentric layering. Hail is associated with thunderstorms that have strong updrafts and relatively great moisture content.

**hailstreak** Accumulation of hail in a long, narrow path along the ground.

**hair hygrometer** An instrument used to monitor relative humidity by measuring the changes in the length of human hair that accompany humidity variations.

**halo** Ring of light about the sun (or moon) caused by refraction of sunlight by tiny ice crystals suspended in the upper troposphere.

**heat** The total kinetic energy of the atoms or molecules composing a substance.

**heat equator** The latitude of highest mean annual surface air temperature; at about latitude 10 degrees N.

**heat lightning** Light reflected by clouds from distant thunderstorms occurring beyond the horizon.

**heating degree-days** A measure of space heating needs on days when the average air temperature falls below 65 °F (18 °C); computed by subtracting the day's average temperature from 65 °F.

**heterosphere** The atmosphere above 80 km (50 mi) where gases are stratified, with concentrations of the heavier gases decreasing more rapidly with altitude than concentrations of the lighter gases.

**hoarfrost** Fernlike crystals of ice formed by deposition of water vapor on twigs, tree branches, and other exposed structures.

**homosphere** The atmosphere up to 80 km (50 mi) in which the proportions of principal gaseous constituents, such as oxygen and nitrogen, are constant.

**hook echo** A distinctive radar pattern that often indicates the presence of a severe thunderstorm and perhaps tornadic circulation.

**horse latitudes** Areas of calm winds associated with subtropical anticyclones; near 30 degrees latitude in both hemispheres.

**hot wire anemometer** An instrument that measures wind speed based on the rate of heat loss to air flowing by a sensor.

**hurricane** Intense warm-core oceanic cyclone that originates in tropical latitudes; called *typhoon* in the western Pacific Ocean. Sustained winds are in excess of 119 km (74 mi) per hour.

**hydrologic cycle** Ceaseless flow of water among terrestrial, oceanic, and atmospheric reservoirs.

**hydrostatic equilibrium** Balance between the vertical air pressure gradient force (directed upward) and the force of gravity (directed downward).

**hygrograph** An instrument that provides a continuous trace of relative humidity variations with time.

**hygroscopic nuclei** Tiny particles of matter that have a special chemical affinity for water molecules, so that condensation may take place on these nuclei at relative humidities under 100%.

**hypothesis** A proposed explanation for some observation or phenomenon; tested through the scientific method.

**ice-forming nuclei (IN)** Tiny particles that promote the formation of ice crystals at temperatures well below freezing; include freezing nuclei and deposition nuclei.

**ice pellets** Frozen raindrops that bounce on impact with the ground; also called *sleet*.

**impaction** Removal of aerosols from air through their impact on buildings and other objects at the Earth's surface; a natural atmospheric cleansing mechanism.

**index of continentality** Degree of maritime influence on continental air temperatures; usually based on the contrast between mean summer and mean winter temperatures.

**Indian summer** A period of mild, sunny weather that occurs in autumn over eastern North America; usually after the first freeze.

**infrared radiation** Electromagnetic radiation at wavelengths shorter than microwaves and longer than visible red light; emitted by most objects on Earth.

**insolation** Incoming solar radiation. *See also* diffuse insolation, direct insolation.

**interglacial climate** Conditions that favor the melting of glacial ice (if present).

**intertropical convergence zone (ITCZ)** Discontinuous belt of thunderstorms paralleling the equator and marking the convergence of the Northern and Southern Hemisphere surface trade winds.

**ion** An electrically charged atom or molecule.

**ionosphere** Region of the upper atmosphere from 80 to 900 km (50 to 600 mi) that contains a relatively high concentration of ions (electrically charged particles).

**isobars** Lines on a weather map joining localities reporting the same air pressure.

**isothermal** Constant temperature.

**isotherms** Lines on a weather map joining locations having the same air temperature.

**ITCZ** *See* intertropical convergence zone.

**January thaw** A period of relatively mild weather around 20–23 January that occurs primarily in New England; an example of a singularity in the climatic record.

**jet maximum** An area of accelerated air flow within a jet stream; same as a jet core.

**jet stream** Relatively narrow corridor of very strong winds embedded in the planetary winds aloft.

**katabatic wind** Downslope flow of cold, dense air under the influence of gravity.

**kinetic energy** Energy possessed by any object in motion.

**La Niña** A period of strong trade winds and unusually low sea-surface temperatures in the central and eastern tropical Pacific; opposite of ENSO.

**lake breeze** A relatively cool mesoscale surface wind directed from a lake toward land in response to differential heating between land and lake; develops during the day.

**lake-effect snow** Snowfall, often highly localized, that occurs along the lee shore of a large lake when cold air flows across a long fetch of relatively mild open water; the cold air becomes less stable and more humid, and uplift along the shoreline triggers cloud development and snowfall.

**land breeze** A relatively cool mesoscale surface wind directed from land to sea or from land to lake in response to differential cooling between land and water body; develops at night.

**latent heat of fusion** Heat released when water changes phase from liquid to solid; 80 calories per gram.

**latent heat of melting** Heat required to change the phase of water from solid to liquid; 80 calories per gram.

**latent heat of vaporization** Heat required to change the phase of water from liquid to vapor; 540 to 600 calories per gram, depending on the temperature of the water.

**latent heating** Transport of heat from one place to another within the atmosphere as a consequence of phase changes of water. Heat is supplied for evaporation and sublimation of water at the Earth's surface, and heat is released with condensation and deposition (cloud formation) within the atmosphere.

**law of energy conservation** Energy is neither created nor destroyed but can change from one form to another; same as the first law of thermodynamics.

**law of reflection** The angle of incident radiation is equal to the angle of reflected radiation.

**lifting condensation level (LCL)** The altitude to which air must be lifted so that expansional cooling leads to condensation (or deposition) and cloud development; corresponds to the base of clouds.

**lightning** A flash of light produced by an electrical discharge in response to the buildup of an electrical potential between cloud and ground, between clouds, or within a single cloud.

**lightning detection network (LDN)** System that provides real-time information on the location and severity of lightning strokes.

**Little Ice Age** A period of relatively cold conditions in many regions of the globe from about 1400 to 1850.

**low-level jet stream** A surge of maritime tropical air in the lower troposphere northward from the Gulf of Mexico.

**macroburst** A downburst that affects a path longer than 4.0 km (2.5 mi).

**magnetosphere** The region of the upper atmosphere encompassed by the Earth's magnetic field; deflected by the solar wind into a teardrop-shaped cavity.

**mammatus clouds** Clouds that form on the underside of a thunderstorm anvil and exhibit pouchlike, downward protuberances; may indicate turbulent air.

**maritime polar (mP) air** Cool, humid air masses that form over the cold ocean waters of the North Pacific and North Atlantic.

**maritime tropical (mT) air** Warm, humid air masses that form over tropical and subtropical oceans.

**mature stage (thunderstorm)** The middle and most intense phase of the life cycle of a thunderstorm cell; begins when precipitation reaches the Earth's surface and is characterized by both updrafts and downdrafts.

**Maunder minimum** A 70-year period from 1645 to 1715 when sunspots were rare.

**McIDAS (Man-computer Interactive Data Access System)** A real-time weather data management system developed at the University of Wisconsin–Madison.

**mercurial barometer** A mercury-filled tube used to measure air pressure; the standard barometric instrument having great precision.

**meridional flow pattern** Flow of westerlies in a series of deep troughs and sharp ridges; westerlies exhibit considerable amplitude.

**mesocyclone** A stage in the development of a tornado; consists of a spinning cylinder of air 10 to 20 km (6.2 to 12.4 mi) in diameter within the updraft of a severe thunderstorm.

**mesopause** Narrow zone of transition between the mesosphere below and the thermosphere above; top of the mesosphere.

**mesoscale convective complex (MCC)** A nearly circular cluster of many interacting thunderstorms covering an area of many thousands of square kilometers.

**mesoscale systems** Weather phenomena operating at the local scale; include thunderstorms and sea breezes, for example.

**mesosphere** Thermal subdivision of the upper atmosphere in which air temperature declines with altitude; situated above the stratosphere and below the thermosphere.

**meteorology** Scientific study of the atmosphere and atmospheric processes.

**microburst** A downburst that affects a path that is 4.0 km (2.5 mi.) or shorter.

**microscale system** The smallest spatial subdivision of atmospheric circulation; a tornado is an example.

**microwaves** Electromagnetic radiation having wavelengths in the 0.1 to 300 mm range; used in weather radars.

**midlatitude westerlies** Planetary-scale prevailing west-to-east winds in the mid- and upper-troposphere between about 30 and 60 degrees of latitude.

**Milankovitch cycles** Systematic changes in three elements of Earth-sun geometry: precession, axial tilt, and orbital eccentricity; likely governed large-scale fluctuations of the Earth's ice sheets during the Ice Age.

**mist** Very thin fog in which visibility is greater than 1.0 km (0.62 mi).

**mistral** A katabatic wind that flows from the Alps down the Rhone River Valley of France to the Mediterranean coast.

**mixing depth** Vertical distance between the ground and the altitude to which convection currents reach; the thickness of the mixing layer.

**mixing layer** Surface layer of the troposphere in which air is thoroughly mixed by convection. Mixing depth is the thickness of the mixing layer.

**mixing ratio** Mass of water vapor per mass of dry air; expressed as grams per kilogram.

**moist adiabatic lapse rate** A variable rate of cooling applicable to saturated air parcels (cloudy air) that are ascending within the atmosphere. This rate is less than the dry adiabatic lapse rate because some of the expansional cooling is compensated for by the release of latent heat that accompanies the phase change of water vapor.

**molecular diffusion** Mixing that takes place at the molecular scale; a relatively slow process.

**molecular viscosity** Frictional resistance arising from the interactions of molecules composing a fluid such as air or water.

**monsoon active phase** A generally cloudy period with frequent deluges of rain.

**monsoon circulation** Characterizes regions where seasonal reversals of winds cause wet summers and dry winters.

**monsoon dormant phase** A generally sunny and hot period that interrupts rainy monsoon episodes.

**mountain breeze** A shallow, gusty downslope flow of cool air that develops at night in some mountain valleys.

**mountain wave** A stationary wave situated downwind of a prominent mountain range and caused by the disturbance of the wind by the mountain range.

**mountain-wave clouds** Stationary clouds situated downwind of a prominent mountain range and formed as a consequence of the disturbance of the wind by the mountain range.

**nacreous clouds** Rarely seen clouds that form in the upper stratosphere; may be composed of ice crystals or supercooled water droplets. Also called *mother-of-pearl* clouds.

**National Climate Data Center (NCDC)** Houses archives of climatic data of the United States; located in Asheville, North Carolina.

**National Hurricane Center (NHC)** Responsible for forecasting tropical storms and hurricanes along the East and Gulf coasts and over the eastern Pacific Ocean; located in Coral Gables, Florida.

**National Oceanic and Atmospheric Administration (NOAA)** Within the U.S. Department of Commerce, the administrative unit that oversees activities of the National Weather Service.

**National Severe Storms Forecast Center (NSSFC)** Issues watches for severe thunderstorms and tornadoes; located in Kansas City, Missouri.

**National Weather Service (NWS)** The agency of NOAA responsible for weather data acquisition, data analysis, forecast dissemination, and storm watches and warnings.

**NWS Cooperative Observer Network** Consists of more than 10,000 weather stations across the United States that record data for hydrologic, agricultural, and climatic purposes.

**neutral air layer** An air layer in which an ascending or descending air parcel always has the same temperature as its surroundings.

**Newton's first law of motion** An object at rest or in straight-line, unaccelerated motion remains that way unless acted upon by a net external force.

**Newton's second law of motion** A net force is required to cause a unit mass of a substance to accelerate (or decelerate); force = mass × acceleration.

**nimbostratus** Low, gray, layered clouds that resemble stratus clouds but are thicker and yield more substantial precipitation.

**noctilucent clouds** Occur in the upper mesosphere, are rarely seen, and are probably composed of meteoric dust.

**Norwegian cyclone model** The original description of the structure and life cycle of a midlatitude low-pressure system, first proposed during World War I by researchers at the Norwegian School of Meteorology at Bergen.

**nuclei** Tiny solid or liquid particles of matter on which condensation or deposition of water vapor takes place.

**numerical model** One or more mathematical expressions that approximate the behavior of a system.

**occluded front** A front formed when a cold front overtakes a warm front; represents the final stage in the life cycle of a midlatitude cyclone. *See also* cold-type occlusion and warm-type occlusion.

**occlusion** The final stage in the life cycle of a midlatitude cyclone. Occlusion occurs when the cold front overtakes the warm front.

**orographic lifting** The forced rising of air up the slopes of a hill or mountain.

**overrunning** The process whereby less dense air displaces more dense air by flowing up and over the denser air; characterizes a warm front.

**ozone hole** A thinning of stratospheric ozone ($O_3$) over Antarctica during the Southern Hemisphere spring; recent deepening of the ozone hole appears linked to CFCs.

**ozone shield** Ozone ($O_3$) within the stratosphere that filters out potentially lethal intensities of ultraviolet radiation (UV) from the sun.

**Pacific air** A North American air mass originating over the Pacific Ocean that travels through the western mountains and emerges on the Great Plains warmer and drier than in its source region.

**parhelia** Two bright spots of light appearing on either side of the sun; each is separated from the sun by an angle of 22 degrees. Parhelia are caused by refraction of sunlight by ice crystals; also called *mock suns* and *sundogs*.

**penumbra** Outer, lighter area of a sunspot.

**perihelion** The time of year when the Earth's orbital path brings it closest to the sun (about 3 January).

**persistence** Tendency for weather episodes to continue for some period of time.

**pH scale** A measure of the range of acidity and alkalinity of different substances. A pH of 7 is neutral; acids have pH values less than 7, and alkaline substances have pH values greater than 7.

**photodissociation** The process by which radiation breaks down molecules into their components.

**photosphere** The visible surface of the sun.

**photosynthesis** The process whereby plants use sunlight, water, and carbon dioxide to manufacture their food.

**physical model** A miniaturized version of a real system.

**planetary albedo** The fraction of solar radiation that is scattered and reflected back into space by the Earth–atmosphere system.

**planetary-scale systems** The largest spatial scale of atmospheric circulation; includes the global wind belts and semipermanent pressure systems.

**polar amplification** Refers to the tendency of a major temperature change to increase with latitude.

**polar front** Transition zone between cold polar easterlies and mild midlatitude westerlies.

**polar front jet stream** A jet stream situated in the upper troposphere between the midlatitude tropopause and the polar tropopause and directly over the polar front.

**polar high** Cold anticyclone that originates in a source region for continental polar air.

**polar-orbiting satellite** Satellite in relatively low orbit that travels over or near the geographical poles on meridional trajectories.

**poleward heat transport** Flow of heat from tropical to middle and high latitudes in response to latitudinal imbalances in radiational heating and cooling. Poleward heat transport is accomplished by air mass exchange (primarily), storms, and surface ocean currents.

**precipitable water** The depth of water produced when all the water vapor in a column of air is condensed; usually the column of air reaches from the Earth's surface to the *top* of the atmosphere.

**precipitation** Water in solid or liquid form that falls to the Earth's surface from clouds.

**pressure gradient force** A force operating in the atmosphere that accelerates air parcels away from regions of high pressure toward regions of low pressure in response to an air pressure gradient.

**primary air pollutants** Substances that are pollutants immediately upon entering the atmosphere.

**psychrometer** An instrument used to determine relative humidity based on the difference in readings between a dry-bulb thermometer and a wet-bulb thermometer.

**pyranometer** The standard instrument for measuring solar radiation incident on a horizontal surface; calibrates the temperature response of a special sensor in units of radiation flux.

**radar (radio detection and ranging)**   An instrument that sends and receives microwaves for the purpose of determining the location and movement of areas of precipitation.

**radar echo**   Microwaves scattered by distant rain or snow back to a receiver where they are displayed as bright spots on a cathode ray tube.

**radiation**   Energy transport via electromagnetic waves; capable of traveling through a vacuum.

**radiation fog**   A ground-level cloud formed by nocturnal radiational cooling of a humid air layer so that its relative humidity approaches 100%.

**radiational temperature inversion**   Cooling of a surface air layer by loss of infrared radiation, so that the coldest air is adjacent to the Earth's surface and the air temperature increases with altitude.

**radio waves**   Long-wavelength, low-frequency electromagnetic waves.

**radiosonde**   A small balloon-borne instrument package equipped with a radio transmitter that measures vertical profiles of temperature, pressure, and relative humidity in the atmosphere.

**rain**   A form of precipitation consisting of liquid water drops having diameters between 0.5 and 5.0 mm (0.02 and 0.2 in.).

**rainbow**   An arc of concentric colored bands formed by refraction and internal reflection of sunlight by falling raindrops. Observer must be looking at a distant rain shower with the sun at his/her back.

**rain gauge**   A device—usually a cylindrical container—for collecting and measuring rainfall.

**rain shadow**   A region situated downwind of a high mountain barrier and characterized by descending air and, as a consequence, a relatively dry climate.

**rawinsonde**   A radiosonde tracked from the ground by a direction-finding antenna to measure variations in horizontal wind direction and wind speed with altitude.

**reduction to sea level**   An adjustment applied to surface air pressure readings in order to eliminate the influence of station elevation.

**reflection**   The process whereby a portion of the radiation that is incident on a surface is reflected by that surface.

**refraction**   The bending of a light ray as it passes from one transparent medium to another (e.g., from air to water). The bending is due to the differing speeds of light in the two media.

**relative humidity**   A measure of how close air is to saturation at a specific temperature; expressed as a percentage.

**return stroke**   A positively charged electrical current that emanates from the ground and meets a downward-moving stepped leader; part of a lightning stroke.

**rime**   An opaque, granular layer of ice formed by the rapid freezing of supercooled water.

**roll cloud**   A low, cylindrically shaped, and elongated cloud occurring behind a gust front; associated with but detached from a cumulonimbus cloud.

**Rossby waves**   Series of long-wavelength troughs and ridges that characterize the planetary-scale westerlies; also called *long waves*.

**Saffir–Simpson Hurricane Intensity Scale**   Hurricane intensity scale based on central pressure, wind speed, and storm surge and damage potential; 1 is minimal, 5 is most intense.

**Santa Ana wind**   A hot, dry (chinook-type) wind that blows downward from the desert plateaus of Utah and Nevada toward coastal southern California.

**saturation mixing ratio**   Maximum concentration of water vapor in a given volume of air at a specific temperature.

**saturation vapor pressure**   The maximum vapor pressure in a sample of air at a specific temperature.

**scattering**   Process whereby small particles disperse radiation in all directions.

**scavenging**   Removal of gases and aerosols from the atmosphere by precipitation; the most effective natural cleansing process operating within the troposphere.

**scientific method**   A systematic form of inquiry that involves observation, speculation, and testing of hypotheses.

**scientific model**   An approximation or simulation of a real system; omits all but the most essential variables of the system.

**sea breeze**   A relatively cool mesoscale surface wind directed from sea toward land in response to differential heating between land and sea; develops during the day.

**secondary air pollutants**   Pollutants generated by chemical reactions occurring within the atmosphere.

**secondary cyclone**   A storm that develops following the occlusion of another cyclone; usually forms at the triple point.

**second law of thermodynamics**   All systems tend toward disorder.

**semipermanent pressure systems**   Persistent cyclones and anticyclones that are components of the planetary-scale circulation. These pressure cells exhibit some seasonal changes in location and in mean surface pressures.

**sensible heating**   The transport of heat from one location or object to another via conduction, convection, or both.

**severe thunderstorms**   Thunderstorms accompanied by locally damaging surface winds, frequent lightning, or large hail.

**sheet lightning**   Bright flashes across the sky due to cloud-to-cloud electrical discharges.

**shelf cloud**   A low, wedge-shaped, and elongated cloud that occurs along a gust front; associated with and attached to a cumulonimbus cloud.

**short waves**   Relatively small ripples (troughs and ridges) superimposed on long waves in the planetary westerlies.

**single-station forecasts** Weather forecasts based on observations at one location.

**singularity** A weather event that occurs on or near a certain date with unusual regularity; the January thaw is an example.

**snow** A type of precipitation consisting of an assemblage of ice crystals in the form of plates, columns, or flakes.

**snowbelts** Areas to the lee of large lakes such as the Great Lakes that are subject to frequent lake-effect snows.

**snow grains** Frozen form of precipitation consisting of particles of white ice having diameters less than 1 mm; originates in the same way as drizzle.

**snow pellets** Frozen form of precipitation consisting of soft conical or spherical white ice particles having diameters of 1 to 5 mm.

**solar altitude** The angle of the sun above the horizon.

**solar constant** The flux of solar radiational energy falling on a surface positioned at the top of the atmosphere and oriented perpendicular to the solar beam when Earth is at its average distance from the sun.

**solar flares** Gigantic disturbances on the sun that emit to space high-velocity streams of electrically charged subatomic particles.

**solar irradiance** The rate of total radiational energy output of the sun.

**solar wind** A stream of charged subatomic particles (mainly protons and electrons) flowing out into space from the sun.

**solstices** When the sun is at its maximum poleward locations relative to the Earth (23 degrees 30 minutes North and South); first days of summer and winter.

**soundings** Continuous altitude measurements that provide profiles of such variables as temperature, humidity, and wind speed.

**southern oscillation** Opposing swings of surface air pressure between the eastern and western tropical Pacific Ocean.

**specific heat** Amount of heat required to raise the temperature of 1 gram of a substance by 1 Celsius degree.

**specific humidity** The mass of water vapor per mass of air containing the water vapor; usually expressed as grams of water vapor per kilogram of air.

**split flow pattern** Westerlies to the north have a wave configuration that differs from that of westerlies to the south.

**Spörer minimum** A period of reduced sunspot activity from 1450 to 1550.

**squall line** A line of intense thunderstorms occurring parallel to and ahead of a fast-moving, well-defined cold front.

**stable air layer** Air layer characterized by a vertical temperature profile such that air parcels return to their original altitudes following upward or downward displacements.

**standard atmosphere** The mean vertical profiles of temperature, pressure, and density within the atmosphere.

**station model** A conventional representation on a weather map, using standard symbols, of weather conditions at some locality.

**stationary front** A nearly stationary narrow zone of transition between contrasting air masses; winds blow parallel to the front but in opposite directions on the two sides of the front.

**steam fog** The general name for fog produced when cold air comes in contact with relatively warm water; has the appearance of rising streamers.

**Stefan–Boltzman law** The total energy radiated by a blackbody at all wavelengths is directly proportional to the fourth power of the absolute temperature (kelvins) of the body.

**stepped leaders** The initial electrical discharges in a lightning flash; consist of negative electrical charge that travels from a cloud base to within 100 m (328 ft) of the ground.

**storm surge** A hurricane-induced rise in sea level that reaches the shoreline ahead of the storm; caused by strong winds and low air pressure.

**stratiform clouds** Layered clouds, such as altostratus, often produced by air mass overrunning.

**stratocumulus** Low clouds consisting of large, irregular puffs or rolls arranged in a layer.

**stratopause** Transition zone between the stratosphere and the mesosphere.

**stratosphere** The atmosphere's thermal subdivision situated between the troposphere and mesosphere; primary site of ozone formation. Within the stratosphere, air temperature is constant in the lower portion and then increases with altitude.

**stratus** Low clouds that occur as a uniform gray layer stretching from horizon to horizon. They may produce drizzle, and where they intersect the ground, are classified as fog.

**streamlines** The flow pattern of air moving horizontally; on a map, lines that are drawn everywhere parallel to wind direction.

**sublimation** Process by which water changes phase from a solid into a vapor without first becoming liquid.

**subpolar lows** High-latitude, semipermanent cyclones marking the convergence of planetary-scale surface southwesterlies of midlatitudes with surface northeasterlies of polar latitudes; Icelandic low and Aleutian low are examples.

**subsidence temperature inversion** A temperature inversion that develops aloft as a result of air gradually sinking over a wide area and being warmed by adiabatic compression; occurs on the eastern flanks of the subtropical anticyclones.

**subtropical anticyclones** Semipermanent warm-core, high-pressure systems centered over subtropical latitudes of the Atlantic, Pacific, and Indian oceans.

**subtropical jet stream**    A zone of unusually strong winds situated between the tropical tropopause and the midlatitude tropopause.

**sundogs**    *See* parhelia.

**sunspot**    Relatively large, dark blotch that appears on the face of the sun.

**synoptic-scale systems**    Weather phenomena operating at the continental or oceanic spatial scale; includes migrating cyclones and anticyclones, air masses, and fronts.

**teleconnection**    A linkage between weather changes occurring in widely separated regions of the globe.

**temperature**    A measure of the average kinetic energy of the individual atoms or molecules composing a substance.

**temperature gradient**    Temperature change with distance.

**temperature inversion**    An extremely stable air layer in which temperature increases with altitude, the inverse of the usual temperature profile in the troposphere.

**terminal velocity**    Constant downward-directed speed of a particle within a fluid due to a balance between gravity and fluid resistance.

**thermal stability**    Resistance to a change in temperature.

**thermograph**    A recording instrument that provides a continuous trace of temperature variations with time.

**thermometer**    An instrument used to measure temperature.

**thermosphere**    Outermost thermal subdivision of the atmosphere in which temperatures increase with altitude.

**thunder**    Sound accompanying lightning; produced by violent expansion of air due to intense heating by a lightning discharge.

**thunderstorm**    A mesoscale weather system produced by strong convection currents that reach to great altitudes within the troposphere. Consists of cumulonimbus clouds accompanied by lightning and thunder and, often, locally heavy rainfall (or snowfall) and gusty surface winds.

**tipping-bucket rain gauge**    A device that collects rainfall in increments of 0.01 in. by containers that alternately fill and empty (tip).

**tornado**    A small mass of air that whirls rapidly about an almost vertical axis. The tornado is made visible by clouds and by dust and debris sucked into the system.

**tornado alley**    Region of maximum tornado frequency in North America; a corridor stretching from eastern Texas northward into Oklahoma, Kansas, and Nebraska.

**trade winds**    Prevailing planetary-scale surface winds in tropical latitudes; blow from the northeast in the Northern Hemisphere and from the southeast in the Southern Hemisphere.

**transpiration**    Process by which water vapor escapes from plants through leaf pores.

**triple point**    The point of occlusion of a cyclone where the cold, warm, and occluded fronts all come together.

**Tropic of Cancer**    Latitude 23 degrees 27 minutes N; a solstice position of the sun.

**Tropic of Capricorn**    Latitude 23 degrees 27 minutes S; a solstice position of the sun.

**tropical depression**    An early stage in the development of a hurricane; winds are at least 37 km (23 mi) per hour but less than 63 km (39 mi) per hour.

**tropical disturbance**    A region of convective activity over tropical seas with a detectable center of low pressure; the initial stage in the development of a hurricane.

**tropical storm**    A tropical cyclone having wind speeds of 63 to 119 km (39 to 74 mi) per hour; pre-hurricane stage.

**tropopause**    Zone of transition between the troposphere below and the stratosphere above; top of the troposphere.

**troposphere**    Lowest thermal subdivision of the atmosphere in which air temperature normally decreases with altitude; site of most weather.

**turbidity**    Dustiness of the atmosphere.

**turbulence**    Irregular, random motions of a fluid such as air or water.

**typhoon**    A hurricane that forms in the western tropical Pacific Ocean.

**ultraviolet radiation (UV)**    Short-wave, energetic electromagnetic radiation that is emitted by the sun; much of the solar ultraviolet radiation is absorbed in the stratosphere where it is involved in the formation and destruction of ozone.

**umbra**    Central dark area of a sunspot.

**unstable air layer**    An air layer characterized by a vertical temperature profile such that air parcels accelerate upward or downward and away from their original altitudes.

**upslope fog**    Ground-level cloud formed as a consequence of the expansional cooling of humid air that is forced to ascend a mountain slope.

**upwelling**    The upward circulation of cold, nutrient-rich bottom water toward the ocean surface.

**urban heat island**    The relative warmth of a city compared with surrounding areas.

**UVB**    The biologically effective portion of solar ultraviolet radiation, in the 0.28 to 0.32 micrometer wavelength range; can cause sunburning and skin cancer.

**valley breeze**    A shallow, upslope flow of air that develops during the day within mountain valleys.

**vapor pressure**    That portion of the total air pressure exerted by the water vapor in a sample of air.

**variables of state**    Temperature, pressure, and density of air.

**virga**    A shaft of rain or snow falling from a distant cloud much of which vaporizes before reaching the ground.

**viscosity**   Friction within fluids such as air and water.

**visible radiation**   Electromagnetic radiation having wavelengths in the range of about 0.40 (violet) to 0.70 (red) micrometer.

**wake**   A region of turbulent (irregular) airflow that develops to the lee (downwind) of an obstacle such as a building.

**warm air advection**   Flow of air from a relatively warm locality to a relatively cool locality.

**warm cloud**   A cloud at a temperature above 0 °C (32 °F); composed of liquid water droplets only.

**warm-core anticyclones**   High-pressure systems occupying a thick column of subsiding warm, dry air; subtropical anticyclones are examples.

**warm-core cyclone**   A surface, synoptic-scale stationary cyclone that develops as a consequence of intense solar heating of a large, relatively dry geographical area; same as a thermal low.

**warm front**   A narrow zone of transition between relatively warm air that is advancing and relatively cool air that is retreating.

**warm sector**   Refers to the region of relatively high surface temperatures to the southeast of a mature midlatitude cyclone.

**warm-type occlusion**   A front formed when a cold front overtakes a warm front and the air behind the front is warmer than the air ahead of the front.

**waterspout**   A tornadolike disturbance that forms over a large body of water; usually much weaker than a tornado but associated with a cumulonimbus cloud.

**wave cyclone**   A low-pressure system that develops along the polar front; cyclonic circulation causes a wave to form along the front.

**wave frequency**   Number of crests or troughs of a wave that pass a given point in a specified period of time, usually 1 second.

**wavelength**   Distance between successive crests or successive troughs of a wave.

**weather**   The state of the atmosphere at some place and time described in terms of such variables as temperature, cloudiness, precipitation, and radiation.

**weather forecast offices (WFOs)**   Offices of the National Weather Service charged with providing regional weather reports and forecasts.

**weather modification**   Any change in weather that is induced by human activity, either intentionally or unintentionally. Cloud seeding and fog dispersal are examples of intentional weather modification, while enhancement of rainfall by air pollution is an example of unintentional weather modification.

**weather warning**   Issued when hazardous weather is observed.

**weather watch**   Issued when hazardous weather is considered possible based on current or anticipated atmospheric conditions.

**weighing-bucket rain gauge**   A device that is calibrated so that the weight of rainfall is recorded directly in terms of rainfall in millimeters or inches.

**wet-bulb depression**   On a psychrometer, the difference in readings between the wet-bulb thermometer and the dry-bulb thermometer; used to determine relative humidity.

**Wien's displacement law**   The higher the temperature of a radiating object, the shorter is the wavelength of maximum radiation intensity; applies to blackbodies.

**wind**   Air in motion relative to the Earth's surface.

**wind shear**   The change in wind speed or direction with distance.

**wind sock**   A large, conical, open bag designed to indicate wind direction and relative speed; usually used at small airports.

**wind vane**   An instrument used to monitor wind direction by always pointing into the wind.

**windchill equivalent temperature (WET)**   An air temperature index that attempts to gauge the sensible heat loss from exposed skin resulting from the combined effect of low air temperature and wind.

**World Meteorological Organization (WMO)**   Coordinates weather data collection and analysis by more than 145 member nations; based in Geneva, Switzerland.

**World Weather Watch (WWW)**   International weather-monitoring network coordinated by the World Meteorological Organization.

**X-rays**   Highly energetic, short-wavelength electromagnetic radiation.

**Younger Dryas**   A relatively cold period from around 12,900 to 11,600 years ago.

**zonal flow pattern**   Flow of the planetary-scale westerlies almost directly from west to east; westerlies exhibit little amplitude.

**zonda**   A foehn-type wind that flows into the Andes mountain valleys of Argentina.

# Index

# Psychrometric Tables

As explained in Chapter 6, a psychrometer is a standard instrument for measuring how close the air is to saturation. It consists of two thermometers mounted side by side: a dry-bulb thermometer and a wet-bulb thermometer. The dry bulb gives the actual air temperature while the wet bulb gives the temperature produced by evaporative cooling. The difference between the dry-bulb and the wet-bulb readings is the wet-bulb depression.

The relative humidity and the dew point can be obtained from the dry-bulb temperature and the wet-bulb depression. Use Table A to obtain the relative humidity: find the dry-bulb temperature in the left column and then read across to the relative humidity that corresponds to the wet-bulb depression. Follow the same procedure to determine the dew point in Table B.

## TABLE A
**Relative Humidity (Percent)**

| Dry-bulb Temp. (°C) | Wet-bulb Depression (°C) | | | | | | | | | | | | | | |
|---|---|---|---|---|---|---|---|---|---|---|---|---|---|---|---|
| | 0.5 | 1.0 | 1.5 | 2.0 | 2.5 | 3.0 | 3.5 | 4.0 | 4.5 | 5.0 | 7.5 | 10.0 | 12.5 | 15.0 | 17.5 |
| −10.0 | 85 | 69 | 54 | 39 | 24 | 10 | — | — | — | — | — | — | — | — | — |
| −7.5 | 87 | 73 | 60 | 48 | 35 | 22 | 10 | — | — | — | — | — | — | — | — |
| −5.0 | 88 | 77 | 66 | 54 | 43 | 32 | 21 | 11 | 0 | — | — | — | — | — | — |
| −2.5 | 90 | 80 | 70 | 60 | 50 | 41 | 31 | 22 | 12 | 3 | — | — | — | — | — |
| 0.0 | 91 | 82 | 73 | 65 | 56 | 47 | 39 | 31 | 23 | 15 | — | — | — | — | — |
| 2.5 | 92 | 84 | 76 | 68 | 61 | 53 | 46 | 38 | 31 | 24 | — | — | — | — | — |
| 5.0 | 93 | 86 | 78 | 71 | 65 | 58 | 51 | 45 | 38 | 32 | 1 | — | — | — | — |
| 7.5 | 93 | 87 | 80 | 74 | 68 | 62 | 56 | 50 | 44 | 38 | 11 | — | — | — | — |
| 10.0 | 94 | 88 | 82 | 76 | 71 | 65 | 60 | 54 | 49 | 44 | 19 | — | — | — | — |
| 12.5 | 94 | 89 | 84 | 78 | 73 | 68 | 63 | 58 | 53 | 48 | 25 | 4 | — | — | — |
| 15.0 | 95 | 90 | 85 | 80 | 75 | 70 | 66 | 61 | 57 | 52 | 31 | 12 | — | — | — |
| 17.5 | 95 | 90 | 86 | 81 | 77 | 72 | 68 | 64 | 60 | 55 | 36 | 18 | 2 | — | — |
| 20.0 | 95 | 91 | 87 | 82 | 78 | 74 | 70 | 66 | 62 | 58 | 40 | 24 | 8 | — | — |
| 22.5 | 96 | 92 | 87 | 83 | 80 | 76 | 72 | 68 | 64 | 61 | 44 | 28 | 14 | 1 | — |
| 25.0 | 96 | 92 | 88 | 84 | 81 | 77 | 73 | 70 | 66 | 63 | 47 | 32 | 19 | 7 | — |
| 27.5 | 96 | 92 | 89 | 85 | 82 | 78 | 75 | 71 | 68 | 65 | 50 | 36 | 23 | 12 | 1 |
| 30.0 | 96 | 93 | 89 | 86 | 82 | 79 | 76 | 73 | 70 | 67 | 52 | 39 | 27 | 16 | 6 |
| 32.5 | 97 | 98 | 90 | 86 | 83 | 80 | 77 | 74 | 71 | 68 | 54 | 42 | 30 | 20 | 11 |
| 35.0 | 97 | 93 | 90 | 87 | 84 | 81 | 78 | 75 | 72 | 69 | 56 | 44 | 33 | 23 | 14 |
| 37.5 | 97 | 94 | 91 | 87 | 85 | 82 | 79 | 76 | 73 | 70 | 58 | 46 | 36 | 26 | 18 |
| 40.0 | 97 | 94 | 91 | 88 | 85 | 82 | 79 | 77 | 74 | 72 | 59 | 48 | 38 | 29 | 21 |